TROPHIC ORGANIZATION *in* COASTAL SYSTEMS

Frontispiece

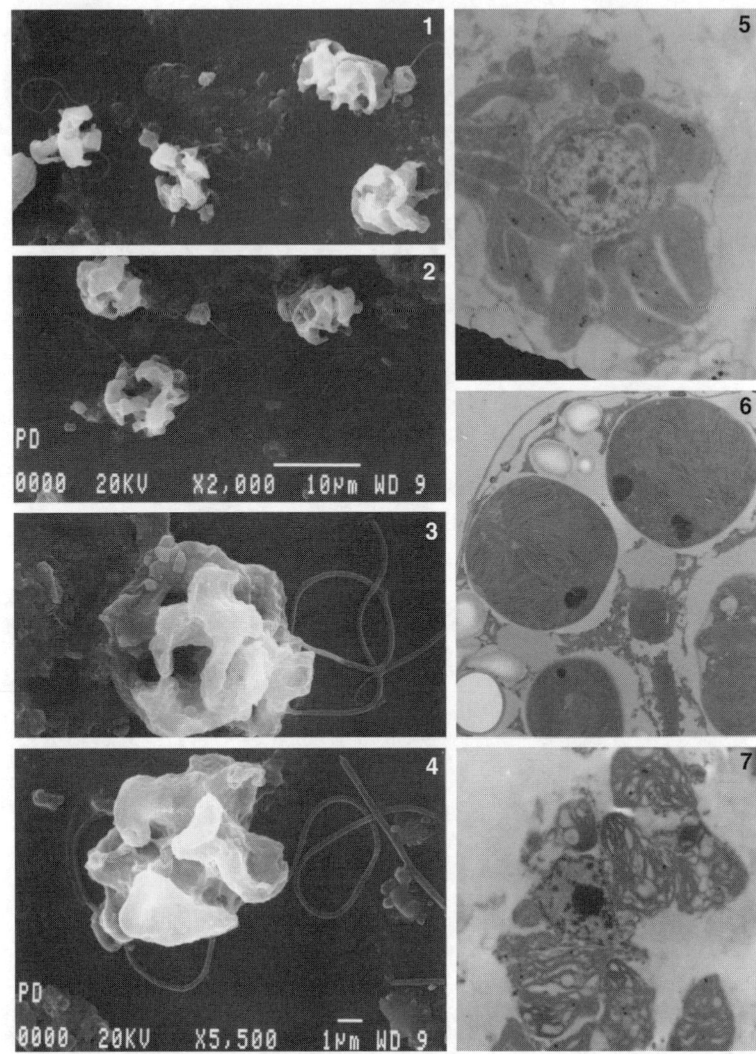

The raphidophyte flagellate, *Heterosigma akashiwo* (Hurlburt) Taylor, one of the bloom-forming species of the NE Gulf of Mexico. Scanning electron micrographs of air-dried material (1 to 4) showing biflagellate condition (3, 4) and the presence of chloroplasts surrounding the central nucleus (4). Transmission electron micrographs of ultrathin sections showing the cellular organization (5 to 7). (Photomicrographs courtesy of A. K. S. K. Prasad.)

TROPHIC ORGANIZATION *in* COASTAL SYSTEMS

Robert J. Livingston

Director
Center for Aquatic Research and Resource Management
Florida State University
Tallahassee

CRC PRESS

Boca Raton London New York Washington, D.C.

The cover design was created by Julia B. Livingston and Robert J. Livingston.

Senior Editor: John B. Sulzycki
Project Editor: Christine Andreasen
Project Coordinator: Pat Roberson
Marketing Manager: Nadja English

Library of Congress Cataloging-in-Publication Data

Livingston, Robert J.
Trophic organization in coastal systems / Robert J. Livingston
p. cm. -- (CRC marine science series)
Includes bibliographical references (p.).
ISBN 0-8493-1110-1
1. Coastal ecology--United States. 2. Food chains (Ecology)--United States. I. Title. II. Marine science series.

QH104 .L6 2002
577.5′1′0973--dc21 2002034916

Direct all inquiries to CRC Press LLC, 2000 N.W. Corporate Blvd., Boca Raton, Florida 33431.

International Standard Book Number 0-8493-1110-1
Library of Congress Card Number 2002034916
Printed in the United States of America 1 2 3 4 5 6 7 8 9 0
Printed on acid-free paper

The CRC Marine Science Series is dedicated to providing state-of-the-art coverage of important topics in marine biology, marine chemistry, marine geology, and physical oceanography. The series includes volumes that focus on the synthesis of recent advances in marine science.

CRC MARINE SCIENCE SERIES

SERIES EDITOR

Michael J. Kennish, Ph.D.

PUBLISHED TITLES

Artificial Reef Evaluation with Application to Natural Marine Habitats, William Seaman, Jr.

The Biology of Sea Turtles, Volume I, Peter L. Lutz and John A. Musick

Chemical Oceanography, Second Edition, Frank J. Millero

Coastal Ecosystem Processes, Daniel M. Alongi

Ecology of Estuaries: Anthropogenic Effects, Michael J. Kennish

Ecology of Marine Bivalves: An Ecosystem Approach, Richard F. Dame

Ecology of Marine Invertebrate Larvae, Larry McEdward

Ecology of Seashores, George A. Knox

Environmental Oceanography, Second Edition, Tom Beer

Estuary Restoration and Maintenance: The National Estuary Program, Michael J. Kennish

Eutrophication Processes in Coastal Systems: Origin and Succession of Plankton Blooms and Effects on Secondary Production in Gulf Coast Estuaries, Robert J. Livingston

Handbook of Marine Mineral Deposits, David S. Cronan

Handbook for Restoring Tidal Wetlands, Joy B. Zedler

Intertidal Deposits: River Mouths, Tidal Flats, and Coastal Lagoons, Doeke Eisma

Marine Chemical Ecology, James B. McClintock and Bill J. Baker

Morphodynamics of Inner Continental Shelves, L. Donelson Wright

Ocean Pollution: Effects on Living Resources and Humans, Carl J. Sindermann

Physical Oceanographic Processes of the Great Barrier Reef, Eric Wolanski

The Physiology of Fishes, Second Edition, David H. Evans

Pollution Impacts on Marine Biotic Communities, Michael J. Kennish

Practical Handbook of Estuarine and Marine Pollution, Michael J. Kennish

Practical Handbook of Marine Science, Third Edition, Michael J. Kennish

Seagrasses: Monitoring, Ecology, Physiology, and Management, Stephen A. Bortone

Preface

In 1969, fresh out of graduate school and newly employed as a lowly assistant professor at Florida State University, my first effort to organize a research program in aquatic ecology involved a trip to the major federal agencies and research foundations in Washington, D.C. The purpose was to obtain funding for a program that would entail long-term, interdisciplinary research to study regional coastal systems. However, this was a period that favored increasing specialization and reductionism in the ecological sciences. There were considerable funding opportunities in various fields of the environmental subdisciplines, but there was relatively little interest in long-term, interdisciplinary analyses of aquatic systems. Needless to say, I was not successful in getting funding for ecosystem-level research in coastal areas, and the explanations for the rejections were virtually unanimous: we need answers now, and we don't have the appropriate bureaucratic system to carry out long-term research.

The emphasis was on "experimental," hypothesis-testing research. Of course, there is nothing wrong with this in the mainstream scientific effort. However, ecosystem research, as the misunderstood crazy uncle of the scientific family, did not fit in with the timeworn assumptions of most scientists. The collection of long-term field data was beneath any right-thinking ecologist. Our modelers would take care of the synthesis of the myriad research programs that were under way in aquatic ecology. During the 1970s, as a member of the Science Advisory Board of the fledgling U.S. Environmental Protection Agency (EPA), I joined a distinguished group of ecologists who attempted to encourage the EPA to establish a comparative field research program in aquatic systems around the country. We thought that comparable field analyses of different coastal systems would eventually give us a better understanding of how to manage and regulate such areas. However, the EPA went for the quick answer provided by short-term bioassays that would tell us everything we needed to know about anthropogenous impacts on aquatic systems.

Now, after nearly four decades of field research, there is a growing realization that we really have very little knowledge of the processes that drive coastal ecosystems. In addition, it has become too obvious that some of the most important drainage systems in North America — the Florida Everglades–Florida Bay–Florida Keys system, Chesapeake Bay, San Francisco Bay, and many other important river–estuarine systems — have been severely damaged by multiple human activities in the form of urban and agricultural development and industrial waste disposal. Much of this loss has been accompanied by relatively little effective input by the research community.

The holistic, ecosystem approach to aquatic research has been rationalized, ridiculed, and even mimicked. There is still a tendency to substitute philosophy and unverified models for hard data. The ecosystem concept is historically based on food web ecology. It follows, therefore, that, if we are to understand how drainage systems actually function, we should base our studies on the following ideas:

- Food webs are central to an understanding of how aquatic systems function.
- The organization of plant communities is a critical factor in the definition of aquatic food webs.
- Food gathering is a primary determinant of the distribution of animals in space and time.
- Both quantity and quality of food types are associated with population maintenance and growth.
- Many aquatic species undergo ontogenetic changes in their feeding habits that ultimately define their habitat needs in complex ways that preclude simplistic assumptions and models of population distributions in space and time.
- There is a hierarchy of cycles of differing periods that is associated with the formation and development of aquatic food webs.

There is, of course, a rich scientific literature concerning trophodynamic processes in coastal areas. Predator–prey relationships, feeding habits of individual species, competitive interactions among coexisting populations, optimization of feeding characteristics, population dynamics with respect to trophic interactions, isotopic determination of food web interactions, modeling of food web development, and myriad other subdisciplines have been constructed around the theme of food web ecology. However, there is still an acknowledged shortage of long-term, interdisciplinary field information that is consistent with the natural periodicities of key processes of coastal systems, and that is based on how food webs interact with habitat features to define long-term trends of system-level responses to natural and anthropogenous driving factors.

In 1970, I set out to establish a research program that would explain some of the long-term phenomena that shape coastal ecosystems. My research group carried out interdisciplinary studies in a series of coastal areas that included nutrient loading and limitation, delineation of habitat variability, quantitative changes of key coastal populations from microbes to fishes, and detailed food web determinations. For almost two decades, our teams of highly trained technicians, graduate students, and principal investigators examined fish and invertebrate stomachs to define ontogenetic feeding habits of the primary populations along the NE Gulf of Mexico. Expert teams of phytoplankton and macrophyte experts established a long-term database of species-specific changes of aquatic plant associations in these Gulf coastal systems. The biological studies were accompanied by synoptic water and sediment quality analyses. Thus, together with chemists, physical oceanographers, geologists, hydrologists, engineers, experimental biologists, and taxonomists, we created a multidisciplinary database of changes in a series of Gulf coastal systems.

The model on which this research program is built required continuous funding, central organization, and cooperative efforts of diverse scientific interests. This approach has been largely rejected as unworkable by most ecologists today as it takes too long and is too costly. It is not based on the "proven" pattern of individual projects organized by subdiscipline with most units working independently of each other. There are some important interdisciplinary projects that have been developed over the years, but these represent the exception, not the rule. Anyone who wants to start a long-term, interdisciplinary research project in the aquatic sciences still faces very dismal prospects for success due to (1) the bureaucratic need for instant results, (2) an antiquated academic environment dominated by a reductionist philosophy, (3) the need to publish more and more about less and less in research journals, and (4) a political establishment that uses scientific ambivalence to continue the incremental destruction of coastal resources. If Pope Leo had enlisted the National Science Foundation to paint the ceiling of the Sistine Chapel, the end result would have been an elaborate patch-quilt composed of individual projects with no overlying theme, organization, or meaning.

The central focus of this book is on the food web organization of coastal systems and the relationship of trophodynamic processes to long-term changes in a series of coastal systems.

Author

Robert J. Livingston, Ph.D., is currently Professor of Biological Science and Director of the Center for Aquatic Research and Resource Management at Florida State University (Tallahassee, Florida, U.S.A.). His interests include aquatic ecology, pollution biology, field and laboratory experimentation, and long-term ecosystem-level research in freshwater, estuarine, and marine systems. Past research includes multidisciplinary studies of lakes and a series of drainage systems on the Gulf and East Coasts of the United States. Over the past 32 years, Dr. Livingston's research group has amassed more than 70 field-years of data in various freshwater and coastal systems. Dr. Livingston has directed the programs of 49 graduate students who have carried out research in behavioral and physiological ecology with individual aquatic populations and communities in lakes, rivers, and coastal systems. He is the author of more than 140 scientific papers and five books on the subject of aquatic ecology, and he has been the principal investigator for more than 100 projects since 1970.

Dr. Livingston's primary research program has been carried out largely in the southeastern United States. The areas studied include the Apalachicola drainage system, the Choctawhatchee drainage system, the Perdido drainage system, the Escambia Bay system, the Blackwater River–estuary, the Escatawpa–Pascagula drainage system, the Mobile River–estuary, the Winyah Bay system (including the Sampit River), Apalachee Bay (the Econfina and Fenholloway River–estuaries), and a series of north Florida lake systems. This work has included determinations of the impact of various forms of anthropogenous activities on a range of physical, chemical, and biological processes. A sophisticated computer system has been developed to aid in the analysis of the established long-term databases. The validation or verification of bioassay results with field data from lakes, rivers, and coastal areas has been an integral part of the research effort. In addition to several ongoing field programs, Dr. Livingston is currently involved in the analysis, modeling, and publication of the established data in peer-reviewed journals and in a series of books and atlases. This research has been, and continues to be, funded by a wide variety of private and public granting agencies.

Acknowledgments

Many individuals have contributed to the research effort over the past 32 years. It is impossible to acknowledge them all by name. Literally thousands of undergraduate students, graduate students, technicians, and staff personnel have taken part in the long-term programs. Various individuals have been involved in CARRMA research for two to three decades. Dr. A. K. S. K. Prasad carried out phytoplankton taxonomy and systematics. Mr. R. L. Howell IV ran the field collections. Mr. G. C. Woodsum designed the overall database organization and, together with Mr. P. Homann, ran database management and day-to-day data analysis of the research information. Dr. F. G. Lewis III was instrumental in the development of analytical procedures and statistical analyses. Dr. D. A. Flemer designed the overall approach for analytical chemistry. Dr. D. A. Meeter, Dr. X. Niu, and Ms. L. E. Wolfe performed statistical analyses. The taxonomic consistency of the field collections was carried out with the help of world-class systematists that include Dr. J. H. Epler (aquatic insects), Dr. C. R. Gilbert (freshwater fishes), and Dr. C. J. Dawes (submerged aquatic vegetation).

The long-term program has benefited from the advice of a distinguished group of scientists. The early work was reviewed by Dr. John Cairns, Jr., Mr. Bori Olla, Dr. E. P. Odum, Dr. J. W. Hedgpeth, Dr. O. Loucks, Dr. K. Dickson, Dr. F. J. Vernberg, and Dr. Ruth Patrick. Dr. D. M. Anderson, Dr. A. K. S. K. Prasad, Mr. G. C. Woodsum, Dr. M. J. Kennison, Dr. K. Rhew, Dr. M. Kennish, Dr. E. Fernald, and Ms. V. Tschinkel have reviewed various aspects of the recent work. Other reviews have been provided by Dr. C. H. Peterson, Dr. S. Snedaker, Dr. R. W. Virnstein, Dr. D. C. White, Dr. E. D. Estevez, Dr. J. J. Delfino, Dr. F. James, Dr. W. Herrnkind, and Dr. J. Travis. Dr. Eugene P. Odum gave me the good advice to write up our results in a series of books, which turned out to be the only way our work could be adequately treated.

Over the years, we have had a long list of excellent co-investigators who have participated in various projects. Recent collaborative efforts in the field have included the following scientists: Dr. D. C. White (microbiological analyses), Dr. D. A. Birkholz (toxicology, residue analyses), Dr. G. S. Brush (long-core analysis), Dr. L. A. Cifuentes, (nutrient studies, isotope analyses), Dr. K. Rhew (microalgal analyses), Dr. R. L. Iverson (primary productivity), Dr. R. A. Coffin (nutrient studies, isotope analyses), Dr. W. P. Davis (biology of fishes), Dr. M. Franklin (riverine hydrology), Dr. T. Gallagher (hydrological modeling), Dr. W. C. Isphording (marine geology), Dr. C. J. Klein III (aquatic engineering, estuarine modeling), Dr. M. E. Monaco (trophic organization), Dr. R. Thompson (pesticide analyses), Dr. W. Cooper (chemistry), Mr. D. Fiore (chemistry), and Dr. A. W. Niedoroda (physical oceanography).

We have had unusually strong support in the statistical analyses and modeling of our data, a process that continues to this day. These statistical determinations and modeling efforts have been carried out by the following individuals: Dr. T. A. Battista (National Oceanic and Atmospheric Administration), Dr. B. Christensen (University of Florida), Dr. J. D. Christensen (National Oceanic and Atmospheric Administration), Dr. M. E. Monaco (National Oceanic and Atmospheric Administration), Dr. T. Gallagher (Hydroqual, Inc.), Dr. B. Galperin (University of South Florida), Dr. W. Huang (Florida State University), Dr. D. A. Meeter (Florida State University), Ms. L. E. Wolfe (Florida State University), Dr. X. Niu (Florida State University), and Mr. G. C. Woodsum (Florida State University).

Experimental field and laboratory ecology has been carried out by the following individuals: Dr. B. M. S. Mahoney (experimental biology), Dr. K. Main (experimental biology), Dr. Frank Jordan (biology of fishes), Dr. K. Leber (experimental biology), Dr. W. H. Clements (trophic analyses), Dr. C. C. Koenig (fish biology), Dr. P. Sheridan (trophic analyses), Dr. J. Schmidt (aquatic macroinvertebrates), Dr. A. W. Stoner (trophic analyses), Dr. R. A. Laughlin (trophic analyses), Dr. J. L. Luczkovich (experimental biology), Dr. J. Holmquist (experimental biology), Dr. D. Bone (experimental biology), Dr. B. MacFarlane (biology of fishes), Ms. C. Phillips (experimental biology), Mr. M. Kuperberg (submerged aquatic vegetation), Ms. T. A. Hooks (experimental biology), Mr. J. Ryan (trophic analyses), Mr. C. J. Boschen (trophic analyses), Mr. G. G. Kobylinski (experimental biology), Mr. C. R. Cripe, (experimental biology), Mr. T. Bevis (biology of fishes), Mr. P. Muessig (experimental biology), Ms. S. Drake (experimental biology), Mr. K. L. Heck, Jr. (experimental biology), Ms. H. Greening (aquatic macroinvertebrates), Mr. D. Cairns (experimental biology), Ms. K. Brady (larval fishes), Mr. B. McLane (aquatic macroinvertebrates), Ms. B. Shoplock (zooplankton), Ms. J. J. Reardon (phytoplankton, submerged aquatic vegetation), and Ms. J. Schmidt-Gengenbach (experimental biology).

Taxonomy is at the heart of a comprehensive ecosystem program, which is something I learned from my major professors (Dr. C. Hubbs, Dr. C. Richard Robbins, Dr. A. Myrberg) while I attended graduate school. Our attention to systematics and natural history has benefited from the efforts of a long line of graduate students, postdoctoral fellows, and trained technicians, including the following: Dr. A. K. S. K. Prasad and Dr. K. Rhew (phytoplankton, benthic microalgae), Dr. F. Graham Lewis III (aquatic macroinvertebrates), Dr. J. Epler (aquatic macroinvertebrates), Dr. R. D. Kalke (estuarine zooplankton), Dr. G. L. Ray (infaunal macroinvertebrates), Dr. K. R. Smith (oligochaetes), Dr. E. L. Bousfield (amphipods), Dr. F. Jordan (trophic analyses), Dr. C. J. Dawes (submerged aquatic vegetation), Dr. R. W. Yerger (fishes), Mr. R. L. Howell III (infaunal and epibenthic invertebrates and fishes), Mr. W. R. Karsteter (aquatic macroinvertebrates), and Mr. M. Zimmerman (submerged aquatic vegetation).

A number of phycologists have given their time to the taxonomy and nomenclature of phytoplankton species over the years. For diatom taxonomy the following individuals are acknowledged: Dr. C. W. Reimer (Academy of Natural Sciences of Philadelphia), Dr. G. A. Fryxell (University of Texas at Austin), Dr. G. R. Hasle (University of Oslo, Norway), Dr. P. Hargraves (University of Rhode Island), Professor F. Round (University of Bristol, England), Mr. R. Ross, Ms. P. A. Sims, Dr. E. J. Cox, and Dr. D. M. Williams (British Museum of Natural History, London, U.K.), Dr. L. K. Medlin and Dr. R. M. Crawford (Alfred-Wegener Institute, Bremerhaven, Germany), Professor T. V. Desikachary (University of Madras, India), Dr. J. A. Nienow (Valdosta State University, Georgia), Dr. P. Silva (University of California at Berkeley), Dr. M. A. Faust (Smithsonian Institution), Dr. R. A. Andersen (Bigelow Laboratory for Ocean Sciences, Maine), Dr. D. Wujeck (Michigan State University), Dr. M. Melkonian (University of Köln, Germany), Dr. C. J. Tomas (Florida Department of Environmental Regulation, St. Petersburg), Professor T. H. Ibaraki (Japan), Dr. Y. Hara (University of Tsukuba, Japan), and Dr. J. Throndsen (University of Oslo, Norway). Technical assistance with transmission and scanning electron microscopes was provided by Ms. K. A. Riddle (Department of Biological Science, Florida State University). Mrs. A. Black and Mr. D. Watson (Histology Division, Department of Biological Science, Florida State University) aided in specimen preparation and serial thin sectioning.

The curators of various International Diatom Herbaria have aided in the loan of type collections and other authentic materials: Academy of Natural Sciences of Philadelphia (ANSP), California Academy of Sciences (CAS) (San Francisco), National Museum of Natural History, Smithsonian Institution (Washington, D.C.), Harvard University Herbaria (Cambridge, Massachusetts), The Natural History Museum (BM) (London, U.K.), and the F. Hustedt Collections, Alfred-Wegener Institute (Bremerhaven, Germany).

Others have collaborated on various aspects of the field analyses. Mr. W. Meeks, Mr. B. Bookman, Dr. S. E. McGlynn, and Mr. P. Moreton ran the chemistry laboratories. Mr. O. Salcedo, Ms. I. Salcedo, Mr. R. Wilt, Mr. S. Holm, Ms. L. Bird, Ms. L. Doepp, Ms. C. Watts, and Ms. M. Guerrero-Diaz provided other forms of lab assistance. Field support for the various projects was provided by Mr. H. Hendry, Mr. A. Reese, Ms. J. Scheffman, Mr. S. Holm, Mr. M. Hollingsworth, Dr. H. Jelks, Ms. K. Burton, Ms. E. Meeter, Ms. C. Meeter, Ms. S. Solomon, Mr. S. S. Vardaman, Ms. S. van Beck, Mr. M. Goldman, Mr. S. Cole, Mr. K. Miller, Mr. C. Felton, Ms. B. Litchfield, Mr. J. Montgomery, Ms. S. Mattson, Mr. P. Rygiel, Mr. J. Duncan, Ms. S. Roberts, Mr. T. Shipp, Ms. A. Fink, Ms. J. B. Livingston, Ms. R. A. Livingston, Ms. J. Huff, Mr. R. M. Livingston, Mr. A. S. Livingston, Mr. T. Space, Mr. M. Wiley, Mr. D. Dickson, Ms. L. Tamburello, and Mr. C. Burbank.

Project administrators included the following individuals: Dr. T. W. Duke, Dr. R. Schwartz, Dr. M. E. Monaco, Ms. J. Price, Dr. N. P. Thompson, Mr. J. H. Millican, Mr. D. Arceneaux, Mr. W. Tims, Jr., Dr. C. A. Pittinger, Mr. M. Stellencamp, Ms. S. A. Dowdell, Mr. C. Thompson, Mr. K. Moore, Dr. E. Tokar, Dr. D. Tudor, Dr. E. Fernald, Ms. D. Giblon, Ms. S. Dillon, Ms. A. Thistle, Ms. M. W. Livingston, and Ms. C. Wallace.

The great tragedy of Science — the slaying of a beautiful hypothesis by an ugly fact.

—*Biogenesis and Abiogenesis,* by Thomas Henry Huxley

A foolish consistency is the hobgoblin of little minds, adored by little statesmen and philosophers and divines.

—*Essays: First Series,* by Ralph Waldo Emerson

Between the idea
And the reality
Between the motion
And the act
Falls the shadow....

This is the way the world ends
Not with a bang but a whimper

—*The Hollow Men,* by T.S. Eliot

A land ethic for tomorrow should be as honest as Thoreau's Walden, and as comprehensive as the sensitive science of ecology. It should stress the oneness of our resources and the live-and-help-live logic of the great chain of life. If, in our haste to "progress," the economics of ecology are disregarded by citizens and policy makers alike, the result will be an ugly America.

—*The Quiet Crisis,* by Stewart Lee Udall

Contents

CHAPTER 1

Introduction

This book is based on a series of long-term studies of the northeastern Gulf of Mexico. These systems have been described in detail by Livingston (2000), and follow a pattern of river–estuarine areas from those that are almost completely free of human influence in eastern sectors (Apalachee Bay, Apalachicola Bay) to increasingly disturbed bay systems to the west (Choctawhatchee Bay, the Pensacola Bay system, Perdido Bay), where human activities have been associated with various forms of adverse impacts. I outline the basis of our food web studies, with the development of the trophic unit as a characteristic form of feeding population in the ontogenetic development of a given fish or invertebrate species. This unit is qualified in terms of both spatial and temporal variability. The ontogenetic trophic units of most of the fishes and invertebrates in the NE Gulf are then applied to the organization of food webs in the various study areas. The structural organization of these food webs is then defined and quantified in terms of spatial distribution along gradients of freshwater runoff and temporal aspects of habitat effects and associated predator–prey interactions.

The relationships of food web organization and phytoplankton community development are then outlined with specific attention to the impact of plankton blooms as a product of anthropogenous nutrient loading patterns. Seasonal and interannual bloom progressions are quantitatively associated with specific nutrient loading characteristics, and these progressions, in turn, are statistically correlated with significant adverse effects on specific parts of associated food webs. Specific differences in food web organization are outlined as part of a series of comparisons of natural and adversely affected food webs in the river–bay systems of the NE Gulf. The gradients of secondary production in areas ranging from natural systems in the east to the deteriorated bays to the west are delineated. The importance of phytoplankton organization in determination of food web organization and secondary production is emphasized, with specific reference to the failure of state and federal agencies to regulate the subtle but significant effects of anthropogenous nutrient loading on such areas. In addition, the lack of attention to food web organization is related to regulatory failures in limiting the effects of toxic agents such as mercury on river-dominated bay systems. The ecosystem approach that underlies these studies is then defined in the application of such studies to the successful management of estuarine resources in one of the last highly productive alluvial systems in the country, the Apalachicola River and Bay system. I end with a suggested outline of how to carry out ecosystem studies in coastal areas with respect to scientific applications for management of estuarine resources. It should be noted that this approach, based on food web organization as part of an ecosystem-level research effort, is not the only way to carry out such studies but has proved successful in the actual application of science to questions related to resource management in coastal areas.

I. FOOD WEBS AND THE ECOSYSTEM PARADIGM

According to O'Neill (2001), the ecosystem concept (Tansley, 1935) actually harkens back to the idea noted by Marsh (1864) that nature was in a constant state with periodic disruptions that are repaired in time. The ecosystem concept had an early expression in the work of Thienemann (1918) as defined in extensive studies of lakes in North Germany. The food chain concept, introduced by Elton (1927), was then developed to define a series of interacting monophagous consumers. Elton (1930) conceived of the ecosystem as a series of plant and animal successions, which was then broadened into a group of interacting units that included polyphagous consumers — thus was the food web concept conceived. The holistic view of the ecosystem was formulated by Tansley (1935) into the concept of a physical/chemical/biological entity with interactions that were unified in space and time. Lindeman (1942) broadened the definitions of patterned feeding organizations with the food web now defined in terms of the energy transfer. The so-called trophodynamic concept of ecosystem functions was related to autotrophic plants that "produce" energy; consumers then take up this energy in the form of organic matter, with the cycle completed by the saprotrophic organisms that break down the organic matter into inorganic nutrients for subsequent renewal of the autotrophic process. In this way, the universal basis for trophodynamic processes was developed.

Lindeman (1942) defined the eutrophication of water bodies as a series of stage equilibria whereby there existed a theoretical "dynamic state of continuous utilization and generation of chemical nutrients in an ecosystem." Hutchinson (1957) further refined productivity as a "food cycle" having "successively different energy contents." He added to the ecosystem concept with ideas of biological efficiency and a broader organization of the system into trophic levels. Thus was born the concept of the interconnectedness of the ecosystem through trophodynamic processes that followed distinctive temporal patterns in accordance with the first and second laws of thermodynamics. The food web now represented a summary of the countless predator–prey interactions that occurred in a given area over a specified time period.

The ecosystem as a natural unit was defined by E. P. Odum (1953) whereby both living and nonliving factors interacted through "circular paths." Eventually, the machine analogy was brought forth (Holling, 1973) with holistic overtones of nutrient cycling. However, as pointed out by O'Neill (2001), the ecosystem was not defined as an empirical entity. Rather, the ecosystem was a paradigm that focused on certain properties of natural systems. Recently, there has been a backlash against the ecosystem concept that is based on the ambiguities associated with the concepts of stability and the mythical ideas of an integrated, homeostatic system that is in a constant state of equilibrium. These criticisms run counter to current conceptual models that view the ecosystem as open, hierarchical, disequilbrial, and patterned in space and time (O'Neill, 2001). Some ecologists accept the notion that species lists define the ecosystem. However, species substitutability is now well recognized, and problems associated with spatial and temporal scales are often ignored or explained away in unrealistic terms. The effects of disturbance on the "stability" of the system have been the subject of many papers. Theoretical ecologists have hotly debated the terminology of temporal changes of community organization for decades, even though tests of the validity of such concepts have been rare. Stability models are often based on short-term experiments that ignore most of the important properties of natural systems. Human intervention has been considered as an external disturbance with serious consequences relative to the so-called self-regulating processes attributed to the ecosystem. However, the details of such disturbance are usually lost in elaborate verbiage and untested assumptions of the proffered models. Statistical elegance and expansion of jargon have outstripped the empirical basis for such modeling efforts.

O'Neill (2001) offered a set of principles that purported to form a foundation for a new paradigm for the ecosystem concept. These included definitions of spatial and temporal scaling and dispersal criteria, explanations of population interactions (as formed through competition and dominance relationships), and sets of stability criteria within scale ranges from local, short-term events to geologic time intervals. In this context, humans are viewed as a keystone species that changes the system stability through alteration of the rate processes, biotic structures, and environmental constraints. O'Neill (2001) considered the new paradigm to address key problems related to spatial patterns, temporal variability, scales of patterns as they relate to stability, and the importance of natural selection and internal feedback mechanisms in the definition of system dynamics. However, as is the case with most theoretical discourses, the actual application of the definition of ecosystem processes to the real world was lost in the elegance of the presentation.

Food webs can be represented as extremely complex diagrams where detail obscures pattern (Raffaelli, 2002). Grouping species into feeding types or trophic levels has been the usual method for representing biomass pyramids (Elton, 1927). Another way of representing food webs is the use of abstract representations as mathematical models (Raffaelli, 2002). The emphasis of these models is on complex predator–prey interactions. Neutel et al. (2002) relate such models to real food webs where loop (i.e., a closed chain of trophic links) lengths are associated with stability. The authors showed that complex ecosystems with intricate food webs retain stability, contrary to theoretical models. Longer-lived predators at the top of the food web are generally characterized by higher energy efficiencies than those lower on the trophic order. Raffaelli (2002) related food web complexity to web stability. However, the author pointed out that freshwater and marine systems often do not have biomass pyramids where primary producers are low. Such systems would thus not fit into the theoretical models that have been proposed by various ecologists to link food web complexity to stability. In shallow coastal systems, the general lack of integration of detailed aquatic food webs with benthic processes that are often dominated by detritivores and intermediary energy paths through microbial activity simply do not fit the general models that have been dominated by oversimplified terrestrial food webs. Trophic feedback, both positive and negative, of various forms of microorganisms in the plankton communities of coastal systems is also usually ignored in the definition of such models. Phytoplankton communities are not well understood even though such associations are a major source of organic matter for coastal food webs. In this way, phytoplankton can have major effects on the stability of associated food webs (Livingston, 2000). The effects of toxic blooms on secondary production remain virtually undetermined due to the complexity of the trophic interactions in space and time. However, the potential effects of such blooms could be one of the most underestimated factors in changes that are currently taking place in coastal systems worldwide.

Instead of rehashing the various arguments associated with stability, resilience, and equilibrium, I approach this problem of definition by addressing the essential components of ecosystem processes that are centered on actual food webs in a series of coastal systems in the NE Gulf of Mexico. I attempt to define processes associated with short- and long-term trophic responses to altered forms of primary production. Included in this empirical approach are definitions of spatial/temporal cycling and feedback phenomena that are crucial to an understanding of how such systems work and how human interventions can play a crucial role in the destabilization processes that ultimately lead to a breakdown of the observed food web structure.

II. TROPHODYNAMIC ASPECTS OF ECOSYSTEM PROCESSES

The considerable number of recent scientific publications concerning food web ecology in aquatic systems is a tribute to the earlier recognition of the ecosystem concept with its emphasis on trophodynamics as a fundamental focal point for a comprehensive understanding of how systems function. In any given ecosystem, the food web remains the central feature around which the basic ecological processes are organized. Food web research reflects various approaches to determination of trophodynamic processes. There is the direct analysis of stomach contents of individual organisms and populations. A more recent approach that is related to such studies includes use of stable isotope techniques to determine qualitative and quantitative aspects of food web dynamics and structure. Another related group of studies uses experimental techniques to evaluate predator–prey interactions and dynamic changes in energy transfer and flow. This includes energy flow determinations and associated models that relate production of organic carbon to associated food webs.

There are numerous modeling procedures for food web definition. These models include both theoretical and applied algorithms to determine food web structure in various aquatic and terrestrial habitats. However, most such models lack adequate empirical information, and most are not field verified. Many of the current food web models also lack definition of spatial/temporal variation, and consequently are not predictive, especially with respect to the effects of disturbance phenomena. The lack of comprehensive databases that define long-term trophodynamic trends in coastal systems is directly related to the general lack of predictive models. Journals that take great pride in the precision and accuracy of statistical approaches to problems involved in modeling the complex trophic interactions of a given system often are vague in the definition of the adequacy of the databases that are used for such studies. This does not diminish the potential importance of food web analyses to evaluations of impacts of toxic substances and nutrient loading; it merely qualifies the current state of understanding of such processes.

The emphasis on studies related to the trophodynamic conception of ecosystem functions does not automatically validate the premises on which many food models depend. Early on, a number of theories concerning how trophodynamic processes operate were advanced (Murdoch, 1966; Rigler, 1975). However, there were problems of interpretation of such models, which were codified by Peters (1977). The original Trophodynamic Concept (Lindeman, 1942) depends on categorization of organisms in a given food chain according to their distance from the ultimate energy source, the sun. However, the real world is based on complex food webs that often cross the artificially set trophic boundaries (Peters, 1977). Adequate definition of quantitative and qualitative aspects of primary productivity also presents problems of interpretation. Not all plankton are primary producers; some utilize organic compounds as a source of primary production. Others are predators on other plankton. There are relatively few studies in coastal systems that actually identify phytoplankton, and the problems associated with the direct use of indicators of phytoplankton activity such as chlorophyll *a* are just now being evaluated (Livingston, 2000). Herbivores also present problems of categorization because many are not strictly vegetarian in their feeding habits (Peters, 1977).

Another weak link in the traditional food web definition involves detritivores that often remain largely undefined in terms of position in the food web. The interactions of bacteria and other microorganisms as intermediaries in the movement of energy through the food web constitute a major complicating factor. The extremely complex transfer relationships between algal producers, bacterial decomposers, and protozoans, herbivores, and predators are just now being addressed in various studies. The importance of such interactions has been well established (White et al., 1977, 1979a,b; Fenchel and Jorgensen, 1978; Wetzel, 1984; Wetzel and Likens, 1990; White, 1983). Detritivores thus represent another source of uncertainty in a given trophodynamic model in that there is no neat category that defines their exact position in the food web. Likewise, omnivores do not fit neatly into a single trophic category, a problem recognized by various authors (Darnell, 1961, 1967; Livingston, 1984a). The use of intermediate trophic levels for omnivores (Kercher and Shugart, 1975) does not conform with the usual approaches used in most food web studies, and certainly is not consistent with modern concepts of food web modeling. At the higher levels of carnivory, there are also problems associated with the taking of prey from different levels of the food web. In addition, there is some flexibility in the feeding habits of predators that is related in some ways to the adventitious nature of what is available as food.

The generally accepted use of the biological species as the unit of trophic definition is also flawed in that many species do not remain at the same level in the food web throughout their life history. For example, Livingston (1982a, 1984b), Stoner (1979a,b, 1980a,b), and Stoner and Livingston (1980a, 1984b) studied the life history feeding habits of the pinfish, *Lagondon rhomboides*, one of the numerically dominant coastal fishes in the eastern Gulf of Mexico. This species actually passed through various trophic stages in its developmental history, including an early planktivorous phase, subsequent benthic carnivory, two distinct stages of omnivory, and adult herbivory. Relatively few feeding studies and derivative models of the food webs take into account the ontogenetic shifts of feeding habitats of coastal invertebrates and fishes. In addition, many such studies view the simplified feeding habits of adults as fixed in space and time, without verification of such assumptions. Once again, the quality of the data on which food web models are constructed should be uppermost in the validation of such models. In any definition of coastal food webs, these problems of interpretation should be addressed so that our assumptions are consistent with our generalizations. This requires that databases on which assumptions are based should be consistent with the many complications of the trophodynamic processes in coastal systems.

Identification of the sources of production for coastal food webs has been well developed over the past few decades. The use of stable isotopes has expanded our understanding of food resource questions and interactions among the various food web components. Determination of food web responses to various forcing functions can include the use of food web information in determination of the effects of anthropogenous sources of nutrients and toxic agents. Earlier work (Hackney and Haines, 1980) related food sources for marsh animals by analyzing $\delta^{13}C$; they established that significant amounts of terrestrial plant matter entered the food web of a Mississippi estuary through river flows. Numerous related studies have been carried out (Haines, 1979; Haines and Montague, 1979) that determined food origins of coastal areas through isotope analyses. These studies identified the relative importance of sea grasses, benthic algae, phytoplankton, and C-4 photosynthetic plants to various estuarine food webs. Other studies (Fry, 1983) used stable C, N, and S isotope ratios to identify fish and shrimp feeding grounds. Overlapping isotopic values obscured exact relationships of *Spartina* marshes and the open bays. However, such studies showed that different shrimp species had convergent feeding patterns as juveniles, but became less convergent as adults, indicating ontogenetic differences in species-specific progressions through time.

The importance of life history patterns cannot be overemphasized in food web analyses. A specific problem associated with stable isotopes has been the unexplained variability of $\delta^{13}C$ (Finlay, 2001). Feeding types play a role in the determination of food origins (Finlay, 2001), and the use of isotope analyses in conjunction with quantitative studies of organic matter dynamics and detailed food web (stomach content) analyses is considered to be important in the identification of the trophodynamic processes in aquatic systems. There are questions of variability of the trophic fractionization that need to be answered. Also, there have been very few long-term analyses of changes in food web dynamics in coastal areas with time. Such analyses, in combination with long-term food web determinations through stomach content analyses, would be an important addition to what is currently known regarding trophic relationships in coastal areas.

A series of food web models has been developed based on energy flow dynamics (Ulanowicz, 1987). These models depend on energy budgets using ECOPATHii software (Christensen and Pauly, 1992) that estimates production and consumption by various compartments of the system. Carbon flows and characteristics of the overall trophic structure of the system (utilizing NETWRK software; Ulanowicz, 1987) are then integrated into the model, with output definition described by Ulanowicz and Kay (1991). Various indices such as guild biomass, production to biomass ratios, diet matrices, and detrital inputs are developed from empirical data and serve as input for the model. In this way, networks of energy flows can be estimated and projected (Monaco, 1995). Using this modeling system, Monaco and Ulanowicz (1997) determined carbon exchanges in three mid-Atlantic estuaries (Narragansett, Delaware, and Chesapeake Bays). Using analyses of cycling structures (magnitude of flows, average carbon cycles lengths) and the organization of the carbon flows (system production:biomass ratios, and harvest rates), the authors determined that the Delaware and Chesapeake systems were more stressed than the Narragansett system. Using another series of measures (system efficiency, cycling structure, food web connectivity), Monaco and Ulanowicz (1997) found that the Delaware system was less affected by pollution and had more potential ability for mitigation of impacts to its food webs than the Chesapeake system. The model depends on determination of guild (trophic compartment) organization and trophic specificity that involves a suite of assumptions in the absence of comprehensive empirical data. Determination of the trophic level of fisheries production would also be sensitive to overfishing effects that are difficult to quantify. Bioenergetic models have been used in various ways to determine the relative roles of top predators (Hartmann, 1993). These predators are largely pelagic feeders that are tied temporally to benthic feeding, which has been shown to be very important to both juvenile and adult piscivores. These models are currently used to address specific management questions in coastal systems.

A modeling effort by the University of British Columbia (Christensen and Pauly, 1992; Pauly et al., 2000) has been developed into a software system for applications of food web functions to static, mass-balanced snapshots of a given system (Ecopath), time-dynamic modeling for policy exploration (Ecosim), and spatial/temporal dynamic models for analyzing impacts and the placement of protected areas (Ecospace). Models are based on biomass pools of guild associations that can be split into ontogenetic groups for analyses of stock assessments, ecological processes, biomass estimates, mortality evaluations, diet compositions, and fisheries evaluations. Relatively simple models are based on linear equations (Ecopath). Data generated can be used for dynamic simulations (Ecosim) with built-in time-series capabilities. The models can be used to track recovery from pollution such as oil spills (Okey and Pauly, 1999). Fisheries applications include estimation of mortality information for determinations of sustainable yields. Modeling efforts have delineated the effects of overfishing on marine food webs. Overfishing of primary species can lead to fishing down the food web with unexpected results that could include declines in overall trophic levels of fish catches and eventual collapse of the food webs and the fisheries that depend on them (Pauly et al., 1998). Removal of top predators that leads to expansion of the predators' competitors can result in increased populations of noncommercial organisms such as jellyfish. Top-down problems with overfishing have been shown in the Black Sea, the Georges Bank, and numerous inland fisheries in the Northern Hemisphere.

Scientific determination of aquatic food webs depends on relatively complete knowledge of the quantity and quality of food available (i.e., phytoplankton and benthic microphytes/macrophytes), a complete list of the predator species, and detailed determinations of their feeding habits through exhaustive stomach content analyses using relatively large numbers of subjects. It is rare that all three components are available in any given study. The considerable complexity of coastal food webs precludes a definitive examination of the myriad trophic interactions that are integral to the definition of a given food web. Most of the published literature regarding stomach contents of fishes and invertebrates is related to single species examinations or to the determination of a set of closely related species. The temporal aspects of food web dynamics in coastal systems have not been well documented. Most of the published literature involves what Schoenly and Cohen (1991) call "cumulative webs" or data collected in specific areas over a series of time periods as opposed to "time-specific" food web development. However, the lack of understanding of food quality in the determination of food web dynamics would qualify such model. Whereas omnivores often select food on the basis of size, the need for identification of food quality is particularly apparent at the autotroph-herbivore level.

Overall, it is apparent that realistic and predictive food web modeling in the future will depend on integrated, long-term data-collection efforts whereby the collection of empirical data takes into consideration the quality of the food sources, the temporal aspects of food web dynamics, and the changes that take place in the complex trophodynamic processes through varying time periods.

III. FOOD WEB ECOLOGY IN COASTAL SYSTEMS

Coastal marine systems encompass a biological continuum from upland, freshwater portions of individual drainage basins through the interface estuary to the open ocean. Within each drainage basin, there are complex combinations of habitats that form one of the most productive ecosystems on Earth. High nutrient levels, multiple sources of primary and secondary production, shallow depths, organically rich sediments, energy inputs from wind and tidal currents, and freshwater inflows combine to establish the high natural productivity of near-shore areas. Although the same general patterns of change occur in many temperate estuaries, the unique combination of habitats in different systems leads to broad variations of trophodynamic processes in different coastal systems (Livingston, 2000). Different combinations of controlling factors, together with specific patterns of inshore–offshore migration of marine forms and offshore movements of euryhaline species, contribute to the area-specific characteristics of any given inshore system. Most coastal areas combine both autochthonous and allochthonous sources of nutrients and resulting high primary productivity to form the basis of high secondary production that translates into high levels of marine sports and commercial fisheries.

The basic characteristics of a given coastal system are determined by physiographic conditions (depth, surface area, connections to the open ocean) and freshwater input with accompanying loading of dissolved and particulate substances (inorganic and organic) (Livingston, 2000). Seasonal cycles of temperature and rainfall add a dimension of recurrent changes of primary and secondary productivity. The particular combination of habitat features and temporal biological responses to such conditions ultimately determines the specific ecological attributes of any given system. Inputs from freshwater and saltwater wetlands (allochthonous and autochthonous), *in situ* phytoplankton productivity, and submerged aquatic vegetation all contribute to the rich coastal soup. With the exception of few species that spend their entire life cycles in such systems (i.e., shellfish, some fishes), most estuarine species migrate into inshore coastal areas as larval stages and juveniles after spawning offshore. Many commercially important species, such as oysters, penaeid shrimp, blue crabs, and various finfish, are euryhaline and eurythermal, and are thus capable of living in rapidly changing habitat conditions. These species utilize the abundant food resources of coastal systems while remaining relatively free from predation by the stenohaline (salinity-restricted) offshore marine forms. Thus, coastal systems are often physically stressed but remain as a highly productive sanctuary for developing stages of offshore forms, many of which are used directly and indirectly by humans. Gradients of physical variables, together with forms of food availability and utilization, provide the basic components for the highly complex food webs of the inshore marine systems. Predator–prey interactions thus combine with a dynamic habitat to produce highly productive coastal food webs.

The rapidly changing physical conditions and intermittent (seasonal, interannual) cycling of organic production are important determinants of the form of the trophic organization in coastal areas. However, the final food web structure of any given system is determined by complex interactions of nutrient loading, corresponding primary production, interacting habitat conditions, and biological modifying factors such as predator–prey interactions and competition. High primary and secondary production, high dominance, and low species richness prevail under such conditions. Along the increasing salinity gradient, biological features become more important as determinants of community structure. With increasing salinity, population distribution becomes more even, with a corresponding decrease in relative dominance and an increase of species richness and diversity. Seasonal changes of predation pressure as well as recruitment characteristics of species with high abundance and/or biomass may modify this general condition; however, at any given time, coexisting estuarine assemblages reach different levels of equilibrium that depend largely on the exact temporal sequence of key habitat factors. The highly adaptable forms of coastal species often prevent direct (linear) relationships with specific driving functions. In this way, nonlinear (i.e., biological) processes are thought to contribute to what may often appear as chaotic conditions even though there is often an underlying organization that is grounded largely in the trophic responses of existing species to a never-ending series of physical/chemical interventions.

Empirical demonstrations of aquatic food webs (Odum and Heald, 1972) have been successfully applied to universal management schemes (Odum et al., 1982). Unfortunately, the successful application of modern methods of food web analysis to research management questions in coastal areas has been inconsistent at best. The extreme complexity of aquatic food webs has led to accepted generalizations concerning the definition of different food classes without corresponding verification of the trophic model as a universal entity. In this way, many studies translate limited data into generalized trophic models without adequate reference to the inherent variability of trophodynamic processes from system to system. The determination of specific criteria for a detailed delineation of trophic organization has proved to be less important than the consistency of the data on which the trophic models have been based. Much of the theoretical scientific literature concerning trophic organization thus remains overgeneralized. In addition, the dependence of trophic response on both qualitative and quantitative aspects of primary production (i.e., phytoplankton assemblages) remains almost completely ignored in most ecosystem-level analyses (Livingston, 2000). Thus, there is a basic gap in our understanding of how coastal systems function at a time when such knowledge is in great demand. The generation of food web information and the application of such information to coastal management questions may thus provide a crucial link to the preservation of what remains of once highly productive coastal ecosystems.

IV. TROPHIC STUDIES IN THE NE GULF OF MEXICO

An in-depth, multidimensional analysis has been carried out in the Perdido drainage system since 1988 (Livingston, 2000). Coffin and Cifuentes (1992, 1999) conducted stable isotope analyses of carbon cycling ($\delta^{13}C$) and nitrogen cycling ($\delta^{15}N_{NO3}$) in Perdido Bay to evaluate carbon sources and nutrient cycling. The authors measured carbon isotope values of suspended particulate matter (SPM), dissolved organic and inorganic matter, and bacteria. Potential sources of organic matter and nutrients included terrestrial runoff (forests, agricultural lands), urban runoff, municipal and industrial wastes, and water from coastal regions through advection from the mouth of the estuary. Terrestrial organic matter was the primary carbon source assimilated by bacteria in the system. Some inorganic nitrogen originated from the coastal region due to water movement through a pass on the Gulf so that the offshore coastal region was considered to be a significant source of nitrogen to the bay. This conclusion was consistent with results from nutrient limitation experiments carried out in Perdido Bay (Flemer et al., 1997). An important source of SPM with unusually light isotopic ratios was considered autochthonous. There was evidence that primary production was linked to an important fraction of the ^{13}C-depleted SPM, and that phytoplankton activity was responsible for such production. This production was high during spring months based on phytoplankton assimilation rates of ^{13}C-depleted dissolved inorganic carbon (DIC); this finding was consistent with analyses of the seasonal distribution of phytoplankton numbers and biomass that peaked during spring months (Livingston, 2000; also, see below).

Although analyses of $^{13}C_{DOC}$ and $^{13}C_{POC}$ in Perdido Bay indicated that a primary source of organic carbon was terrestrial, phytoplankton appeared to be an important source of the organic carbon pool in Perdido Bay. Macauley et al. (1995) found that phytoplankton production in Perdido Bay was an important source of organic carbon to associated food webs. The studies in Perdido Bay thus indicated that there was a direct connection between phytoplankton in the water column and benthic processes that were important to the overall food web of the bay. Interactions of freshwater flows, nutrient loading, light, and temperature were all important factors that were related to phytoplankton production and bloom generation in the bay (Livingston, 2000). These factors are universal in phytoplankton dynamics in coastal areas (Gilbert et al., 1995, 2001).

A recent study in the Apalachicola Bay system (Chanton and Lewis, 1999) indicated that, although there were inputs of large quantities of terrestrial organic matter, net heterotrophy was not dominant relative to net autotrophy during a 3-year period. Chanton and Lewis (2002), using $\delta^{13}C$ and $\delta^{13}S$ isotope data, found that there were clear distinctions between benthic and water column feeding types. They found that the estuary was dependent on river inflows to provide floodplain detritus (Mattraw and Elder, 1982; Elder and Cairns, 1984; Livingston, 1984b) during high-flow periods, and dissolved nutrients for estuarine primary productivity during low flows.

Floodplain detritus was significant in the important East Bay nursery area, thus showing that peak flows were important in washing floodplain detritus into the estuary. The data indicated that altered river flow, especially during low-flow periods, could adversely affect overall bay productivity. These studies indicated that phytoplankton productivity was an important component of estuarine food webs along the Gulf coast and that a combination of river-derived organic matter and autochthonous organic carbon provided the resources for consumers in Gulf coast river-dominated estuaries.

The use of holistic reconstructions of bay habitat represents another element in the time-based factors that determine food web structure in coastal systems (Abood and Metzger, 1996). These authors showed the importance of historical reviews to an understanding of human impacts on the Hudson River system. Brush (1991) carried out a long-core analysis of the Perdido system. The historical data recorded in long cores sampled in Perdido Bay sediments indicated that salinity increases in the Perdido estuary were a sharp deviation from the natural (i.e., freshwater) conditions of the estuary at the turn of the 19th century. The artificial opening of Perdido Pass during the early 1900s was an important modifying factor in the current state of the bay by introducing saline water to a previously freshwater system. Brush (1991) also indicated changes that were possibly related to non-point sources such as timbering, agriculture, and municipal development, but very few if any changes could be specifically related to the paper mill activity in the upper basin.

CHAPTER 2

The Northeast Gulf of Mexico

I. BACKGROUND

The Gulf coastal zone of north Florida extends from the alluvial Perdido River–Bay system in eastern Alabama and the western Florida Panhandle to the blackwater rivers of Apalachee Bay (the Big Bend area: Figure 2.1). The drainage systems that have been part of the long-term studies of our research group are given in Figure 2.2. Also shown are the permanent sampling stations that have been used for these studies. The Panhandle landscape is the result of stream and river flows and wave action that has acted on the land surface over the past 10 to 15 million years. This region is dominated by beach ridges, barrier islands, spits, cliffs, swales, sloughs, dunes, lagoons, and estuaries along a relatively flat upland configuration. Western bay systems are characterized by a series of alluvial rivers with relatively restricted coastal plain areas. The Apalachicola and Apalachee Bay basins are part of broad coastal plains that include extensive marsh areas. Barrier islands start in Apalachicola Bay, extending west to the Pensacola and Perdido Bay systems. On the eastern end of the Panhandle coast (Apalachee Bay), there is no barrier island development. Here, the coast is dominated by coastal swamps and marshlands. This area is characterized by shallow, sloping margins, the lack of wave action, and an inadequate supply of sand (Tanner, 1960) for barrier island development.

The coastal area from the Ochlockonee drainage to the southern reaches of Apalachee Bay is characterized by the flat, Karst topography of Gulf coastal lowlands punctuated by a series of small streams that flow through freshwater swamps and salt marshes into a shallow marine area bounded by a broad, marine shelf (Tanner, 1960). The Apalachicola River provides alluvial input for the barrier islands (St. Vincent, St. George, and Dog) that envelop the southern end of the estuary. From western Apalachicola Bay to the Choctawhatchee Bay area, alluvial and shelf sediments are channeled into deeper water via westward drift. There are no true barrier islands in the region between the Apalachicola and Choctawhatchee systems where moderate wave action forms a series of beaches. The Choctawhatchee River provides sand for an adjoining barrier system to the west including Santa Rosa Island. Downdrift from the Choctawhatchee Bay and from the Continental Shelf supplies sediments for Santa Rosa Sound to the Pensacola Bay system. Perdido Key encloses the Perdido system at its southern terminus.

The climate of the NE Florida Gulf coast is south temperate, with mild temperatures and a humid atmosphere (Fernald and Patton, 1984; Wolfe et al., 1988). Winter temperature changes occur due to cold fronts. Mean annual temperatures in the region approximate 20°C, with some variation depending on elevation and proximity to the coast (Livingston, 1984a). Annual rainfall along the NE Gulf coast of Florida is usually bimodal on an annual basis, with a major peak during summer–early fall months (June–September) and a secondary peak during winter–early

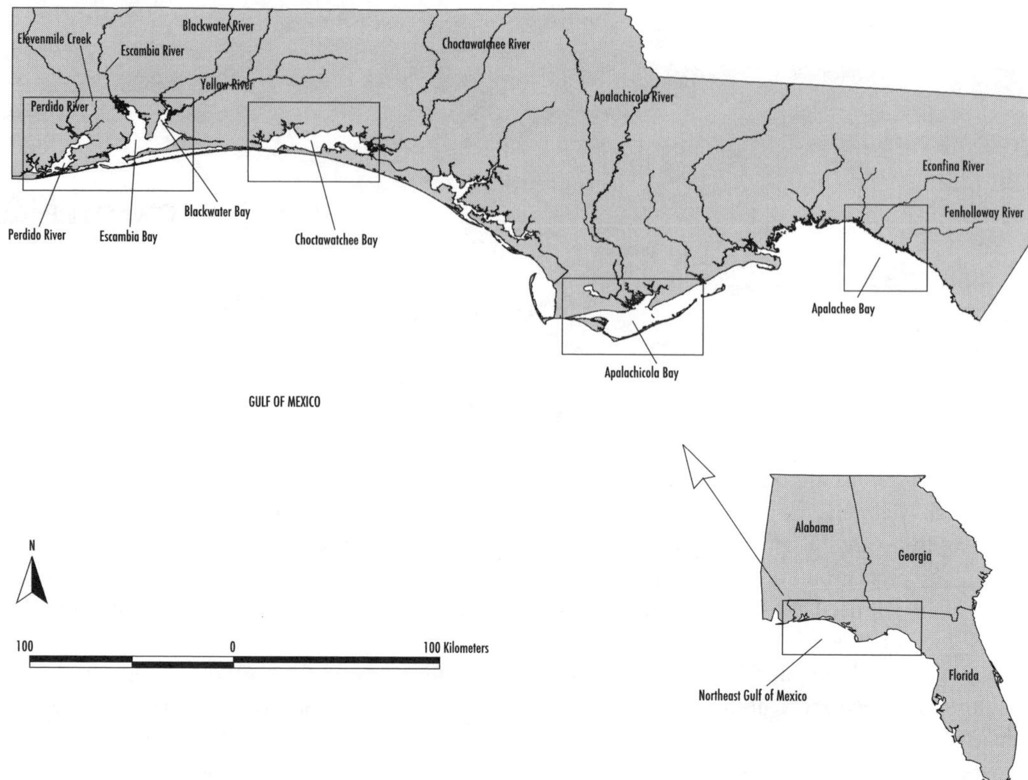

Figure 2.1 The Florida Panhandle showing the primary river basins and distribution of associated river–estuaries and coastal systems. This figure is a composite of information provided by the National Oceanic and Atmospheric Administration, the Florida Department of Environmental Protection, and the ESRI Corporation.

spring periods (January–April). Mean annual rainfall in the region approximates 152 cm, which ranges from 163 cm in western areas to 142 cm in the east (Wolfe et al., 1988). Alluvial river flows in the region are influenced by Georgia and Alabama rainfall patterns that approximate equal summer and winter peaks. Winter rains usually occur with frontal systems from the north, and are generally of longer duration than summer storms. There is heavy rainfall associated with periodic storms and hurricanes that are relatively frequent. Gulf air circulation is mainly anticyclonic around a high pressure region during periods from March through September (Wolfe et al., 1988). Winds tend to be highest from January through March. In spring (March–May), winds come mainly from the south or southeast depending on the region. During summer (June–August), the prevailing winds are mixed, coming from the southeast (in eastern sections of Panhandle Florida) or the southwest (in western sections). In shallow estuarine areas, winds can have a major influence on tidal current patterns. Such effects are due to the orientation, physiography, and spatial dimensions of a given river-estuarine or coastal system.

Livingston (2000) described habitat distribution and the ecological features of the study area (NE Gulf of Mexico: Figure 2.3). The Florida Panhandle is one of the least developed coastal areas in the continental United States. This region has relatively low human populations (Figure 2.4). The Apalachicola and Apalachee Bay drainage basins are among the least populated coastal areas in the United States. Coastal urban areas include Pensacola (Escambia Bay–Pensacola Bay), Destin/Fort Walton Beach/Niceville (western Choctawhatchee Bay), and Panama City (St. Andrews Bay) (Figure 2.4). The distribution of sewage treatment facilities, hazardous

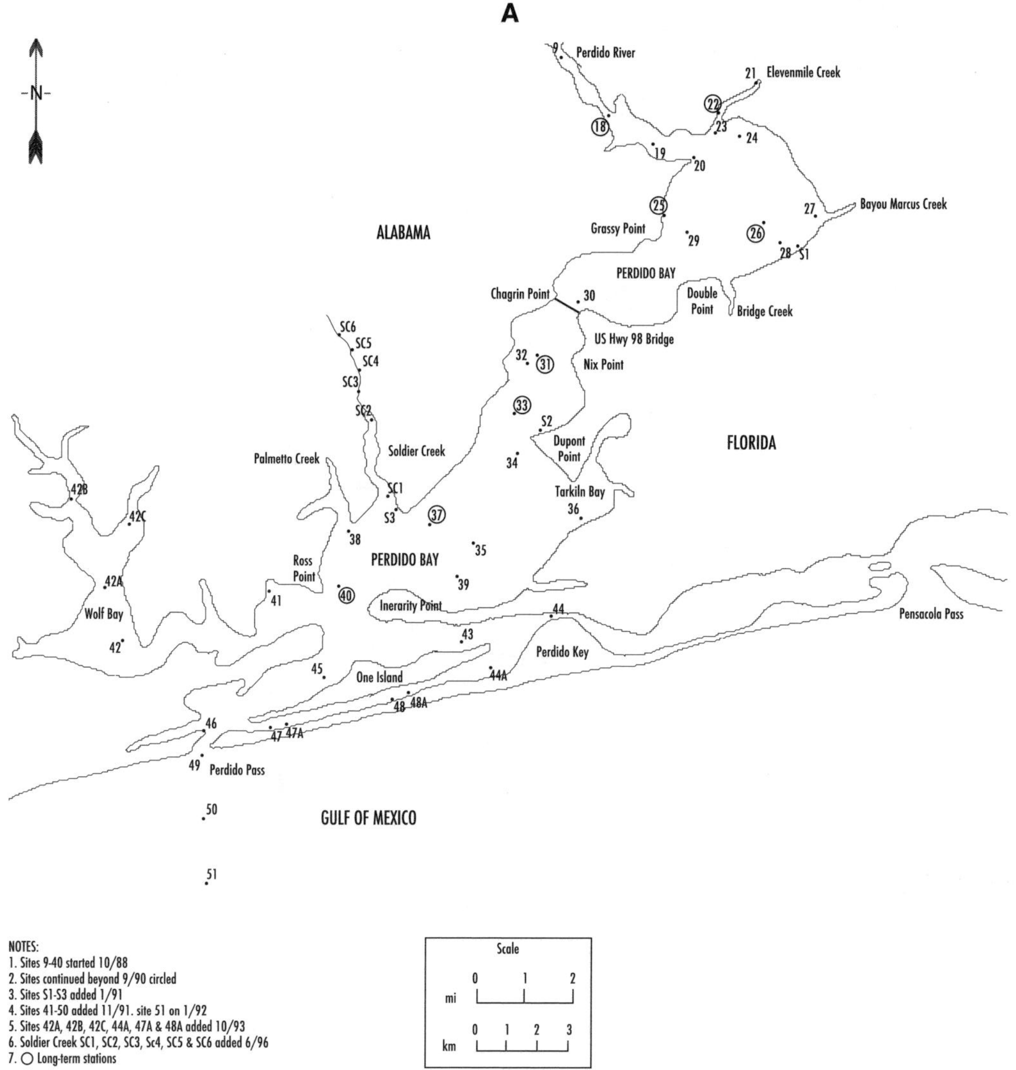

Figure 2.2 (A) The Perdido River–estuary showing sampling stations for long-term synoptic monitoring from October 1988 through June 2002. (continued)

waste sites, and NPDES permit sites follows closely the distribution of human populations along the NE Gulf coast (Figure 2.5). Outstanding sources of pollution loading in the Perdido Bay system include a pulp mill in the upper bay, and agricultural and urban runoff in the lower bay. The highest concentration of such pollution sources occurs in the Pensacola Bay system with secondary increases in western sections of Choctawhatchee Bay and St. Andrews Bay. Apalachicola Bay and Apalachee Bay remain relatively free of such discharges and are among the least polluted such coastal systems in the conterminous United States. In the Apalachicola system, the single most important pollution source is a sewage treatment plant in Apalachicola that discharges into a creek north of the city. In Apalachee Bay, the primary source of pollution is a pulp mill on the Fenholloway River, with discharges near the inland city of Perry, Florida. In later chapters, these east-to-west gradients of human activities are related to differences of food web structure and associated losses of secondary productivity in river–bay systems of the NE Gulf coast.

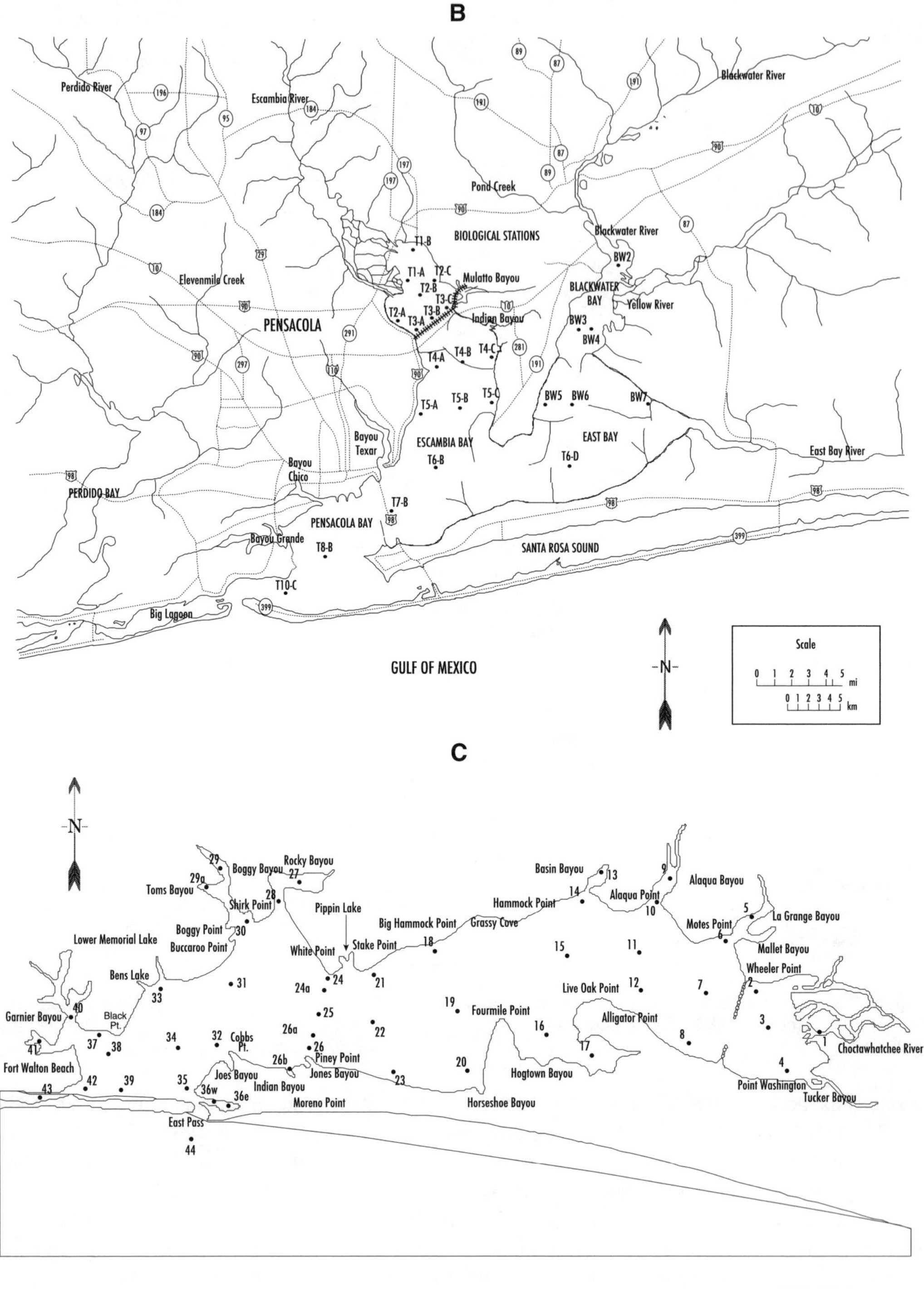

Figure 2.2 (continued) (B) The Pensacola Bay system (Escambia, Blackwater, Yellow) showing sampling stations for studies from May 1997 through October 1998. (C) The Choctawhatchee River and Bay system showing sampling stations for studies from September 1985 through 1986. (continued)

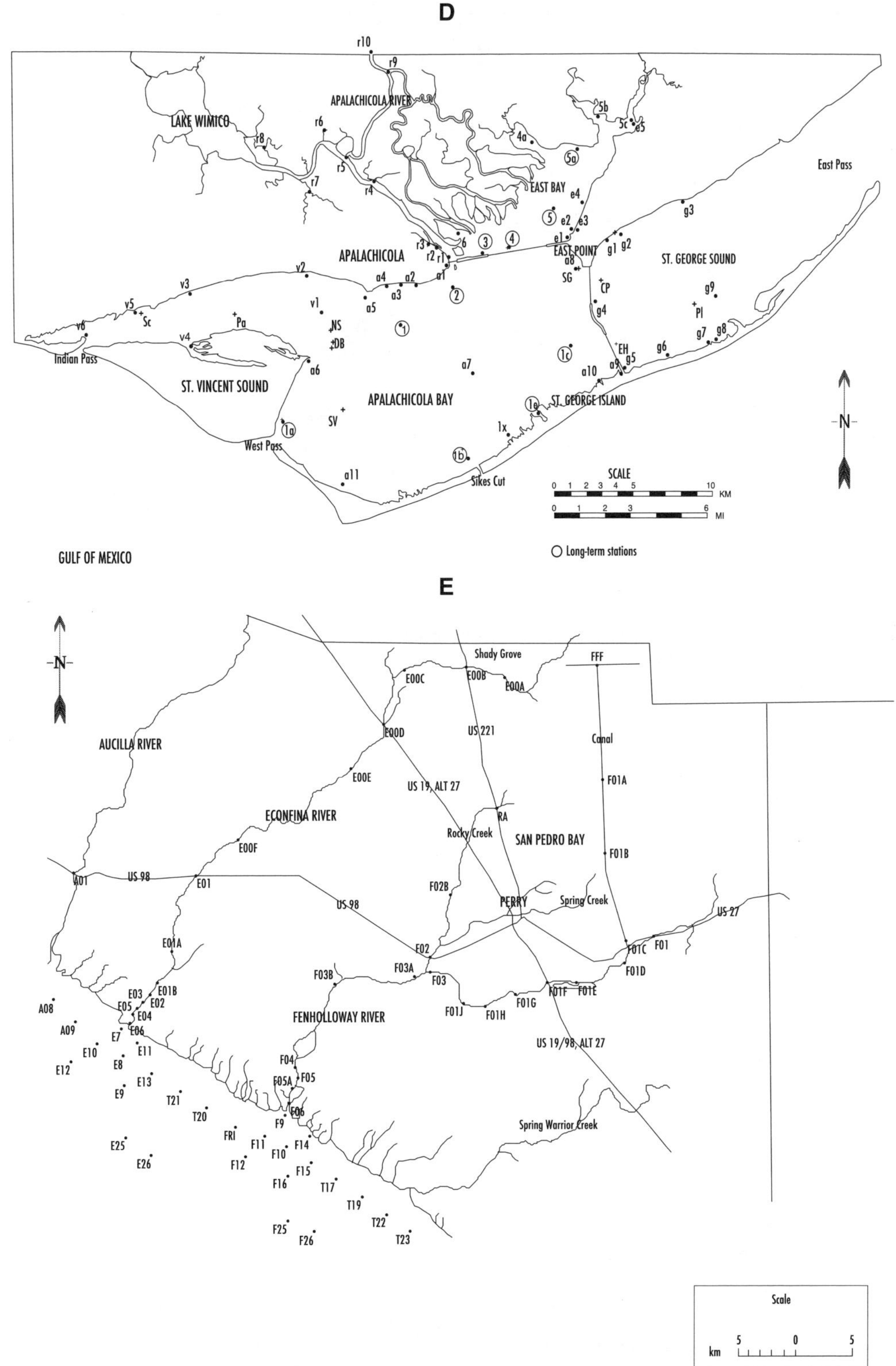

Figure 2.2 (continued) (D) The Apalachicola Bay system showing sampling stations for studies from March 1971 through December 1991. (E) Apalachee Bay showing the Econfina and Fenholloway estuaries with sampling stations for studies from June 1971 through June 2002.

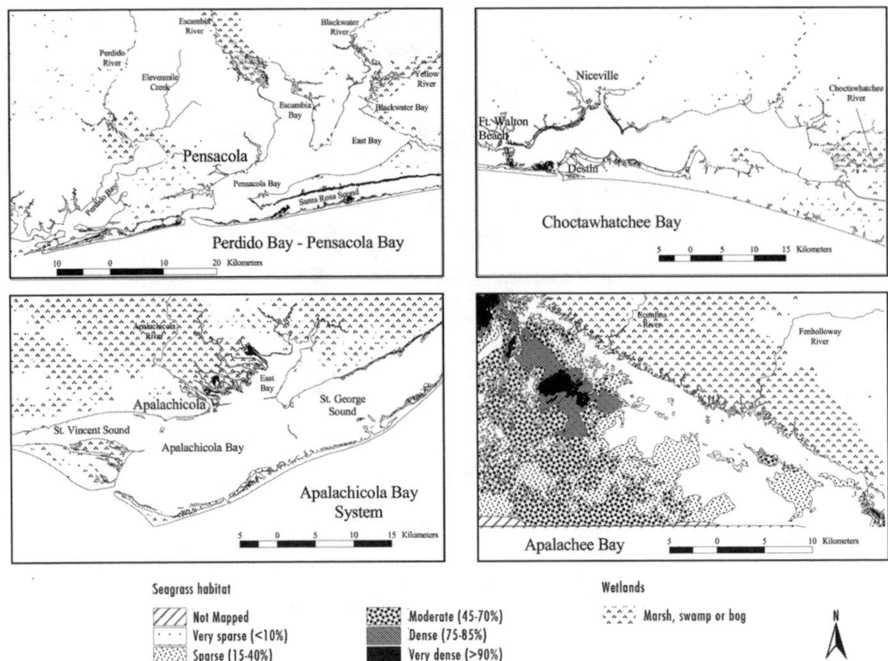

Figure 2.3 Sea grass and wetland habitat features of bay systems in Northwest Florida. This figure is a composite of information provided by the National Oceanic and Atmospheric Administration, the Florida Department of Environmental Protection, and the ESRI Corporation.

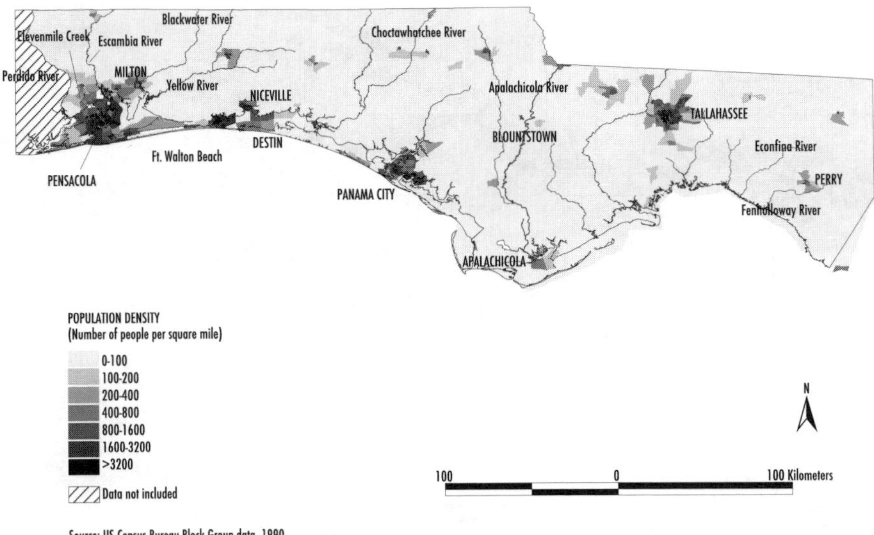

Figure 2.4 Population density in the Northwest Florida area of study. This figure is a composite of information provided by the National Oceanic and Atmospheric Administration, the Florida Department of Environmental Protection, and the ESRI Corporation. (From U.S. Census Block Group data, 1990.)

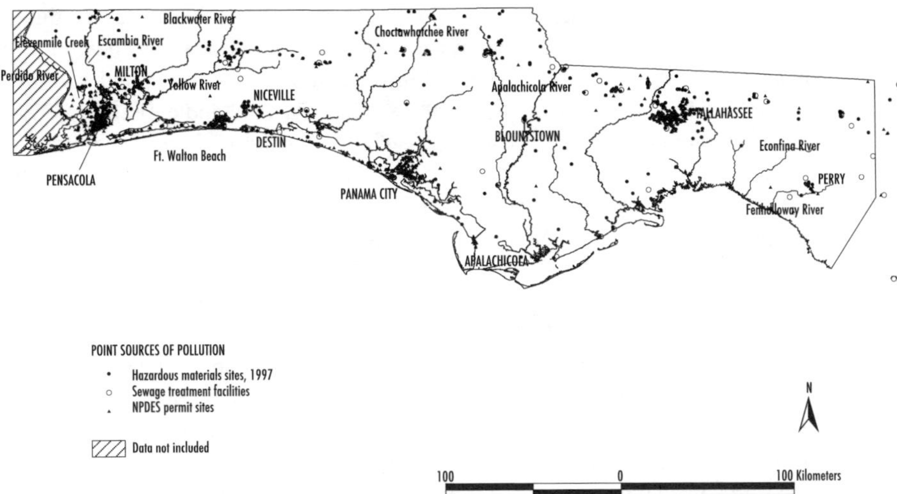

Figure 2.5 Distribution of point sources of discharges to coastal waters in the Northwest Florida study area. This figure is a composite of information provided by the National Oceanic and Atmospheric Administration, the Florida Department of Environmental Protection, and the ESRI Corporation.

II. DRAINAGE BASINS

Upland watersheds for coastal regions of the NE Gulf are located in Alabama, Florida, and Georgia, in an area approximating 135,000 km^2 (Figure 2.6). Associated estuarine/coastal systems are characterized by habitats that are largely controlled by the upland freshwater drainage basins. The salinity regimes of these areas are variously affected by major river systems or, as is the case in Apalachee Bay, by a series of small rivers and groundwater flows. The combined drainage of springs and streams contributes about 1 billion gallons of fresh water per day to Apalachee Bay, which is the only study area not affected by a major alluvial river system (Livingston, 1990). Nine

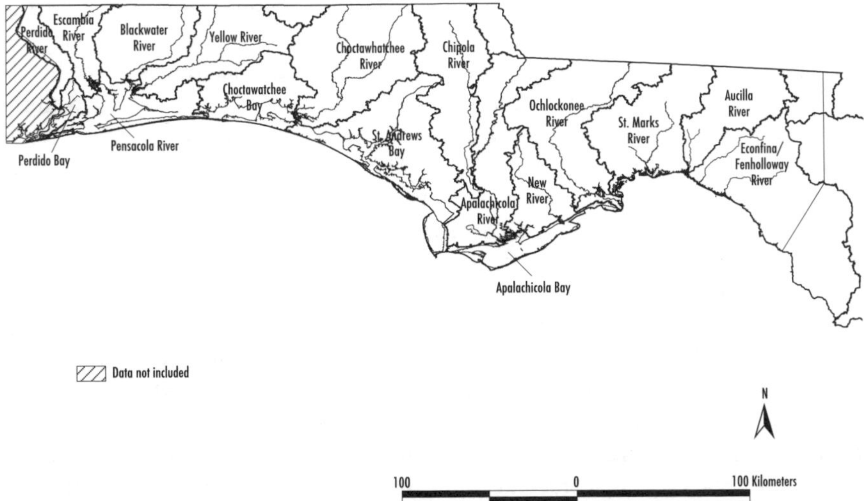

Figure 2.6 Hydrologic basins of drainages in the Northwest Florida study area. This figure is a composite of information provided by the National Oceanic and Atmospheric Administration, the Florida Department of Environmental Protection, and the ESRI Corporation.

of the 12 major rivers and five of the seven major tributaries of Florida occur in this region. Intersecting habitats of the coastal zone include saltwater marshes, sandy beaches, tidal creeks, intertidal flats, oyster reefs, sea grass beds, subtidal unvegetated soft bottoms, and various transitional areas. The major alluvial rivers of the northwest Florida Panhandle (Perdido, Escambia, Choctawhatchee, Apalachicola) have their headwaters in Georgia and Alabama (see Figure 2.1). A series of smaller streams along the Panhandle coast include the Blackwater and Yellow Rivers of the Pensacola Bay system, the Chipola River (part of the Apalachicola drainage), and the Ochlockonee River on Apalachee Bay. Farther down the coast, a series of small streams (St. Marks, Aucilla, Econfina, and Fenholloway) with drainage basins in Florida, flow into Apalachee Bay.

In the Panhandle region of Florida, groundwater and surface water bodies are often directly connected by rock channels (Wolfe et al., 1988). The Karst topography that is so prevalent in this area is characterized by porous limestone that is penetrated by various solution channels and sandy substrates. The connections to the surface are typically in the form of sinkholes and springs, which can act as direct connections between the surface waters and groundwater aquifers. Rainwater drains rapidly into the Surficial Aquifer, which is composed mainly of coarse-to-fine sands intermixed with surface gravels. The Surficial Aquifer, which is recharged locally, is often separated from the underlying Floridan Aquifer by a clay-confining (impermeable) layer. The condition of this Surficial Aquifer, in terms of previous (recent) rainfall conditions, has an effect on the surface flows in the stream systems of the region. Some areas, such as the Perdido drainage basin, have relatively flashy flows in response to precipitation incidents. Periodic flash flooding occurs in the non-alluvial streams such as Elevenmile Creek.

III. COASTAL HABITATS

Habitat conditions in Gulf coastal areas of north Florida are determined by different factors. Astronomical tides in the western Panhandle from the Apalachicola to the Perdido systems are diurnal, with small amplitudes between 0.37 and 0.52 m (Livingston, 1984a). Tides from Apalachee Bay to the Apalachicola estuary are mixed and semidiurnal, with two (unequal) highs and lows each tidal day. Tides in this region range from 0.67 to 1.16 m. Wave height and energy density are moderate (12.25 to 306.25 J/m^2) between the Ochlockonee and Perdido estuaries (Fernald and Patton, 1984). This Gulf coastal region has low energy conditions (1.96 to 12.25 J/m^2), but is dominated by relatively high freshwater flows from rivers and groundwater input. The physical dimensions of a series of upland basins are a key element in the determination of the distribution and nature of coastal aquatic habitats. The size and depth of a given bay along with the location of passes to the Gulf are also important factors in habitat definition through determination of stratification potential and salinity distribution. The combination of freshwater influxes, basin characteristics (depth), and areal dimensions thus determines the spatial/temporal dimensions of habitat characteristics of the Gulf coastal systems of north Florida.

The inshore marine habitats of the northwest Florida coast (see Figure 2.2) represent a diverse series of coastal systems with relatively high primary and secondary productivity. Multiple sources (allochthonous, autochthonous) of nutrients together with more or less continuous disturbance by winds and tides provide a continuous supply of organic matter for inshore food webs (Livingston, 1984a). Freshwater inflows, saltwater wetlands, *in situ* phytoplankton productivity, and submerged aquatic vegetation all contribute to the coastal food webs. These areas are physically stressed, but provide a highly productive nursery area for eurytopic developing stages of offshore forms, many of which form the basis of the sports and commercial species in the region. In a functional sense, these coastal systems encompass a biological continuum from upland, freshwater drainage basins to the offshore Gulf that is characterized by certain common physical, chemical, and biological attributes. Each coastal system represents a unique combination of these habitats. Primary productivity differs widely both within a given system and among the various coastal areas. This unique identity is also supported by individual adaptive responses of indigenous species from phytoplankton to fishes. Seasonal and interannual cycles of controlling factors, together with intermittent storms and hurricanes, provide the highly variable background for the area-specific characteristics of these coastal systems.

There are significant differences in the development of emergent and submergent vegetation in the subject coastal systems of the NE Gulf (see Figure 2.3). Perdido Bay and the Pensacola system have moderate concentrations of marshes whereas the Choctawhatchee system is practically devoid of emergent vegetation. The Apalachicola and Apalachee Bay systems have extensive and well-developed marsh systems. The big alluvial systems have limited development of submerged aquatic vegetation whereas Apalachee Bay is characterized by one of the highest concentrations of sea grass beds in the Northern Hemisphere. The Econfina and Fenholloway estuaries are distinguished from the alluvial river–bay systems to the west by the relatively low flow/watershed and flow/open water ratios (see Figure 2.3), and by the well-developed marsh areas relative to the flow rates of the contributing rivers to Apalachee Bay. The shallowness of the bay, together with the relatively low freshwater flows and considerable development of fringing coastal wetlands, contributes to the sea grass beds as the dominant offshore habitat along the Big Bend area of north Florida.

IV. ESTUARINE/COASTAL SYSTEMS OF THE NE GULF

The following is a brief description concerning background information on the primary study sites on which this book is based. A more detailed evaluation is presented by Livingston (2000).

A. Perdido Bay System

The Perdido Bay system (see Figure 2.2A) lies in an area of submergence on the north flank of the active Gulf coast geosyncline, and is a shallow to moderately deep (average depth; 2.2 m) inshore water body oriented along a northeast–southwest axis. The bay can be divided into four distinct geographic regions: the lower Perdido River, upper Perdido Bay (north of the route 98 bridge), lower Perdido Bay (bounded to the south by a line between Ross Point and Inerarity Point), and the Perdido Pass complex (see Figure 2.2). The bay system has a length of 53.4 km and an average width of 4.2 km. The primary source of freshwater input to the estuary is the Perdido River system that flows southward about 96.5 km, draining an area of about 2937 km^2 (Livingston, 1998, 2000). The Elevenmile Creek system (including Eightmile Creek; see Figure 2.2A) is a small drainage basin (about 70 km^2) that receives input from a municipal waste system and a paper mill. The relatively small Bayou Marcus Creek drains a residential area of western Pensacola with input from urban storm water runoff and a sewage treatment plant (recently diverted to adjoining marshes). The Gulf Intracoastal Waterway (GIWW) runs through the lower end of Perdido Bay about 5.6 km northeast of Perdido Pass. The U.S. Army Corps of Engineers (USACOE) maintains the Perdido Pass channel at about 4 m as part of the GIWW (U.S. Army Corps of Engineers, 1976). As in the Choctawhatchee Bay system, there is a shelf that can extend up to 400 m in width around the periphery of the Perdido estuary. This shelf usually does not exceed 1 m in depth. The upper bay is relatively shallow, and depth tends to increase southward with the deepest parts of the estuary located at the mouth of the Perdido River and in the lower bay off Ross Point.

Prior to the opening of Perdido Pass in the early 1900s, Perdido Bay was a largely freshwater system, covered with freshwater plants (Brush, 1991). Access to the Gulf was restricted by the shallow, shifting body of fresh water. At the time of an early survey (1767), the pass had a depth of ~2 m. During an outbreak of malaria in the early 1900s, the mouth of the bay was opened to the Gulf, making it the saltwater system it is today. This action resulted in the creation of the Ono Island/Old River complex. The history of Perdido Bay is thus comparable to that of the Choctawhatchee Bay system (Livingston, 1986a,b), which was also opened at the mouth by another group of citizens with shovels in 1929. Since the turn of the 19th century, there has been a steady increase in the human population in the Perdido basin. By the 1980s, Perdido Key was undergoing rapid residential and commercial development. Problems of sewage treatment and urban storm water runoff remain unresolved in various residential areas around the bay to the present day. Agricultural runoff, based largely in Alabama, contributes to nutrient loading in Wolfe Bay at the lower end of the basin.

Detailed, long-term analyses of the Perdido drainage system have been carried out for the past 14 years (Livingston et al., 1998a; Livingston, 2000) that include long-term, baywide synoptic analyses at fixed stations (see Figure 2.2A). Previous information (Schropp et al., 1991) indicated that concentrations of dissolved organic carbon, Kjeldahl nitrogen, nitrate-nitrite, orthophosphate, particulate carbon, total nitrogen, total phosphorus, and total suspended solids were uniformly higher in Elevenmile Creek than in the other tributary streams. Isphording and Livingston (1989) found that wind events, salinity stratification, and temperature conditions were associated with observed changes in water quality and hydrological conditions. Low bottom dissolved oxygen was associated with high temperature and salinity stratification. The Perdido River represented the primary source of fresh water and sediments to the bay (Isphording and Livingston, 1989). Metal concentrations (Cd, Co, Cu, Cr, Ba, Zn, Fe, Va, Al) in sediments of Perdido Bay were comparable to those found in other Gulf areas. Relatively low organic carbon and high percentages of coarser-grained sediments indicated only limited sites for the accumulation and attachment of contaminants that enter the bay. During most times and in most portions of the bay, chlorophyll *a* was relatively low in the Perdido Bay system compared with other reference bay systems such as Apalachicola Bay and Choctawhatchee Bay.

According to a U.S. Fish and Wildlife Service Report (1990), submerged aquatic vegetation (SAV) was largely concentrated in the lower bay. Historically, SAV has decreased by more than half from 1940–1941 to 1979. Dredging of the GIWW in the 1930s and continuous pass enlargement and open water spoil disposal have been postulated as factors in the decline of SAV in the lower bay (Bortone, 1991). SAV development has been restricted to Grassy Point on the west side of the upper bay with *Vallisneria americana* as the dominant species. Field/laboratory experiments (Livingston, 1992) indicated that light is the chief limiting factor with grass development not extending deeper than 0.8 to 1.0 m. Based on the success of previous grass bed transplant experiments, Davis et al. (1999) concluded that *V. americana* beds in upper Perdido Bay were recruitment limited rather than constrained by water quality, toxic substances, light inhibition, or unsuitable substrate.

Overall, the Perdido Bay system has been adversely affected by pulp mill effluents in the upper bay, the dredged opening to the Gulf in mid- to lower bay areas, and urbanization and agricultural runoff in the lower bay.

B. Pensacola Bay System

The Pensacola Bay system includes five interconnected estuarine components: Escambia Bay, Pensacola Bay, Blackwater Bay, East Bay, and Santa Rosa Sound (see Figure 2.2B). This system is dominated by a series of river basins (the Escambia, Blackwater, and Yellow). The alluvial Escambia River system (10,880 km^2) extends northward about 386 km from the north end of Escambia Bay. The Blackwater River (2230 km^2) and the Yellow River basin (3535 km^2) contribute to the Blackwater/East Bay system. Escambia Bay is located east of the City of Pensacola (see Figure 2.2B). The primary source of fresh water to the bay is the Escambia River; other sources include the Pace Mill Creek and Mulatto Bayou drainage basins in the upper bay, and the Bayou Texar, Bayou Chico and Bayou Grande basins in the lower bay. Pensacola Bay receives flows from Escambia Bay, East Bay, and Bayou Grande, and is bordered to the north by the City of Pensacola and to the south by Santa Rosa Island. Pensacola Bay empties into the Gulf of Mexico through a pass at its southwestern terminus. Santa Rosa Sound is a lagoon between the mainland and Santa Rosa Island that connects Pensacola Bay with Choctawhatchee Bay to the east. The individual estuarine basins are characterized by a relatively shallow shelf, a slope area, and a deeper coastal plain similar to that noted in the Perdido and Choctawhatchee systems.

The history of Escambia Bay reflects long-term adverse impacts due to urban runoff, nutrient loading from point and non-point sources, and toxic waste loading from industrial sites. From the 1950s to the present, the system has been seriously affected by nutrient loading. During the late 1960s to early 1970s, the Pensacola system underwent extensive fish kills, habitat loss, and the almost complete destruction of sea grass beds and a once-thriving oyster industry due to hyper-eutrophication (Livingston, 1989, 1990, 1997a, 1999, 2000; Collard, 1991a,b). Upper Escambia Bay was particularly affected due to reduced circulation and a combination of natural and anthropogenous factors. Extensive fish kills led to restrictions on nutrient loading during the early 1970s that were accompanied by partial recovery of water quality and secondary production. However, non-point-source pollution from the City of Pensacola and the Naval Air Station, together with major pollution of the bayous (Texar, Chico, Grande) in this region, has led to continued low secondary production in this system (Livingston, 1999). Non-point-source pollution and sewage discharges continue to cause serious water quality problems in local areas to this day. Storm water outfalls are particularly numerous in western sections of Escambia Bay, and septic tank numbers are high in the lower bay.

Using quantitative whole-water phytoplankton data, Livingston (1997a, 1999) found that Escambia Bay was in a moderately eutrophic state. Subdominant phytoplankton populations (known to be noxious bloom species in Perdido Bay) were found in mid and lower Escambia Bay. Urban storm water runoff was considered a factor in the maintenance of these populations (Livingston, 1997a, 1999). Habitat factors, such as sediment type, dissolved oxygen, and salinity distribution, while important determinants of the general distribution of organisms in the bay system as a whole, had effects that could only be understood within the context of the distribution of primary (phytoplankton) production and the trophic organization of the bay (Livingston, 1997a, 1999). When compared with unpolluted systems such as Apalachicola Bay, secondary production in upper Escambia Bay was relatively low (Livingston, 2000).

The Pensacola Bay system (see Figure 2.2B) is one of the most heavily polluted bays in Florida in terms of toxic substances (Seal et al., 1994; Livingston, 2000). The bayous on western parts of Escambia Bay collect urban runoff and have a long history of poor water and sediment quality with accompanying fish kills (Moshiri, 1976, 1978, 1981; Moshiri and Crumpton, 1978; Moshiri et al., 1978, 1979, 1980, 1987). High levels of metal contamination have been found in Bayou Grande (Cd, Pb, Zn) and Bayou Chico (Cr, Zn). Bayou Chico has also been contaminated with polynucleated aromatic hydrocarbons (PAHs) and polychlorinated biphenyls (PCBs). In Pensacola Bay, PAHs and PCBs were found in sediments close to shore and in central parts of the bay. Phenolic compounds were found at one site near Pensacola Harbor. Long et al. (1997), in a study of sediment toxicity in four bays of the Florida Panhandle, found that the greatest toxicity of all the bays analyzed occurred in Bayou Chico, with the other developed bayous in the Pensacola area showing "relatively severe toxicity." Toxicity of the Pensacola Bay sediments was attributed to high-molecular-weight PAHs, zinc, DDD/DDT isomers, total DDT, and dieldrin, a highly toxic organochlorine pesticide.

C. Choctawhatchee Bay System

The Choctawhatchee Bay system (see Figure 2.2C) is a drowned river plain surrounded by shallow shelf-slopes and inshore bayous (Livingston, 1986a,b), and is aligned along an east–west axis. Freshwater input is primarily associated with the Choctawhatchee River with secondary inflows from a series of bayous located primarily in northern sections of the bay. Saltwater input comes from the Gulf through the dredged East Pass, a channel opened by a group of citizens during particularly heavy freshwater flooding in 1929. The "blow-out" of East Pass was accompanied by increased salinity throughout Choctawhatchee Bay. High salinities were associated by local observers with the loss of emergent (marsh, swamp) and submergent (sea grass) vegetation throughout the bay. Choctawhatchee Bay currently has one of the least well-developed fringing (emergent) vegetation zones of the various Gulf estuaries in the region (Reyer et al., 1988; see Figure 2.3). The artificial creation of the channel across Santa Rosa Island has also contributed to erosional and depositional patterns in areas of the Choctawhatchee Bay system.

The Choctawhatchee River, the third largest in Florida in terms of freshwater discharge, is formed by a series of tributaries in Alabama and Florida. Mean stream flow rates range between 156 and 198 $m^3 s^{-1}$ with low flows approximating 96 $m^3 s^{-1}$ and high flows of 736 $m^3 s^{-1}$ (U.S. Geological Survey, Tallahassee, Florida; unpublished data). Choctawhatchee Bay (see Figure 2.2C) has average depths of ~3 m in eastern sections and ~8 m in western sections. Over the past 30 to 35 years, there has been a major increase of urbanization in western parts of the bay (Livingston, 1986a,b, 2000; see Figure 2.4). Eastern sections remain relatively undeveloped. The opening of East Pass has led to increased salinity stratification that, in turn, has been associated with periodic hypoxia and anoxia at depth in various parts of the system. Stabilization of the water column has also been associated with extensive development of liquid mud or nephlos (i.e., suspensions of fine particulate matter above the sediment/water interface) (Livingston, 1986a,b). The combination of salinity stratification in central and western sections of the bay, together with low turnover rates and associated habitat deterioration due to hypoxia and liquid mud development, has led to relatively low secondary productivity in the main stem of the system (Livingston, 1986a,b). This situation is similar to that described in Perdido Bay to the west.

A complete survey of the Choctawhatchee River and Bay (see Figure 2.2C) was conducted during the mid-1980s (Livingston, 1986a,b, 1987a,b, 1989, 1991a,b). The Choctawhatchee Bay system represents a series of complex habitats that are strongly influenced by freshwater input (river, bayou drainages) and salinity stratification. Various human activities associated with urbanization in the western Choctawhatchee basin (e.g., storm water runoff and sewage spills) have contributed to the deterioration of natural aquatic resources in receiving bay areas. According to Livingston (1987), discharges of sewage and storm water associated with unrestricted land development in western sections of Choctawhatchee Bay were responsible for water quality deterioration in a series of bayou areas in the northern sections of the bay and in Destin Harbor (Old Pass Lagoon) to the south. The combination of nutrient loading together with a lack of flushing contributed periodically to severe water quality problems (hypoxia, anoxia) and associated reductions of natural productivity in the lagoon. This situation was indicated by high nutrients (N, P) in the water column, silt-laden sediments (especially in eastern sections of the lagoon), high phytoplankton productivity, low numbers of herbivores (zooplankton, meroplankton), a depauperate infaunal invertebrate community through central/deep portions of the lagoon, and relatively few epibenthic invertebrates and fishes. It was suggested (Livingston, 1987c) that actions be taken to reduce input from sewage and urban storm water that accounted for the loading of nutrients, organic matter, and pollutants to the lagoon and some of the more polluted northern bayous.

Recent anecdotal information (K. Spencer, *Destin Log,* 1999–2000) indicated that uncontrolled urban development in western sections of the bay was associated with increased runoff and sewage spills. The anecdotal data were consistent with established signs of advanced hypereutrophication (i.e., hypoxia at depth, fish kills throughout the river–bay system, reduced secondary production; Livingston, 2000). These conditions were accompanied by extensive "red tide" (*Gymnodinium breve*) activity both in the bay and in offshore Gulf areas. There were also associated problems in the form of the deaths of sea turtles and "bottlenose dolphins" in the Choctawhatchee system and along the coast between Choctawhatchee Bay and St. Andrews Bay (K. Spencer, *Destin Log*, 1999–2000). About 121 dead dolphins were noted in this area between August 1999 and mid-January 2000. The deaths of fish-eating terrestrial mammals such as raccoons and foxes in the region were also observed in various parts of the western Choctawhatchee Bay area. Although there are no definitive answers regarding causation of the mammal deaths, researchers found red tide toxins in the dolphins' systems (K. Spencer, *Destin Log*,

1999–2000). Compared with an estimated annual death rate of three to five dolphins per year, the current die-offs were considered unusual. Lack of scientific information prevented an evaluation of causative factors relating to the blooms, and there is little evidence of changes in the phytoplankton assemblages that led to the red tide incidents.

D. Apalachicola Bay System

The Apalachicola Bay system (see Figure 2.2D) is a lagoon-and-barrier-island complex oriented on an east–west axis, parallel to the Gulf of Mexico. The Apalachicola River, the primary source of fresh water to the bay, is one of the last major free-flowing, unpolluted alluvial systems in the conterminous United States. The river drains a basin area of 45,405 km^2. The upper, tidally influenced reaches of the estuary form an extensive delta region as the river enters the bay through a series of tributaries. As an extension of the Apalachicola River, the estuary represents a shallow, width-dominated coastal plain system (Livingston, 1984a). The bay system, composed of East Bay, Apalachicola Bay, St. Vincent Sound, and St. George Sound, is shallow (mean depth ~2 m). Three barrier islands form the Gulf-ward extent of the bay. There are four natural openings to the Gulf: Indian Pass, West Pass, East Pass, and a pass to the East. A human-made opening (Sikes Cut) was established in the western portion of St. George Island in 1954. Sand, silt, and shell components that have accumulated over tertiary limestones and marls of alluvial origin with some recent modifications due to hurricanes dominate sediments in this estuary.

The alluvial Apalachicola Bay system remains in a relatively unaltered state with freshwater flow as the primary controlling variable of bay habitat conditions (Livingston et al., 1974, 1997; Livingston, 1983a, 1984a). Other driving forces of the estuary include seasonal fluctuations of rainfall and Apalachicola River flow, the physiographic structure of the receiving basin, wind speed (duration and direction), and exchanges between the estuary and the Gulf (Livingston, 1984a; Livingston et al., 1997, 2000). Hydrodynamic processes are further complicated by bathymetric modifications due to dredging activities that include the opening and maintenance of Sikes Cut and the Intracoastal Waterway that extends from the mouth of the river to St. George Sound to the east. Meeter and Livingston (1978) and Meeter et al. (1979) showed a strong correlation of Apalachicola River flow with the spatial and temporal distribution of salinity throughout the bay. The Apalachicola Bay system is the major oyster production area for the entire region. By coupling hydrodynamic modeling with descriptive and experimental biological data, Livingston et al. (2000) found that high salinity, relatively low velocity current patterns, and the proximity of a given oyster bar to entry points of saline Gulf water into the bay were important factors that contributed to increased oyster mortality. The highest oyster densities and maximum overall bar growth were found in the vicinity of the confluence of high salinity water moving westward from St. George Sound and river-dominated (low-salinity) water moving south and eastward from East Bay. By influencing salinity levels and current patterns throughout the bay, the Apalachicola River was important in controlling oyster mortality due to predation and disease.

Macroinvertebrate and fish production of the Apalachicola estuary is affected by complex, long-term (interannual) changes in river flow and the bay response to drought and flood conditions (Livingston et al., 1997). At specific thresholds of drought-induced reductions of fresh water, there was clarification of the normally turbid and highly colored river–estuarine system that led to increased phytoplankton productivity. These changes were associated with alterations of trophic organization of the system in the form of initial increases of herbivore/omnivore abundance. However, there was a dichotomous response of the estuarine trophic organization to drought–flood periodicity. Herbivores and omnivores were responsive to river-dominated physical/chemical factors, whereas the various levels of carnivory responded to biological factors such as prey availability. Trophic response time was measured in months to years from the point of the initiation of low-flow conditions. During later stages of the drought, there were reductions of secondary production throughout the bay. Reduced nutrient loading during the drought was postulated as a cause of the loss of productivity during and after the drought period. Recovery of secondary productivity with resumption of increased river flows was also a long-term event. There was a range of cyclic periodicity among the various trophic levels in the bay. Responses to regular interventions such as droughts varied among the different food web components with lag periods measured from months to years (Livingston et al., 1997).

E. Apalachee Bay System

The region along the upper Gulf coast of peninsular Florida from the Ochlockonee River to the Suwannee River is characterized by a series of drainage basins that include the Aucilla, Econfina, and Fenholloway Rivers. These streams drain into Apalachee Bay (see Figure 2.2E), which is the northern extension of a broad, shallow shelf area that occurs along the entire Gulf coast of peninsular Florida. The smaller basins are wholly within the coastal plain as part of a poorly drained region that is composed of springs, lakes, ponds, freshwater swamps, and coastal marshes. The Econfina and Fenholloway river–estuaries both originate in the San Pedro Swamp (see Figure 2.2E). This basin has been affected, in terms of water flow characteristics, by long-term physical modifications through forestry activities. However, most of Apalachee Bay remains in a relatively natural state due to the almost complete lack of human development in the primary drainage basins. The dominant habitat feature of Apalachee Bay is an extensive series of sea grass beds that extends to Florida Bay in the south to Ochlockonee Bay in the north (Iverson and Bittaker, 1986). The one area of significant anthropogenous effects in an otherwise pristine system is the Fenholloway River–estuary, where pulp mill discharges have caused adverse effects due to high dissolved organic carbon (DOC) and water color, high biochemical oxygen demand (BOD), low dissolved oxygen (D.O.), and high nutrient loading (ammonia and orthophosphate) (Livingston 1980a, 1982a, 1984d, 1985a,b, 1988a; Livingston et al., 1998b). The Econfina River remains one of the most natural blackwater streams along the coast, and has been used as a reference area for studies in the Fenholloway system since 1971. Both the Econfina and Fenholloway drainages (see Figure 2.2) share a common meteorological regime. River flow characteristics in the two drainages are comparable in rate and seasonal variation.

Phytoplankton productivity in Apalachee Bay is generally low compared with the alluvial systems to the west (Livingston et al., 1998b). Reductions of the Fenholloway sea grass beds have been associated with altered food web patterns in terms of species shifts and feeding alterations relative to reference areas (Livingston, 1975a, 1980a; Livingston et al., 1998b). Feeding habits of the dominant fish species were different in the offshore Fenholloway system from those found in similar areas of the Econfina drainage. These differences were traced to basic changes in habitat due to the reduced sea grass beds in the Fenholloway system (Clements and Livingston, 1983). In offshore Fenholloway areas, there was increased phytoplankton production as dilution increased light transmission in areas of increased nutrient loading from the paper mill. Subsequent improvement in water quality in 1974 was associated with some recovery of the fish trophic organization in outer areas of the Fenholloway system (Livingston, 1982a, 1984d; Livingston et al., 1997b). Livingston (1984d) showed that altered SAV distribution led to changes in the trophic organization of the Fenholloway system. During earlier periods of untreated mill effluents, plankton-feeding fishes in areas affected by discharges replaced grass bed species. Subsequent water quality improvement due to the implementation of secondary treatment of the mill effluents was accompanied by trophic shifts that followed habitat changes in the outer portions of the Fenholloway estuary.

CHAPTER 3

Long-Term Studies: Northeastern Gulf of Mexico

I. INTRODUCTION

Over the past 32 years, a series of field studies has been carried out in coastal areas of the Gulf of Mexico and the South Atlantic. An outline of field effort is given in Table 3.1. The research effort is based on written, peer-reviewed protocols for all field and laboratory operations. Water quality methods and analyses and specific biological methods have been continuously certified through the Quality Assurance Section of the Florida Department of Environmental Protection (Comprehensive QAP 940128 and QAP 920101). These methods have been published in the reviewed scientific literature, including the following: Livingston et al., 1974, 1976b, 1977, 1997, 1998a,b; Livingston, 1975a,b,c, 1976a, 1979, 1980a, 1982a, 1984a,b, 1985a,b, 1987c, 1988b, 1992, 1997a; Flemer et al., 1997. Ontogenetic feeding units have been determined from a series of detailed stomach content analyses carried out with the various epibenthic invertebrates and fishes in the region (Sheridan, 1978, 1979; Laughlin, 1979; Sheridan and Livingston, 1979, 1983; Livingston, 1980a, 1982a, 1984b, unpublished data; Stoner and Livingston, 1980, 1984; Stoner, 1980c, 1982; Stoner et al., 1982; Laughlin and Livingston, 1982; Clements and Livingston, 1983, 1984; Leber, 1983, 1985). An outline of such methods is given in Appendix I.

Subject coastal areas under study are shown in Figures 2.1 and 2.2, and include both alluvial (Perdido, Escambia, Choctawhatchee, Apalachicola) and blackwater (Blackwater, Econfina, Fenholloway, Amelia, Nassau) systems. As shown (Table 3.1), the long-term (i.e., >14 years) databases have been collected in two alluvial systems (Apalachicola, Perdido) and two blackwater systems (Econfina, Fenholloway). We have conducted nutrient loading determinations for more than three decades in these areas. Long-term analyses of phytoplankton communities have been carried out in the Perdido, Econfina, Fenholloway, Nassau, and Amelia systems. Shorter-term phytoplankton analyses have been made of the Escambia, Choctawhatchee, and Blackwater systems, with full databases (including phytoplankton community structure) for the Escambia and Choctawhatchee River–estuaries. Long-term studies of the Apalachicola River–estuary provide a base line of information for one of the last major unpolluted systems of its kind in the conterminous United States. A complete description of these systems, data collection efforts and field/laboratory operations, protocols for all data-collection and experimental methods, and methods of statistical analysis and modeling is given by Livingston (2000).

II. FIELD PROGRAMS

A general outline of the integrated field descriptive and field/laboratory experimental program carried out in the different coastal systems from 1971 to the present is given in Table 3.1 and Figure 3.1. This approach includes the use of continuous, long-term field descriptive data combined with

Table 3.1 Long-Term, Interdisciplinary Sampling Effort of the Center for Aquatic Research and Resource Management (Florida State University) from 1971 through 2002

System	Years of Field Data												
	MET	PC	LC	POL	PHYTOPL	SAV	ZOOPL	INF	INV	FISH	FW	NL	NR
Perdido Bay[a,b]	50[c]	13[a]	13[a]	3	10.5[a]	3[d]	3[d]	14[a]	13[a]	13[a]	13[a]	13[a]	1[d]
Apalachicola Bay[a,b]	85[c]	14[d]	2[d]	2	1	1	1	8[d]	14[d]	14[d]	8[d]	2[d]	ND
Apalachee Bay (Econfina, Fenholloway)[d]	50[c]	20[d]	10[a]	2	3[d]	14[d]	3[d]	2	14[d]	14[d]	14[d]	3[d]	1[d]
Escambia/Pensacola Bay[d]	30[c]	1.5[d]	1.5[d]	1	1.5[d]	ND	1.5[d]	1.5[d]	1.5[d]	1.5[d]	1.5[d]	1.5[d]	ND
Blackwater/East Bay[d]	30[c]	1.5[d]	1.5[d]	1	ND	1	ND	1.5[d]	1.5[d]	1.5[d]	1.5[d]	1.5[d]	ND
Choctawhatchee Bay[d]	30[c]	2[d]	2[d]	2	2[d]	1[d]	2[d]	2[d]	2[d]	2[d]	2[d]	2[d]	ND
Nassau/Amelia Estuaries[d]	3[d]	3[d]	3[d]	2	3[d]	3[d]	3[d]	2[d]	1[d]	1[d]	2[d]	ND	1[d]

[a] Monthly, quarterly.
[b] Long term.
[c] Daily.
[d] Monthly.

Abbreviations: MET = river flow, rainfall; PC = salinity, conductivity, temperature, dissolved oxygen, oxygen anomaly, pH, depth, Secchi; LC = NH_3, NO_2, NO_3, TIN, PON, DON, TON, TN, PO_4, TDP, TIP, POP, DOP, TOP, TP, DOC, POC, TOC, IC, TIC, TC, BOD, SiO_2, TSS, TDS, DIM, DOM, POM, PIM, NCASI color, turbidity, chlorophyll *a, b, c,* sulfide; POL = water/sediment pollutants (pesticides, metals, PAH); PHYTOPL = whole water and net phytoplankton; SAV = submerged aquatic vegetation; ZOOPL = net zooplankton; INF = infaunal macroinvertebrates taken with cores and/or ponars; INV = invertebrates taken with seines in freshwater areas around the edges of the bay and with otter trawls in the bay; FISH = fishes taken with seines in freshwater areas around the edges of the bay and with otter trawls in the bay; FW = food web transformations; NL = nutrient loading; NR = nutrient limitation experiments; ND = no data.

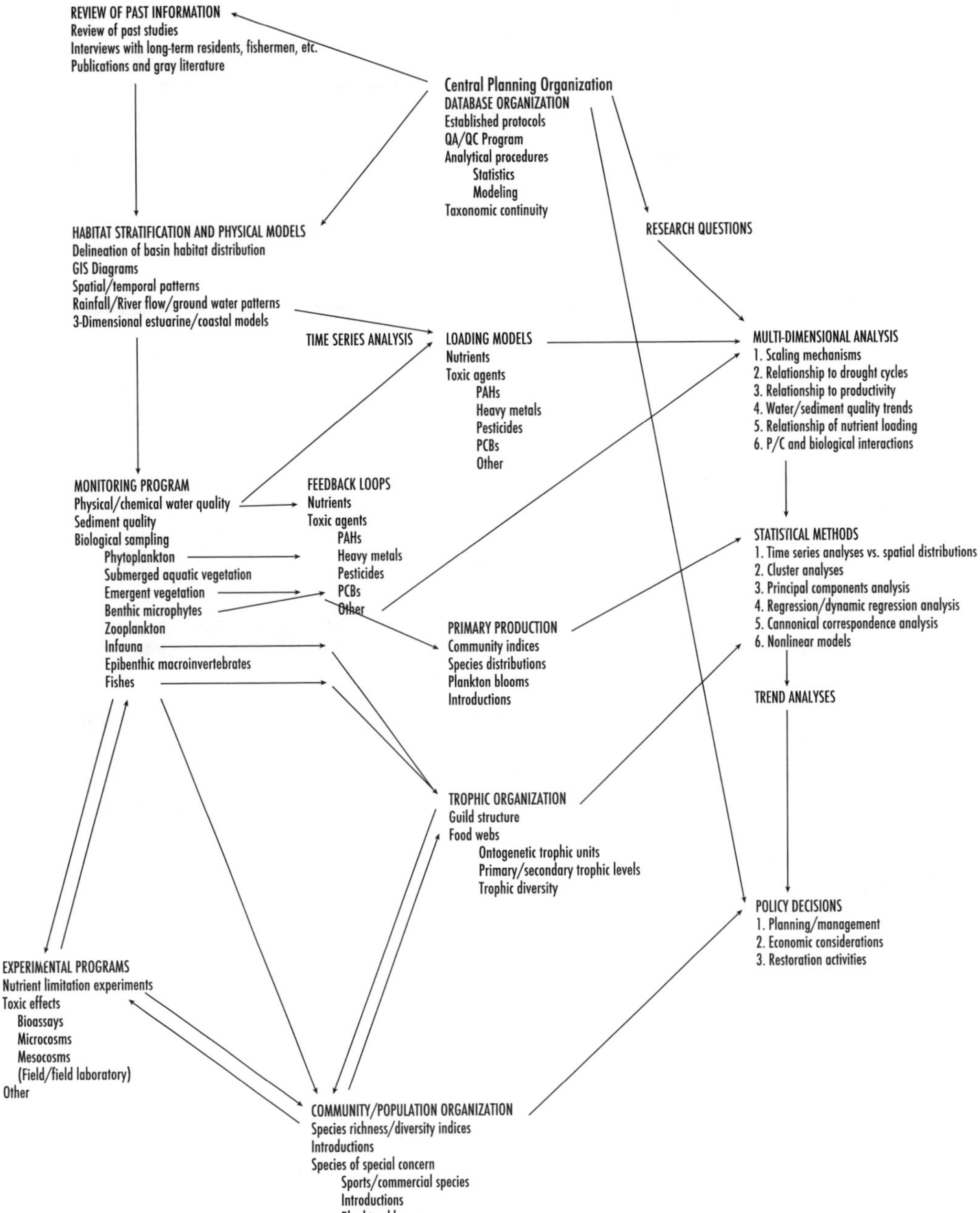

Figure 3.1 Outline of the interdisciplinary program of study by CARRMA in coastal areas of the NE Gulf of Mexico from 1970 to the present.

detailed trophic analyses of fishes and invertebrates of our region and an active laboratory and field experimental program.

A complete description of these systems, data collection efforts and field/laboratory operations, protocols for all data collection and experimental methods, and methods of statistical analysis and modeling is given by Livingston (2000). The combination of the various forms of research involved in this program is shown in Figure 3.1. Long-term analyses of phytoplankton communities have been carried out in the Perdido, Econfina, and Fenholloway systems. Shorter-

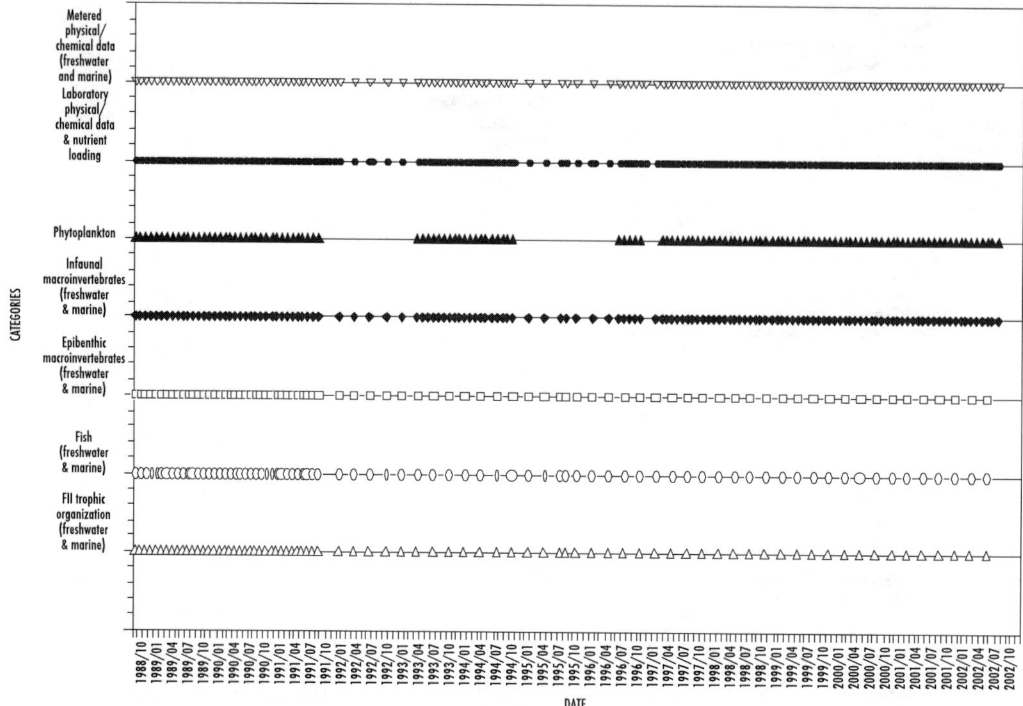

Figure 3.2 Outline of the Perdido River and Bay field database taken from October 1988 through June 2002.

term analyses have been made of the Escambia, Choctawhatchee, and Blackwater systems, with full databases (including phytoplankton community structure) for the Escambia and Choctawhatchee River–estuaries. A 14-year program has been carried out in the Perdido Bay system (Figure 3.2) to determine the impact of a pulp mill on the Elevenmile Creek/Perdido Bay system. During this period, nutrient loading from the mill has caused a series of plankton blooms that have had various effects on the structural elements of the phytoplankton assemblages and on the food webs of the receiving areas (Livingston, 2000).

By using a long-term multidisciplinary database that includes detailed (species-specific) phytoplankton data collected in a series of coastal systems in the NE Gulf of Mexico, I evaluate the influence of nutrient loading and plankton blooms on the food web structure of these systems. Use of the extensive database taken in these areas has permitted detailed determinations of bloom origin in addition to delineation of seasonal and interannual bloom sequencing and successions. In this book, I analyze the effects of these blooms on associated food webs. By monitoring recovery as the pulp mill reduced its nutrient loading and then relating the changes in the Perdido system to other bays in the NE Gulf, I identify physical/chemical and biological indicators of the bloom sequences and of the trophic responses to progressive changes of the plankton assemblages. By analysis of the multiple causes of the blooms (anthropogenic nutrient loading) and the modifying factors that facilitated bloom proliferation (drought sequences and related habitat features of the Perdido system), I define the elements that have contributed to the occurrence and impact of plankton blooms on secondary production in a series of Gulf coastal systems.

CHAPTER 4

Trophic Organization in Gulf Coastal Systems

I. INTRODUCTION

Feeding habits of coastal organisms have been extensively studied in habitats such as mangroves (Odum and Heald, 1972), sea grass beds (Kikuchi, 1974; Stoner, 1976; Brook, 1976, 1977; Livingston, 1982a), coral reefs (Hobson, 1973, 1974, 1975), river-dominated estuaries (Livingston, 1997c), and the open ocean (Ross, 1977, 1978). Shallow, coastal areas are highly productive yet physically unstable (Livingston et al., 1997) with extreme spatial/temporal variability of features such as temperature, land runoff, and associated water quality conditions (e.g., salinity, color, turbidity, light penetration). However, many coastal biological components remain temporally stable from year to year (Livingston and Loucks, 1978, Livingston et al., 1978, 1997; Dugan and Livingston, 1982). Questions remain concerning the temporal stability of coastal food webs.

Trophic organization of fishes in coastal areas has been widely assumed to involve adventitious use of nonlimiting resources. Most published analyses of trophic relationships remain limited in scope, usually determining feeding patterns of individual species or groups of related species. Early studies (Darnell, 1958, 1961) defined the "opportunism" of coastal consumers within the context of ontogenetic progressions of "distinct nutritional stages." Subsequent work has delineated distinct ontogenetic changes of basic dietary patterns in a number of invertebrate and fish species (Carr and Adams, 1972, 1973; Adams, 1976; Sheridan, 1979; Sheridan and Livingston, 1979; Livingston, 1980a, 1982a, 1984d, 1997b; Stoner, 1980a; Leber, 1983). However, these ontogenetic changes have often been left out of most trophic modeling efforts due, in part, to the complexity involved and to the lack of reliable information.

II. APPROACHES TO THE STUDY OF TROPHODYNAMICS

Trophic studies are still being published within a broad range of contexts. Feeding habits of coastal organisms have been related to productivity, spatial habitat variability, resource partitioning, and optimal foraging theory (Livingston, 2000). Habitat complexity is widely recognized as one of the key determinants of trophic diversity. Estuarine and coastal marine areas encompass a biological continuum from upland, freshwater parts of individual drainage basins through estuarine habitats to offshore marine systems. The basic characteristics of a given drainage system are defined by physiographic conditions (depth, area, access to freshwater runoff, and open ocean conditions) and freshwater input (i.e., dilution of seawater), with accompanying loading of dissolved nutrients (inorganic and organic) and particulate organic matter. Seasonal cycles of temperature and rainfall add a dimension of recurrent, time-related changes, with interannual cycles of meteorological conditions contributing a long-term dimension to the organization of coastal food webs. The complex relationships that determine trophic diversity have contributed to general acceptance of simplified food webs as the bases for modeling and theoretical definitions of food web characteristics. This simplification has led to extrapolation from general conditions to specific food web processes, which in many ways is antithetical to a comprehensive evaluation of trophic organization in coastal systems.

Food web relationships in estuarine and coastal systems are invariably delineated at the species level in most published studies. Resource partitioning among fishes and invertebrates is evident in various aquatic habitats (McEachran et al., 1976; Chao and Musick, 1977; MacPherson, 1981). However, niche breadth of a given species can be so extensive that quantitative determinations of significant ecological processes may be difficult to make unless the ontological progression of such niches is delineated. Without adequate recognition of the complex ecological stages that characterize coastal invertebrates and fishes, definitions of trophic organization risk oversimplification. There is a growing need to associate process-oriented food web relationships with key driving factors such as meteorological events, habitat variability, and biological interactions such as competition and predation. There should also be particular emphasis on the qualitative and quantitative aspects of primary productivity. Somehow, qualitative aspects of food resources in the form of phytoplankton community organization have received little attention.

A. Field Collections

Determinations of the feeding habits of aquatic organisms in the NE Gulf study areas have been carried out as an integral part of the long-term field collections (see Figure 3.1). Infaunal/epibenthic macroinvertebrates and fishes have been quantitatively collected at fixed stations over long-term periods, with detailed stomach-content analyses carried out for most of the dominant and subdominant species in the NE Gulf of Mexico coastal region (see Figure 2.2). Methods of collection and taxonomic identification were standardized over the entire study period (Livingston, 2000; see Appendix I).

B. The Trophic Unit

Ontogenetic feeding units were determined for epibenthic invertebrates and fishes in the nearshore Gulf regions such as the Apalachicola River and Bay system (Sheridan, 1978, 1979; Laughlin, 1979; Sheridan and Livingston, 1979, 1983); Apalachee Bay (Econfina and Fenholloway drainages) (Livingston, 1980, 1982, 1984b, unpublished data; Stoner and Livingston, 1980, 1984; Laughlin and Livingston, 1982; Stoner, 1982; Clements and Livingston, 1983, 1984; Leber, 1983, 1985); and the Choctawhatchee River and Bay system (Livingston, 1986a,b, 1989; Livingston et al., 1991). It was determined that an adequate sample for a given trophic designation (size class, species, location) required processing at least 15 fish or invertebrate stomachs (Livingston, unpublished data). For example, 4129 blue crab (*Callinectes sapidus*) stomachs were processed from the Apalachicola system (Laughlin, 1979). Other representative species analyzed included penaeid shrimp (*Penaeus duorarum*; 1115 stomachs), pinfish (*Lagodon rhomboides*; 4915 stomachs), spot (*Leiostomus xanthurus*; 4091 stomachs), and pigfish (*Orthopristus chrysoptera*; 8634 stomachs).

The development of the trophic unit concept has been outlined in a series of studies of the pinfish (*Lagodon rhomboides*), a numerical dominant in shallow inshore waters of the NE Gulf. (Livingston, 1982, 1984; Stoner, 1979a,b, 1980: Stoner and Livingston, 1984). A generalized outline of the ontogenetic feeding patterns of pinfish is given in Figure 4.1. A list of the general food types found in fishes of Apalachee Bay is given in Table 4.1. There was a well-ordered progression of changes in food preferences of this species in the unpolluted Econfina system. Young-of-the-year recruits from offshore spawning grounds (<20 mm standard length [SL]) were planktivorous, feeding mainly on calanoid and cyclopoid copepods and fish eggs. With growth (21 to 35 mm SL), there was gradual transition to benthic carnivory in the form of

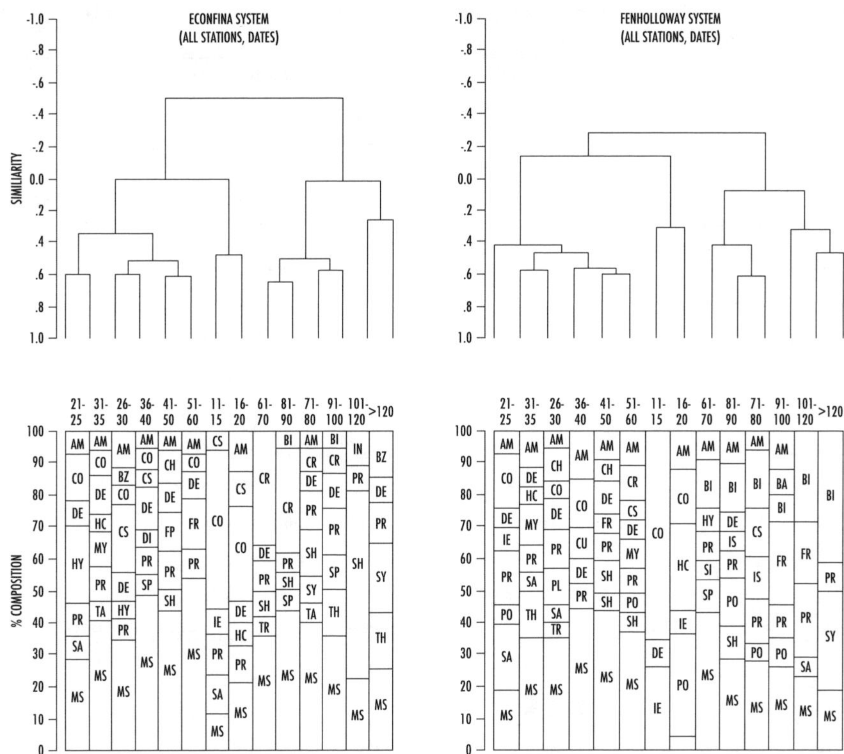

Figure 4.1 Ontogenetic changes in the diet of *Lagodon rhomboides* taken in the Econfina and Fenholloway coastal areas from 1971 to 1977. Histograms represent the relative proportions of the major dietary components (dry weight) and dendograms represent cluster analyses of the diet similarity among size classes. Codes for the food items are given in Table 4.1.

amphipods, mysids, harpacticoids, and other small animals. With growth to 60 mm SL, there was an increased preference for amphipods, shrimp, plant remains, and detritus in the pinfish diet. Plant matter consisted largely of microalgae. Fish ranging from 61 to 120 mm SL then feed on larger benthic invertebrates such as crabs, shrimp, and bivalve mollusks; plant matter changed with such growth from microalgae to pieces of *Syringodium* and *Thalassia.* Pinfish that exceeded 120 mm SL generally fed on plant matter in the form of *Syringodium* and *Thalassia* and were largely herbivorous in the Econfina system. This species thus passed through a planktivorous phase with a shift through benthic carnivory to two distinct stages of omnivory and ended in herb ivory in fishes >120 mm SL.

A systematic analysis of the dominant fishes in Apalachee Bay (Livingston, 1982a) indicated that various grass bed fishes (i.e., *Diplodus holbrooki,* spottail pinfish; *Orthopristis chrysoptera,* pigfish; *Bairdiella chrysura,* silver perch; *Centropristis melana*, Gulf black sea bass; *Leiostomus xanthurus,* spot) followed species-specific and age-specific feeding patterns, whereas other fishes (i.e., *Anchoa mitchilli,* bay anchovy; *Paraclinus fasciatus,* banded blenny) did not follow distinct ontogenetic feeding stages. Transitions from one feeding stage to the next were often gradual and not necessarily distinct, but were, in general, related to growth stages in the ontological development of each species. Plankton feeders, herbivores, infaunal macroinvertebrate feeders, and benthic predators of various types went through species-specific, ontological changes in feeding habits. The feeding progressions were associated with ontological changes in morphology as it related to locomotion (body shape), mouth dimensions, dentition, and gut dimension (Stoner and Livingston, 1984).

Table 4.1 List of General Food Types Found in Fish Stomachs and Codes Used to Describe Them in the Presentation of the Data

AM	Amphipods	HO	Holothuroidia
AN	Annelids	HY	Hydracarina
AP	Appendicularians	IE	Invertebrate eggs
AR	Animal remains	IL	Insect larvae
BA	Barnacles (ad, juv)	IN	Insects
BI	Bivalves (ad, juv)	IP	Insect pupae
BN	Barnacles (larvae)	IS	Isopods
BR	Branchiopods	MA	Microalgae
BZ	Bryozoans	MS	Miscellaneous
CA	Cambaridae	ML	Molluscan larvae
CD	Cephalochordates	MY	Mysids
CE	Cephalopoda	NE	Nematodes
CH	Chaetognaths	NM	Nemerteans
CI	Ciliophora	OL	Oligochaeta
CL	Cladocerans	OP	Ophiuroids
CN	Cnidaria	OR	Organic remains
CO	Copepods (calanoid, cyclopoid)	OS	Ostracods
CR	Crabs (ad, juv)	PL	Polychaetes (larvae)
CS	Crustacean remains	PO	Polychaetes (ad, juv)
CT	Chitons	PR	Plant remains
CU	Cumaceans	PY	Pycnogonida
DE	Detritus	RA	Radiolarians
DI	Diatoms	RP	Reptilia
DL	Decapod larvae	SA	Sand grains
EC	Echinoderms	SH	Shrimp (ad, juv)
EG	Egg cases	SI	Sipunculids
FE	Fish eggs	SP	Sponge matter
FL	Fish larvae	ST	Stomatopods
FO	Foraminiferans	SY	*Syringodium filiforme*
FP	Fecal pellets	TA	Tanaids
FR	Fish remains (ad, juv)	TH	*Thalassia testudinum*
GA	Gastropods	TM	Trematodes
HC	Copepods (harpacticoid)	TN	Tunicates

Note: "Plant remains" represents living plant matter and "detritus" represents dead organic matter. "Miscellaneous" (MS) represents all items that comprise <3% of the total mass; ad = adult; juv = juvenile.

The organization of fishes in the Apalachee sea grass beds could be divided into three major groups (Livingston, 1982a). There were planktivorous forms that included the early stages of various species such as spot, pinfish, the filefishes (*Monacanthus* spp.), and the mojarras (*Eucinostomus* spp.). Included in this group were the water column feeders such as the bay anchovies (*Anchoa* spp.). Copepods were a prime source of food for the planktivores. A second group included benthic carnivory and omnivory. Each species went through a distinct serial transition of feeding. Species that were spatially and temporally sympatric (i.e., pinfish and spottail pinfish) had relatively little dietary overlap. This level included the later growth stages of pinfish, filefishes, and spottail pinfish. The third group included crustacean feeders that tended to specialize in amphipods, shrimp, and crabs: early growth stages of pigfish, silver perch, and the syngnathids (pipefishes). Species such as the banded blenny, the black sea bass, and adult forms of pigfish and silver perch fed largely on crustaceans. Species with relatively complex feeding progressions tended to be trophic specialists whereas species with more simple trophic organization were often generalists. Ontogenetic feeding strategies, life history characteristics, and habitat preferences were closely associated. Similar ontogenetic feeding units were determined in the Apalachicola (Sheridan, 1978, 1979; Laughlin, 1979; Sheridan and Livingston, 1979, 1983) and Choctawhatchee River–bay systems (Livingston, 1986a,b, 1989).

Decapod crustaceans, in the form of shrimp and crabs, are the numerical and biomass dominants in coastal areas of the NE Gulf (Livingston et al., 1974, 1997; Livingston, 1976a, 1984d, 1985a, 1986a, 1989, 1992; Hooks et al., 1976; Greening, 1980; Dugan and Livingston, 1982; Greening and Livingston, 1982). There have been relatively few systematic analyses of the ontogenetic feeding stages of the decapods. Odum et al. (1982) hypothesized that penaeid and caridean shrimp and crabs represented a major link between detritus production in wetlands and coastal food webs. Laughlin (1979), Laughlin and Livingston (1982), and Leber (1983, 1985) found ontogenetic feeding patterns in the dominant (decapod) invertebrates along the NE Gulf coast that were similar to those described for fishes. In a study of blue crabs (*Callinectes sapidus*) in the Apalachicola system, Laughlin (1979) found that there were three trophic stages in developing blue crabs (Figure 4.2). The smaller crabs (<30 to 40 mm carapace width) fed on bivalve mollusks, detritus,

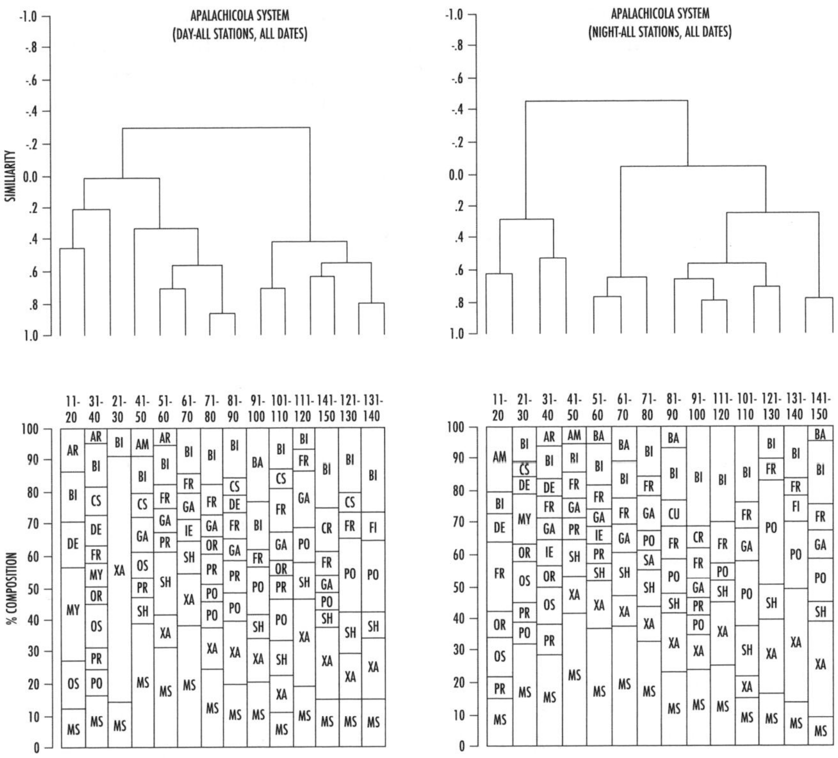

Figure 4.2 Ontogenetic changes in the diet of blue crabs (*Callinectes sapidus*) taken in the Apalachicola Bay system during the day and at night monthly from January 1978 through December 1978. Histograms represent the relative proportions of the major dietary components (dry weight) and dendograms represent cluster analyses of the diet similarity among size classes. Codes for the food items are given in Table 4.1.

xanthid crabs, ostracods, and plant matter. The larger crabs (40 to 90 mm carapace width) ate bivalves in increasing proportions along with shrimp, xanthid crabs, detritus, and plant matter. The largest crabs (>90 mm carapace width) fed on bivalve mollusks, xanthid crabs, shrimp, and fishes. In addition, this group was cannibalistic, eating significant numbers of their own species.

Leber (1983) defined the ontogenetic feeding habits of 14 decapod species in Apalachee Bay. Penaeid shrimp (*Penaeus duorarum* or pink shrimp) were carnivorous, feeding on small crustaceans such as amphipods and harpacticoid copepods at early stages of development. With growth, they went through two feeding stages, going from shrimp and amphipods (stage 2) to a mixture of shrimp, amphipods, bivalve mollusks, gastropod mollusks, and polychaete worms (stage 3). *Sicyonia laevigata* fed mainly on amphipods, harpacticoids, and plant matter at early stages with increasing dependence on amphipods and polychaetes during the largely carnivorous second stage of feeding in the older organisms. The caridean *Ambidexter symmetricus* went from polychaete/crustacean feeding in the smaller carnivorous individuals to omnivorous feeding in the second growth stage with increasing dependence on detritus and red algae. The caridean *Processa bermudensis* had only one stage of feeding and was largely carnivorous, feeding mainly on polychaetes and amphipods. The alpheid green snapping shrimp (*Alpheus normanni*) was almost completely herbivorous, feeding mainly on sea grass (*Thalassia testudinum*) and filamentous red algae. The palaemonid shrimp *Palaemon floridanus* fed mainly on caridean shrimp, amphipods, and filamentous epiphytes during its early growth stages, changing to a diet of caridean shrimp, amphipods, and bivalve mollusks during its later stages of development. This progression was from omnivory to carnivory in this species. The xanthid crab *Neopanope packardii* was mainly herbivorous, feeding on coralline red algae, *T. testudinum*, and plant detritus. Larger forms in a second stage became more omnivorous, adding caridean shrimp and amphipods to the diet. The xanthid *N. texana*, however, was more omnivorous, feeding mainly on various invertebrates and plant matter during early developmental stages and gradually becoming herbivorous at larger sizes, feeding mainly on *T. testudinum,* encrusting algae, and small invertebrates. The majid spider crab (*Libinia dubia*) was largely herbivorous at both stages of feeding, going from small epiphytes in the early growth stages to sea grasses (*T. testudinum*, *Syringodium filliforme*) in the larger animals. The majid crab *Epialtus dilatatus* was an herbivore with no real ontogenetic feeding progression with growth. The crab *Podochela riisei* was an omnivore at early growth stages, changing to an herbivore with growth. The crab *Metoporhaphis calcarata* was an omnivore at all size classes, whereas the majid crab *Pitho anisodon* was mainly herbivorous, going through two stages of plant eating. The predominant forms of decapod crustaceans thus went through ontogenetic feeding patterns similar to those described for fishes.

The largely herbivorous brachyuran crabs were contrasted to the largely carnivorous penaeid and caridian shrimp with omnivory occasionally represented in specific ontogenetic feeding groups. As with the fishes, there was considerable overlap in the dietary stages of invertebrates so that the feeding units were generally not discrete. However, there were definitive feeding trends with time in most species so that feeding groupings of a given species often were more strongly associated with similar groupings of another species than with succeeding ontogenetic feeding units of the same species.

Overall, the trophic organization of both invertebrates and fishes indicated the following:

1. The diets were generally well mixed with little dependence on one form of food.
2. The presence of plant matter took many forms, and was often size specific, ranging from smaller epiphytes to the larger forms of plant matter such as sea grasses.
3. Strict herbivory was relatively rare compared with the numbers of omnivores and carnivores.
4. There were definite ontogenetic feeding trends in most species.
5. There was little evidence of purely opportunistic (i.e., random, unspecialized) feeding patterns.
6. The progressive feeding stages within a given species indicated changes in microhabitat usage, with specificity of feeding dictating adherence to specific habitat features.

7. There was a common interaction of habitat structural complexity with biological features such as prey avoidance as modified by complex predator–prey interactions (Leber, 1983).
8. The direct use of living plants by invertebrate herbivores and omnivores was more important than previously indicated. This was also found to be true of fishes (Stoner, 1980; Clements and Livingston, 1983).

Fish and invertebrate feeding patterns cannot be generalized, and should be viewed as species-specific progressions of often-disparate feeding stages that are usually more closely related to trophic stages of other species than to their own ontogenetic feeding stages.

III. FEEDING VARIABILITY

The issue of spatial and temporal variability of feeding in coastal invertebrates and fishes is rarely addressed in most trophic evaluations. However, such variation is usually taken for granted as an example of the adventitious feeding behavior of aquatic organisms. However, although many researchers agree with the concept of broad ontogenetic habitat preferences along well-developed habitat gradients, the interrelationships of the biological processes that determine trophic organization in coastal systems remain unclear even though there is an extensive scientific literature concerning fish feeding habits in coastal areas. Early findings addressed the issue of whether or not the presence of a given coastal population was a function of physical habitat distribution or trophic response to overland runoff even though these factors are not mutually exclusive. In fact, habitat gradients and spatial distribution of primary productivity in coastal areas are closely linked, and there is considerable evidence that the spatial and temporal distribution of many coastal populations are closely associated with both features (Livingston, 2000).

In a comprehensive study of the blue crab in the Apalachicola estuary, Laughlin (1979) determined the nature of ontogenetic feeding progression of this species as it was affected by spatial and temporal (diurnal, seasonal) habitat distributions. In the analysis of thousands of blue crab stomachs, Laughlin (1979) compared spatial variation as well as the diurnal and seasonal aspects of feeding by this species. The author found that spatial habitat distribution affected the prey that, in turn, was the chief controlling factor in the distribution of the predator. Environmental variation thus had an indirect rather than a direct effect on blue crab distribution in space and time in the Apalachicola system. This same conclusion was reached within a very different context in studies of the trophic organization of fishes in the Apalachicola system (Livingston et al., 1997). These authors found that, within certain ranges, Apalachicola River flow was a key element in the control of the biological organization of the receiving estuary. This control was both direct (affected by the distribution of salinity and light penetration) and indirect (related to species-specific predator–prey relationships). Complex biological variables were physically forced in the highly variable estuarine environment when compared with the more biologically controlled systems outside of the main influence of the river. Trophic components directly linked to phytoplankton and benthic algal production (herbivores, omnivores) were more immediately affected by changes in river flow, whereas the higher trophic levels were more biologically controlled as predation was the paramount factor in the definition of the various carnivore groups in space and time. Livingston et al. (1997) demonstrated this dichotomy in quantitative terms with respect to both seasonal and interannual trophic relationships. Herbivore/omnivore populations were associated with river-dominated habitat phenomena related to primary productivity, whereas predators followed patterns of prey distribution.

The physical instability of the Apalachicola system was thus associated with a relatively stable trophic organization within specific ranges of seasonal river flow because of the ascendancy of trophic relationships. Periodic drought conditions increased light transmission through the water column, thereby stimulating primary productivity and initiating increased herbivore activity. Light penetration, mediated by physical influences of the river on the estuary, thus emerged as a primary factor in the biological organization of the bay, and the seasonal and interannual changes of the trophic organization followed a well-ordered progression. The interaction of the natural life history progressions of numerous estuarine species with short- and long-term changes in the qualitative and quantitative cycles of primary production thus led to the observed long-term trophic responses of fishes in the Apalachicola system.

A. Spatial Features of Trophic Response

There is considerable habitat variation in the sea grass meadows of Apalachee Bay. Feeding habitats of fishes and invertebrates in the relatively unpolluted Econfina drainage was diverse. Pulp mill effluents in the Fenholloway system simplified the sea grass habitat with increased open sediment areas and reduced submerged aquatic vegetation (SAV) cover. Brady (1981) analyzed the distribution of fish eggs and larvae in Apalachee Bay, showing that ichthyoplankton were defined by high diversity but relatively low abundance in unpolluted areas. There were two primary reproductive types: a cold-water fauna composed of off-shore, winter spawners that moved into near-shore waters as postlarvae, and a warm-water assemblage that included resident species spawning during spring/summer months. Cold-water spawners entered inshore areas as demersal juveniles, suggesting nonrandom dispersal of the early life stages into SAV-dominated habitats. The cold-water faunal group was composed of postlarval stages of migratory species that were relatively large (>9.0 mm SL) and well developed, and able to swim. This group of species was dominated numerically by pinfish and spot. A collective spring spawning peak, described by Johannes (1978), and observed in this and other surveys of near-shore habitats of the Gulf of Mexico (Hoese et al., 1968; Blanchet, 1979), was associated with a replacement of the cold-water assemblage by resident species that spawn near shore. Eggs and larvae of the warm-water spawners thus followed the known distributional patterns of adults, with the exception of the engraulids that produced pelagic eggs and followed local patterns of spawning activity. The sea grass ichthyofauna of the Apalachee Bay sea grass meadows depended heavily on recruitment from offshore areas, but the distribution of juveniles in the nursery areas followed a complex combination of individual spawning behavior, species-specific life history traits and survivorship, and ontogenetic patterns of habitat choice (Brady, 1982).

Postlarvae of off-shore spawners such as *Lagodon rhomboides* begin to search for high-density sea grass at very early stages. Alternatively, survivorship may be greatest in dense sea grass areas even for postlarvae. Pinfish juveniles are abundant in coastal creeks and salt marshes (Weinstein and Heck, 1979). Low salinity excludes many piscivorous fishes. Brady (1981) noted that anchovy (*Anchoa mitchilli*) eggs were most abundant in the nutrient-enriched inshore areas of the Fenholloway system, following patterns described in river-dominated estuaries such as Apalachicola Bay (Blanchet, 1979). In the largely unpolluted Apalachee Bay, primary production is dominated by benthic macrophytes, and water column plankton are relatively low in abundance compared to river-dominated systems where water column feeders such as anchovies are numerous. Alluvial river–estuaries are also characterized by unvegetated soft sediment areas, which accounts for the dominance of benthic grubbers such as spot and croaker (*Micropogonius undulatus*) and invertebrates such as penaeid shrimp and blue crabs (Livingston, 1984d). Laughlin (1979) found that bivalve mollusks, located in river-dominated areas of low salinity and high phytoplankton productivity, constituted a major form of food of blue crabs. Habitat thus determined the distribution of food types that, in turn, was associated with blue crab distribution. Seasonal shifts in food preferences also followed prey availability (in terms of abundance and size) with such species considered as opportunistic (i.e., number maximizers rather than energy maximizers) in a physically unstable environment. Habitat instability reduced predation pressure so that eurytopic populations adapted to changing environmental conditions were able to take full advantage of the highly productive estuarine environment.

A comparison of the unpolluted Econfina system and the polluted Fenholloway system provided a basis of comparison for the influence of spatial habitat and productivity differences on coastal food webs. After a pollution abatement program undertaken in 1974, there was a 30% reduction of water color (an indicator of the pulp mill pollution in the Fenholloway system). Outer stations in the Fenholloway offshore areas (see Figure 2.2E) were not adversely affected by mill effluents. However, there was relatively little water quality improvement in the inshore Fenholloway estuary. Transition areas between the unaffected offshore areas and inshore polluted areas reflected the cessation of mill effluent effects, as illustrated by reduced color. Previous habitat conditions related to mill impacts had caused a disjunct trophic organization that was directly related to altered prey availability. The general replacement of pinfish by plankton-feeding anchovies and worm-eating sciaenid fishes (spot) was complemented by increased activity by predators such as silver perch (*Bairdiella chrysura*). These alterations reflected prey associations with productivity trends and habitat distribution. Increased light penetration in the Fenholloway system was followed by increased SAV distribution in transition areas. Livingston (1980, 1982) outlined the trophic response in the Fenholloway system with a return of sea grass habitat and a reversion of the trophic order to that noted in the reference Econfina area (low phytoplankton activity in an SAV-dominated habitat). Carnivorous and omnivorous trophic units of the dominant pinfish that had been altered due to the loss of sea grass beds in the Fenholloway system were restored. Plankton feeding populations in the Fenholloway were reduced accordingly, whereas species that depended on SAV for habitat and food (i.e., pinfish) increased. There was thus a spatial pattern to the differentiation of pinfish feeding habits between the polluted and unpolluted estuaries, which was based, in large part, on the distribution of grass beds in the respective study areas (Livingston, 1980a).

Hierarchical analyses (Livingston, 1975a) indicated that fishes such as anchovies and spot, which feed largely at the river mouths, did not differentiate among general food types on the basis of the individual estuaries; rather, feeding patterns were ordered along inshore–offshore habitat and productivity gradients that tended to follow specific water quality patterns (e.g., salinity). The mojarras (*Eucinostomus* spp.) followed a somewhat similar pattern, with spatial feeding associations grouped by habitat (mudflats, grass beds). With these species, the inshore areas of both estuaries tended to be associated with the preferred food. Thus, fishes that were feeding predominantly on harpacticoid copepods and polychaete worms were not affected by the water quality improvements as there was relatively little habitat change in these inshore areas relative to the reference site.

The habitat–food preference relationships of the grass bed fishes that were benthic omnivores and carnivores were very different in the two study areas. Although there was some overlap in areas of grass bed recovery (stations Fl2, Fl6; see Figure 2.2), fish feeding habits in the heavily polluted areas (F9, Fl1), which showed little or no recovery of benthic macrophytes with time, remained quite different from those in the undisturbed grass bed areas offshore. Crustacean feeders (syngnathids; pigfish; silver perch; black sea bass, *Centropristis striata*) and the banded blenny (*Paraclinus fasciatus*) also reflected habitat changes in the Fenholloway transitional areas (i.e., increased grass bed distribution). These changes were most evident in the feeding habitats of syngnathids and black sea bass and were less prevalent among the pigfish and silver perch, which were less likely to have habitat-specific food habits. Plant biomass, dominated by *Thalassia* in the Econfina estuary and *Syringodium* in the Fenholloway estuary, showed divergent trends before and after the pollution abatement program. There was also a general (natural) decline of plant biomass in the Econfina system in 1974, whereas there were indications of some SAV recovery in the Fenholloway.

The plankton and polychaete feeders (Group I, Table 4.2) were consistently more dominant in the Fenholloway estuary with the exception of two periods of high river flow (1973 and 1976–1977), at which times Group I fishes were also dominant in the Econfina estuary. This result is consistent with the habitat/productivity conditions at the time of flooding and the trophic relationships of the planktivorous populations. Group II, the benthic omnivores and carnivores (Table 4.2; Livingston 1982a), were dominant in the Econfina grass beds except during the two periods of high river flow. The relative abundance of these fishes did not change appreciably over time, which is consistent with previous observations concerning the trophic requirements of such fishes. The Group III fishes,

Table 4.2 Fish Feeding Types in the Econfina and Fenholloway Systems as Noted over 7 Years of Observation (1971 to 1979)

Group I. Plankton, Harpacticoid Copepod, and Polychaete Feeders
Leiostomus xanthurus (spot)
Eucinostomus gula (silver jenny)
Eucinostomus argenteus (spotfin mojarra)
Anchoa mitchilli (bay anchovy)
Group II. Benthic Omnivores, Carnivores
Lagodon rhomboides (pinfish)
Diplodus holbrooki (spottail pinfish)
Monacanthus ciliatus (fringed filefish)
Stephanolepis hispidus (planehead filefish)
Chilomycterus schoepfii (striped burrfish)
Opsanus beta (Gulf toadfish)
Arius felis (sea catfish)
Synodus foetens (inshore lizardfish)
Rhinoptera bonasus (cownose ray)
Group III. Crustacean Feeders
Syngnathus floridae (dusky pipefish)
Syngnathus scovelli (Gulf pipefish)
Orthopristis chrysoptera (pigfish)
Bairdiella chrysoura (silver perch)
Centropristis striata (black sea bass)
Paraclinus fasciatus (banded blenny)

the crustacean feeders (Table 4.2; Livingston 1982a), tended to be more prevalent, on a percentage basis, in the Fenholloway estuary. Overall, the Econfina estuary was composed largely of sea grass species that occasionally reflected temporary habitat changes (i.e., low salinity, high color, high nutrient loading) in their relative abundance. The Fenholloway estuary was dominated by anchovies and spot, which reflected the productivity and habitat structure of the polluted system. It is clear that spatial habitat differences affect fish feeding habits.

Another important question involves the comparability of feeding habits of a given species that occurs in different coastal systems. The plankton-feeding bay anchovy (Figure 4.3) was primarily a copepod feeder in all systems. However, there were some differences of the relative food composition of this species in the different alluvial (Apalachicola, Choctawhatchee) and blackwater systems (Econfina, Fenholloway). In the Apalachicola system, there was a tendency for increased food diversification in the more mature stages. This trend was not apparent in the Choctawhatchee estuary where copepods were the primary form of food throughout the life history cycle of the fish. Although copepods were the dominant form of food in Apalachee Bay, there was a higher level of anchovy food diversity in both the Econfina and Fenholloway systems; detritus and plant remains were generally found in higher proportions than those found in the alluvial systems. Thus, although the overall pattern of feeding for this species was similar in all systems, there were specific differences that were probably related to differences in the quality and distribution of the habitats within each estuary.

Spot (*Leiostomus xanthurus*) are bottom feeders that forage at river mouths. This species generally does not differentiate among general food types on the basis of the individual estuaries; rather, the feeding patterns appeared to be ordered along inshore–offshore gradients that tended to follow specific water quality patterns (e.g., salinity). A comparison of spot feeding habits in different coastal areas (Figure 4.4) indicated some differences in feeding patterns among these systems. Copepods appeared to be more dominant as food in the Choctawhatchee system. In most of the

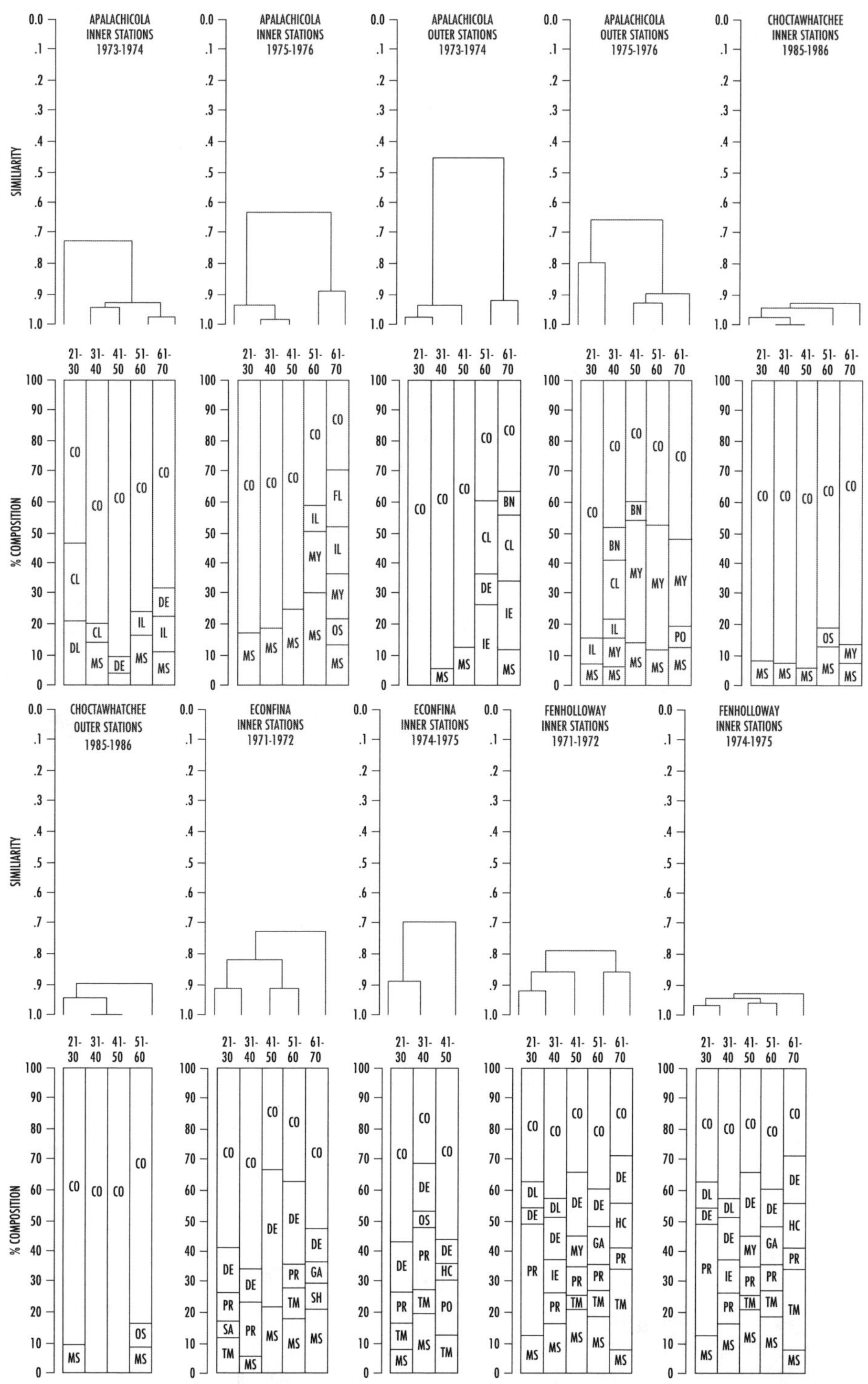

Figure 4.3 Ontogenetic changes in the diet of bay anchovies (*Anchoa mitchilli*) taken in the Apalachicola Bay system (inner and outer stations, 1973–1974; 1975–1976), Choctawhatchee Bay (inner and outer stations, 1985–1986), the Econfina system (inner stations, 1971–1972; 1974–1975), and the Fenholloway system (inner stations 1971–1972; 1974–1975). Histograms represent the relative proportions of the major dietary components (dry weight), and dendograms represent cluster analyses of the diet similarity among size classes. Codes for the food items are given in Table 4.1.

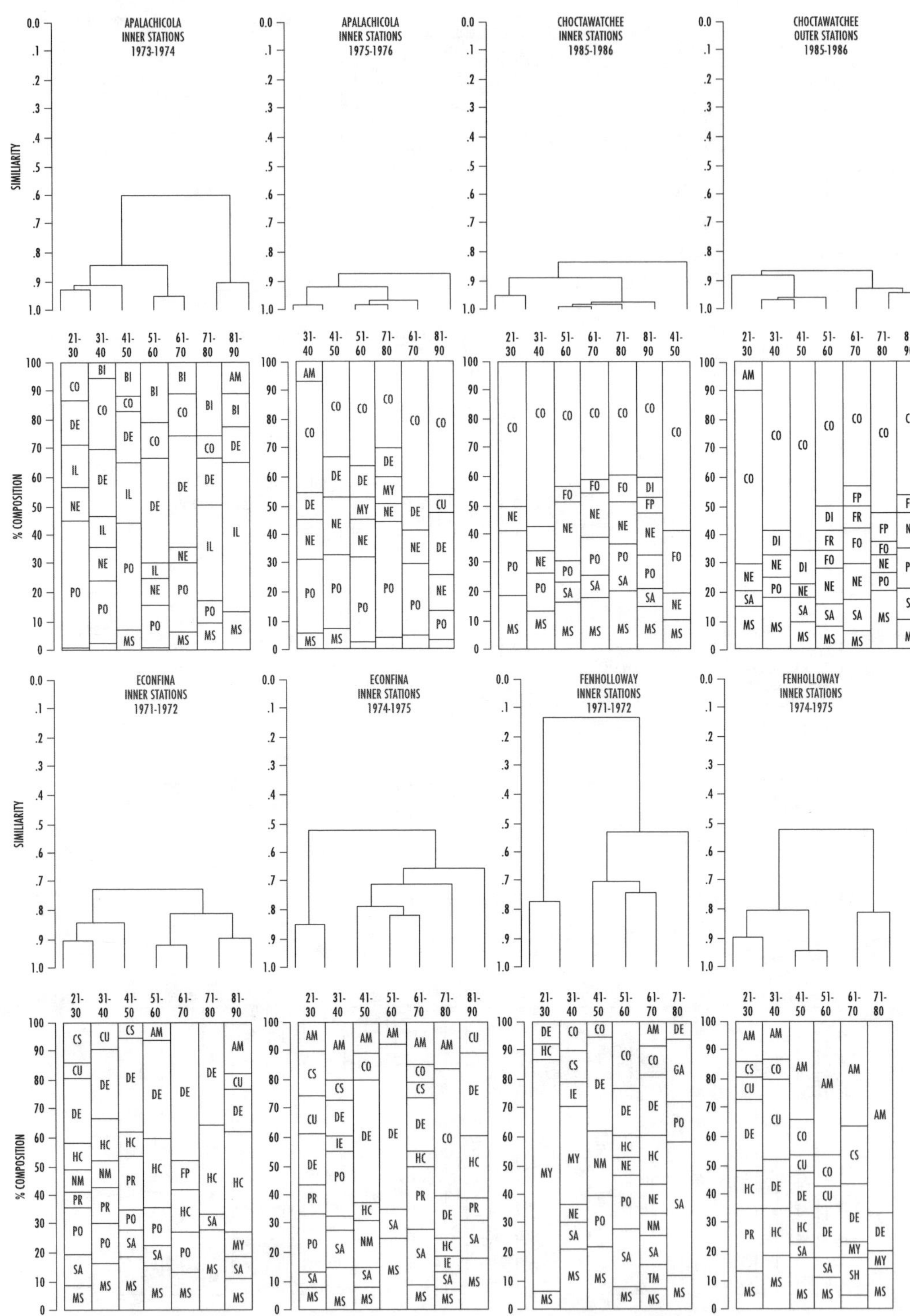

Figure 4.4 Ontogenetic changes in the diet of spot (*Leiostomus xanthurus*) taken in the Apalachicola Bay system (inner stations, 1973–1974; 1975–1976), Choctawhatchee Bay (inner and outer stations, 1985–1986), the Econfina system (inner stations, 1971–1972; 1974–1975), and the Fenholloway system (inner stations, 1971–1972; 1974–1975). Histograms represent the relative proportions of the major dietary components (dry weight), and dendograms represent cluster analyses of the diet similarity among size classes. Codes for the food items are given in Table 4.1.

estuaries, polychaetes were a major form of spot feeding during early stages of development. With growth, detritus became more important in most areas. In the Econfina (unpolluted) and Fenholloway (polluted) systems, there were differences in feeding during the period of maximum pollution (1971–1972). The Econfina estuary was composed largely of sea grass species, which occasionally reflected temporary habitat changes such as low salinity in fish relative abundance. The Fenholloway estuary was dominated by anchovies and spot, which reflected the productivity and habitat structure of the polluted system where soft sediment areas replaced the sea grass beds in areas closely associated with the pollution (i.e., inner stations). Thus, although species-specific patterns of feeding are generally similar among different coastal systems, differences of productivity and habitat can affect the ontogenetic patterns of such feeding. Pollution and habitat alteration can cause changes in associated trophodynamic feeding patterns, and can be a source of restricting population distribution through habitat loss and breaks in the ontogenetic feeding progressions.

B. Temporal Features of Trophic Response

1. Diurnal Variation

Trophic generalists such as blue crabs, penaeid shrimp, and the dominant fishes of the coastal waters of the Gulf are characterized by combinations of omnivory, detritivory, and even cannibalism. These species pose significant problems with respect to their position in coastal trophic organization. The situation is further complicated by species-specific ontogenetic progressions that are influenced by temporal (day/night, seasonal, interannual) periodicity.

Approximately one third to one half of the species from most fish communities are nocturnally active (Helfmann, 1978). Notable exceptions can be found in some kelp beds (13%; Ebeling and Bray, 1976) and in temperate sea grass beds in North Carolina (Adams, 1976). Diel activity patterns among fish communities have been documented for coral reefs (Hiatt and Strasburg, 1960; Hobson, 1965, 1968, 1973, 1974; Starck and Davis, 1966; Collette and Talbot, 1972; Harmelin-Vivien and Bouchon, 1976), kelp forests (Ebeling and Bray, 1976; Hobson and Chess, 1976), mesopelagic waters (Marshall, 1954; Merrett and Roe, 1974; Clarke, 1978), and shallow estuaries (Hoese et al., 1968; Livingston, 1976a; Ogren and Brusher, 1977). Relatively little is known about diel patterns of fish feeding in sea grass areas. Previous studies indicated that sea grass fishes are predominantly diurnal in their feeding activities (Hatanaka and Iizuka, 1962; Reid, 1967; Randall, 1967; Carr and Adams, 1973; Kikuchi, 1974; Adams, 1976; Brook, 1977; Stoner, 1979b, 1980c; Orth and Heck, 1980). However, such studies usually do not address day/night variability of feeding.

Ryan (1981) studied temporal feeding relationships of grass bed fishes inhabiting Apalachee Bay over a 14-month period, and he outlined temporal ontogenetic feeding patterns for numerically dominant nocturnal predators in Apalachee Bay. Stomach content analyses showed that 17 of the 59 species (28.8%) collected by otter trawl engaged in some form of nocturnal feeding activity. These nocturnal fishes were characterized by large mouths and eyes. Most fed on crustaceans. Many species engaged in a dual feeding behavior whereby fishes were largely consumed during the day and crustaceans at night. Crepuscular piscivores were much more generalized in their prey selection than their nocturnal counterparts. Nocturnal forms were represented by both primitive and more recently evolved percoid species, whereas day-feeders were characterized by species of more recent origin. As invertebrate fauna becomes increasingly active at night in sea grasses, it is curious that there is currently no well-adapted nocturnal fish fauna to exploit the available food supply at night.

Diurnal predators such as *Lagodon rhomboides*, *Leiostomus xanthurus*, and *Orthopristis chrysoptera* were taken in greater numbers at night, although no feeding activity was observed (Ryan, 1981; Ryan and Livingston, unpublished data). Although no fishes were collected exclusively during the day, several burrowing and/or demersal species such as the spotted worm eel (*Myrophis punctatus*), bank cusk eel (*Ophidion beani*), Gulf flounder (*Paralichthys albigutta*), tonguefish

(*Symphurus plagiusa*), and sea robins (*Prionotus scitulus* and *P. tribulus*) were taken only at night. Nocturnal trammel collections were dominated by bonnethead sharks (*Sphyrna tiburo*), Atlantic sharks (*Rhizoprionodon terranovae*), longnose gar (*Lepisosteus osseus*), and Atlantic stingray (*Dasyatis sabina*). *Bairdiella chrysoura* (silver perch) were the most numerous of the nocturnal predators collected.

Recently spawned silver perch young-of-the-year (<20 mm) enter Apalachee Bay from the Gulf in late winter–early spring months (Brady, 1981) and grow to approximately 120 mm by the time they emigrate from the bay in the fall. Silver perch showed the greatest ontogenetic, temporal, and spatial trophic variability of any of the nocturnal specimens examined (Ryan, 1981). Trophic analyses of silver perch (Figure 4.5) taken at night in the Econfina system indicated that Group I young of the year (21 to 40 mm SL) preyed heavily on small crustaceans (copepods, mysids, tanaids, and caridean shrimp). Based on the species composition of these prey items, the smallest silver perch are benthic generalists. The second trophic group (41 to 80 mm SL) preyed on shrimp and amphipods. Fish from Group III (81 to 120 mm SL) fed on penaeid and caridean shrimp, which represented 80% of their diet. Diets of fish larger than 141 mm SL were composed primarily of shrimp, crabs, and fish remains. Silver perch switch from being trophic generalists to trophic specialists with age. Spatial variation was indicated. Silver perch taken at Fenholloway stations at night had a greater percentage of shrimp, and the diet of such fish was more simplified than that of fish taken at Econfina reference stations. Silver perch were found to feed sporadically during the day as a number of the diurnally collected specimens contained well-digested shrimp remains in the Econfina system. The presence of digested fish remains in nocturnal specimens suggests that some silver perch are diurnal or crepuscular feeders. However, larger silver perch fed almost exclusively at night in vegetated areas that contributed to the much-simplified daytime diet. By comparison, fish collected at unvegetated Fenholloway sites fed equally during the day and at night (Figure 4.5). The diet was more diverse during the day, with shrimp and crustacean remains dominating in the smaller fish, and amphipods and fish remains more prevalent in the larger fish.

There was considerable flexibility in silver perch diets relative to seasonal fluctuations in prey populations (Ryan, 1981). Larger fish fed heavily on shrimp through fall months just prior to their migration from the estuary. By spring, mature specimens as well as young-of-the-year had switched to more abundant prey, apparently in response to reduced shrimp abundances during these months (Greening, 1980). Group II silver perch preyed heavily on amphipods and mysids at this time, whereas Group III fish changed their diet to include mysids and juvenile fishes that had recently entered the estuary. There was also considerable cannibalism at this time, primarily on smaller young of the year (<30 mm SL). As shrimp populations began to increase in numbers during the summer, Group III again began foraging on shrimp. Fish <80 mm SL did not start regularly feeding on shrimp until fall. Thus, trophic analyses of the silver perch showed the complexity of feeding habits when delineated by habitat, time of the day, and season.

The southern sea bass (*Centropristis striata*) was present during most of the study period. This species employs an ambush strategy already reported for numerous serranids (Hobson, 1974; Harmelin-Vivien and Bouchon, 1976). Stomach analyses of 322 sea bass indicated a dual (day/night) feeding pattern over the diel cycle (Figure 4.6). Although prey composition was roughly the same for sea bass taken during the day and at night at Econfina stations, the relative proportions of food items (based on biomass) were consistently greater for nocturnal specimens (Ryan, 1981). Amphipods and crabs (particularly xanthids) were more abundant in the stomachs of the smaller classes of nocturnal feeders, and these foragers took much greater proportions of fishes and shrimp (particularly penaeids). There were also indications of crepuscular feeding in this species. Larger sea bass (121 to 140 mm SL) preyed almost exclusively on shrimp at night and on crab remains during the day. This reflected a tendency to feed on the most available crustaceans during the diurnal cycle. In the Fenholloway system, shrimp were taken by the first two trophic units (31 to 80 mm SL) at night, whereas during the day the diet was more varied. The larger sea bass took crabs, fishes, and ophiuroids at night, whereas, during the day, these size classes took mainly crabs,

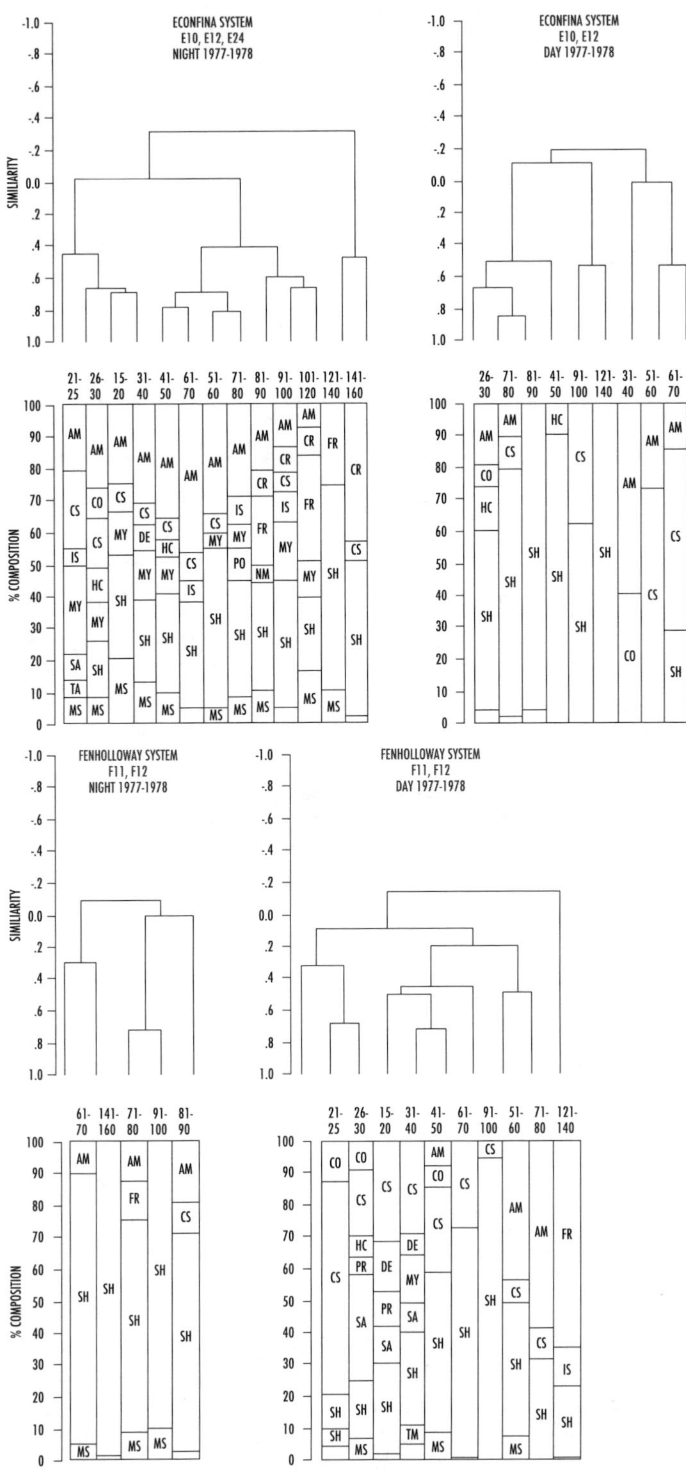

Figure 4.5 Ontogenetic changes in the diet of silver perch (*Bairdiella chrysoura*) taken in the Econfina and Fenholloway systems (inner stations, 1977–1978) at night and during the day. Histograms represent the relative proportions of the major dietary components (dry weight), and dendograms represent cluster analyses of the diet similarity among size classes. Codes for the food items are given in Table 4.1.

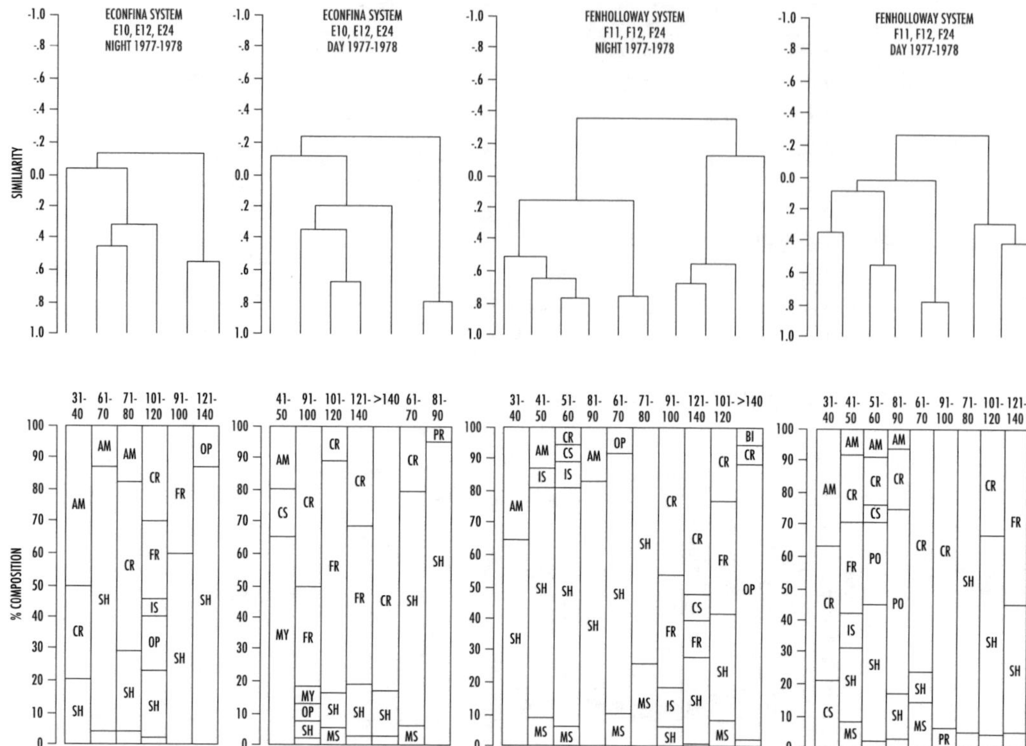

Figure 4.6 Ontogenetic changes in the diet of southern sea bass (*Centropristis striata*) taken in the Econfina and Fenholloway systems (inner stations, 1977–1978) at night and during the day. Histograms represent the relative proportions of the major dietary components (dry weight), and dendograms represent cluster analyses of the diet similarity among size classes. Codes for the food items are given in Table 4.1.

shrimp, and fishes. There was a strong seasonal component in the food selection of this species (Ryan, 1981; Ryan and Livingston, unpublished data). Smaller sea bass fed heavily on shrimp throughout the summer and winter months, but switched to fishes and crabs during the spring after shrimp populations declined (Greening, 1980). Thus, sea bass specialize on relatively few prey organisms although distinct dietary shifts were evident during 24-h periods and at different times of the year. Both spatial and temporal differences in prey selectivity of *Centropristis striata* were evident with habitat and prey availability differences, together with the generally nocturnal feeding habits of this species, contributing to considerable variability in food preferences.

Gulf toadfish (*Opsanus beta*) were present in the grass beds of Apalachee Bay throughout the year (Ryan, 1981). Gravid females appeared in May collections and recently spawned young-of-the-year were taken during late summer. As with most ambush predators, toadfish exhibit a dual feeding periodicity. Juveniles (<30 mm) fed largely at night on amphipods and shrimp. Larger toadfish (>30 mm) had a more pronounced feeding periodicity, taking fishes and crustaceans during the day, and switching to a crab diet at night. Unlike the previously mentioned nocturnal feeders, toadfish showed little ontogenetic, spatial, and seasonal trophic variability. Another nocturnal predator, *Urophycis floridanus* (southern hake), occasionally feeds during the day. However, a large number of empty stomachs were found in diurnal collections, and this species is largely crepuscular and nocturnal in its feeding habits. Southern hake passed through three distinct ontogenetic trophic stages with the diet of the first two groups (31 to 90 mm SL) differing only by percent composition of amphipods, mysids, and shrimp eaten. Larger hake (91 to 140 mm SL) preyed more heavily on shrimp, although juvenile fishes (pinfish, spot, silver perch, and grunts) became increasingly important prey with growth. Spatial differences in feeding habits were noted for fishes inhabiting the two inshore stations (E10 and F11), where juveniles of several fish species were included in the diet. Shrimp consumption by hake was high during all months at the vegetated sites. However, more than 70% of prey organisms taken at F11 (unvegetated) was composed of fishes and amphipods, Greening (1980) found that shrimp population densities were greatest during winter months at all stations except F11, thus again indicating habitat and prey availability as determinants of the diet of the southern hake.

Sphyrna tiburo (bonnethead sharks) were the most abundant fish taken by trammel nets at night (Ryan, 1981). Greatest numbers were caught during summer months when water temperatures had reached 25°C. All specimens contained decapod crustaceans, 85% of which consisted of blue crabs (*Callinectes sapidus*). The only other food organism encountered was the pink shrimp (*Penaeus duorarum*). Both prey species are known to exhibit strong nocturnal activity patterns (Laughlin, 1979; Greening, 1980; Ryan, 1981). Given this increased nocturnal prey activity and the large quantities of sea grass encountered in stomachs, it appears that this shark searches for these decapods as it swims just above the grass blades. All of the specimens of *Rhizoprionodon terranovae* (Atlantic sharpnose sharks) were taken in the Econfina grass beds during summer–early fall months. Specimens of this species ranged from 500 to 710 mm (total length, TL). Most contained fresh pink shrimp, blue crabs, and fish remains (including *Lagodon rhomboides* and *Syngnathus* spp.). Well-digested fish remains indicated that sharpnose sharks are piscivorous during crepuscular periods, switching to invertebrate prey after dark. Both sand and spotted sea trout (*Cynoscion arenarius* and *C. nebulosus*, respectively) showed a dual feeding pattern. Spotted sea trout were taken regularly in vegetated areas, whereas *C. arenarius* was more abundant at the sparsely vegetated sites (especially at F11). Both species fed on fishes and a variety of crustaceans (crabs, shrimp, isopods, and amphipods) during the day, whereas those that foraged at night took shrimp almost exclusively (primarily *P. duorarum*). Both species changed their diet with increased size (Ryan, 1981). Amphipods and small caridean shrimp were found in the stomachs of small (<70 mm) *C. nebulous*. Larger spotted sea trout took increasingly larger prey in the form of crabs, fishes, and larger shrimp. The eight-fingered threadfin (*Polydactylus octonemus*), another transient species, were taken only at Fenholloway stations and showed a nocturnal foraging pattern in which amphipods, cumaceans, and shrimp comprised a large portion of the diet. Flatfishes (*Paralichthys lethostigma*, *Symphurus plagiusa,* and *Etropus crossotus*) fed largely on caridean shrimp and amphipods at night. Sea robins (*Prionotus scitulus* and *P. tribulus*) foraged after dark on shrimp and amphipods. Speckled worm eel (*Myrophis punctatus*) and bank cusk eel (*Ophidion beani*) are burrowers during the day, hunting actively at night, for amphipods, shrimp, and cumaceans.

Overall, the feeding periods observed in various fish communities may be closely related to peaks in prey activity patterns (Starck and Davis, 1966; Helfmann, 1978; Hall et al., 1979; Miller, 1979; Ryan, 1981). Seasonal dietary shifts are related to differential prey abundances throughout the year and from year to year. Various species enter the grass beds during colder months when food densities are greatest (Greening, 1980; Stoner, 1980a,b) and prey heavily on amphipods, shrimp, mysids, and juvenile fishes. Shrimp abundances drop sharply by the time juvenile silver perch and pinfish move into Apalachee Bay (Greening, 1980) during late winter, and seasonal dietary shifts reflect such changes in prey densities. These juveniles feed primarily on amphipods and copepods, whose densities remain high throughout the spring. By summer, shrimp field abundances increase once again, with reflected increases in the diets of various fish species. Of the nocturnal specimens examined, only the requiem sharks showed no seasonal dietary changes. Bonnethead sharks fed almost exclusively on blue crabs and pink shrimp from March to December. Trophic variability thus appeared to be related to differences in habitat structure and prey abundance, both within the grass beds and between grass bed and unvegetated areas (i.e., the Fenholloway).

Only the ambush strategists failed to exhibit spatial feeding variations, possibly because such predators sit and wait for defensive lapses by preferred prey (Ryan, 1981). Various studies have indicated that both the intensity of predation and total prey consumption are inversely related to habitat complexity (Glass, 1971; Ware, 1972; Vince et al., 1976; Van Dolah, 1978; Stoner, 1980a). Stoner (1980a–c) found that feeding efficiency of diurnal pinfish dropped markedly with increased macrophyte densities in Apalachee Bay. However, no such pattern was evident for nocturnal feeders (Ryan, 1981). Structural differences in habitat complexity may thus be perceived much differently by nocturnal predators. For example, Group III silver perch consumed proportionately greater numbers of shrimp at the most heavily vegetated sites, whereas at E10, where large amounts of drifting red algae (primarily *Laurencia poitei*) are present, amphipods dominate the stomach contents. It appears that windrows of drift algae provide a greater refuge for certain shrimp at E10 at night than denser grasses at E12 (Greening, 1980; Ryan, 1981). Additionally, amphipod densities may be extremely high in drift algae, and fishes may simply be feeding on these crustaceans as they swarm into the water column at night (F. G. Lewis III, personal communication).

Diet breadth in nocturnal fishes may follow food abundance patterns. Food densities were generally highest at the three vegetated stations, and the degree of trophic specialization reflected habitat-related changes in prey abundance. Greening (1980) found fewer invertebrate species at the unvegetated sites. Nocturnal predators foraging at the vegetated sites took a much narrower range of prey taxa than those fishes feeding on bare bottom substrata at F11. These results are compatible with several previously published reports that have correlated dietary specialization with increased food supply (Ivlev, 1961; Zaret and Rand, 1971; Werner and Hall, 1974; Pyke et al., 1977; Keast, 1979; Stoner, 1979a, 1980c). Consequently the feeding strategy of a particular species may differ markedly over relatively small distances (2 km in this case). Differences in prey availability over the diel cycle also appear to influence feeding periodicity of many grass bed predators. Whereas *Bairdiella chrysoura* fed almost exclusively at night in vegetated areas, fish at station F11 were characterized by an arrhythmic feeding periodicity. Such trophic flexibility would be highly adaptive since, at lowered food densities, silver perch could switch to different prey organisms as they became available throughout the day or night. Plasticity in maintaining activity periods that coincide with those of suitable prey organisms appears to be widespread among fishes and invertebrates, allowing them to shift their activity rhythms according to their ecological requirements (Muller, 1970, 1978; Ericksson, 1978; Grossman et al., 1980).

2. *Interannual Variation*

There is little scientific information regarding interannual variability of trophic relationships of coastal organisms. Livingston (1980a, 1984d) indicated changes of fish feeding habits over a 7-year period. The basis of long-term changes in the trophic relationships of fishes in the Econfina and Fenholloway systems can be illustrated by trends of trophodynamically complex ontogenetic feeding progressions of the pinfish (Figure 4.7). There was a spatial pattern to the differentiation of pinfish feeding habits between the polluted and unpolluted estuaries that was based, in part, on the distribution of grass beds. Plant biomass, dominated by *Thalassia* in the Econfina estuary and *Syringodium* in the Fenholloway estuary, showed divergent trends over the 7-year period of observation. There was a general decline of plant biomass in the Econfina system, while there were indications of a limited recovery in peripheral areas of the Fenholloway estuary. The general patterns of pinfish feeding were similar in both the inner and outer areas of the Econfina system over the 7-year study period. Early growth stages depended largely on copepods, amphipods, and other small invertebrates with a gradual gradation into different levels of omnivory. Further growth culminated in a largely herbivorous existence. These patterns were similar in both areas where the well-developed grass beds remained relatively stable in time. A stable habitat led to little interannual change in diet.

The sequential feeding progression with growth was not noted with pinfish taken at the inner (most polluted) Fenholloway stations during the study period (1971 to 1974). The absence of sea grasses in these areas was reflected in the relative paucity of those stages of pinfish (>120 mm) that were mainly herbivorous in reference sea grass beds. In later years, there was still limited recovery

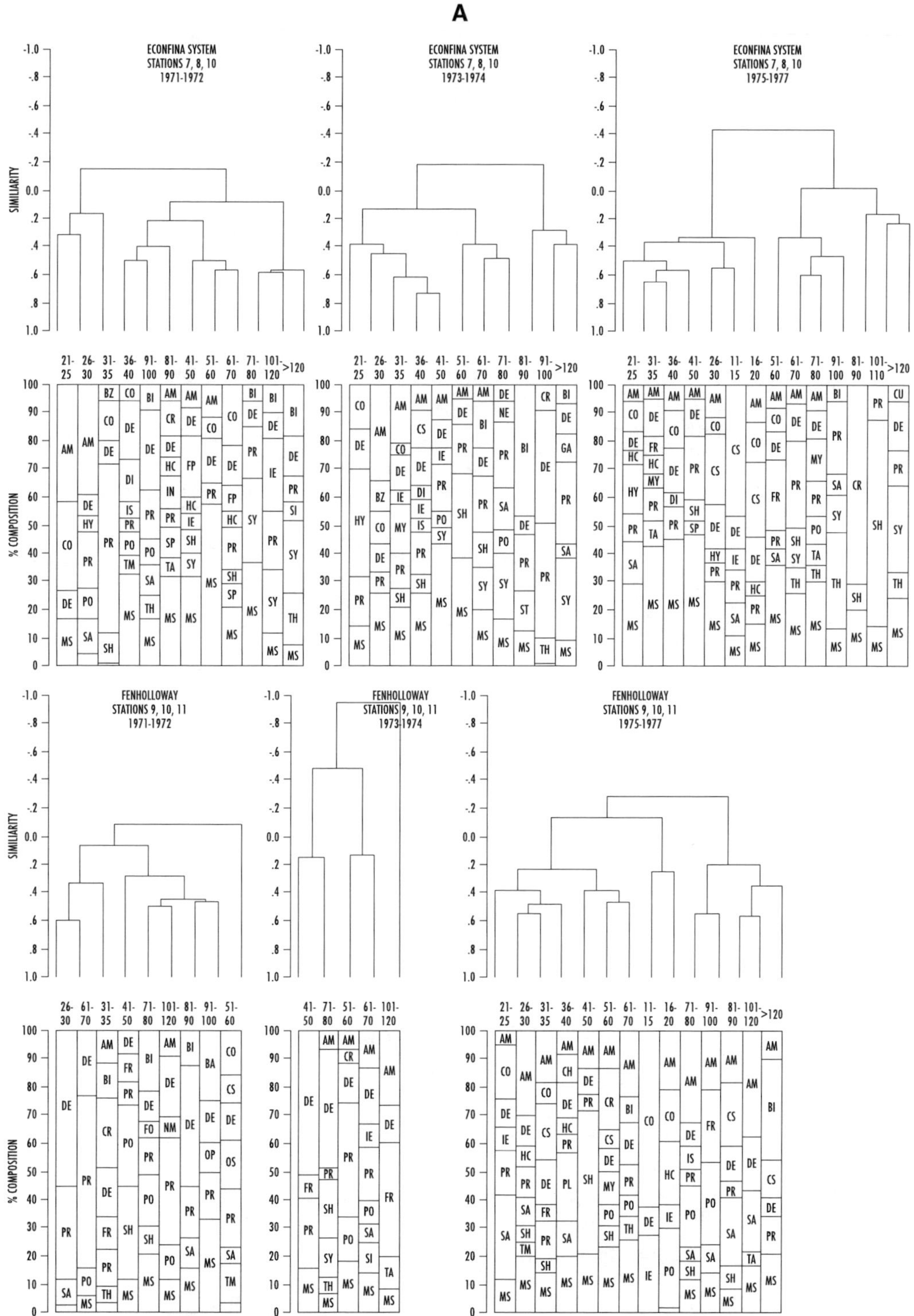

Figure 4.7 Ontogenetic changes in the diet of *Lagodon rhomboides* taken in the Econfina and Fenholloway coastal areas from 1971–1977. Histograms represent the relative proportions of the major dietary components (dry weight), and dendograms represent cluster analyses of the diet similarity among size classes. Codes for the food items are given in Table 4.1. (A) Inner stations of the Econfina and Fenholloway systems with feeding patterns shown at three intervals: 1971–1972, 1973–1974, 1975–77. (continued)

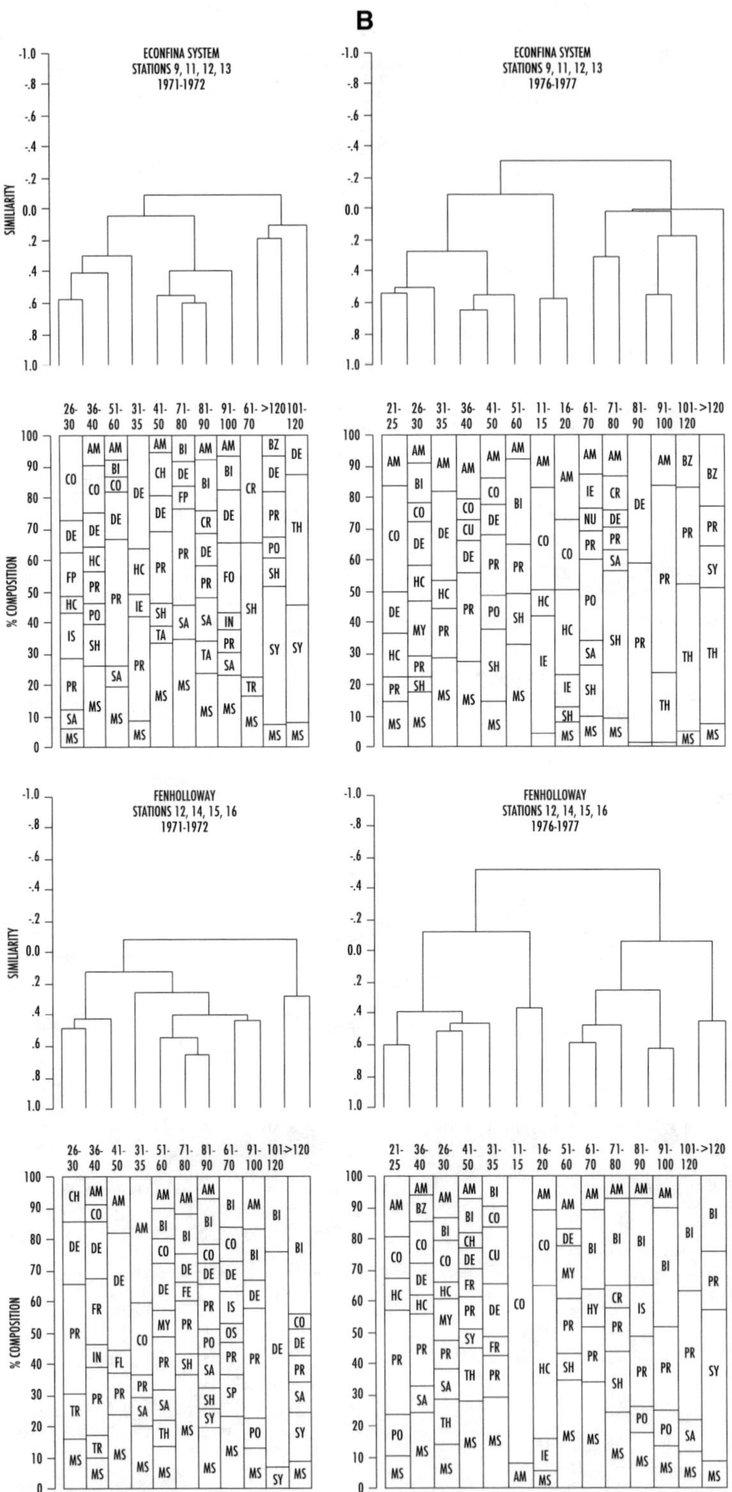

Figure 4.7 (continued) Ontogenetic changes in the diet of *Lagodon rhomboides* taken in the Econfina and Fenholloway coastal areas from 1971–1977. Histograms represent the relative proportions of the major dietary components (dry weight) and dendograms represent cluster analyses of the diet similarity among size classes. Codes for the food items are given in Table 4.1. (B) Outer stations of the Econfina and Fenholloway systems with feeding patterns shown at two intervals: 1971–1972, 1976–1977.

of SAV in the inner Fenholloway areas with the larger fishes eating mainly bivalve mollusks and various form of dead organic matter. However, in outer areas of the Fenholloway, where there was less pollution and some improvement of the sea grass beds with time, there was a different pattern (see Figure 4.7). During 1971–1972, the earlier stages of pinfish tended to resemble those in the reference area in terms of general food distribution. However, the larger fishes were mainly dependent on bivalves and various forms of organic detritus during this time. By 1976–1977, however, there was a recovery of pinfish feeding that closely resembled that in the Econfina system, with a return to increased dependence on *Syringodium* in the larger fishes. The trends of trophic organization followed pinfish numbers that were always higher in the Econfina system, but which tended to increase in the Fenholloway system with time at the outer (not inner) stations. Thus, in areas with relatively stable levels of habitat organization and productivity (Econfina), there was relatively little alteration of feeding habits of pinfish over interannual periods. In the Fenholloway system, however, changes in habitat and productivity were followed by qualitative changes in the feeding habits of this species that were consistent with the increased levels of habitat quality and productivity. These changes of trophodynamic processes were followed by increased production by this species.

As noted by Livingston (1984d), various grass bed fishes followed regular seasonal, age-specific feeding patterns, which did not change substantially in terms of qualitative food composition in the unpolluted estuary over the 7-year period of observation. This feeding behavior helped to explain temporally conservative cycles of relative abundance despite extreme (natural) habitat changes. Anthropogenous habitat alterations in the Fenholloway system, though seemingly slight, were associated with reductions in benthic macrophyte distribution, enhanced phytoplankton productivity, and changes in the relative dominance and numerical abundance of associated fish assemblages. Grass bed species were replaced by plankton-feeding fishes, and disruption of feeding habits of various species was apparent in the affected area relative to the unpolluted estuary. Subsequent water quality improvement over time was associated with shifts in the age-specific dietary patterns of various species toward those observed in the unaffected estuary, although such recovery varied from species to species according to habitat utilization and trophic needs. From these results, it is clear that a relatively complex coastal sea grass system exposed to periodic, extreme natural disturbance is relatively resilient to such changes in terms of relative dominance and food web structure. However, apparently slight water quality changes due to pollution that are outside the evolutionary experience of coastal populations can cause serious disruptions of basic habitat structure, energy flow, and community composition at various levels of biological organization.

C. Inadequacy of Species-Level Designations

Fish and invertebrate feeding patterns vary widely, according to habitat structure, qualitative and quantitative aspects of primary productivity, season, and fish size, thus complicating any comprehensive evaluation of trophodynamic processes in coastal areas. Ontogenetic patterns of feeding add another dimension of complexity to the problem. Most food web studies and models are based on species-specific trophic designations (Peters, 1977; Stoner, 1979a, 1980c; Livingston, 1980). In the overwhelming majority of ecological papers concerning trophic relationships in coastal systems, the central focus is the species. Where the basic life history characteristics of the individual species are the subject of concern, such emphasis is appropriate. However, when complex fish and invertebrate assemblages in shallow coastal areas are the subject of the study, the use of the species as a unit of measure may not be suitable. This is especially true in attempts to associate process-oriented relationships with forcing factors such as meteorological events, habitat variables, or biological interactions. Diel feeding rhythms are flexible, and fishes can adjust their trophic preferences to varying environmental conditions. However, such adaptations are still limited by species-specific behavior. Thus, the ontogenetic feeding patterns of individual species, within determined boundaries of both spatial and temporal variability, should be taken into consideration when evaluating trophodynamic processes in coastal systems.

Livingston (1988) examined the utility of the common use of species-level designations in the development of food web associations in coastal areas. Resource partitioning among fish species has been shown in various habitats (McEachran et al., 1976; Chao and Musick, 1977; Ross, 1977). Changes in feeding patterns can be used as indicators of natural and anthropogenous disturbance phenomena as well as recovery of water and sediment quality. Even relatively slight water quality changes have been associated with basic changes in food availability, which, in turn, are directly related to shifts in the trophic assemblages. Species-specific feeding trends have been identified for the numerically dominant fishes and invertebrates in the near-shore sea grass beds of Apalachee Bay (Econfina and Fenholloway estuaries), Apalachicola Bay, and Choctawhatchee Bay. In Apalachee Bay, a sea grass–dominated system, young-of-the-year fishes and invertebrates move into these inshore coastal habitats during different times of the year. Some species, such as spot (*Leiostomus xanthurus)*, undergo relatively smooth transitions based primarily on changes in the size of the food. Others, such as the pinfish (*Lagodon rhomboides)*, have more distinctive trophic stages in addition to the obvious food-size designations.

Seasonal progressions of feeding activity are integral to food web dynamics in coastal areas of the NE Gulf of Mexico. During winter–early spring periods characterized by high numbers of copepods, epibenthic invertebrates, amphipods, and polychaetes, young pinfish and spot move into inshore grass beds. As the spring peaks of phytoplankton and submerged aquatic vegetation occur, anchovies (*Anchoa* spp.), pigfish (*Orthopristis chrysoptera)*, and spotted pinfish (*Diplodus holbrooki)* become dominant. By early summer, anchovies and the species that feed on crustaceans (*Bairdiella chrysura*, *Stephanolepis hispidus*) become numerous. During summer, species that feed on invertebrates such as the mojarras (*Eucillostomus* spp.), syngnathids (*Syngnathus* spp.), and *Centropristis striata* are numerically important. By early fall, later feeding stages of pinfish. spottail pinfish, mojarras, monacanthids, and syngnathids are prevalent. At this time, the top predators such as lizardfish (*Synodus foetens*), toadfish (*Opsanus beta*), sea trout (*Cynoscion* spp.), catfish (*Arius felis*), rays, and sharks are active in inshore feeding grounds. As water temperature drops precipitously during the procession of atmospheric cold fronts in November, *Stephanolepis hispidus* and *Monacanthus ciliatus* become abundant (Clements and Livingston, 1983). Stoner and Livingston (1984) showed that the sparids (*Lagodon rhomboides* and *Diplodus holbrooki*), although spatially and temporally sympatric, have distinct trophic stages and little dietary overlap. This ontogenetic variation was correlated with differences in locomotor attributes, mouth dimensions, and ontogenetic changes in dentition. Thus, complex combinations of spawning periods, migratory behavior, habitat and food availability, growth rates, and ontogenetic morphological and physiological changes lead to the observed feeding patterns of the various numerically dominant sea grass fishes in systems along the NE Gulf coast.

There is a considerable range of feeding adaptations among coastal fishes and invertebrates, from the broad span of trophic stages of pinfish to the relatively specialized feeding style of the banded blenny (Stoner and Livingston, 1980). Various species undergo significant developmental stages that span a range of habitats and trophic niches. The relatively distinct within-species trophic identities of such species are accompanied by highly differentiated habitat utilization. Consequently, the assignment of niche characteristics based on species designations represents an oversimplification of the actual role that such organisms play in coastal areas. In an attempt to quantify the relationships of fish trophic organization to habitat and resource changes, Livingston (1988) developed a station-by-time matrix that consisted of physical and chemical variables, productivity determinations (phytoplankton, submerged aquatic vegetation), macrophyte features (standing crop biomass, species richness, and species diversity), epibenthic invertebrate features (numerical abundance, species richness, and diversity), and fish data (numerical abundance, trophic unit organization). Multiple and stepwise regression analyses were run with these variables; correlation coefficients and r^2 values were uniformly low, reflecting high temporal variability and possible nonlinear relationships of trophic organization. Various species (e.g., monacanthids, pinfish, silver perch, pigfish, spotted pinfish) had relatively strong and consistent associations with different aspects of the submerged aquatic vegetation (productivity, standing crop biomass, species richness). Species with complex trophic histories and distinct ontogenetic feeding stages (i.e., pigfish, spotted pinfish, pinfish) were distinguished by close relationships between individual trophic units and habitat characteristics, whereas species-level designations of these fishes, spanning broader ranges of niche characteristics, were subject to very different statistical relationships. For instance, early pinfish stages were not associated with the sea grass beds as were later stages. Analyses with the species as a whole did not reflect such changes in habitat utilization. In fact, statistical analyses carried out with species-level data led to results that did not correspond to any specific habitat associations. At the same time, various trophic stages of different species were often more similar than the serial stages of the same species. Overall, statistical analyses using trophic units came closer to identification of the ecological relationships of a given species succession than similar analyses carried out with the species as the unit of measure.

The species concept remains the fundamental unit of systematists, and, as such, is useful as a descriptor of basic natural history. However, from early papers (Thienemann, 1918) to more recent discussions (Peters, 1977; Livingston, 2000), there has been an underlying discontent with the largely theoretical association of the species with a distinct and stable niche. Lindeman (1942) proposed a more bioecological species-distributional approach based in large part on the trophic relationships within the "community-unit." Based on the results of the analysis of the trophic organization of Gulf coastal fishes, the use of species-level data in the quantification of habitat relationships was not possible because of the complexity of the life history stages as they relate to feeding relationships. Models based on the species as the central unit of measure thus run the risk of oversimplification and overgeneralization when used for quantification of ecological associations. Although it has long been known that many coastal fishes and invertebrates undergo complex ontogenetic changes in their feeding habits and their use of available habitats, relatively little has been done to use such data in a way that allows quantification of the ecological relationships of such species. There is a range of such responses among various species from relatively simple life history strategies (e.g., anchovies) to the highly complex progressions that encompass a complex series of distinct trophic levels (e.g., pinfish). The identification of multispecies assemblages with similar food habits would allow a more comprehensive understanding of the underlying ecological processes by relating the relevant biological unit of organization to process-oriented factors (i.e., food web functions). For questions relating to process-oriented trophic organization, there is thus good reason to look at subspecific units based on life history stages rather than the species as a whole. Under such circumstances, quantitative ecological studies that concentrate on ontological trophic entities and food web organization will be more realistic than the continued dependence on the taxonomic species as the primary unit of measure.

Overall, the studies of trophodynamics in Gulf coastal systems indicate that the species is not always appropriate as a unit of measure when used in quantitative ecological studies. In many instances, more substantial ecological differences exist among life stages of a given species than among similar trophic units of different species. The use of a species in quantitative ecological studies can lead to problems of interpretation concerning the relationships of coastal organisms to complex habitats. Use of the species as a convenient unit of measure substitutes a basically taxonomic entity for ecologically relevant life history stages. Niche breadth of a given species can be so extensive that quantitative determinations of significant ecological processes are difficult to make. Migratory coastal fishes undergo diverse ontogenetic trophic transformations. Without adequate recognition of the complex ecological stages that characterize such organisms, the interpretation of ecosystem-level data will remain incomplete.

IV. TROPHIC LEVEL DESIGNATIONS

A. Trophic Unit Transformation

All biological data (as biomass$\cdot$m^{-2} mo^{-1} of the infaunal, epibenthic macroinvertebrates and fishes) were transformed from species-specific data into a new data matrix based on trophic organization as a function of ontogenetic feeding stages of the species found in the various coastal systems over the multiyear sampling program. Ontogenetic feeding units were determined from a series of detailed stomach content analyses carried out with the various epibenthic invertebrates and fishes in the region (Sheridan, 1978, 1979; Laughlin, 1979; Sheridan and Livingston, 1979, 1983; Livingston, 1980a, 1982a, 1984d; Stoner and Livingston, 1980, 1984; Laughlin and Livingston, 1982; Stoner, 1982; Clements and Livingston, 1983, 1984; Leber, 1983, 1985). Based on the long-term stomach content data for each size class, the fishes and invertebrates were reorganized first into their trophic ontogenetic units using cluster analyses. Infaunal macroinvertebrates were also organized by feeding preference based on a review of the scientific literature (Livingston, unpublished data). A detailed listing of the trophic organization is given in Appendix II. The field data were then reordered into trophic levels so that monthly changes in the overall trophic organization of the various systems could be determined. We assumed that feeding habits (at this level of detail) did not change over the period of observation; this assumption is based on previous analyses of species-specific fish feeding habits that remained stable in a given area over a 6- to 7-year period (Livingston, 1980).

Because of the relatively high level of spatial and temporal variability of the trophic successions of a given species, a more generalized approach was used to organize the long-term databases into a derived database founded on life history factors of the infaunal and epibenthic macroinvertebrates and fishes. This reconstruction included the definition of a trophic database founded on the trophic units determined by grouped data as outlined above (see Appendix II). The long-term field data from the various coastal systems were transformed as a function of the individual trophic units of the organisms taken (infauna, epibenthic invertebrates, fishes = FII indices) into an empirical food web. Life history factors were reorganized into guild associations that transcended taxonomic boundaries. The similar trophic units were grouped into a definitive trophic order that included various levels of specificity. The most generalized reorganization included the following: herbivores (feeding on phytoplankton and benthic algae), omnivores (feeding on detritus and various combinations of plant and animal matter), primary carnivores (feeding on herbivores and detritivorous animals), secondary carnivores (feeding on primary carnivores and omnivores), and tertiary carnivores (feeding on primary and secondary carnivores and omnivores). All data were given as ash-free dry mass m^{-2} mo^{-1} or as percent ash-free dry mass m^{-2} mo^{-1}. Unless otherwise indicated, the data were presented as monthly means of data taken at fixed stations (identified by Global Positioning System [GPS] designations) or grouped stations. In this way, the long-term database of the various collections of infauna and epibenthic macroinvertebrates and fishes was reorganized into a quantitative and detailed trophic matrix based not solely on species (Livingston, 1988) but on the complex ontogenetic feeding stages of the various organisms in each coastal area.

CHAPTER 5

Structural Components of Trophic Organization

I. INTRODUCTION

The distribution of organisms in aquatic systems is strongly influenced by feeding preferences. Ontogenetic shifts in diet define habitat needs as each species follows relatively conservative morphological and physiological transitions. Consequently, the trophic needs of a given species may vary considerably in time, leading to a close connection between habitat variables and related prey distributions. Livingston (1987b) outlined the spatial/temporal scaling problems associated with the definition of population and community variability in dynamic coastal ecosystems. There are families of spatial and temporal scaling phenomena that define interspecific trophic interactions. The dimensions of variation change along spatial/temporal gradients of salinity, habitat complexity, and productivity with different responses at different levels of trophic organization. There is thus a continuum of scaling dimensions that defines the population distributions in coastal systems.

So-called background noise is an inherent characteristic of ecological systems; such variation is particularly high in physically unstable coastal areas. Disequilibria induced by physical disturbance are often specific to particular levels of biological organization within a given system so that, although many coastal systems are considered to be physically unstable, the persistence, inertia, and resilience (Dayton et al., 1992) of the biological components of such systems remain less prone to rapid (habitat-induced) changes. Regularity and predictability of community structure cannot depend solely on homogeneous or typical areas. The complexity and highly variable nature of estuarine assemblages have precluded definitive experimental demonstrations of the processes that underlie the constant shifts of coastal populations. Dayton (1979) alluded to the many scales of disturbance (physical and biological) that continually intergrade into a series of short-term, seasonal, and interannual episodic events. These changes, when superimposed over highly variable recruitment patterns, contribute to the difficulty of explaining, in a quantitative sense, the mechanisms that drive spatial and temporal variation of estuarine assemblages. The fact that various populations are interchangeable in terms of their feeding needs at various stages of their development is related to the contradictory relationship of high variability of population distribution with a relatively time-stable trophic organization.

Most of the dominant forms of macroinvertebrates and fishes in the highly productive shallow coastal waters of the Gulf of Mexico are benthic feeders. To define in more detail the trophic interactions involved with the primary food webs of the region, a series of field descriptive and experimental analyses was carried out with an emphasis on the infaunal macroinvertebrates as the chief link to the primary food webs of the Apalachicola estuary. Weekly monitoring of water quality and infaunal variation was joined with predator-exclusion and toxic treatment experiments. These tests were designed to evaluate the possible importance of trophic organization as an indicator of the natural variability of coastal habitats and the influence of toxic agents on trophodynamic processes.

The Apalachicola system (see Figure 2.2D) is controlled to a considerable degree by freshwater input from the Apalachicola River (Livingston et al., 1997). The most obvious effects of river runoff include changes of salinity and other physical/chemical factors (Schroeder, 1978). Riverine loadings of nutrients and organic carbon, together with the physiographic characteristics of the estuary, combine to create the conditions that lead to extremely high primary and secondary production in the Apalachicola system (Livingston, 1984a). At the same time, the rigors of estuarine conditions in terms of rapid habitat changes reduce predation pressure (Livingston et al., 2000), thus allowing the rich nursery environment to contribute to the rapid growth and productivity of eurytopic populations that are adapted to estuarine conditions (Livingston, 1984a, 1991b; Livingston et al., 1998b). Salinity plays an important role in the determination of which species are able to exist in a given estuarine habitat; this is particularly important in low-salinity estuaries where tidal freshwater and marine associations come together.

Determinations of biological interactions along salinity gradients have not been well documented in terms of the relationship of habitat to food web organization over prolonged periods. Nichols (1985) found that periods of prolonged low river flow in northern San Francisco Bay led to changes in the relationships of benthic food webs relative to planktonic ones; drought conditions led to shifts from pelagic to benthic food webs, whereby energy was passed directly from the water column primary producers to the benthos. Cloern et al. (1983) attributed such changes to altered positioning of suspended particulate maxima in the bay, with phytoplankton composition determined by river discharge. Livingston et al. (1997) found that prolonged drought conditions in the Apalachicola estuary were associated with alterations in the benthic food webs, and that changes in productivity as a result of a drought were translated into long-term alterations of the overall estuarine food web structure. Montagna and Kalke (1992), working in two Texas estuaries, showed that increased river flow rates led to increased macrofaunal production. Low freshwater inflow and high salinity led to increased macrofaunal diversity. Flint (1985) reported increased infaunal abundance and biomass following higher river flow into Corpus Christi Bay, Texas. The above-noted studies showed that estuarine food webs are intricately tied to variables such as river flow input, primary productivity, habitat quality, and biological factors such as predator–prey relationships.

A number of generalizations concerning estuarine biological relationships have been derived from small-scale studies of the interactions of climatological conditions, physical/chemical factors, nutrients and organic carbon, predators, parasites, competitors, and pathogens. Previous studies indicate that predation can be a controlling factor of unvegetated benthic communities in temperate and subtropical estuaries (Virnstein, 1977; Holland et al., 1977, 1980; Reise, 1978). The question of the relative importance of predator control vs. primary producer influence on food webs has been discussed by various authors (Hunter and Price, 1992). The theoretical underpinnings of pattern and scale in organism distribution have been thoroughly reviewed (Levin, 1984). Application of food web linkage to such questions, however, remains largely undeveloped. Despite a plethora of natural history studies in coastal systems, there is very little information concerning how the natural variation of coastal assemblages can be explained relative to basic interactions of physical controls, predator–prey relationships, and food web structure.

Experiments have been carried out to determine population response to natural disturbance and predator–prey effects in estuarine/coastal habitats. Peterson (1991) compared various aspects of physiological stress (high in river-dominated estuaries), physical and biological disturbance (potentially high), competition (not a dominant process in soft sediments with the possible exception of large-scale food limitation), and predation (often a factor in shaping infaunal communities). Disturbance is a natural part of many temperate estuaries, and there is ample evidence that various forms of disturbance can be followed by successional changes in estuarine/marine infaunal associations (Pearson and Rosenberg, 1978; Santos and Simon, 1980a,b; Thistle, 1981). Levin (1984) related the effects of a given disturbance to the specific life history characteristics of infaunal species. Zajac and Whitlach (1982a) found that estuarine infauna showed variable responses to disturbance with no spatial or temporal (seasonal) pattern of recolonization. Zajac and Whitlach

(1982b) indicated that the timing of a given disturbance affected subsequent species successions. Livingston et al. (1999) noted the same phenomenon in storm effects on oyster populations. Factors involved in responses to disturbances thus include the timing of the disturbance event, ambient habitat characteristics at the time of disturbance, reproductive periodicity of existing infaunal populations, the dynamics of individual populations, and available resources of space and food. Whitlach and Zajac (1985) found that, while opportunistic estuarine infaunal species may be affected by biotic interactions such as inhibition due to intraspecific settlement and competition for food and resources, the actual form of interaction usually depended on the species in question and existing habitat conditions. No characteristic type of biological interaction was found that defined or predicted soft-bottom infaunal successions.

The Apalachicola Bay system is an unpolluted, highly productive river-dominated estuary (Livingston et al., 1974, 1997; Livingston, 1976a, 1984a, 1991c; Sheridan and Livingston, 1979, 1983). Major parts of the estuary are characterized by soft sediments. Bay food webs are dominated by relatively few dominant infaunal macroinvertebrates, epifaunal macroinvertebrates, and fishes (Livingston et al., 1997). Interactions of these groups in estuaries are often mediated by predator–prey relationships (Virnstein, 1977; Peterson, 1982, 1992; Livingston, 1984b; Whitlach and Zajac, 1985). A combined descriptive and experimental study was carried out in the Apalachicola estuary along a series of habitat and productivity gradients related to the river flows. Analytical support for the following analyses was provided by Dr. F. Graham Lewis III, Mr. Glenn C. Woodsum, Dr. Xufeng Niu, and Dr. Gary L. Ray. These colleagues should be considered co-authors of this chapter.

II. INFAUNAL MACROINVERTEBRATES

The upper, tidally influenced reaches of the Apalachicola estuary form an extensive delta region as the Apalachicola River enters the bay. East Bay (stations 3 and 5a: Figure 5.1) represents a primary nursery of the headwater system. Previous studies (Livingston, 1984a) suggest an interaction among several dynamic physical mechanisms operating at different timescales that control habitat conditions in the receiving estuary. Seasonally fluctuating Apalachicola River flow, local rainfall, the physiographic structure of the receiving basin, wind speed (duration and direction), and exchanges between the estuary and the Gulf of Mexico are thought to be the primary physical controlling factors (Livingston, 1984a; Livingston et al., 1997). Tides generally have a localized effect on salinity and stratification in the estuary (Livingston, 1997a); these effects are less important on baywide salinities than the influence of freshwater discharge, local precipitation, and wind events (Livingston et al., 2000). St. George Sound (station ML: see Figure 5.1), on the other hand, remains largely outside of the influence of the Apalachicola River. This part of the estuary has relatively constant high salinities and is usually well mixed.

Previous studies indicate that both phytoplankton production (Myers and Iverson, 1977, 1981) and river-derived allochthonous detritus (Livingston et al., 1977) are higher in East Bay than in areas removed from direct river flow. Within East Bay, station 3 can be characterized by seasonal increases in allochthonous detritus derived from upland river-wetlands areas, whereas station 5a had lesser quantities of such organic matter (Livingston et al., 1977). Peak levels of macrodetrital accumulation occur during winter/spring periods of high river flow (Livingston, 1981a). These periods are coincident with increased infaunal abundance (McLane, 1980). Four of the five dominant infaunal species at river-dominated stations are detritus feeders. A mechanism for the direct connection of increased infaunal abundance has been described by Livingston (1983a, 1984a), whereby microbial activity at the surface of the detritus (Federle et al., 1983a) leads to microbial successions (Morrison et al., 1977) that then provide food for a variety of detritivorous organisms (White et al., 1979a,b; Livingston, 1984a). The disturbance of predation on microbes increases microbial production (Federle et al., 1983a,b, 1986). The transformation of nutrient-rich particulate organic

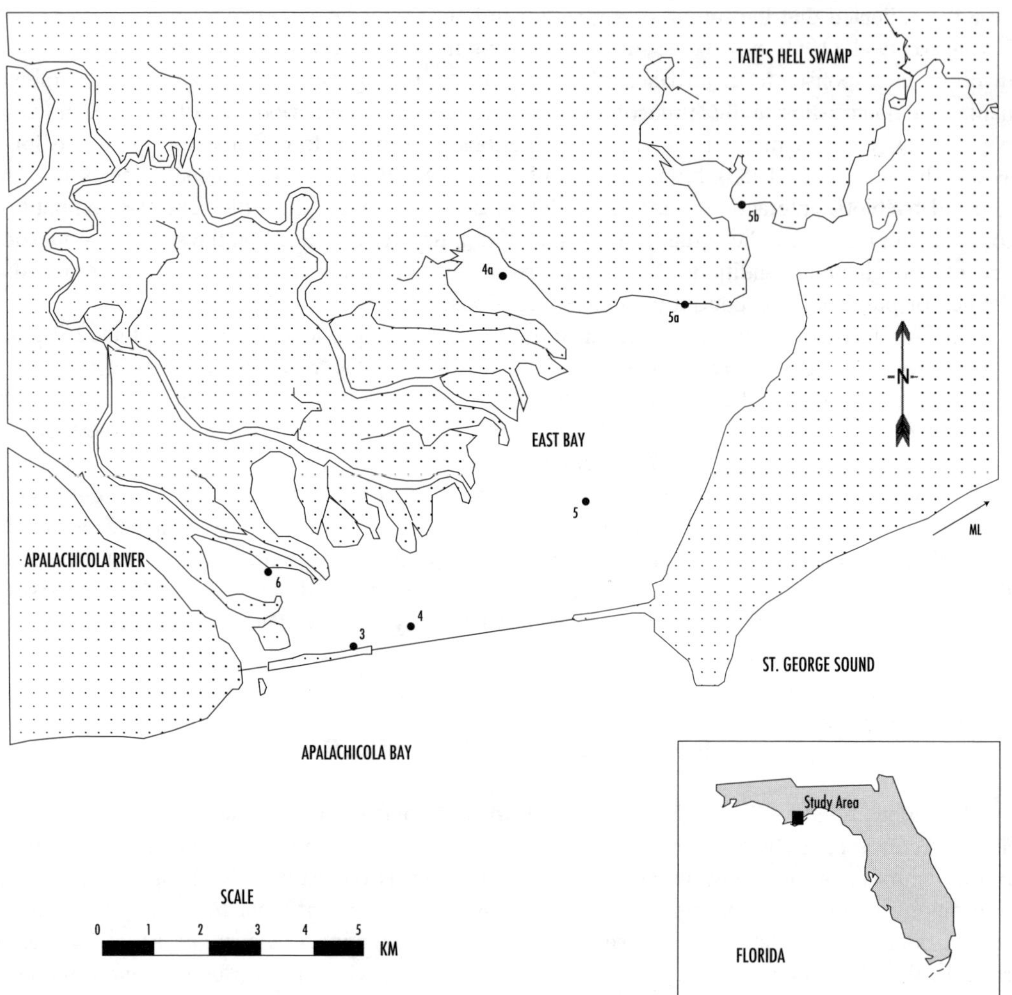

Figure 5.1 The Apalachicola Bay system, showing stations used in the long-term field analyses and experimental studies.

matter from periodic river-based influxes of dissolved and organic detritus coincides with the abundance peaks of the detritus-based (infaunal) food webs of the Apalachicola system (Livingston and Loucks, 1979) during periods of increased river flooding. Chanton and Lewis (2002) provide analytical support for these observations. These processes help to explain the differential primary and secondary production levels among the various study sites (stations 3, 5a, ML).

A. Descriptive Field Data

Detailed methods for the field collections of physical/chemical data are given by Livingston (1979, 1981b, 1984a) and Livingston et al. (1974, 1976). Long-term (weekly to monthly) field water quality measurements were taken at the three stations (see Figure 5.1). Sediment analyses (top 10 cm of core samples) were run from samples taken at the fixed stations throughout the field descriptive and experimental program. Sediments were analyzed for particle size distribution and organic composition according to methods described by Mahoney (1982) and Mahoney and Livingston (1982) as defined by Galehouse (1971). Infaunal macroinvertebrates were collected using coring devices (7.6 cm diameter, 10 cm deep). Water quality and biological samples were taken monthly at stations 3 and 5a (July 1978 to June 1984) and station ML (November 1981 to August 1986). Weekly collections were taken during the period from November 1981 through November 1983 concurrent with the predation experiments (described below). During October 1981, three sets of 100 cores uniformly distributed on a grid (2 m on a side, 20 cm between core centers) were taken from platforms placed at each of the three established stations to determine small-scale spatial variability and spatial correlations of benthic macroinvertebrates within a given area. For each set of three sampling efforts at each site, the platform was moved about 5 m in such a way that the distribution of the sets was haphazard. Biological data derived from weekly and monthly collections were used to develop the spatial and temporal context within which the results of the field experiments could be interpreted.

B. Habitat Changes

The most obvious difference between the aquatic habitats of East Bay (stations 3 and 5a) and St. George Sound (station ML) is the lack of direct (station 3) or indirect (station 5a) Apalachicola River flow at station ML. Apalachicola River flows varied from 186 to 2219 $m^3 s^{-1}$ over the study period with high flows during late winter–early spring and low flows during fall. The study period included a drought during 1980–1981 when river flows were significantly ($p < 0.05$) below the long-term monthly means for 22 consecutive months. Considerable station-specific differences were noted in the long-term data concerning measured habitat variables (Table 5.1, Figure 5.2). Mean depths were 1.2, 1.4, and 1.8 m at stations 3, 5a, and ML, respectively. Habitat variables varied with distance from the river.

East Bay station 3, which receives direct river flow, had lower salinity, temperature, and dissolved oxygen (DO) anomaly than the other sites. Bottom salinity in East Bay ranged from essentially freshwater conditions during winter/spring floods to salinities between 10 and 20 ppt during the fall low water periods (Figure 5.2). There were rapid changes of salinity at station 3. Station ML was characterized by less salinity variability. During periods of high Apalachicola River flow, salinity was not markedly reduced in St. George Sound, and there was evidence (low salinities during summer months) that local runoff was responsible for the relatively small seasonal differences in salinity in this area of the bay. Salinity remained relatively high at station ML with little in the way of rapid changes as noted at the East Bay stations. Color levels in East

Table 5.1 Long-Term Mean Values (± standard deviations) of Environmental and Infaunal Community Characteristics Summarized from Monthly Collections over the Longest Period of Record at Each Site

Habitat and Community Characteristics	Station 3	Station 5a	Station ML
	Environmental Characteristics		
Salinity	4.3 (4.9)	6.2 (6.4)	28.8 (3.6)
Temperature	21.4 (6.2)	22.0 (6.0)	22.0 (5.9)
Color	57.5 (57.2)	75.5 (71.3)	24.6 (21.9)
DO anomaly	–1.01 (1.31)	–0.52 (1.09)	–0.44 (0.88)
pH	7.25 (0.47)	7.38 (0.69)	7.87 (0.35)
	Community Characteristics		
Density	7830 (10867)	3206 (3363)	3669 (1686)
Number of taxa	9.9 (3.3)	6.2 (3.3)	28.9 (6.0)
Biomass	4.7 (8.1)	6.9 (17.7)	1.7 (1.8)

Note: Longest period of record was 7/78 to 6/84 for stations 3 and 5a, 11/81 to 8/86 for station ML. Environmental characteristics were taken from bottom water samples.

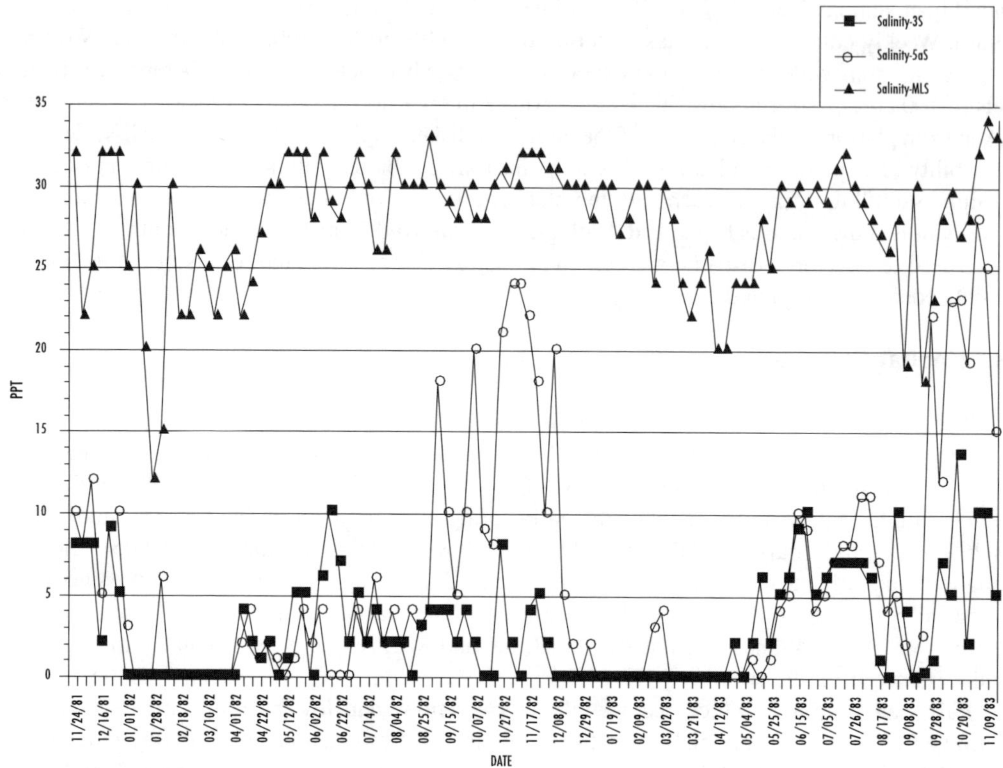

Figure 5.2 Weekly changes of surface salinity (ppt) at stations in the Apalachicola system from November 1991 through November 1993.

Bay generally reflected river flow conditions with relatively high color during winter/spring periods of high flow (Table 5.1, see Figure 5.2). Station 5a was outside of the direct effects of the river; however, this station received periodic highly colored fresh water input from adjacent upland swamps. Long-term pH levels tended to be lower in areas most directly affected by Apalachicola River flow. Overall, primary habitat components varied less at station ML than at the East Bay sites.

Statistical tests used in this chapter are given in Appendix III. A statistical analysis of the weekly habitat data taken during the 2-year experimental period was carried out for all habitat variables (Table 5.2). Differences were noted in both mean values and variability among sites for the various habitat characteristics. Salinity means were significantly different among the three stations for each of the 2 years of observations, with highest values at station ML. Salinity variances were significantly different between stations in East Bay and station ML, with the lowest such values at station ML (Table 5.2). Differences were only significant between stations 3 and 5a during the first year. Water temperature, although significantly different between sites, varied by no more than 1.3°C on average. No consistent trend was noted between stations for the 2 years examined. Variances in water temperature were similar among areas. Variance in oxygen anomaly was similar at all sites. Water color (means and variances) was significantly lower in St. George Sound than that noted at the East Bay stations; highest values were observed at station 5a. Significantly higher pH was observed at station ML during both years, with significantly lower variances noted during the second year.

Sediment characteristics differed slightly among the three sites, with mean grain size varying between 0.11 mm at station 3 and 0.17 mm at station ML. Station ML was the sandiest of the sites and was also characterized by the least amounts of organic matter.

Table 5.2 Spatial Differences in Environmental Characteristics

Environmental Characteristics	Year	Means 3	Means 5a	Means ML	Variances 3	Variances 5a	Variances ML
Salinity	1982	3.9	6.2	29.4	5.1	7.2	2.6
	1983	5.2	7.4	28.3	7.1	8.2	3.4
Temperature	1982	21.8	22.6	23.1	6.6	6.2	6.4
	1983	21.5	22.1	21.7	6.3	6.4	6.7
DO anomaly	1982	−1.59	−1.17	−0.15	0.98	1.03	1.00
	1983	−1.36	−0.76	−0.83	0.88	1.01	0.89
Color	1982	79	108	18	51	80	12
	1983	88	111	22	56	89	21
pH	1982	7.2	7.2	7.7	0.35	0.38	0.44
	1983	7.3	7.3	7.9	0.36	0.31	0.21

Note: Differences between sites in the means and variances (shown in table as standard deviations) for various environmental characteristics collected weekly over the experimental period from 11/81 to 11/83. Means are compared with a Wilcoxon paired difference test; variances are compared with a modified Levene's test. Values within a year are significantly different ($p < 0.01$), except for those connected by underlining.

C. Spatial Variation

1. *Within-Site Variation*

In all, 166 taxa were collected at station ML compared with 47 and 40 at stations 3 and 5a, respectively, over the long-term monthly sampling program (see Appendix III). The dominant organisms (i.e., those species making up the top 90% of density) at each site comprised 9 (station 3), 10 (station 5a), and 28 taxa (station ML). The polychaete *Mediomastus ambiseta* was the numerical dominant at all stations. The three most abundant taxa composed more than 50% of the total faunal density at all three sites. Long-term means of community indices differed among stations over the period of record with the highest density at station 3, the greatest number of taxa at station ML, and the greatest biomass at station 5a (Table 5.1). Analysis of the long-term field data indicated that species richness reached maximal levels in all three sampling areas during fall, a period of relatively low numerical abundance. Variability of the multispecies assemblages of benthic macroinvertebrates was maximal for estimation of spatial relationships.

Fine-scale community variation was examined with the replicated sets of 100 cores taken at each of the three study sites in October 1981. Data from one of the replicate sets are presented graphically in Figure 5.3. Similar results were noted with the other two sets at each site (Livingston, 1987). Infaunal abundance per core was highest at station 3 and lowest at station ML. Species richness per core was highest at station ML and lowest at station 5a. Relative dominance per core was highest at station 3 and lowest at station ML. Station 3 had high numbers per square meter with a relatively low cumulative species richness (21 species). Numerical dominants were the polychaetes *M. ambiseta* and *Streblospio benedicti*. Station 5a was moderately productive in terms of numerical abundance with low cumulative species richness (10 species) and high dominance of the same species noted at Station 3. A total of 72 species was found at station ML with relatively low dominance of the most numerically abundant species. Dominants included *Apoprionospio pygmaea*, *Paraprionospio pinnata,* and *M. ambiseta.*

The spatial distributions of the infauna were evaluated as part of the preparation for a series of predator exclusion experiments. Average sample spatial autocorrelations of numbers, species richness, and the top two dominants differed among stations. Sample *Z*-scores for the average spatial autocorrelations of the above biological characteristics at the three stations are shown in Table 5.3.

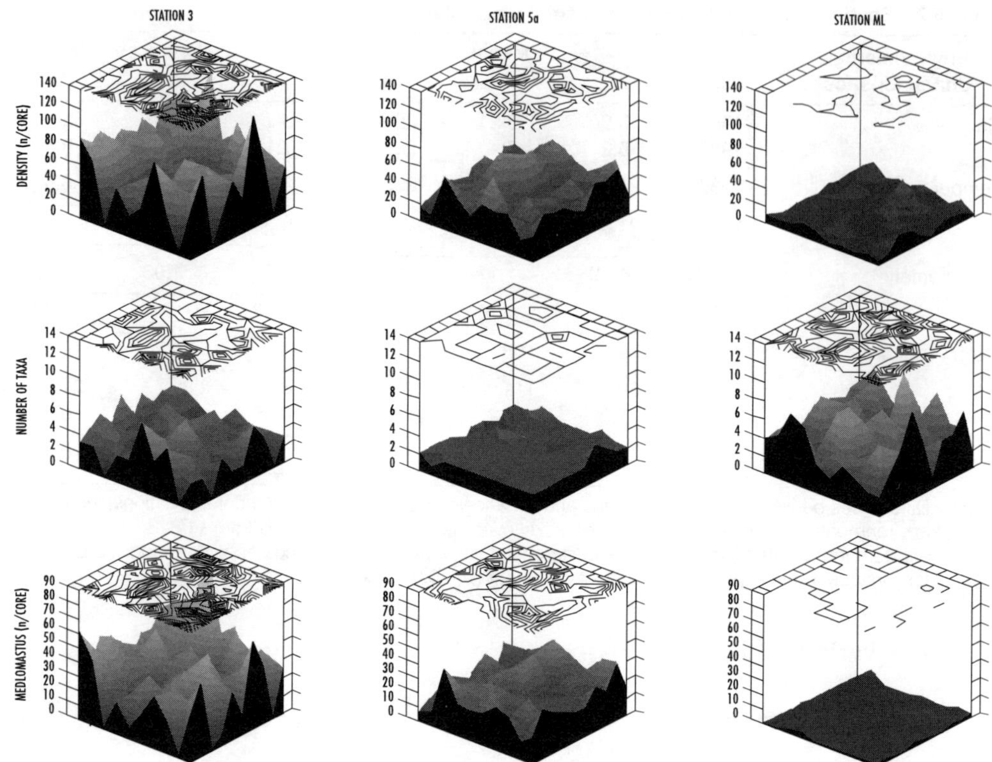

Figure 5.3 Three-dimensional views of numbers of individuals, species richness, and the top numerical dominants of infaunal macroinvertebrates taken in the 100-core samplings at stations 3, 5a, and ML during October 1981. Only one of three replicates is shown.

Table 5.3 Sample *Z*-Scores for Testing the Average Spatial Autocorrelation among Adjacent Cores of Varying Distance Apart for Number of Individuals, Number of Taxa, and Densities of the Two Dominant Taxa

Factor	Station	Order 1	2	3	4	5	6	7	8
No. of	3	5.69*	4.16*	3.66*	3.10*	0.94	−1.20	−1.06	−1.71
individuals	5a	5.29*	2.08*	3.28*	2.89*	4.67*	−0.09	0.36	−1.37
	ML	−0.57	0.62	−0.76	0.43	1.09	0.74	0.18	−0.51
No. of	3	2.85*	2.85*	0.54	0.01	1.87	1.58	−0.17	−0.68
taxa	5a	0.90	−1.31	0.01	−1.28	1.87	0.74	1.07	−0.34
	ML	1.06	0.93	−0.07	1.18	1.25	−1.39	−1.06	−0.34
Abundance	3	5.86*	3.93*	3.13*	1.50	−1.86	−2.03	−1.24	−0.51
First dominant	5a	4.15*	2.23*	2.90*	2.99*	4.82*	0.47	0.01	−1.37
	ML	0.01	−0.69	−0.84	−1.17	−0.62	−0.92	1.07	0.01
Abundance	3	4.15*	3.16*	3.36*	2.57*	2.65*	−0.28	0.72	−1.89
Second dominant	5a	3.90*	0.46	1.22	2.35*	2.65*	−1.02	−1.42	−1.20
	ML	−0.08	0.31	3.05*	0.96	1.09	−0.37	−0.35	−1.03

Note: Distances among cores are denoted by order of the cores; increasing order is equivalent to increased distance apart. An * indicates a significant correlation ($\rho \neq 0$; $p < 0.05$).

In terms of numerical abundance, the first- through fourth-order spatial autocorrelations of numbers at station 3 were significantly different from zero, which indicated that macroinvertebrates at station 3 had patches of 40 to 50 cm diameter. Macroinvertebrates at station ML had no overt clumping. Species richness patches of about 30-cm diameter were noted at station 3; no such patches were noted at the other two stations. Results concerning the numerical distribution of the top two dominants were similar to those for numerical abundance. Significant autocorrelations were found in East Bay but not in St. George Sound. The spatial autocorrelation results of the 900 core observations thus indicated that for any given sample, cores should be taken at least 60 cm apart in order to have independence. This number was programmed into the protocols for all experimental testing. Overall, the infaunal distribution pattern at station ML differed from that noted at stations 3 and 5a (see Table 5.3).

2. *Between-Site Variation*

Statistically significant differences among stations were observed over the concurrently sampled time period for all of the biological variables (Table 5.4). Infaunal numerical abundance, biomass, and species richness were usually significantly higher at station 3 than at 5a, as was the variability of such factors. Station 5a tended to have significantly lower numbers, biomass, and species richness than station ML, whereas station 3 tended to have higher abundance and biomass variation than station ML. Whereas density was lowest at station 5a during both years, highest abundance alternated between stations 3 and ML depending on year. Mean biomass at station 5a was significantly lower than that at the other stations during both years. There was no significant difference between infaunal biomass taken at stations 3 and ML. Numbers of species and species richness variability were significantly higher at station ML than at the East Bay stations, with greater variability a function of the greater numbers of species.

D. Temporal Variation

To examine temporal differences in various field characteristics, two-way analysis of variance (ANOVA) was run by year and season using monthly values within each season as replicates (for discussion of this procedure, see Livingston et al., 1997). Monthly values within each season were used as replicates even though these factors were not replicates in the true sense. To test the assumptions of this model, an autocorrelation function (ACF) and a Wald–Wolfowitz runs test (Wald and Wolfowitz, 1940) were used to check for serial correlation in the model residuals. The

Table 5.4 Spatial Differences in Community Characteristics

Community Characteristics	Year	Means			Variances		
		3	5a	ML	3	5a	ML
Density	1982	7354	3061	3286	14004	4017	1847
	1983	2921	632	3902	2483	591	1589
Number of taxa	1982	9.7	5.5	25.4	4.5	2.7	6.9
	1983	10.4	4.3	29.3	3.5	1.8	5.6
Biomass	1982	1.72	1.03	1.47	1.53	1.30	0.89
	1983	1.12	0.49	1.27	0.88	0.70	0.60

Note: Differences between sites in the means and variances (shown in table as standard deviations) for various infaunal community characteristics collected weekly over the experimental period from 11/81 to 11/83. Means are compared with a Wilcoxon paired difference test; variances are compared with a modified Levene's test. Values within a year are significantly different ($p < 0.01$), except for those connected by underlining.

Table 5.5 Temporal Differences in Environmental Characteristics

Factor	df	Salinity	Temperature	Color	DO Anomaly	pH
			Station 3			
Year	5	1.10 ns	0.29 ns	3.65**	3.08*	4.14**
Season	3	5.76**	57.39***	7.89***	0.86 ns	3.96*
Year × Season	15	1.16 ns	0.45 ns	1.91*	0.91 ns	4.00***
Residual	48					
			Station 5a			
Year	5	1.76 ns	0.42 ns	2.24 ns	1.53 ns	14.48***
Season	3	23.61***	46.64***	6.36***	3.05*	7.60***
Year × Season	15	0.85 ns	0.96 ns	0.85 ns	1.34 ns	6.59***
Residual	48					
			Station ML			
Year	4	3.28*	1.08 ns	2.74*	2.59 ns	4.86**
Season	3	3.02*	27.57***	0.55 ns	1.70 ns	1.02 ns
Year × Season	12	1.55ns	0.38 ns	2.05*	1.34 ns	1.03 ns
Residual	40					

Note: Two-way ANOVA results (*F*-ratios with significance levels) for various physical/chemical characteristics. ANOVAs were run by year and season for each site using monthly averages within each season as replicates. ns = no significant difference; * $p < 0.05$; ** $p < 0.01$; *** $p < 0.001$.

runs test (with cutoff = 0) determined the possibility of clumping of the positive and negative residuals. In most cases, negative autocorrelations were found at lags 1 and 2 that indicated generally conservative ANOVA results (i.e., error variance was overestimated). Occasionally, positive correlations were found that tend to underestimate the error variance and bias the test outcomes. A lack of significance in the Wald–Wolfowitz run tests on residuals was usually found. *F*-ratios and their associated *p*-values were calculated for the final determinations with pairwise comparison of the effects/interactions via *post hoc* tests. *Post hoc* comparisons used Scheffe's test.

A statistical analysis of seasonal and interannual variations of the long-term (monthly) habitat data is given in Table 5.5. Seasonal differences of habitat characteristics were most pronounced at East Bay sites. All variables differed significantly on a seasonal basis at station 5a and all but DO anomaly differed significantly at station 3. Water color and pH had significant seasonal differences only at the East Bay stations, a reflection of the influence of river flow on the immediate receiving areas of the bay. Only salinity and temperature had significant seasonal components at station ML, although differences were not as strong as those in East Bay. Interannual trends, on the other hand, were least pronounced at East Bay station 5a due, in part, to the high variability noted. Three of five habitat variables had strong interannual differences at stations 3 and ML; only station ML had significant interannual salinity differences. Except for annual differences in salinity at station ML, the most pronounced significant seasonal or annual differences in habitat characteristics, as indicated by the magnitude of the *F*-values (see Table 5.3), occurred at East Bay sites.

Considerable seasonal and interannual variability was observed in monthly density, number of taxa, and biomass at the three stations over the collection period. Relative overall stability was observed at the ML site for the three faunal indices. Abundance of *Mediomastus ambiseta, Streblospio benedicti, Oligochaeta* spp., and *Grandidierella bonnieroides* increased significantly during the first years of collection, followed by a striking decline; abundances of the dominants at both East Bay stations dropped to generally less than 50 individuals/m during the 1982–1984 collection period. No such changes were found in the dominant taxa at station ML where relatively

Table 5.6 Temporal Differences of Infaunal Community Characteristics

Factor	df	Density	No. of Taxa	Biomass
		Station 3		
Year	5	28.60***	18.11***	26.75***
Season	3	7.01***	1.45 ns	3.74*
Year × Season	15	2.39*	1.95*	1.24 ns
Residual	48			
		Station 5a		
Year	5	70.14***	28.60***	20.92***
Season	3	14.24***	4.54**	1.55 ns
Year × Season	15	3.27***	2.17*	2.89**
Residual	48			
		Station ML		
Year	4	1.55 ns	3.93**	0.64 ns
Season	3	6.38**	2.15 ns	7.37***
Year × Season	12	1.77 ns	3.01**	2.92**
Residual	40			

Note: Two-way ANOVA results (*F*-ratios with significance levels) for infaunal density, number of taxa, and biomass. ANOVAs were run by year and season for each site using monthly averages within each season as replicates. ns = no significant difference; * $p < 0.05$; ** $p < 0.01$; *** $p < 0.001$.

constant densities were observed over the 5-year collection period. Significant differences were noted among collection years for each of the infaunal characteristics at both East Bay sites, whereas no such differences were noted at station ML for density and biomass (Table 5.6). Significant annual differences were noted only for species richness at station ML. Seasonal patterns of the three infaunal characteristics differed by station. Density and number of taxa at East Bay sites were generally bimodal with peaks in the winter and summer/fall. Seasonal density patterns at station ML were less distinct with highest values in the spring and fall; number of taxa exhibited little pattern. Winter and fall biomass peaks occurred at stations 3 and ML, with a single spring peak at station 5a. There were significant differences between seasons for all characteristics except number of taxa at stations 3 and ML and biomass at station 5a (see Table 5.6). East Bay stations clearly exhibited the greatest variability over time, with density varying as much as three orders of magnitude (station 3) and biomass varying as much as 5 orders of magnitude (station 5a) over the period of observation.

Similar trends to those observed during the short-term weekly collections were noted among infaunal community characteristics from the long-term monthly records with the exception of biomass. A disparity was noted in biomass at the three stations. Biomass was highest at station 5a over the long-term monthly period of record. However, biomass at station 5a was lowest when compared using the short-term weekly data. Infaunal densities, number of taxa, and biomass were generally lower during the experimental period (November 1981 to November 1983) than over the entire, long-term collection period. Density and biomass were significantly lower (ANOVA multiple comparisons between years, $p < 0.05$) at East Bay stations 3 and 5a during the short-term experimental period than in other years; biomass at station 3 during 1978–1979 was similar to the experimental period. Numbers of taxa and biomass at East Bay sites were significantly greater during 1980–1981 than in all other years (ANOVA multiple comparisons between years, $p < 0.05$). Faunal characteristics at station ML during the experimental period were similar to those observed over the long-term collections (ANOVA multiple comparisons between years, $p > 0.05$).

1. Predation Experiments

A series of predation experiments was carried out in East Bay (station 3) and St. George Sound (station ML) to repeat and complement previous experiments in East Bay at stations 3 and 5a (Mahoney and Livingston, 1982). Experiments were carried out in unvegetated, soft-bottom sediments in large cages (2 m on a side) using previously described methods (Mahoney and Livingston, 1982). No significant caging effects were noted at the East Bay stations (Mahoney and Livingston, 1982). Infaunal macroinvertebrates were taken from gridded platforms set on the cages in such a way that a coring device inserted into one of the holes in the grid was directed to a precise position beneath the platform. Samples were taken and analyzed using methods described above. Experimental treatments included controls (no mesh), exclusion cages (mesh-enclosed to exclude fishes and epibenthic macroinvertebrates), and inclusion cages (mesh-enclosed cages into which were placed fishes and invertebrates that reflected the dominant species, i.e., same size and density, present during the experimental period). The trophic organization of the bay was an important factor in the establishment of the experimental protocols as we used specific feeding data (Sheridan and Livingston, 1979, 1983a) and background population information to establish the types and numbers of predators for the various experiments. A randomized block design was used with three replicate cages for each treatment at each site. The cages were placed in such a way that the blocks were perpendicular to the prevailing currents. Twelve core samples were taken randomly within each cage at generally weekly intervals for 5 to 8 weeks. This number of samples was considered representative based on species accumulation curves generated to determine sampling efficiency (as described above). Four sets of experiments were carried out over a 2-year period between November 1981 and November 1983. These experiments were designed to test for potential predation effects of fishes and epibenthic macroinvertebrates on infaunal macroinvertebrates in both an oligohaline portion of East Bay that was directly affected by river flow and a polyhaline habitat in St. George Sound that was largely outside of the influence of the Apalachicola River.

Results of the field predation tests are given in Table 5.7. During the spring 1982 experiment at the high salinity site (station ML), there was a significant increase in the numbers of the primary dominant, *Mediomastus ambiseta*, in the exclusion cages when compared to numbers in the control and inclusion cages. Such differences were significant by the third week of the experiment; by week

Table 5.7 Predation Experiments

Factor	df	Total Numbers	No. of *Mediomastus*	No. of Taxa
		Spring 1982: Station ML		
Treatment	2	4.25^{ns}	9.62^{*}	2.13^{ns}
Location	2	0.50^{ns}	0.23^{ns}	8.67^{*}
Week	4	6.43^{*}	7.85^{**}	1.68^{ns}
Treatment × Week	8	1.70^{ns}	2.80^{*}	0.87^{ns}
		Fall 1983: Station ML		
Treatment	2	2.25^{ns}	1.15^{ns}	1.14^{ns}
Location	2	0.23^{ns}	1.09^{ns}	0.17^{ns}
Week	4	13.48^{***}	22.37^{***}	22.78^{***}
Treatment × Week	8	4.57^{**}	3.77^{*}	0.60^{ns}

Note: Randomized block ANOVA results (*F*-ratios with significance levels) for infaunal macroinvertebrate caging experiments at stations ML and 3 in the Apalachicola Bay system. Treatments included controls, caged exclusion, and caged inclusion with measurements made over time. Location relative to currents was used as the blocking factor. Results are presented for total numbers, numbers of the dominant taxon *Mediomastus ambiseta* and numbers of taxa per core. ns = no significant difference; * $p < 0.05$; ** $p < 0.01$; *** $p < 0.001$.

4, densities reached nearly six times their initial values. This increase was also noted, although to a lesser extent, in the greater total number of individuals in the exclusion cages. This increase, however, was not significantly different because of high variance within treatments. No predation effect was observed for numbers of taxa where there were no significant differences among the three treatments.

During the fall 1983 experiment at station ML, there was also an increase by the end of the experiment in the numbers of *Mediomastus* in the exclusion treatment; densities increased nearly sevenfold over initial conditions and were twice that of controls. These differences among treatments, unlike the spring results, however, were not significant. In the fall, *Mediomastus* densities increased naturally at ML. Although the exclusion cages had the highest densities by the end of the fall 1983 experiment, increases in the other treatments prevented the difference from being considered significant. The spring 1992 increases at ML occurred despite general declines of *Mediomastus*. As with the spring experiment, there were no significant treatment-specific differences in species richness indicating that the release of the dominant species was not accompanied by changes in the general species distributions. During the fall experiment, there was a progressive increase in species richness in all three treatments over the 8-week period of testing. This was due, in large part, to the shift from summer to fall conditions, which was usually accompanied by marked increases in the numbers of species of infaunal macroinvertebrates. Increases of species richness were paralleled by increases of numbers of individuals over the period of experimentation, as indicated by the highly significant week effects in the ANOVA results (see Table 5.7). Significant interaction terms in both spring and fall experiments indicated that the treatments behaved differently in time.

Neither the fall 1982 nor the spring 1983 experiment in East Bay resulted in significant treatment-related differences in overall numbers, numbers of the top dominant (*M. ambiseta*), or species richness. Total numbers and species richness generally increased in all three treatments during the fall 1982 experiment (i.e., significant week effects; see Table 5.7); however, treatments were not significantly different.

E. Physical Habitat Changes, Predation, and Food Web Relationships

The influence of river flow on receiving areas was observed as a gradient of increasing temporal habitat stability with distance from the river. Salinity variation, as well as ambient salinity, at any given time appeared to be primary factors that defined the East Bay habitat. Areas directly (station 3) and indirectly (station 5a) affected by Apalachicola River flow had high salinity variation relative to areas distant from such flow (station ML). Although important habitat variables had high levels of variation within short-term periods at river-dominated sites, such variation was relatively consistent from year to year. Other variables such as water color and oxygen anomaly had significant annual differences at river-dominated stations relative to areas not directly affected by river flow. There were thus fundamental differences of habitat stability between a system that was physically dominated by freshwater runoff (East Bay) and an area that was little influenced by changes in river flow (St. George Sound).

Table 5.8 summarizes the differences in environmental and infaunal community characteristics noted among the three study areas of the Apalachicola estuary. Apalachicola River flow has a dominant influence on the physical setting of areas in close proximity to its mouth. The river is associated with relatively high primary productivity and detrital inputs to East Bay (Chanton and Lewis, 2002); these factors probably account for the higher levels of sediment organic content. East Bay is characterized by low salinity, high color, and low pH, with generally high variability of these factors. The habitat conditions set the stage for the composition of the infaunal community and interactions between the infauna and primary/secondary carnivores. East Bay infaunal assemblages have high densities of relatively few species that tend to aggregate in patches. Densities and biomass vary widely as environmental conditions change. The species present appear well adapted to these highly fluctuating conditions. These adaptations carry over in their response to predation. Biological responses to rapidly and widely fluctuating environmental conditions tend to overwhelm the potential influence of predators, even though their numbers may at times be high (Mahoney and Livingston, 1982; Livingston, 1984a).

Table 5.8 Summary of Environmental and Infaunal Community Characteristics Observed in the Apalachicola Estuary

Characteristics	Station 3	Station 5a	Station ML
	Physical Setting		
Proximity to river	Near	Intermediate	Far
Sediment organic content	High	High	Low
Mean grain size	Low	Intermediate	High
Phytoplankton productivity[a]	High	High	Low
Detrital input [b]	High	Intermediate	Low
	Environmental Characteristics		
Level			
Salinity	Low	Low	High
Temperature	ND	ND	ND
Color	Intermediate	High	Low
pH	Low	Low	High
DO anomaly	Low	Intermediate	High
Variability			
Salinity	High	High	Low
Temperature	ND	ND	ND
Color	Intermediate	High	Low
pH	ND	ND	ND
DO anomaly	ND	ND	ND
	Community Characteristics		
Level			
Density	High	Low	Low
Biomass	Intermediate	High	Low
Species richness	Intermediate	Low	High
Variability			
Density	High	Intermediate	Low
Biomass	Intermediate	High	Low
Species richness	Low	Low	High
Degree of spatial clumping			
Density	High	High	Low
Species richness	High	Low	Low
Influence of predation	Low	Low	High

[a] Myers and Iverson (1977).
[b] Livingston et al. (1977) and Livingston (1981a, 1984a).
Note: Table entries denote relative magnitudes among stations of the levels and variability for the various characteristics; ND = no differences among stations.

The above argument presupposes that opportunists have a relatively higher capacity for growth and reproduction than "non-opportunists." A partial list of the natural history of the bay species is given in Appendix II. Infaunal macroinvertebrate assemblages in the river-affected study areas were dominated by polychaetes such as *Mediomastus ambiseta* (below-surface deposit feeder and detritivorous omnivore) and *Streblospio benedicti* (above-surface deposit feeder and detritivorous omnivore). Oligochaete worms (browser/grazers and detritivores) were also among the primary dominants. Almost 60% of all infauna taken at station 3 was represented by these three types. At station 5a, *M. ambiseta* and *S. benedicti* represented more than 50% of infauna taken. These species have short generation times (4 to 6 weeks) and are well adapted to low salinity and a rapidly changing habitat. Most such species live in or near the sediment–water interface, with most of the trophic organization of the bay dependent on interactions among bottom-living infaunal and epibenthic macroinvertebrates and fishes. Areas distant from the river (station ML) were characterized by reduced numbers of opportunistic species, reduced dominance by such species, and a more diverse trophic community.

The influence of the river is thus considerable in terms of habitat characteristics, primary/secondary production, and the trophic composition of the biological assemblages. Areas distant from the river, as represented by station ML, have less variable conditions with significantly less organic matter (either primary productivity or detrital input) to support high densities at the higher trophic levels. Thus, there is lower overall infaunal density and biomass in such areas. Higher, less variable salinities, relative to East Bay, allow greater numbers of species to coexist at station ML. The relative influence of predation on these assemblages tends to increase as the variability of environmental conditions decreases. If natural sequences of drought and flood could be considered as disturbances, it is possible that pulsed conditions provide a form of natural destabilization that the system needs to maintain the observed relatively high levels of primary and secondary production over the long term. Given adequate raw materials, the high productivity of largely opportunistic species may be part of a basic system response to physical instability, where the mechanism depends, in part, on regular disturbance for the provision of high natural variation (instability). This, in turn, perpetuates conditions favorable for the highly productive opportunists.

In the river-dominated East Bay system, the results of the predation experiments indicated that there was no overt control by predation on various aspects of infaunal assemblages. These results are identical to the results of similar experiments carried out at stations 3 and 5a in East Bay during fall 1978 and fall 1979 (Mahoney and Livingston, 1982). Fall is probably not a time of high predation intensity. During spring, the influx of juvenile sciaenid fishes is high. It is this period in which one might expect to observe an effect if predation is excluded. The background field data and the repeated experiments reported here suggest that, in river-dominated areas, physical factors affected by river flow are more important as controlling factors than biological response to predation. Seasonal changes in infaunal macroinvertebrate indicators within the various experimental treatments at station 3 were far more significant than treatment differences, indicating that natural seasonal fluctuations of biological response are probably physically driven. The experimental results concerning predator exclusion at station ML suggest that exclusion of fish and invertebrate predators released one of the top species, the polychaete *M. ambiseta*. The lack of a response of this species in the predator inclusion treatment further emphasizes the probability that there was some control of the dominant species by predators. Additionally, the lack of significant differences in the number of taxa among the three treatments indicates that the increased numbers of the dominant species in the absence of predation did not influence the numbers of other infaunal species, at least in the short term.

Long-term descriptive field data together with predator exclusion experiments with infaunal macroinvertebrates in the Apalachicola estuary suggest that river flow influences both abundance and species composition of infauna along both spatial and temporal gradients. Physical instability, coupled with high productivity in the upper (river-dominated) parts of the bay system, resulted in high densities of relatively few populations that are adapted to high habitat variation. In areas of the bay that are outside direct riverine influence but are still affected by river fluctuations (station 5a), there are relatively low numbers of species, high numbers of individuals of a few populations, high biomass, and high relative dominance. These physically controlled areas show little in the way of a response to predation due to the adaptive nature of the adventitious species that inhabit the upper estuary. Biological control in the form of predation is thus minimal here. In areas of the bay that are outside riverine influence, there are relatively high numbers of species, low numbers of individuals, low biomass, and low relative dominance. Even in the high-salinity environment, however, salinity and temperature have some degree of control of the biological aspects of infaunal macroinvertebrates in terms of seasonal changes of such assemblages. Experimental results indicated that predation is a factor in the organization of the infaunal assemblages in the high-salinity portions of the Apalachicola system. Predation activities in these areas reduced dominance by individual species. Over time, this process could contribute to the relatively high numbers of species by reducing the influence of dominant species. Increasing habitat stability, coupled with lower primary productivity, leads to reduced numbers of infauna and high species richness that is mediated, in part, by predator control of the top dominants. Resource limitation among dominant forms of deposit-feeders in soft-sediment estuarine communities has not been shown to occur. Mortality as a result of competitive interactions among soft-sediment biota remains relatively uncommon in such systems. According to Whitlach and Zajac (1985), biotic interactions among opportunistic estuarine infaunal species depend on habitat conditions, the species involved, and their density with no characteristic form of biotic process that controls successional dynamics. Peterson (1991), in a comprehensive review of processes that shape soft-sediment assemblages, indicated that competition for space is not usually a factor in the distribution of such populations. Competitive exclusion is relatively rare on sedimentary bottoms. Various studies have indicated that the experimental exclusion of large mobile predators has considerable effects on infaunal associations in soft-sediment communities (Virnstein, 1977; Reise, 1978; Holland et al., 1980). On the other hand, there is evidence that indirect effects of altered physical habitat variables usually are more important than processes such as competition and predation in soft-sediment areas (Peterson, 1991). Peterson (1992) indicated that soft-sediment communities hold promise for testing specific hypotheses relative to the interplay between physical and biological processes in the organization of estuarine communities. Our results indicate that infaunal community structure varies in its relative sensitivity to physical and biological control along gradients of river influence, and that such control is also influenced by the nature of the physicochemical stressors and primary productivity.

Considerable attention has been directed to the relative effects of natural physical and biological disturbance on soft-sediment communities. Turner et al. (1995) analyzed long-term changes of infaunal associations under varying environmental conditions. Despite considerable physical and biological instability, the authors found that the soft-sediment communities exhibited both resistance and resilience to such disturbance. As pointed out by Zajac and Whitlach (1982), estuarine infaunal successions, with dominant forms usually opportunistic, have varying types of recovery after disturbances with no set recovery stages. Similar results have been reported by Santos and Simon (1980a,b). River flow fluctuations in river-dominated estuaries create a highly variable habitat in terms of key environmental determinants leading to the exclusion of many stenohaline species and high relative dominance of eurytopic forms. Our microbial trophic models (Morrison et al., 1977; White et al., 1977, 1979a,b; Bobbie et al., 1978) indicate that river flow, tidal/wind action, and the shallowness of the bay contribute to the high benthic productivity of the portions of the Apalachicola estuary that receive direct river flow (Livingston, 1984a). Factors that favor the numerical dominance of opportunistic eurytopic populations could contribute to the relative resistance of such populations to predation pressure. With distance from the river, physical stress and primary production are reduced. These gradients are associated with increased species richness, reduced levels of overall secondary production, and reduced dominance by opportunistic populations.

The effects of a large alluvial river on its associated estuary thus include a complex series of gradients of primary and secondary production, infaunal species richness and relative dominance, and predator--prey interactions. There is thus a well-defined relationship of physical habitat stress and biological response that can be related to spatial and temporal gradients of river influence on the habitat and productivity of the receiving estuary. This analysis indicates that physical habitat stressors and biological processes such as predation interact with spatial and temporal trends of productivity and the exclusion of stress-susceptible marine species to result in observed responses of infaunal assemblages. The high production of natural river-dominated estuaries is thus a product of physical and biological processes that are directly and indirectly related to seasonal and interannual trends of freshwater input to the estuary.

F. Effects of Toxic Agents on Trophic Relationships

Pentachlorophenol (PCP) is a major industrial chemical that has been used as a biocide throughout the world (Crosby, 1981). This compound has been widely studied, and field tests with PCP have shown that its toxicity to individual species in laboratory tests is similar to that in field tests

(Crossland and Wolff, 1985). However, field experiments indicated an added effect of toxicity to filamentous algae that caused trophic responses that could not be predicted by the laboratory experiments (Crossland and Wolff, 1985).

A series of experiments was carried out in St. George Sound (station ML) to evaluate the effects of PCP on field plots of infaunal macroinvertebrates. Methods used for the field and laboratory tests with PCP have been described in detail by Livingston et al. (1985b), so that only an outline of the methodology will be presented here. A parallel set of identical experiments was carried out with mesocosms and microcosms established in the York River-estuary (Diaz et al., 1985).

Experiments were carried out in unvegetated soft-bottom sediments in large cages (2 m on a side). Infaunal macroinvertebrates were taken from gridded platforms with coring devices as noted above (Section 5.II.D). All experimental procedures followed those described for the predation experiments. Experimental treatments included three controls, three areas dosed with sediments having concentrations of 10 ppm PCP (low dosage), and three areas dosed with sediments having concentrations of 100 ppm PCP (high dosage). The sampling platform remained unscreened for these tests. A randomized block design was used with three replicate areas for each treatment with blocks perpendicular to the prevailing currents. Test treatments were chosen randomly and sediments (controls, 10 ppm PCP, 100 ppm PCP) were spread using a sifting device designed to make even applications of 1-cm depth at each test site. Immediately after treatment, samples were taken (noted as To). After 24 h, cores were also taken for PCP analysis. Chemistry samples were extracted at each weekly sampling time. The PCP samples were analyzed using methylene chloride extractions, with standard gas-liquid chromatography (flame ionization and electron capture detection).

Twelve core samples were taken randomly in each area at weekly intervals for 9 weeks during spring 1985 and fall 1985. Fall represents the recruitment period for most infaunal species in this area. This number of samples was representative, based on species accumulation curves generated to determine sampling efficiency (as described above). The PCP experiments were analyzed using randomized block ANOVA tests (F-ratios with significance levels) with location relative to currents used as the blocking factor. Treatments were compared statistically to controls at weekly intervals. Results were analyzed for total numbers, numbers of dominant taxa, numbers of taxa per core, and a series of guild associations calculated from the data sets. Pattern comparisons between the laboratory and field sets were made subjectively.

A parallel set of laboratory microcosm experiments were run with the same PCP sediment concentrations. The flowing water laboratory microcosms (0.1 m^2) were established with cored samples from the field test area. Nine sets of six microcosms (three sets of controls, three sets of low PCP concentrations, three sets of high PCP concentrations) were established and four cores per sampling time were taken in an identical way as that in the field areas. In both areas, water temperature, salinity, and DO were sampled every third day for the duration of the experiment. Sediment preparation for PCP treatments was identical for the field and laboratory tests.

Guild assignments were made by classifying the various species into functional groups based on (1) feeding mode, (2) trophic level (Appendix II), (3) mobility mode, and (4) reproductive mode. Guild associations, based in part on trophic organization, were found to have certain advantages. Nine separate guild divisions were used in the analysis of test results. The use of functional or process-associated groups of species allowed representative comparisons between the laboratory and field organisms. Guild associations also provide comparisons between areas represented by different species. In addition, the use of guilds was more specific to identifiable modes of exposure to PCP. This use also simplifies the analysis by joining entire groups of species, where high species-specific variability is evened out by the grouping process.

Sediment dose-equivalency between laboratory and field treatments was achieved in both the spring and fall experiments. During the spring 1985 experiment, dose-specific effects on total abundance and species richness indices were noted in both the field and laboratory tests although the laboratory tests were more severe and longer lasting than effects in the field. Deposit-feeding detritivores were significantly reduced by the PCP treatments in the field (first half of the experiment) and in the laboratory (practically eliminated by the end of the experiment). During the fall 1985 experiment, there were again dose-specific reductions of numerical abundance and species richness in laboratory and field with the most pronounced significant effects noted in the laboratory treatments. Trophic simplification was noted in the laboratory PCP treatments. Once again, PCP effects were limited to the deposit-feeding detritivores in both the field and laboratory tests. In both the spring and fall experiments, the main difference between the laboratory and field was the higher recruitment levels in the field that tended to dilute PCP impacts in time. This was because of the small size of the field treatment area relative to the almost limitless areas that provided recruitment populations from adjacent (untreated) areas.

Results of field and laboratory experiments concerning the effects of PCP on infaunal macroinvertebrates in the Apalachicola system emphasized the fact that impacts of toxic substances on estuarine organisms should be interpreted within the context of physicochemical habitat variation, predator–prey interactions, and recruitment patterns of dominant populations. The limitations of laboratory tests in predicting effects of toxic agents include problems of interpretation due to the inability of such tests to account for the effects of these natural processes. The same problem arises when individual species are used as subjects for such tests because laboratory artifacts often have impacts on time-related test results with individual species. The laboratory microcosm and field mesocosm data provided reliable predictions of PCP effects with the community indicators (numerical abundance, species richness) and the guild that represented deposit-feeding detritivores (mobile burrowers, wide dispersers). The selective effects on this guild (no other guild showed a response to PCP) indicate that trophic organization plays an important role in the effects of toxic substances in coastal systems. Feeding modes and trophic interactions are important aspects in the evaluation of the effects of toxic agents on estuarine organisms.

G. Long-Term Changes of Trophic Organization

Human modifications of the flow regime of the Apalachicola River have been the subject of considerable debate in recent years as various interests in the tri-river (Apalachicola, Chattahoochee, Flint) basin vie for the use of fresh water. An evaluation of the implications of reduced river flow on the Apalachicola River–bay system is complicated by the considerable seasonal and interannual variation of river flows. Individual estuarine fish and invertebrate species use the Apalachicola estuary as a nursery ground. The trophic organization responds to changes of river flow. Previous analyses (Livingston, 1984) indicate that physical instability of the estuary is actually a component in the continuation of a biologically stable estuarine system. Multiple responses to river flow cycles reflect species-specific adaptations that allow serial dominance throughout the normal drought/flood cycles.

1. Habitat Background

Riverine input influences various trophic conditions in receiving estuaries. Nutrient loading, advected nutrient-rich offshore water, nutrient outwelling from estuarine wetlands, and rapid recycling of nutrients all contribute to the generally high primary productivity in alluvial systems (Snedaker et al., 1977, Cross and Williams, 1981). Food web responses to such loading and recycling are important in the search for an answer to the freshwater question (Livingston, 1984a, 1991b, 2000; Peterson and Howarth, 1987; Howarth, 1988; Baird and Ulanowicz, 1989).

The relative importance of various sources of both organic carbon (dissolved and particulate) and inorganic nutrients varies from one estuary to the next (Peterson and Howarth, 1987). It is now generally agreed that the significance of organic input, allochthonous or autochthonous, to a given estuary depends on specific habitat-related variables (Haines, 1979; Odum et al., 1979; Nixon, 1980, 1981a,b; Welch et al., 1982; Kemp and Boynton, 1984). Despite an extensive literature concerning the nature of nutrient fluxes between freshwater and estuarine/coastal areas, definitive limits on anthropogenous freshwater diversion for the purpose of protecting estuarine production remain elusive. However, long-term data sets are fundamental to an understanding of the factors that regulate system-level processes in estuaries and the linkages among the various food web components (Smith et al., 1982).

Interacting factors that control population distribution in the Apalachicola Bay system have been described (Livingston, 1976a,b, 1979, 1981a,b, 1984a). Dominant invertebrates include blue crabs (*Callinecties sapidus*) and three species of penaeid shrimp (*Penaeus* spp.). Fish assemblages are numerically dominated by anchovies (*Anchoa mitchilli*), Atlantic croaker (*Micropogonias undulatus*), spot (*Leiostomus xanthurus*), and sand sea trout (*Cynoscion arenarius*). Anchovies are dominant during fall–winter periods, whereas sciaenids such as Atlantic croaker and spot are dominant during winter–spring months. Spot and croaker are in direct competition for food, especially as young populations that are numerically high during the winter–early spring months of high river flow (Sheridan, 1979; Sheridan and Livingston, 1979). The sand sea trout is a piscivorous fish that feeds primarily on anchovies (Sheridan and Livingston, 1979), reaching peak numbers during late spring and early summer. There are indications that long-term changes in population distribution of these species could be related to competition for food (Livingston, 1991c).

Long-term, monthly measurements were taken at stations in East Bay (see Figure 5.1). Detailed descriptions of methods for the field collection of physicochemical data are given by Livingston (1979, 1981a, 1982, 1984a,b) and Livingston et al. (1974, 1976). Statistical analyses for the long-term (14-year) field collections are outlined in Appendix III.

Trends of Apalachicola River flow are shown in Figure 5.4. The seasonal pattern of river flow was similar from year to year with high flows during winter–spring months and low flows during summer–fall months. During the first year of record (1972), there was a minor drought that was broken by increases in river flow during the winter–spring and summer of 1973. This was followed by somewhat drier conditions (1973, 1974) after which there was another period of high river flow (1975). Over the next 5-year period (1976 through 1980), winter–spring flows were comparable. The period from 1976 to 1977 was characterized by relatively low winter flows compared to preceding and succeeding years. Low summer flows became increasingly prolonged from 1975 to 1978. Winter–spring peak flows occurred during the period from 1978 to 1980. From May 1980 through the end of 1981, there was a major drought with substantially lower river flows during the winter–spring of 1981. Flow rates during the 20-month period prior to the winter of 1982 were often less than 50% of that which East Bay usually receives in the way of freshwater runoff from the river. In terms of duration, the 1980–1981 drought was the fourth longest period in this century with below-average flows in consecutive months (Livingston et al., 1997). The combination of individual wet and dry seasons during this period formed the third most severe drought in terms of magnitude (mean flow) over three consecutive seasons. The following 2.3 years were characterized by a general return to the prevailing patterns of Apalachicola River flow as noted during the period from 1978 through the winter of 1980.

According to ANOVA analyses of the long-term data set (Table 5.9), river flow highs during winter–spring periods were significantly different ($p < 0.05$) from summer–fall lows. River flows during 1981 were significantly ($p < 0.05$) different from all other years except 1978. Peak flows in 1973 and 1975 were significantly ($p < 0.05$) different from flows in 1981.

Monthly values of important water quality factors over the study period are shown in Figures 5.5 and 5.6. Overall, the strong influence of Apalachicola River flow on water quality factors in East Bay is evident. Seasonal increases of river flow during winter–spring months were associated with reduced salinity, increased color and turbidity, reduced Secchi depths, reduced oxygen anomaly, and reduced pH. Salinity peaked during late fall low-flow conditions. Oxygen anomaly increased during summer months, at which time pH levels were highest. The oxygen anomaly and pH data indicated other influences such as microalgal productivity. The eigenvectors of the sample correlation matrix and the estimated coefficients of the first six principal components of the PCA analysis are listed in Table 5.10. The first six principal components accounted for nearly 89% of the total

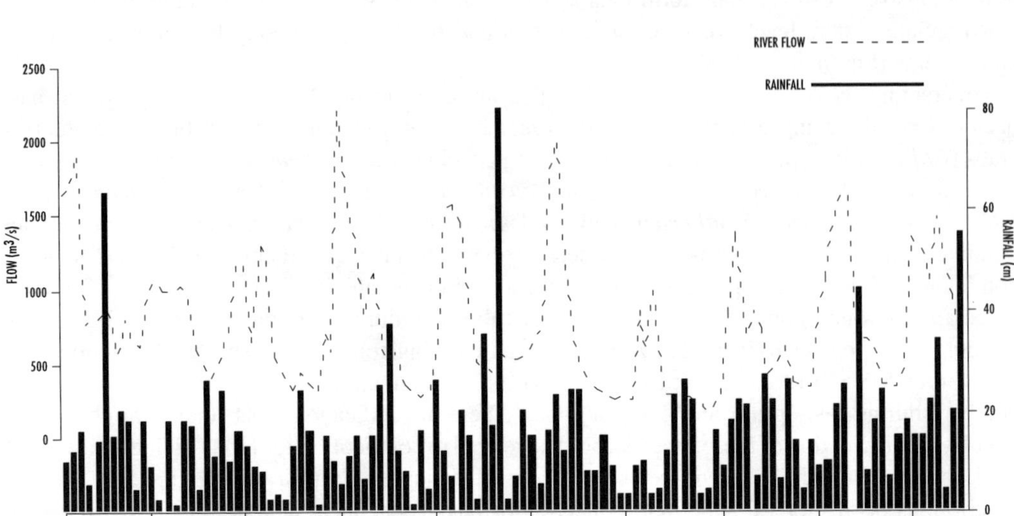

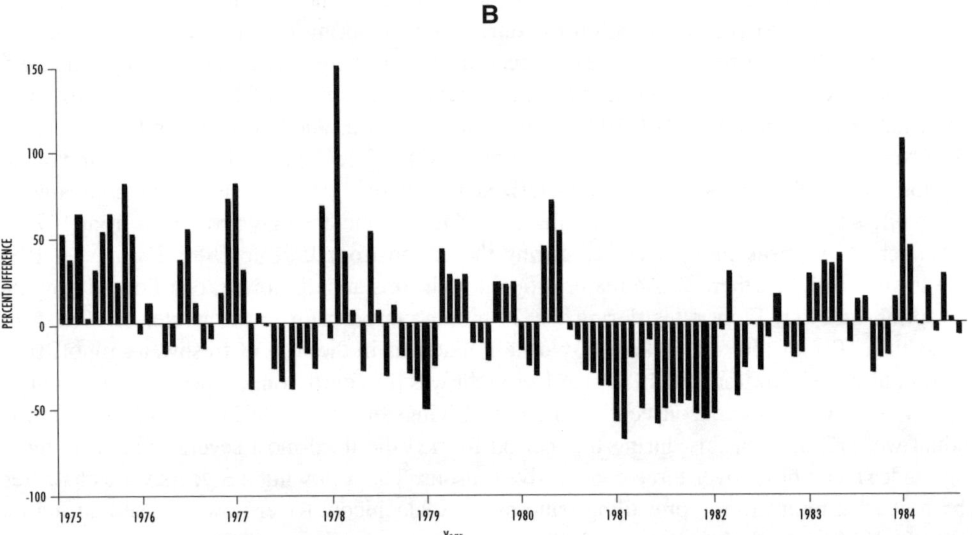

Figure 5.4 (A) Apalachicola River flow and East Bay rainfall shown monthly from February 1975 through July 1984. (B) Monthly Apalachicola River flow expressed as the percent of differences from the long-term (1950 through 1990) monthly means. Values are presented as

$$\frac{(Y_{ij} - \overline{Y}_i)}{\overline{Y}_i} \times 100$$

where Y_{ij} = monthly river flow observation for the ith month and the jth year and

$$\overline{Y}_i = \frac{\sum_{j=1}^{40} Y_{ij}}{40}$$

and where i = month and j = year.

Table 5.9 Results of the Two-Way ANOVA for Physicochemical Variables Taken at East Bay Stations

Factor	Source	df	*F*-Value	*p*-Value
River flow	Year	8	5.26	0.0001
	Season	3	41.98	0.0001
	Year × season	24	1.86	0.0232
	Residual	71		
Salinity	Year	8	3.13	0.0044
	Season	3	29.44	0.0001
	Year × season	24	1.02	0.4582
	Residual	71		
Color	Year	8	3.59	0.0015
	Season	3	13.21	0.0001
	Year × season	24	1.26	0.2247
	Residual	71		
Turbidity	Year	8	6.77	0.0001
	Season	3	5.98	0.0011
	Year × season	24	2.34	0.0031
	Residual	71		
Secchi depth	Year	8	2.61	0.0147
	Season	3	9.16	0.0001
	Year × season	24	0.97	0.5186
	Residual	71		
DO	Year	8	2.17	0.0397
	Season	3	38.41	0.0001
	Year × season	24	0.82	0.6981
	Residual	71		

Note: ANOVAs were run by year and season using monthly values within each season as replicates.

variation, which implies that the water quality variables can be well represented by the first few principal components. The first principal component was dominated by river flow, water color, salinity, and Secchi depth readings. River flow was negatively correlated with salinity and Secchi depth readings and was positively associated with water color. The second principal component represented temperature and its inverse relationship to DO and the oxygen anomaly. The third principal component emphasized pH.

The most outstanding aspect of the bay salinity over the period of study was the sustained high salinities during the drought (1980–1981), although such values were not significantly different due to wide variance of salinity by season. Salinities were lowest during high river flows in 1975 and 1979–1980. Water temperature during summer periods did not vary significantly from year to year. However, winter low temperatures tended to increase from 1977 to 1980 with sustained high winter temperatures occurring from 1980 through 1984. Water color was lowest and Secchi disk readings were highest during the drought, with Secchi readings in 1980 significantly different ($p < 0.05$) from preceding years. Color levels during the drought years were significantly ($p < 0.05$) different from those taken during preceding and succeeding years. Winter peaks of turbidity were lower ($p < 0.05$; Tukey compromise, Scheffe's *S*) during 1975–1976 than readings taken during all succeeding years. The high turbidity levels during the second year of the 1980–1981 drought coincided with low color levels; the slight decrease in Secchi depths reflected such changes in turbidity and color. Periodic high oxygen anomalies were evident during low flow periods from 1975 through 1977. The 1980–1981 drought was also characterized by periodically high positive values. A general reduction of oxygen anomalies began midway into the drought, and sustained low anomalies were noted throughout the period from 1982 to 1984. The 1980–1981 drought was a defining feature of the long-term habitat data.

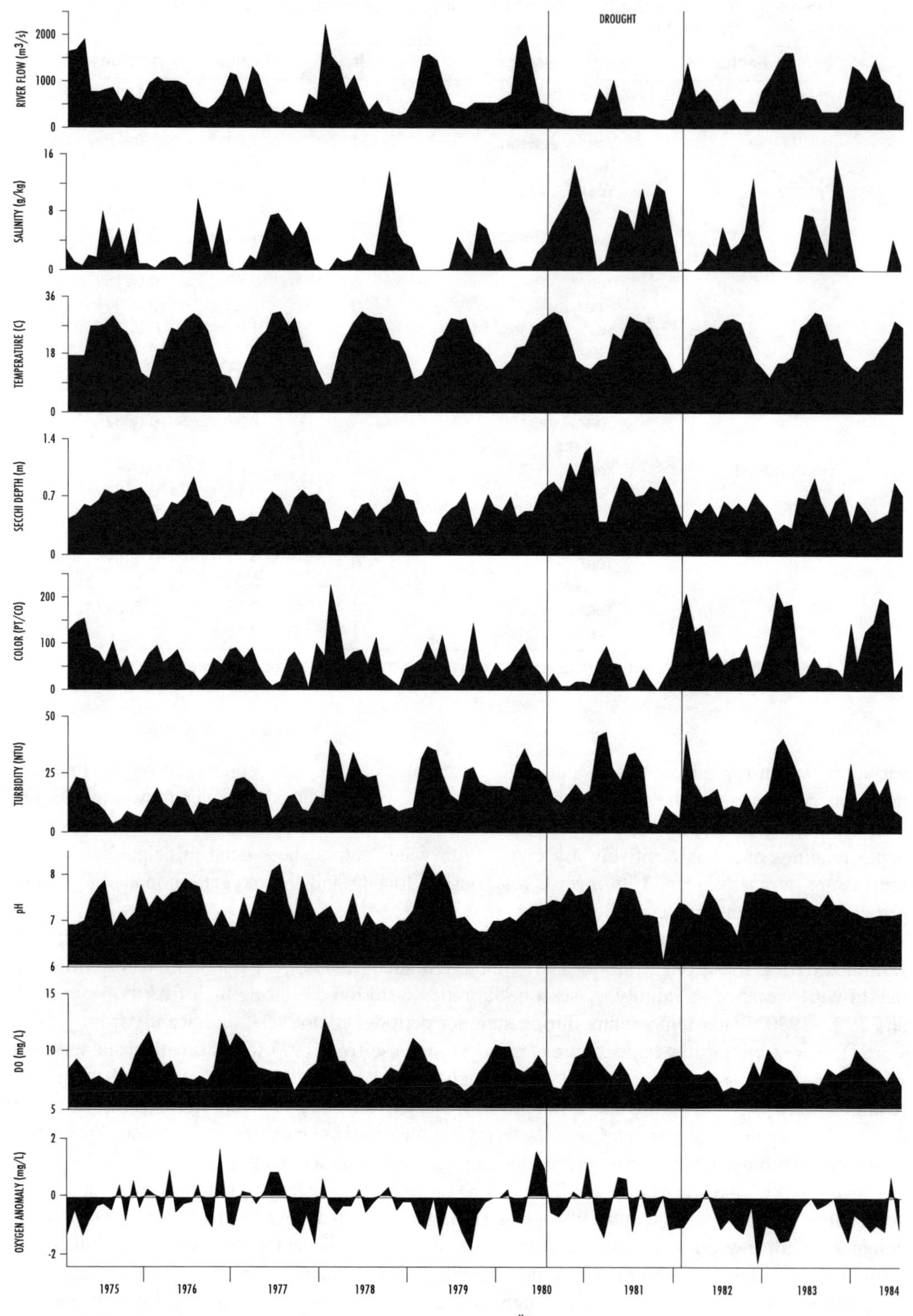

Figure 5.5 Long-term monthly values of river flow and important environmental variables collected from surface waters of East Bay from February 1975 through July 1984.

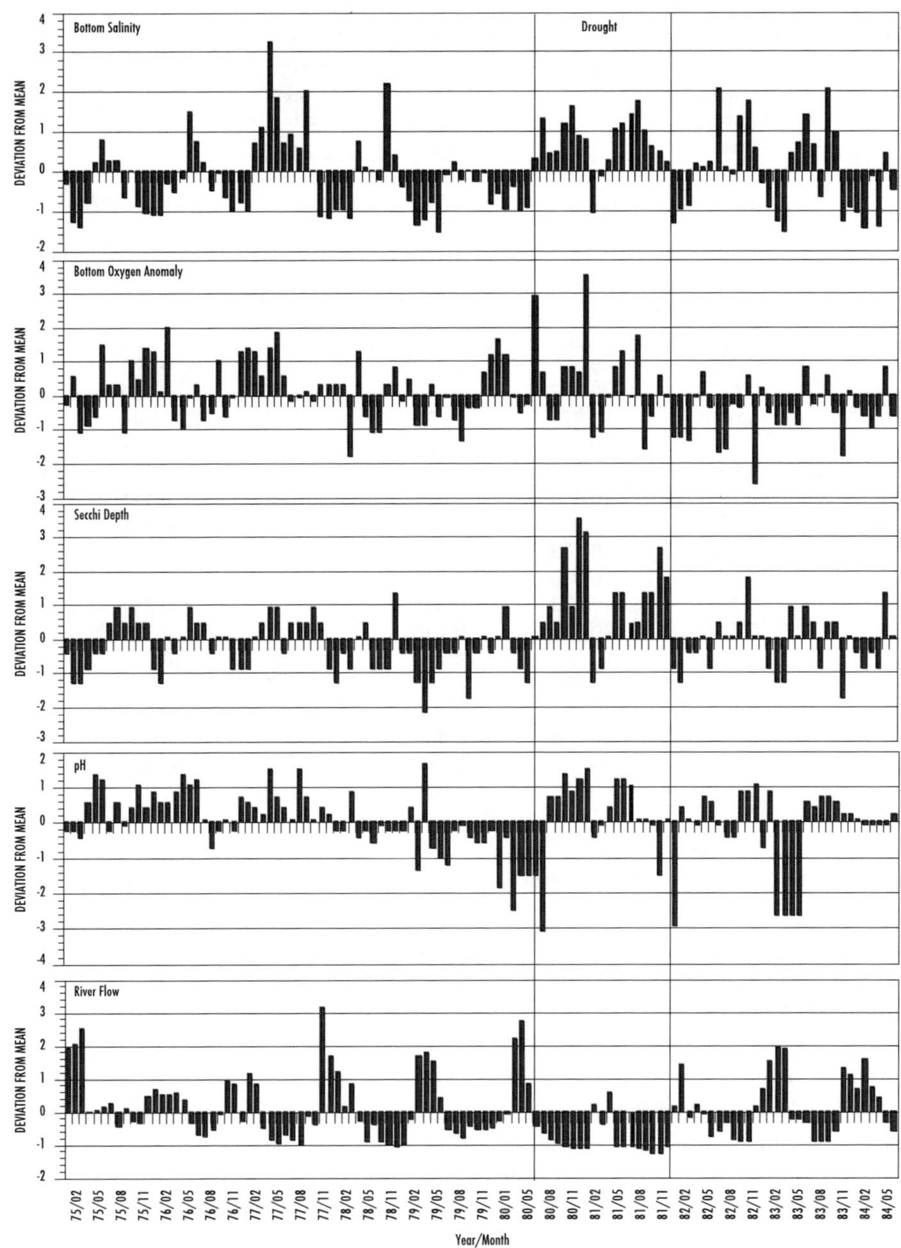

Figure 5.6 Long-term monthly values of river flow and important environmental variables collected from surface waters of East Bay from February 1975 through July 1984 and expressed as deviations from the grand means.

Table 5.10 Results of Principal Component Analysis Based on Monthly Environmental Data Taken in Surface Waters of the Apalachicola System

Eigenvalues of the Correlation Matrix

Axis No.	Eigenvalue	Difference	Proportion	Cumulative Percent
P_1	6.272	2.747	0.348	0.348
P_2	3.525	1.716	0.196	0.544
P_3	1.809	0.115	0.100	0.645
P_4	1.694	0.302	0.094	0.739
P_5	1.392	0.131	0.077	0.816
P_6	1.261	0.630	0.070	0.886

Eigenvectors

	Principal Component Axis					
Variable[a]	P1	P2	P3	P4	P5	P6
APR4	0.329	0.030	0.022	0.062	0.090	–0.224
APRAIN	–0.067	–0.267	0.176	–0.446	0.389	0.205
EBRAIN	–0.066	–0.285	0.097	–0.396	0.446	0.219
SECCHI	0.316	0.093	–0.023	–0.152	–0.018	–0.006
COL-S	–0.320	–0.090	0.180	0.045	0.077	–0.274
DO-S	–0.161	0.440	–0.001	–0.128	0.111	0.084
OXAN-S	0.174	0.288	–0.223	0.128	0.479	–0.150
PH-S	0.127	0.108	0.609	0.282	0.108	0.094
SAL-S	0.339	0.039	0.003	–0.037	–0.058	0.224
TEMP-S	0.226	–0.334	–0.091	0.247	0.238	–0.203
TURB-S	–0.252	–0.046	–0.158	0.366	0.134	0.443
COL-B	–0.317	–0.085	0.177	0.077	0.059	–0.267
DO-B	–0.152	0.412	0.041	–0.167	0.124	0.074
OXAN-B	0.131	0.350	–0.174	0.071	0.442	–0.185
PH-B	0.153	0.116	0.602	0.241	0.108	0.078
SAL-B	0.338	0.025	0.012	–0.068	–0.083	0.214
TEMP-B	0.230	–0.330	–0.094	0.242	0.235	–0.199
TURB-B	–0.217	–0.057	–0.196	0.382	0.123	0.506

[a] Variable names include APR4 = Apalachicola river flow, APRAIN = Apalachicola rainfall, EBRAIN = East Bay rainfall, SECCHI = Secchi depth, COL = color, DO = dissolved oxygen, OXAN = dissolved oxygen anomaly, PH = pH, SAL = salinity, TEMP = temperature, TURB = turbidity; S = surface, B = bottom.

Note: Eigenvalues and eigenvectors are presented only for the first six component axes, which together account for nearly 89% of the total variation observed.

2. Biological Trends

Biological samples (infauna, epibenthic macroinvertebrates, fishes) were taken monthly at fixed stations in East Bay (see Figure 5.1). Long-term trends of infaunal biomass and species richness are shown in Figure 5.7. Infaunal biomass was highest during the middle of the drought period with a sharp reduction by spring 1982, at which time the lowest biomass of the data series was found. Year-by-season trends were significantly different; the annual differences were also highly significant ($p < 0.05$; Tukey compromise, Scheffe's *S*). Infaunal biomass taken during peak years (1980 and 1981) was significantly different from all other years of collection. The lowest such biomass occurred during the years following the drought (1982 and 1983). Winter–spring peaks of infaunal invertebrate biomass were generally higher than numbers found during the warmer periods of the year ($p < 0.05$). During winter 1981, infaunal species richness peaks coincided with the beginning of the biomass buildup. During the years following the drought, infaunal species richness was lower than at any other time of the sampling period. Sustained low river flow was accompanied by initially high numbers of species as *Streblospio benedicti, Mediomastus ambiseta,* and *Macoma*

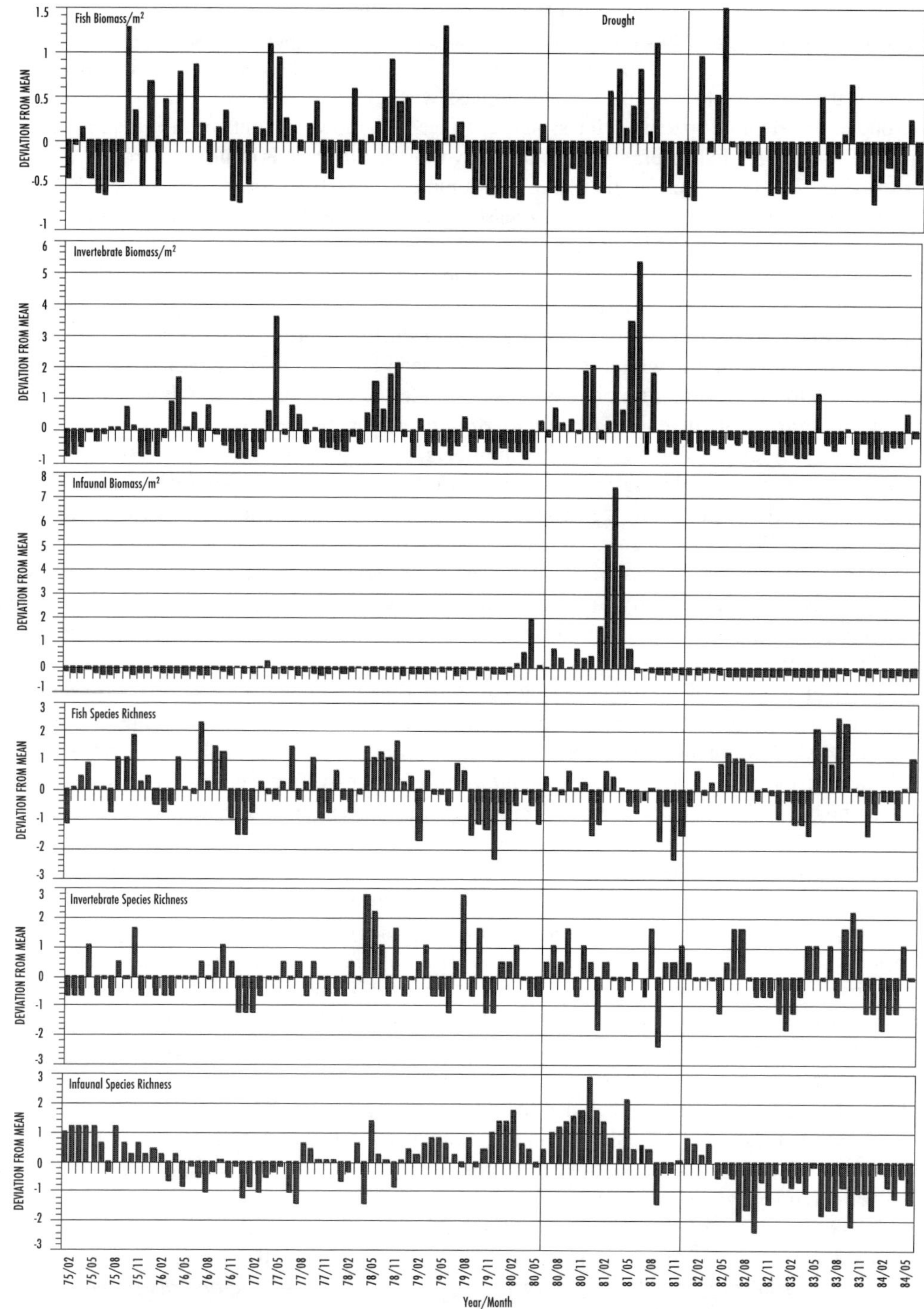

Figure 5.7 Long-term monthly values of the infaunal, macroinvertebrate, and fish biomass m^{-2} and species richness collected from stations in East Bay from February 1975 through July 1984. Each value represents the mean of all stations per month for each variable and is expressed as a deviation from the grand mean.

mitchelli were noted. These numbers decreased dramatically over the course of the drought. These changes in density were accompanied by initially reduced trophic diversity that recovered to previous levels with resumption of high river flow conditions.

Macroinvertebrate biomass was highest ($p < 0.05$) during fall periods (see Figure 5.7). Peaks of invertebrate biomass tended to follow the infaunal biomass peaks noted during the mid to late months of the drought period. Invertebrate biomass was relatively low during the years following the drought. Invertebrate species richness tended to reach the lowest points of the entire data series during 1982 and 1983. Invertebrate trophic diversity followed the trends of infaunal trophic diversity with sustained low levels during the spring and summer of 1981. Recovery was evident by spring 1983. There was evidence of a general decrease of invertebrate trophic diversity during the brief drought of the summer of 1977, although such reduction was not as pronounced as that during the 1981 period. These reductions of invertebrate trophic diversity were significantly ($p < 0.05$) different only during the 1977 and 1981 periods of drought.

Fish biomass was significantly ($p < 0.05$) higher during winter–spring periods, with particularly high peaks during the winter–spring of 1981 (see Figure 5.7). These biomass peaks tended to correspond to the invertebrate biomass peaks during the drought. Initially high fish numbers, comprising in large part juvenile *Leiostomus xanthurus,* during the early months of the drought gave way to progressive decreases in fish numbers (Livingston et al., 1997). Fish abundance returned to former levels during the recovery period of 1983–1984. Fish species richness was very low during the year before the drought and during most of the drought period, with some recovery during 1982. Fish species richness and trophic diversity were usually highest during summer–fall periods. These factors were lowest ($p < 0.05$) during the drought of 1981 with recovery of fish trophic diversity to previous levels by spring 1982. Although fish numbers and trophic diversity were lowest during most of the drought, the recovery of these indices followed somewhat different patterns.

Long-term changes of dominant species in East Bay are shown in Figure 5.8. In terms of frequency of occurrence during the sampling effort, the infaunal macroinvertebrate assemblages in East Bay were dominated by species such as *Mediomastus ambiseta* (below-surface deposit feeder and detritivorous omnivore), *Hobsonia florida* (above-surface deposit feeder and detritivorous omnivore), *Grandidierella bonnieroides* (grazer/scavenger and general omnivore), *Streblospio benedicti* (above-surface deposit feeder and detritivorous omnivore), and *Parandalia americana* (primary carnivore). Biomass of the infauna was influenced by larger types such as *Rangia cuneata* and *Mactra fragilis* (plankton-feeding herbivores). Dominant epibenthic macroinvertebrates in East Bay over the period of study included the palaemonetid shrimp (*Palaemonetes* spp., detritivorous omnivores), xanthid crabs (*Rhithropanopeus harrisi;* primary carnivores), blue crabs (*Callinectes sapidus,* primary carnivores at <30 mm; secondary carnivores at >30 mm), and penaeid shrimp (*Penaeus setiferus, P. duorarum,* and *P. aztecus*, primary carnivores at <25 mm; secondary carnivores at >25 mm). Most of these invertebrate species are browsers, grazers, or seize-and-bite predators.

Dominant fishes in East Bay included the plankton-feeding primary carnivore *Anchoa mitchilli* (bay anchovy) and benthic feeding primary carnivores such as spot (*Leiostomus xanthurus*), hogchokers (*Trinectes maculatus*), young Atlantic croakers (*Micropogonias undulates,* <70 mm), and silver perch (*Bairdiella chrysoura*, 21 to 60 mm). Secondary carnivores among the dominant fishes included larger croakers (>70 mm), Gulf flounder (*Paralichthys albigutta*), and sand sea trout (*Cynoscion arenarius*). Tertiary carnivores in East Bay included the larger spotted sea trout (*C. nebulosus*), southern flounder (*P. lethostigma*), largemouth bass (*Micropterus salmoides*), and gars (*Lepisosteus* spp.). With the exception of the bay anchovies, all of the above species live near the sediment–water interface, with most of the trophic organization of the bay dependent on interactions among bottom-living infaunal and epibenthic macroinvertebrates and fishes.

The drought of 1980–1981 was thus associated with a succession of biomass increases starting with the infauna and ending with peak biomass of invertebrates and fishes. These changes were

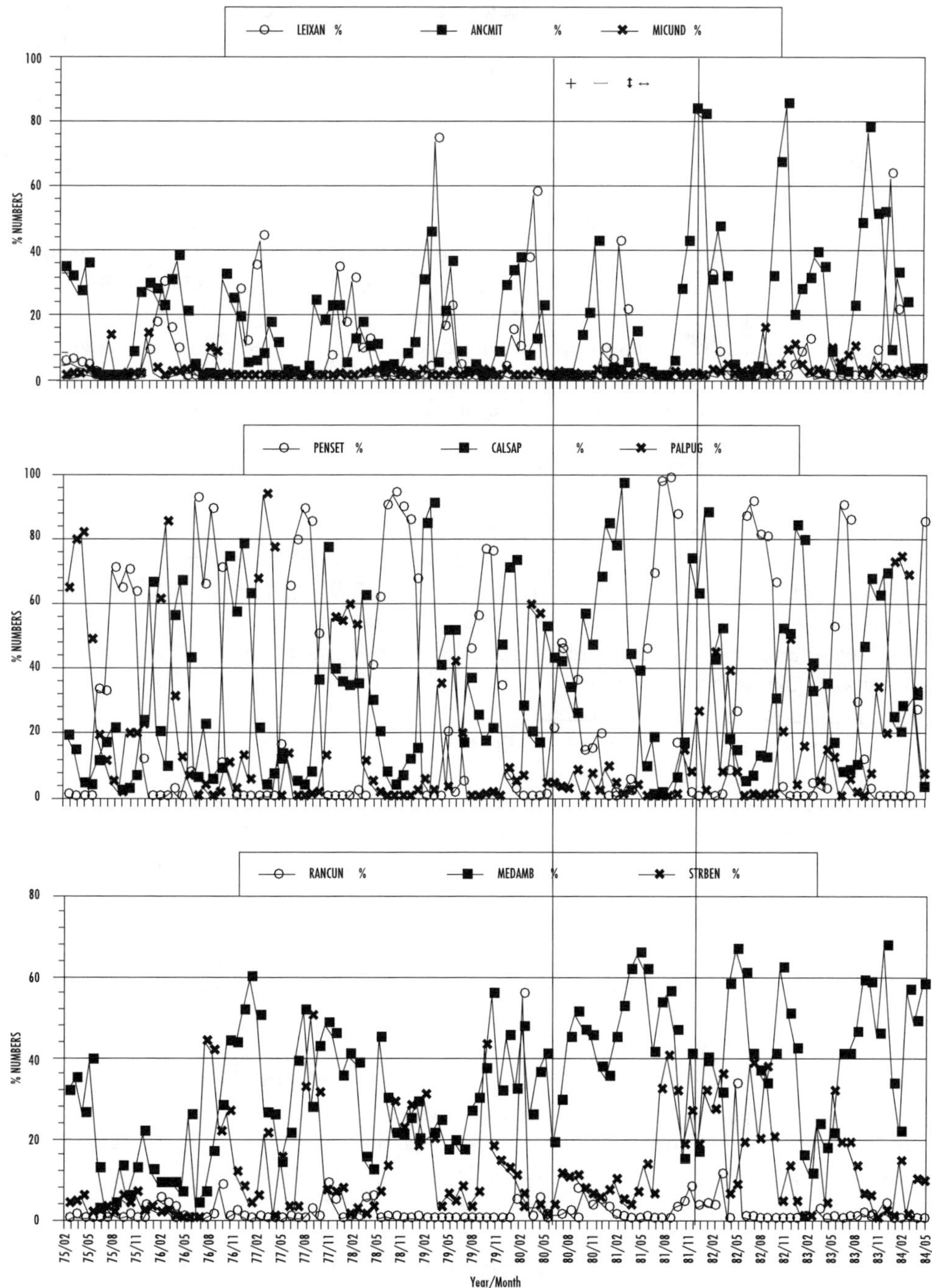

Figure 5.8 Long-term monthly percent of total numerical abundance infauna, invertebrates, and fishes in East Bay taken monthly from March 1975 through July 1984. Biological data are given as means of all stations in East Bay for each month. LEIXAN = *Leiostomus xanthurus*, ANCMIT = *Anchoa mitchilli*, MICUND = *Micropogonias undulatus*, PENSET = *Penaeus setiferus*, CALSAP = *Callinectes sapidus*, PALPUG = *Palaemonetes pugilis*, RANCUN = *Rangia cuneata*, MEDAMB = *Mediamastus ambiseta*, STRBEN = *Streblospio benedicti*.

followed by relatively low biomass of all three groups during the years after the drought. Infaunal species richness reached peaks during the first half of the drought; reductions of this index were noted by the end of the drought. Infaunal species richness did not recover during the 1982–1983 period. However, there was evidence of a recovery of both invertebrate and fish species richness following low numbers of species of both groups during the drought. The recovery of fish and invertebrate trophic diversity also coincided with the return of increased river flow conditions.

3. Trophic Relationships

All biological data (as biomass·m^{-2} mo^{-1} of the infauna, epibenthic macroinvertebrates, and fishes) were transformed from species-specific data into a new data matrix based on trophic organization as a function of ontogenetic feeding stages of the species found in East Bay over the multiyear sampling program (see above). The field data were then reordered into trophic levels so that monthly changes in the overall trophic organization of East Bay could be determined over the study period. The data were summed across all taxonomic lines and translated into the various trophic levels that included herbivores (feeding on phytoplankton and benthic algae), omnivores (feeding on detritus and various combinations of plant and animal matter), primary carnivores (feeding on herbivores and detritivorous animals), secondary carnivores (feeding on primary carnivores and omnivores), and tertiary carnivores (feeding on primary and secondary carnivores and omnivores). Trophic assignments are given in Appendix II.

The evaluation of the monthly mean values of the combined (fishes, infauna, invertebrates; FII) trophic data over the 9.5-year study period (Livingston et al., 1997) indicated that herbivore biomass peaked during winter–spring periods, whereas omnivores tended to peak during fall–winter months. Primary carnivore biomass peaked during summer and early winter, whereas secondary carnivore biomass peaked during late spring and summer months. Tertiary carnivore biomass was highest during spring and late summer–fall months.

The long-term trends of the trophic groupings in East Bay are shown in Figure 5.9. The mean monthly biomass of herbivores over the study period was 2.39 g/m^2 with averages of 0.21 g/m^2 for omnivores, 0.53 g/m^2 for primary carnivores, 0.04 g/m^2 for secondary carnivores, and 0.004 g/m^2 for tertiary carnivores. There was a marked increase in the herbivore biomass during the drought of 1980–1981. Herbivore biomass peaks coincided with low winter river flows, peak turbidity, low water color, and increased Secchi depths (Livingston et al., 1997). The more frequent periods of positive oxygen anomaly during the drought also coincided with the high levels of herbivore biomass. There was a marked decrease of herbivore biomass that started during the second half of the drought. Herbivore biomass virtually collapsed during the following year (see Figure 5.9). There was a general decrease of omnivore biomass during the 2-year drought period. Primary carnivore biomass increased during the fall of 1980 and continued at high levels through the spring of 1981. Carnivore biomass, as shown by deviations from the grand mean (see Figure 5.9), was generally low during the drought, with increases of the primary carnivores during 1982. The biomass deviations of secondary and tertiary carnivores increased during 1983. The tertiary predators were virtually absent during the 1980–1981 drought. This group returned to the bay with the return of winter–spring peak river flows in 1982 and 1983. Overall, the drought of 1980–1981 had a profound effect on the trophic organization of East Bay, with recovery from the drought following very different patterns among the different trophic groupings. The lack of recovery of the FII herbivores during 1982–1983 coincided with increased biomass of primary carnivores, and these patterns could be related by predator–prey interactions. Likewise, the return of the tertiary carnivores during a period of reduced primary carnivore biomass could also have been related to predator effects.

With trophic organization expressed as total biomass m^{-2} yr^{-1}, there was a clear relationship between the mean annual river flow rates and the overall animal (infauna, macroinvertebrate, fish) biomass in East Bay. There were significant ($p < 0.05$) seasonal and annual differences in biomass from year to year (Table 5.11); however, during the first 5 years of sampling, river flow and total

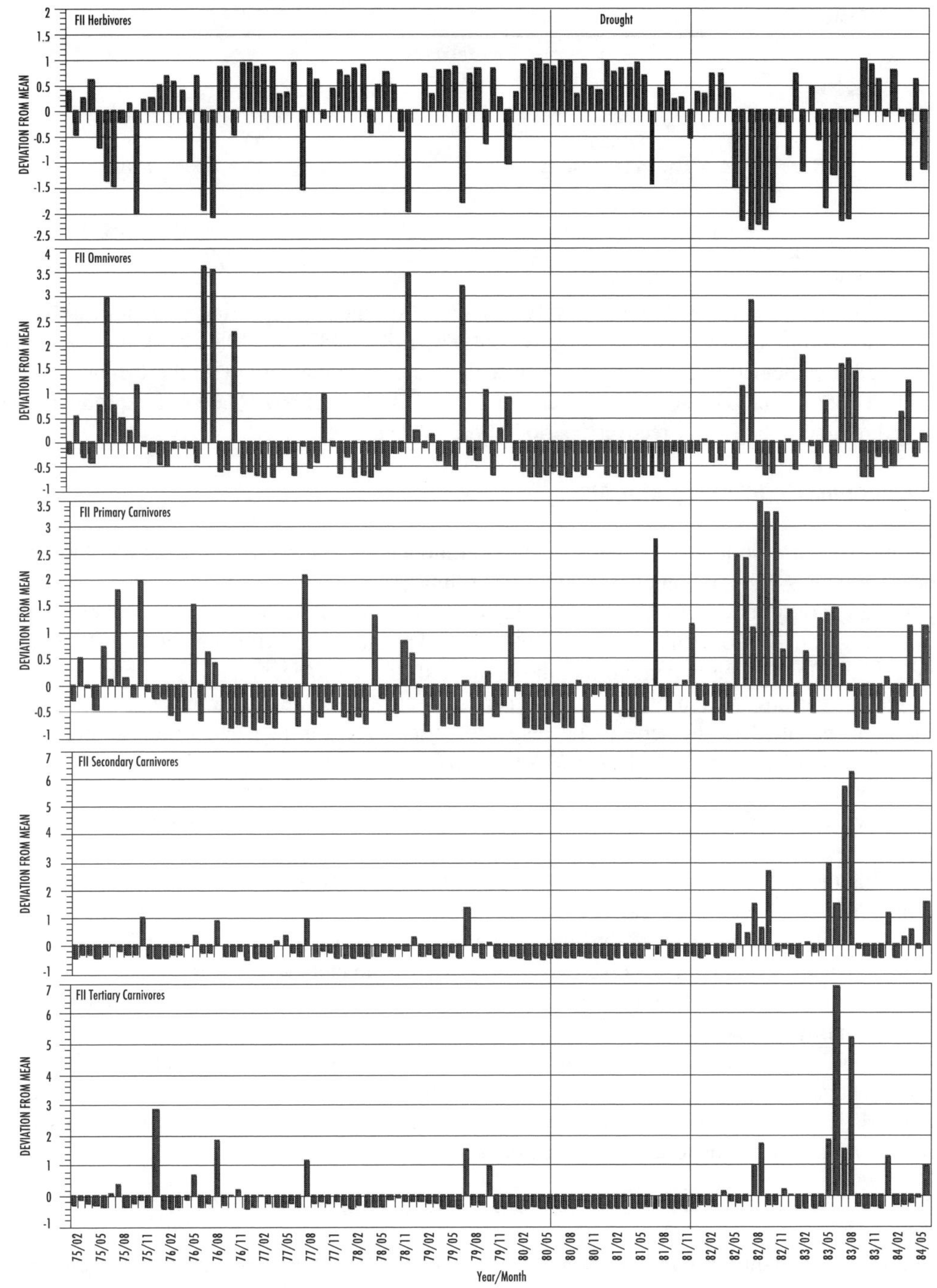

Figure 5.9 Long-term monthly values of the FII biomass m^{-2} of herbivores, omnivores, and primary, secondary, and tertiary carnivores collected from stations in East Bay from February 1975 through July 1984. Each value represents the mean of all stations per month for each variable and is expressed as the deviation from the grand mean.

Table 5.11 Results of the Two-Way ANOVA for the Summed Biological Data Taken in East Bay

Factor	Source	df	*F*-Value	*p*-Value
	Total Fauna Combined			
Biomass/m^2	Year	8	16.85	0.0001
	Season	3	5.90	0.0012
	Year × season	24	1.19	0.2827
	Residual	72		

Note: ANOVAs were run by year and season using monthly averages within each season as replicates.

animal biomass remained within a relatively small level of interannual variance. Significantly ($p < 0.05$) different biomass was noted during years 1980, 1981, 1982, and 1983. Peak biomass years (1980–1981) coincided with reductions in river flow and were due largely to the increases in the herbivore component. The significant decrease in biomass, which began late in the drought, continued throughout the 2-year recovery period (1982–1983).

4. Dynamic Regression Models

Univariate time series and dynamic regression models were developed for the five biological series, herbivores $\{Hb(t)\}$, omnivores $\{Om(t)\}$, primary carnivores $\{_{C1}(t)\}$, secondary carnivores $\{_{C2}(t)\}$, and tertiary carnivores $\{_{C3}(t)\}$, following the well-known Box–Jenkins modeling procedures (Box and Jenkins, 1976; Pankratz, 1983, 1991) (see Appendix III). All parameters in the models were estimated using the maximum likelihood method. If a biological series is serially correlated (i.e., the current observation of the series depends on past observations), a univariate time series model may be fitted to the series. The response series $\{Y(t)\}$ is modeled as an autoregressive integrated moving average model $\{\text{ARIMA } (p, d, q)\}$ with the form:

$$\nabla^d Y(t) - \phi_1 \nabla^d Y(t-1) - \ldots - \phi_p \nabla^d Y(t-p) = \mu + \varepsilon(t) - \theta_1 \varepsilon(t-1) - \ldots - \theta_q \varepsilon(t-q) \quad (5.1)$$

where ∇ is the differencing operator such that $\nabla Y(t) = Y(t) - Y(t-1), \ldots, \nabla^d Y(t) = Y(t) - Y(t-d)$ and the $\varepsilon(t)$ are assumed to be independent and normally distributed with mean zero and variance σ^2. In the model, p is the order of the autoregression (AR) term, d is the order of differencing, and q is the order of the moving-average (MA) term. The parameters ϕ and θ represent the AR and MA coefficients, respectively. Each biological series was modeled using only present and past (i.e., lagged) monthly values of the series.

Dynamic regression models, also called transfer function models (Box and Jenkins, 1976), were used to characterize how the biological variables responded to changes in the environmental characteristics. A linear dynamic regression model describing the relationship between a response series $\{Y(t)\}$ and an input series $\{X(t)\}$ has the form:

$$Y(t) = \mu + \upsilon_0 X(t) + \upsilon_1 X(t-1) + \ldots + \upsilon_k X(t-k) + \xi(t) \quad (5.2)$$

where μ is a constant term and $\xi(t)$ is assumed to be an ARIMA process and independent of the input series $\{X(t)\}$. If $\xi(t)$ follows a stationary AR(p) model, such that $\xi(t) - \phi_1\xi(t-1) - \phi_2\xi(t-2) - \ldots - \phi_p\xi(t-p) = \varepsilon(t)$ and letting B be the backward shift operator such that $B\xi(t) = \xi(t-1), \ldots, Bk\xi(t) = \xi(t-k)$, then $\xi(t)$ can be expressed as $\xi(t) = (1 - \phi_1 B - \phi_2 B^2 - \ldots - \phi_p B^p)^{-1}\varepsilon(t)$ and Equation 5.2 can be written as:

$$Y(t) = \mu + \upsilon_0 X(t) + \upsilon_1 X(t-1) + \ldots + \upsilon_k X(t-k) + (1 - \phi_1 B - \phi_2 B^2 - \ldots - \phi_p B^p)^{-1}\varepsilon(t) \quad (5.3)$$

The dynamic regression model in Equation 5.3 states that the effects of $\{X(t)\}$ on $\{Y(t)\}$ are distributed across several time periods. The coefficients $\{\upsilon_0, \upsilon_1, ..., \upsilon_k\}$ in the model, called υ-weights, indicate how much the response series changes when the past and current values of the input series change one unit. Multiple inputs may be introduced into the model by adding other variables (e.g., $X_1(t)$, $X_2(t)$, $X_3(t)$) along with their appropriate weights for different time lags $\{(t-k)\}$.

In the current analysis, each of the five biological series was modeled using the six principal component axes (representing the environmental variables) as well as the other biological series (i.e., Hb, Om, C1, and C2). Inclusion of the other biological variables as inputs into each model allowed us to examine possible feedback mechanisms among the biological series. Although the tertiary carnivores were modeled for completeness, the C3 series was not included as an input variable in the formation of the dynamic regression models for the lower trophic groups (Hb to C2). Tertiary carnivores are highly motile organisms and were the least well-represented group in our trawl collections.

An F-test was used to compare the two nested models (i.e., the univariate model and the dynamic regression model). RSS_1 and RSS_2 are the residual sum of squares from the two models, respectively. The F-statistic for comparing the two models is defined by:

$$F = \frac{(\mathrm{RSS}_1 - \mathrm{RSS}_2)/f_1}{\mathrm{RSS}_2/f_2}$$

where f_1 is the difference in numbers of parameters in the two models and f_2 is the degrees of freedom for RSS_2. The F-statistic has an approximate F distribution with degrees of freedom f_1 and f_2.

The usefulness of both the univariate time series and the dynamic regression model can be measured by r^2, which is calculated by the formula:

$$r^2 = 1 - \frac{\sum_{t=1}^{n} (Y(t) - \hat{Y}(t))^2}{\sum_{t=1}^{n} (Y(t) - \bar{Y})^2}$$

where $\hat{Y}(t)$ is the fitted value of $Y(t)$ and $\bar{Y}$ is the mean of $Y(t)$.

a. *Herbivores*

The univariate time series model for the herbivore series was identified to be a parsimonious AR(3) process, in which the current biomass of herbivores is positively correlated with changes in herbivore biomass 1 and 3 months earlier (Livingston et al., 1997). The r^2 for the AR(3) model is 0.58, indicating that the serial correlation explains about 58% of the total variation in the herbivore series. The fitted dynamic regression model describes the relationship between the herbivore series and the various input series (i.e., the principal components, representing the environmental variables, and the other biological series). Four of the principal component axes and two of the biological variables contributed significantly to the herbivore biomass model. In particular, the first three principal components as well as the sixth had significant influences on herbivore biomass. The herbivore series responded negatively to changes in the first (with a 12-month lag) and third (with a 6-month lag) principal components and positively to changes in the second, third, and sixth (with 2- and 6-month lags). In addition to these environmental factors, variation in the herbivore series was explained by changes in the omnivore and the primary carnivore series. Herbivore biomass was associated positively with omnivore biomass (no time lag) and negatively with primary carnivores (with a 20-month lag).

The fitted dynamic regression model (using the four principal components, the omnivore and primary carnivore biomasses, and the serial correlation of the herbivore series itself) explained about 82% of the total variation in herbivore biomass. The univariate time series model had fewer parameters than the fitted dynamic regression model but explained significantly less of the herbivore variance (F-value = 16.84 with df = 7, 85).

b. *Omnivores*

The univariate time series model for omnivores was an AR(1) process that explained only about 33% of the total variation of the series (Livingston et al., 1997). The fitted dynamic regression model showed that the omnivore series was significantly associated with only the first principal component, and herbivore and primary carnivore biomasses. Omnivore biomass was correlated with P_1 negatively at 6- and 12-month lags and positively with a 24-month lag. Omnivores were correlated with the herbivore series at lag 0 and 11, and with the primary carnivores at lag 1. The r^2 for the fitted dynamic regression model was 0.65, which was a significant improvement over the univariate time series model (F-value = 12.49 with df = 6, 82).

c. *Carnivores*

The univariate time series model for the primary carnivores was identified as a parsimonious AR(4) process, whereas both secondary and tertiary carnivores were described by AR(1) processes (Livingston et al., 1997). The univariate models explained only 38, 15, and 12% of the variation in the primary, secondary, and tertiary carnivores, respectively. The fitted dynamic regression models for the three carnivore groups indicated that the only significant variables entering the models were those of the other biological series. Specifically, primary carnivores were positively related to the herbivore series with 2- and 8-month lags. Secondary carnivores were positively related to the herbivore series with a 6-month lag and the primary carnivore series with a 1-month lag, but were negatively related to the omnivore series with an 11-month lag. Finally, tertiary carnivores were negatively related to primary carnivores with a 1-month lag and positively correlated with secondary carnivores with no lag. The dynamic regression models explained about 51, 32, and 25% of the variation in the respective trophic series. In all three cases, the dynamic regression models explained significantly more of the variation in the response series than did the univariate models (C1: F = 12.49 with df = 2, 101; C2: F = 7.86 with df = 3, 98; C3: F = 9.55 with df = 2, 109).

5. *Long-Term Trends*

Previous studies have indicated that predation is a dominant force in unvegetated benthic communities of temperate and subtropical coastal areas (Virnstein, 1977; Reise, 1978; Holland et al., 1980; Stoner, 1982; Hunter and Price, 1992). In the Apalachicola system, river flow and rainfall followed recurrent patterns with peaks of river flow highly correlated with reductions of salinity in the receiving estuary. During the first year of the drought of 1980–1981, there was increased salinity, increased water clarity (low color, high Secchi depths), and generally higher oxygen anomalies in East Bay, indicative of a period of high primary productivity during the early stages of the drought. Secchi readings and the relative shallowness of East Bay indicated that light reached the bottom in much of the system during the drought and may have contributed to increased benthic microalgal productivity. Despite some high turbidity levels during the second half of the drought, Secchi depths remained high during this period, probably as a result of very low water color. The declining oxygen anomalies during the last half of the drought indicated possible other biological activities, which could have involved reduced primary production and/or increased secondary production.

Prior to the onset of the drought in East Bay in 1980, there appeared to be a relatively stable series of changes in the trophic organization of the system. Within a certain range of river flow variation, there did not seem to be wide deviations in the trophic structure from year to year. The 1980–1981 drought tended to coincide with increased predominance of the herbivores, followed in quick succession by the various types of carnivores over the succeeding 2 years (i.e., after the drought subsided). After initial unprecedented levels of herbivore biomass, there was a distinct reduction of the herbivores, suggesting a possible nutrient-limited reaction of the system to the drought. During the relatively short period of the drought, extremes of secondary production at various levels of the trophic organization were evident, indicating a rapid succession of changes in the heterotrophic community. These patterns indicated that the postulated extreme autotrophic response to reduced river flow takes 1 to 2 years to make its way through the food webs of the estuary. It is possible that the succession times for the various trophic levels in East Bay could be related to their generation time because such times tend to increase up the trophic scale. The absence of phytoplankton data represented a serious omission, and the various productivity indices such as DO, oxygen anomaly, and pH were used as indicators of phytoplankton trends.

There are other studies that would tend to substantiate the production cycles associated with drought. Observations concerning peaks in primary and secondary production noted in San Francisco Bay during the 1976–1977 drought indicated high primary productivity in portions of the system during the early phase of the drought. High chlorophyll *a* concentrations were statistically associated with increased water clarity and salinity and decreased freshwater inflow (Lehman, 1992). Phytoplankton biomass dropped to very low levels during the course of the drought (Cloern et al., 1983) concurrent with peak densities of various suspension-feeding herbivores and omnivores, primarily bivalves, polychaetes, and amphipods (Nichols, 1985). Two alternative mechanisms were proposed to account for the observed population data: a circulation-induced local plankton maximum controlled by the volume of freshwater inflows (Cloern et al., 1983) and high levels of benthic grazing (Nichols, 1985). Nutrient limitation and zooplankton grazing (other possible alternatives) were not thought to be operational in this case. Although the mechanisms underlying the observed plankton dynamics remained unclear, changes in primary and secondary production resulting from the drought were linked to long-term repercussions in food web dynamics in San Francisco Bay. Such data were consistent with the trends in East Bay trophic structure during the 1980–1981 drought, at which time herbivores and detritivorous omnivores, such as the polychaetes *Mediomastus ambiseta* and *Streblospio benedicti* and the bivalve *Macoma mitchelli,* peaked in abundance. The connection of the benthos with phytoplankton production and the subsequent rapid decline in benthos during the latter portions of the drought indicated a complex trophic coupling of East Bay with long-term trends of Apalachicola River flow.

In East Bay, there was a dichotomy of response of the trophic elements to controlling factors. The herbivores and omnivores were directly linked to physical and chemical controlling factors that were associated implicitly with the primary productivity of the estuary. These observations were supported by the dynamic regression models in which these two feeding groups were coupled to the principal component axes representing the environmental variables. The river, which mediates such factors, was thus directly linked to the response of these trophic components. The primary, secondary, and tertiary carnivores, on the other hand, were associated more closely with biological factors; none of the carnivore groups was related directly to any environmental principal component after taking into account the effects of herbivores and omnivores. This suggested that river flow and primary production were mainly associated with changes in the lower trophic levels, and that the carnivores were associated primarily with the lower animal trophic components.

The preponderance of positive signs associated with the biological coefficients in the dynamic regression models suggested primarily "bottom-up" control of trophic organization in East Bay. Presumably, predation effects (i.e., "top-down" control) at a lower trophic level would be noted with least ambiguity as negative coefficients associated with higher trophic groups over relatively short time lags (e.g., 0 to 2 months). A negative relationship with primary carnivores was observed in the herbivore series, indicating the potential influence of predation at this level. However, the introduction of significant long-term time lags (i.e., a 20-month lag) confounded this interpretation and made the determination of controlling functions more difficult. The general lack of support for top-down control is consistent with previous experimental work in the bay (Mahoney, 1982; Mahoney and Livingston, 1982; see above) where river flow was postulated as the key factor. Thus, the complex interactions involved in the long-term changes in the various trophic groupings could well include both bottom-up (primarily) and top-down (to a limited extent) control with river flow as the dominant controlling factor that directly (herbivores and omnivores) and indirectly (carnivores) determined the long-term response of the system to fluctuations in freshwater inflows.

The preoccupation in the literature with the relative importance of top-down vs. bottom-up control of aquatic food webs is often misdirected. In many cases, both forms of control occur depending on various conditions. A more appropriate question concerns interactions of the various components, physical and biological, that drive the trophic organization of aquatic systems in both the qualitative (species composition) and quantitative (population numbers and biomass) sense. Power (1992) demonstrated the various levels at which top-down and bottom-up limitation combined, and there are specific examples of co-limitation by resources and predators (Mittlebach, 1988; Mittlebach et al., 1988; Arditi and Ginzburg, 1989; McQueen et al., 1989). However, relatively little information has been generated to address the complex aspects of control (Menge, 1992). Separation of the top-down and bottom-up trophic processes by researchers can lead to an oversimplification of the situation since both processes are inextricably interconnected within the sediment–water interactions of shallow estuarine systems.

The increases of the herbivore biomass during the drought were based on increased numbers of bivalve mollusks, a finding consistent with previous analyses of long-term changes of biological components of the Apalachicola estuary. Meeter et al. (1979) found that oyster landings from 1959 to 1977 were correlated negatively with river flow. The highest oyster landings coincided with drought conditions. Wilber (1992), using oyster data from 1960 to 1984, found that river flows were correlated negatively with oyster catch per unit effort within the same year and positively with catch 2 and 3 years later. Highest oyster harvests occurred in 1980–1981, coinciding with the drought years in our study period. Predation on newly settled spat during periods of high salinity was given as a possible explanation of the 2-year time lags between low flow events and subsequent poor production. Mechanisms underlying the contemporaneous inverse correlation were unclear, although both predation and disease accompanied reduced oyster populations with increased salinity (Livingston et al., 2000). Based on our present analysis, a trophic explanation is thus possible at least in part for the observed changes in the phytoplankton-feeding herbivore groups. Observed high oyster production during low flow years could thus be due to increased primary productivity as a function of altered physical conditions in the receiving estuary. Increased productivity contributes to increased growth rates and ultimately increased oyster production. With time, production decreases could follow as nutrient limitation sets in with continued low flows. Extension of this model to postulated permanent anthropogenous reductions of Apalachicola River flow due to projected anthropogenous activities in the tri-river basin indicated serious damage to the highly productive oyster industry in the Apalachicola estuary (Livingston et al., 2000).

Within certain ranges, Apalachicola River flow was a key element in the control of the biological organization of East Bay as a function of both direct (salinity and light penetration) and indirect (trophic) effects. Complex biological variables were more physically forced in the highly variable estuarine environment as compared with the more biologically controlled systems outside the main influence of the river. Trophic components directly linked to phytoplankton and benthic algal production (herbivores, omnivores) were thus immediately affected by changes in river flow, whereas the higher trophic levels were more biologically controlled, with predation the key factor in the definition of the various carnivore groups in space and time. Below a certain level of river flow, water quality changes occurred that increased light transmission through the water column, thereby stimulating primary productivity and initiating a series of trophic responses. The data thus lead to the hypothesis that light penetration and nutrient loading, mediated by physical influences of the river on the estuary, emerge as the primary determinants of the biological organization of East Bay.

The specific criteria for the detailed delineation of trophic organization proved to be less important than the consistency of the data on which the trophic models were based. In East Bay, there was a dichotomy of response of the trophic elements to controlling factors. Most of the estuarine organisms are eurythermal and euryhaline. Physical control of their distribution in space and time is less important than trophic considerations. In other words, the primary ordering of spatiotemporal distribution of coastal populations of carnivores is trophic. The results of the long-term analyses of trophic organization in the Apalachicola system suggest that river flow, nutrient loading, and associated primary productivity are mainly associated with changes in the lower trophic levels, and that the carnivores are associated primarily with the distribution of their prey.

The long-term data from the Apalachicola system indicated that, with a specific reduction of freshwater flow to a level specific for the receiving system, the physically controlled, highly productive river–estuarine system would become a species-rich, biologically controlled bay of substantially reduced productivity. Permanent base flow reductions of even a relatively small magnitude could cause the system to be more vulnerable to droughts of less magnitude than that examined here.

CHAPTER 6

Coastal Phytoplankton Organization

I. FRESHWATER RUNOFF AND PRIMARY PRODUCTION

Freshwater input is an important determinant of the overall productivity of river-dominated estuaries (Snedaker et al., 1977; Cross and Williams, 1981). River flow into estuaries controls many factors that influence habitat structure, productivity, and food web interactions (Schroeder, 1978; Livingston, 1984a, 2000; Randall and Day, 1987; Gallegos et al., 1992; Mallin et al., 1993; Livingston et al., 1997). Basic features such as salinity, light penetration, stratification, dissolved oxygen (DO), sediment quality, nutrient loading, primary and secondary production, and food web structure are controlled in varying degrees by the cyclic changes in river discharges (Livingston, 2000). Freshwater input controls the long-term nutrient dynamics in river-dominated coastal systems (Livingston, 1981b, 1984a; Peterson and Howarth, 1987; Baird and Ulanowicz, 1989). Mallin et al. (1993) noted that primary production and algal bloom periodicity were directly related to rainfall in the watershed of the Neuse River estuary in North Carolina and consequent river influxes into the coastal system. River runoff was associated with eutrophication levels in the estuary to the point that the authors noted that episodic runoff should be taken into consideration in mitigation efforts to reduce nutrient loading to coastal areas. Under natural river flow conditions, the combination of high estuarine primary production and reduction of predation due to low/variable salinity contributes to rapid growth and enhanced productivity of eurytopic populations that are adapted to rapidly changing environmental conditions (Livingston, 1984a, 1991b, 2000). Climatological changes have biological effects on coastal systems that can be followed by population shifts over decades (Finney et al., 2000).

River-dominated estuaries in the NE Gulf of Mexico are highly productive due to factors such as nutrient enrichment from land runoff, the shallow nature of the receiving system, and energy supplements from wind, tidal currents, and thermohaline circulation. However, the processes that determine primary productivity (based largely on phytoplankton activity) and associated food webs in coastal areas vary due to differences in nutrient loading, physiography of the receiving area, and habitat features that include temperature, salinity, stratification characteristics, currents, light transmission, and sediment quality. Nutrient loading is fundamental to the growth of coastal phytoplankton (Livingston, 2000). Light availability is an important determinant of estuarine and coastal phytoplankton communities (Philips et al., 2000). Biological processes (competition, predation) also influence primary and secondary production. Different combinations of the set variables determine the highly individual rates of primary production and food web responses that differentiate one system from another with resulting differences in population dynamics, community structure, and overall secondary production (Livingston, 2000).

The response of phytoplankton to nutrient loading in coastal systems has been well studied (Anderson and Garrison, 1997). Specific effects of nutrient loading on phytoplankton assemblages can be related to currents and salinity distribution (Squires and Sinnu, 1982), the physiography of contributing systems (Marshall, 1982a,b, 1984), and effects of human activities. Industrial wastes such as pulp mill effluents are known to affect phytoplankton (Reddy and Venkateswarlu, 1986) with resulting changes such as increased domination by Cyanophycean types along with the reduction of green algae. Sewage wastes have been associated with *Oscillatoria* spp., *Rhopalodia gibberula,* and *Nitzschia pale*. Blue-green algae (*O. nigroviridis*) are often indicators of waters affected by sewage (Premula and Rao, 1977) although blue-green algae are also abundant in marine areas under natural conditions (Potts, 1980). A *Prorocentrum micans* bloom in a New Zealand estuary was coincident with increased nitrogen from upwelling (Chang, 1988). The problem with many such studies is that there is little in the way of nutrient loading data so that the specific details of phytoplankton response remain undetermined.

II. THE PERDIDO DRAINAGE SYSTEM

The Perdido Bay system (Figure 6.1) is a shallow to moderately deep inshore body of water that is approximately perpendicular to the Gulf of Mexico. The Elevenmile Creek system is a small drainage basin (about 70 km^2) that receives input from a pulp mill. Elevenmile Creek drains into the estuary about 1.6 km east of the entry point of the Perdido River in the northernmost part of the system. The relatively small Bayou Marcus Creek drains a residential area of western Pensacola with input from urban storm water runoff.

A. Seasonal and Interannual River Flow Patterns

There were two basic components of Perdido River flow: winter–early spring highs and late summer–fall lows. Two-way ANOVAs, run by year and season using monthly averages within each season as replicates, indicated significant ($p = 0.05$) differences between winter–early spring and late summer/fall flows. Aperiodic freshwater influxes to the bay due to storm activity accounted for occasional peak flows during most months of the year over the 14-year study period. Periods of low flow occurred during 1988–1989, 1993–1994, and from 1998 to 2002 (Figures 6.2 and 6.3). Droughts were defined by continuous low summer flows and relatively flat river curves during winter–early spring flood periods. The 1998 to 2002 drought was the most prolonged period of low flows over the study period. Relatively heavy river flow events occurred during the winter–spring of 1990, over a prolonged period from spring 1995 through winter 1996, during winter 1998, and during five storm periods. A trend analysis of monthly river flows indicated that storm-related increases in river flows were most noticeable during 1995 and 1997–1998. River flows during drought periods were significantly ($p = 0.05$) different from those during the peak years of 1990 and 1995–1996. With the exception of the 1994 storm, most storm events occurred during summer low flows.

The drought–flood sequences are important determinants of habitat quality in the Perdido estuary (Livingston, 2000). Salinity was inversely correlated with river discharge. Niedoroda (1992) found that, during relatively wet years, surface and bottom salinity of Perdido Bay decreased, but the lower layer decreased the most. The similar flushing times of the middle and lower bay were somewhat less than those of the upper bay. Most of the differences in flushing times were attributed to differences in the width of the bay. The upper layer of the middle bay had flushing times between 0.3 and 4.5 days. Corresponding values for the whole lower layer in the middle bay ranged between 0.3 and 6 days. The computed flushing times for a representative cell in the lower bay ranged between 0.1 and 3 days for the upper layer with much longer values (up to 15 days) for the lower layer. Thus, turnover rates were lowest in sub-haloclinal areas of the lower bay.

Flushing periods in Perdido Bay are minimal during drought periods, and this feature of bay organization is critical to the development of phytoplankton blooms (Livingston, 2000). Moderate

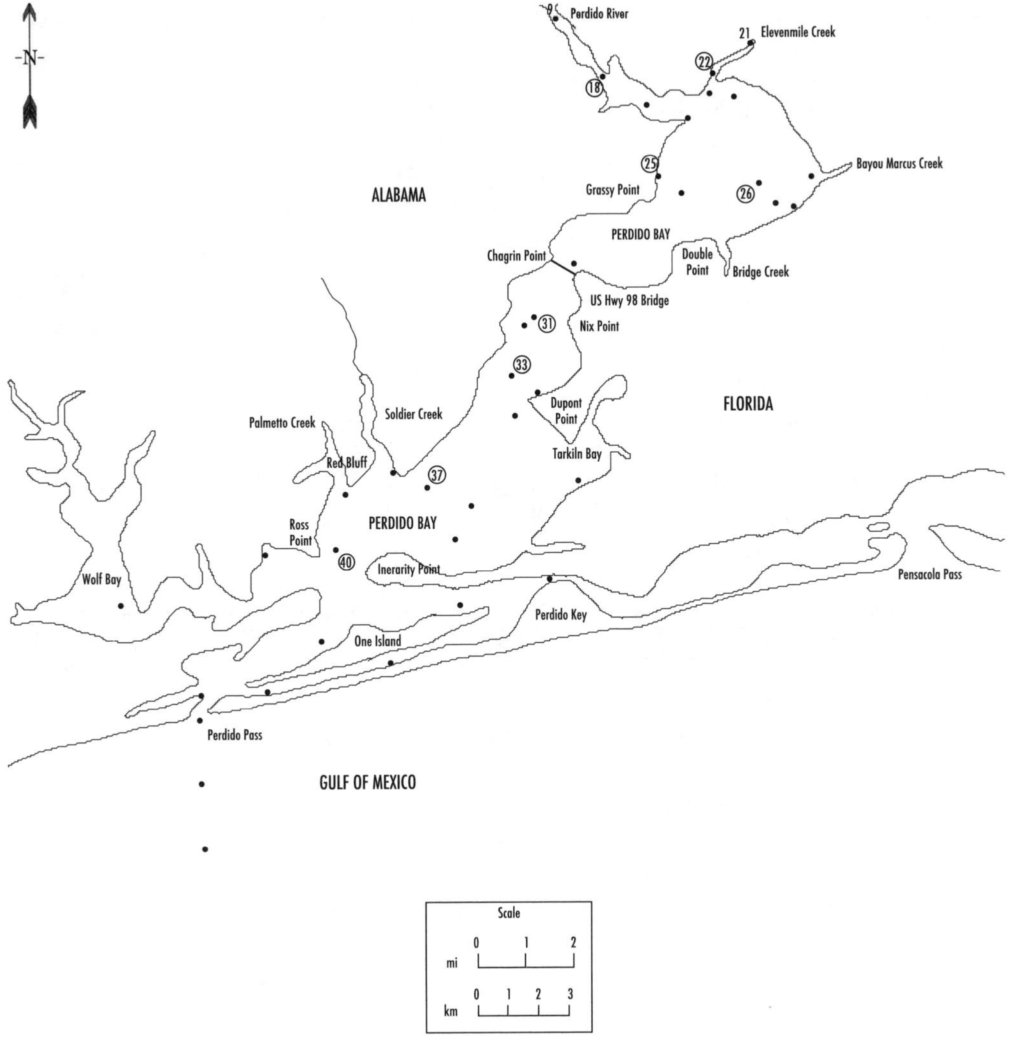

Figure 6.1 The Perdido Bay system showing long-term sampling stations for the long-term (14-year) field program.

discharge events tend to flush out the upper layer of the upper bay while not destroying the lower layer, which leads to increased salinity stratification and sharply reduced entrainment of the lower layer into the upper layer. Major storms flush the entire upper bay, with the return of the bottom salinity to the upper bay usually taking no more than a month (Livingston, 2000). Under prolonged conditions of very low flow, such as the condition during the 1998 to 2002 period, high salinities are carried up the bay by bottom flows under highly stratified conditions throughout the bay. These conditions are conducive to the development of plankton blooms.

The artificially maintained Perdido Pass is the primary source of salt water to Perdido Bay, and the pass controls salinity relationships in Perdido Bay. It is the primary contributor to the relatively strong stratification of the bay (Livingston, 2000) that, in combination with turbulence and vertical turnover rates, represents an important determinant of water quality, phytoplankton production, and sub-haloclinal hypoxia during warm seasons of the year. The relatively narrow bay, with a trending NE to SW axis, also tends to minimize vertical mixing, thus contributing to the relatively strong halocline and warm-season hypoxia at depth. There is also a northward movement of lower bay nitrogen during summer periods in Perdido Bay that is similar to that noted in the Patuxent River

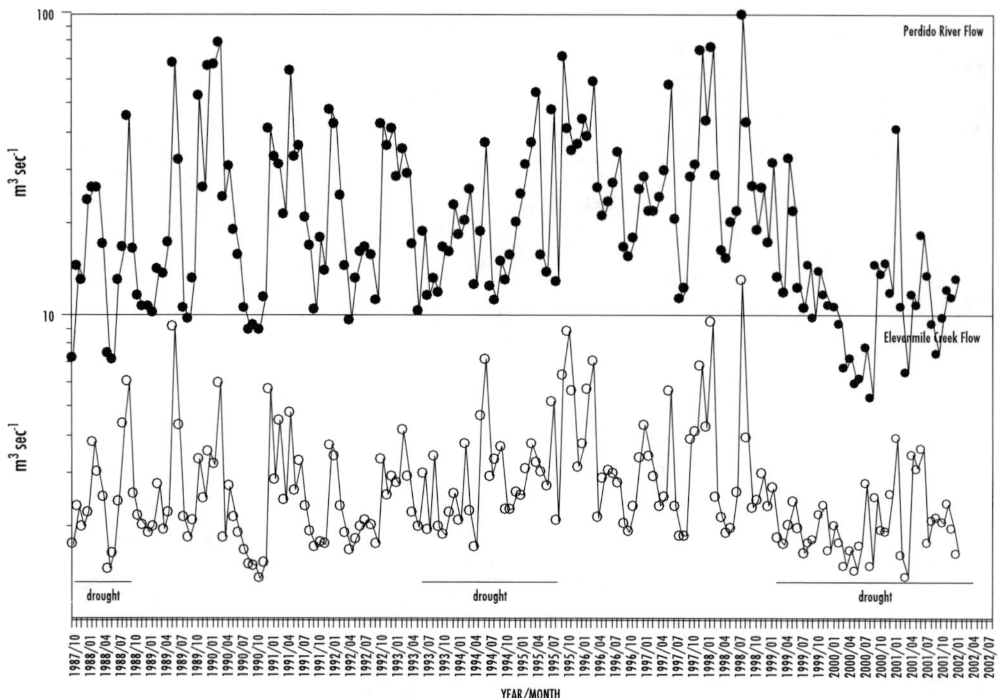

Figure 6.2 Monthly Perdido River and Elevenmile Creek flows (m[3] sec[−1]) shown by month over the 14-year sampling period.

(Hagy, 1996). Low flows to the Patuxent were often associated with high algal biomass in the upper estuary (Hagy, 1996). River flow trends are an important consideration in the development and maintenance of phytoplankton production and bloom formation in river-dominated estuaries such as the Perdido because of both changes in the physical habitat of the bay and the complex relationships of river flow, nutrient loading, and nutrient concentrations.

B. Nutrient Loading

Nutrient loading is an important determinant of the ecological response of coastal systems to freshwater inflow (Flemer et al., 1997). Rabalais (1992) noted that 23 of 58 estuaries in the Gulf of Mexico, including Perdido Bay, have problems or potential problems of overenrichment. Detailed, long-term nutrient loading analyses were carried out in the various river–estuarine systems in the NE Gulf of Mexico (Livingston, 2000). The U.S. Geological Survey (Tallahassee, Florida; M. Franklin, personal communication) provided river flow data for these systems. The models used to calculate nutrient loading were based on a ratio estimator developed by Dolan et al. (1981) that corrects for biases due to specific temporal sampling regimes by using auto- and cross-covariance values of flows and nutrients for correction of loading calculations (Appendix I). Nutrient data for the models were usually taken monthly, whereas river flow data were taken daily. In the Perdido system, program modifications by A. Niedoroda and G. Han (personal communication) provided models for evaluation of nutrient loading from Elevenmile and Bayou Marcus Creeks and the Styx, Blackwater, and Perdido Rivers (usually summed as direct contributions to upper Perdido Bay). Nutrient loading models were also calculated for overland (non-point source) runoff from unmonitored areas in eastern and western parts of the upper bay.

Livingston (2000) noted that the Apalachicola River system had the greatest flow rates of all the systems in the NE Gulf of Mexico (see Figure 2.2D). The next highest flow rates were the

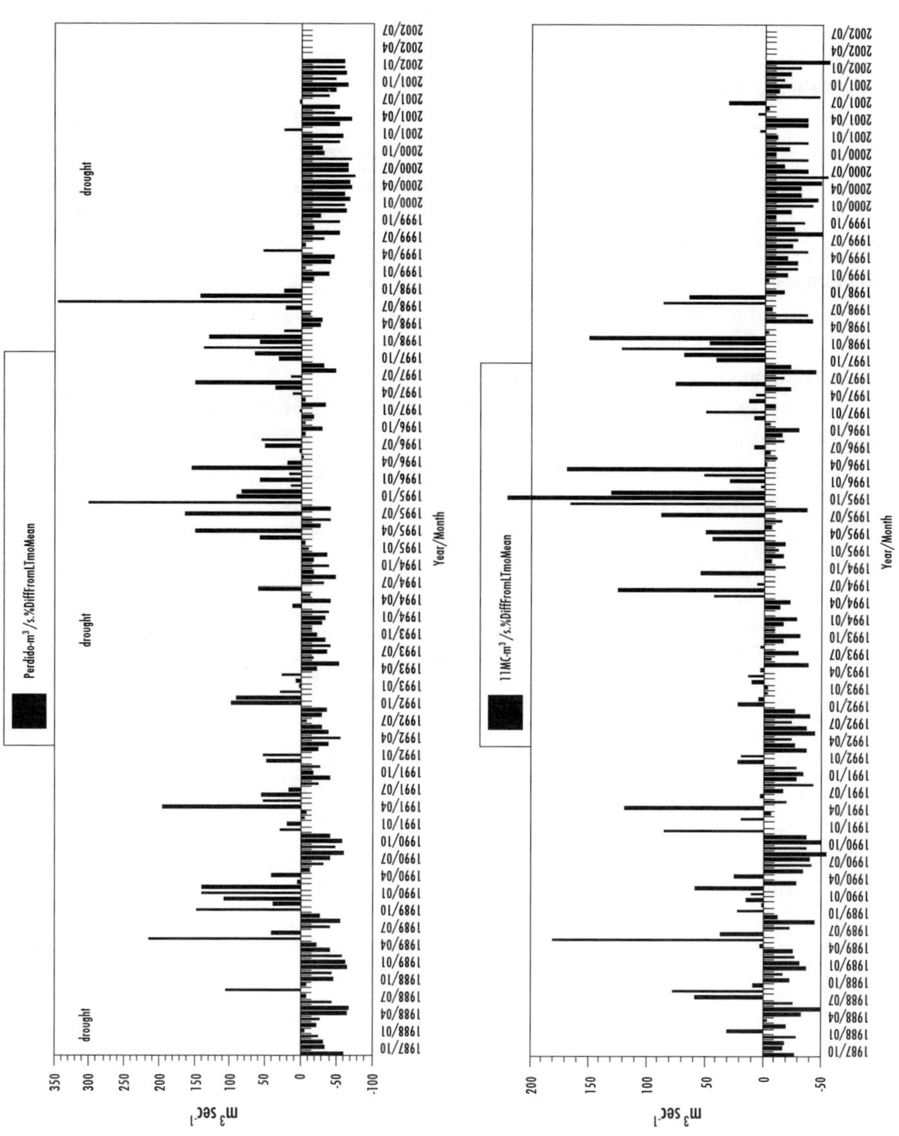

Figure 6.3 Monthly Perdido River and Elevenmile Creek flows (m^3 sec^{-1}) plotted by month over the 14-year sampling period. Data also were calculated as percent differences from the grand mean.

Escambia and Choctawhatchee Rivers. The Perdido River (including the Alabama Blackwater River and the Styx River), the Yellow River, and the Florida Blackwater River (Pensacola Bay system, Figure 2.2B) were somewhat lower in flow rates. In most cases, peak flows were noted during winter–spring periods and seasonal drought periods were usually found during fall. The Fenholloway River, the Econfina River, and Elevenmile Creek were relatively small blackwater steams having considerably lower flow rates than the above-mentioned alluvial rivers.

Orthophosphate loading was highest in the largest alluvial rivers (Apalachicola, Choctawhatchee, and Escambia systems) although such loading was also high in the Fenholloway River and Elevenmile Creek (1994–1995) due to pulp mill activities. Orthophosphate concentrations in the Fenholloway River and Elevenmile Creek were one to two orders of magnitude higher than those found in the other systems. These comparisons indicated that nutrient loading was higher in the alluvial rivers than the nutrient-enriched Fenholloway River and Elevenmile Creek. However, these small streams had significantly higher orthophosphate concentrations than the alluvial rivers. Total phosphorus loadings were comparable among the various rivers despite the flow disparities among these systems. Total phosphorus concentrations were again highest in the Fenholloway River and Elevenmile Creek (Livingston, 2000).

Ammonia loading was comparable among the various systems with the exception of the Econfina, which had relatively low ammonia loading (Livingston, 2000). Ammonia concentrations were an order of magnitude higher in the Fenholloway River and Elevenmile Creek than the other rivers. Thus, as with orthophosphate, ammonia loading in the small, nutrient-enriched streams was comparable to that of the alluvial rivers, whereas ammonia concentrations in the Fenholloway River and Elevenmile Creek were much higher than those in the largely unpolluted alluvial rivers. These differences were due to nutrient loading by kraft pulp mills. Nitrate/nitrite loading appeared to be directly related to river flow characteristics in most of the systems tested. Somewhat higher nitrate/nitrite concentrations occurred in Elevenmile Creek. The lowest nitrate/nitrite measurements were found in the Econfina system.

A review of the annual loading rates of important nutrients into upper Perdido Bay from Elevenmile Creek is shown in Figure 6.4. Compared to total nutrient loading from all sources in the upper bay, these two streams had the highest overall loading values. Of all the nitrogen, phosphorus, and carbon compounds loaded to upper Perdido Bay, only ammonia and orthophosphate loading from Elevenmile Creek (and the pulp mill) were significantly higher than other sources of ammonia and orthophosphate to the bay. Orthophosphate loading from Elevenmile Creek was highest from 1994 to 1996 with apparent transition periods from 1992–1993 and 1997–1998. Overall orthophosphate loading was lowest during the first and third drought periods. Loading of total dissolved phosphate from Elevenmile Creek tended to follow the long-term trend of orthophosphate loading to the bay. Ammonia loading from the creek was particularly high from 1995 to 1998. Ammonia loading was lowest during the three drought periods. The major drought during 1998 to 2002 was associated with significant reductions of total organic nitrogen (TON), total organic phosphorus (TOP), particulate organic carbon (POC), dissolved organic carbon (DOC), total nitrogen (TN), and total phosphorus (TP). Silica loading to the bay tended to follow river flow patterns, although comparability varied somewhat through time. These data indicated that specific patterns of ammonia and orthophosphate loading associated with pulp mill discharges were anomalous relative to other sources of nutrients to upper Perdido Bay. The influence of drought periods on overall nutrient loading (especially the 1998 to 2002 low river flows) was evident.

To understand more fully the pattern of nutrient loading to Perdido Bay over the study period, an analysis was made of the monthly loading of ammonia and orthophosphate from Elevenmile Creek and the Perdido River system. These data were compared to ammonia and orthophosphate loading from Perdido River (Figure 6.5). Mill orthophosphate loading was significantly ($p = 0.0212$, $r^2 = 0.505$) and positively associated with orthophosphate loading from Elevenmile Creek (Livingston, 2000). The mill had temporary increases of orthophosphate loading during 1988 and 1989. Mill orthophosphate loading remained comparable to that of the Perdido River system from the second quarter 1989 through the fourth quarter 1991. There were periodic increases of

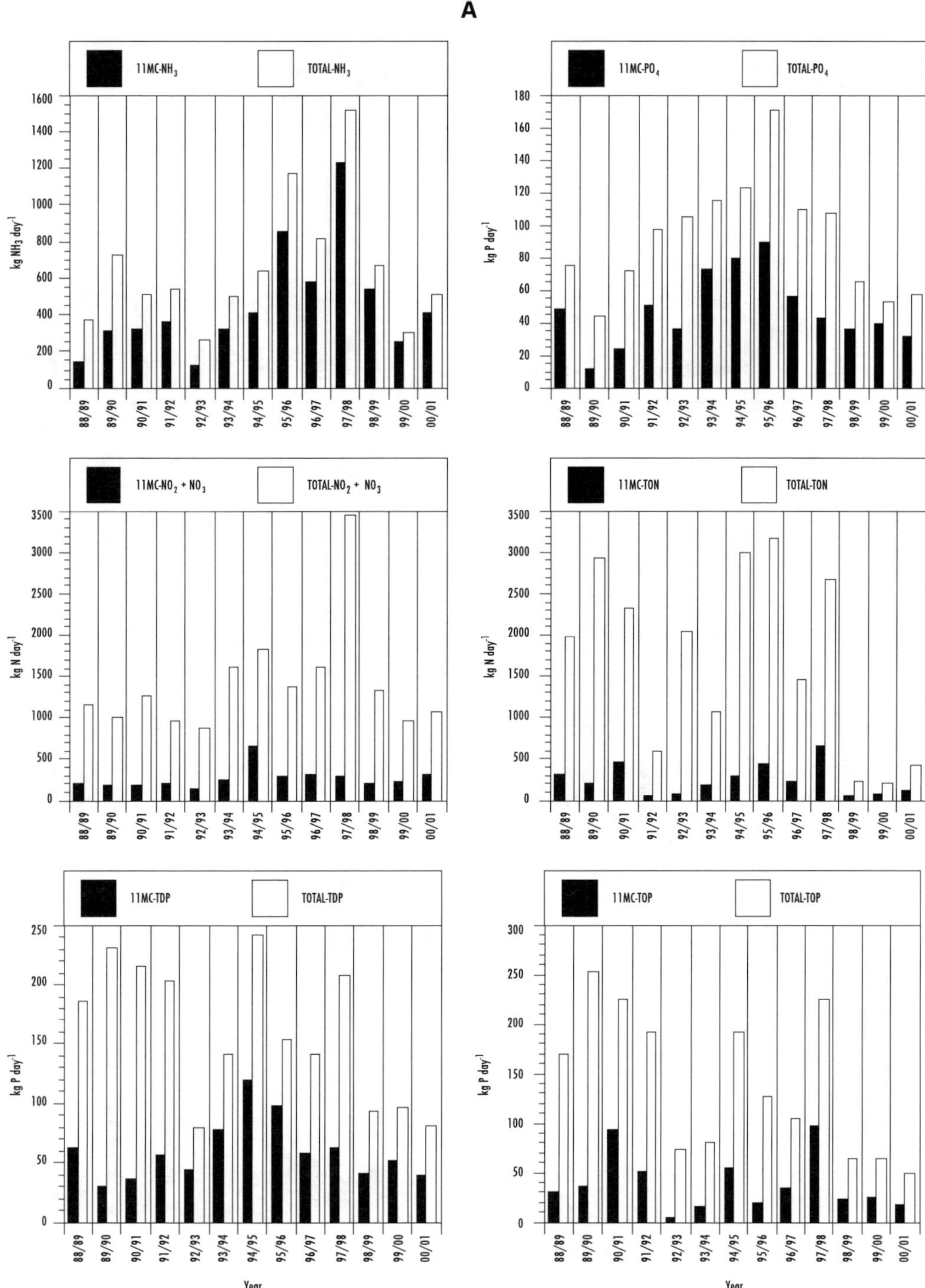

Figure 6.4 Annual loading rates (kg d^{-1}) of Elevenmile Creek vs. total upper bay loading rates (kg d^{-1}) of nitrogen, phosphorus, carbon, and silica in Perdido Bay from 1998 through 2001. (A) Ammonia (NH_3), orthophosphate (PO_4), nitrite + nitrate (NO_2 + NO_3), total organic nitrogen (TON), total dissolved phosphorus (TDP), and total organic phosphorus (TOP). (continued)

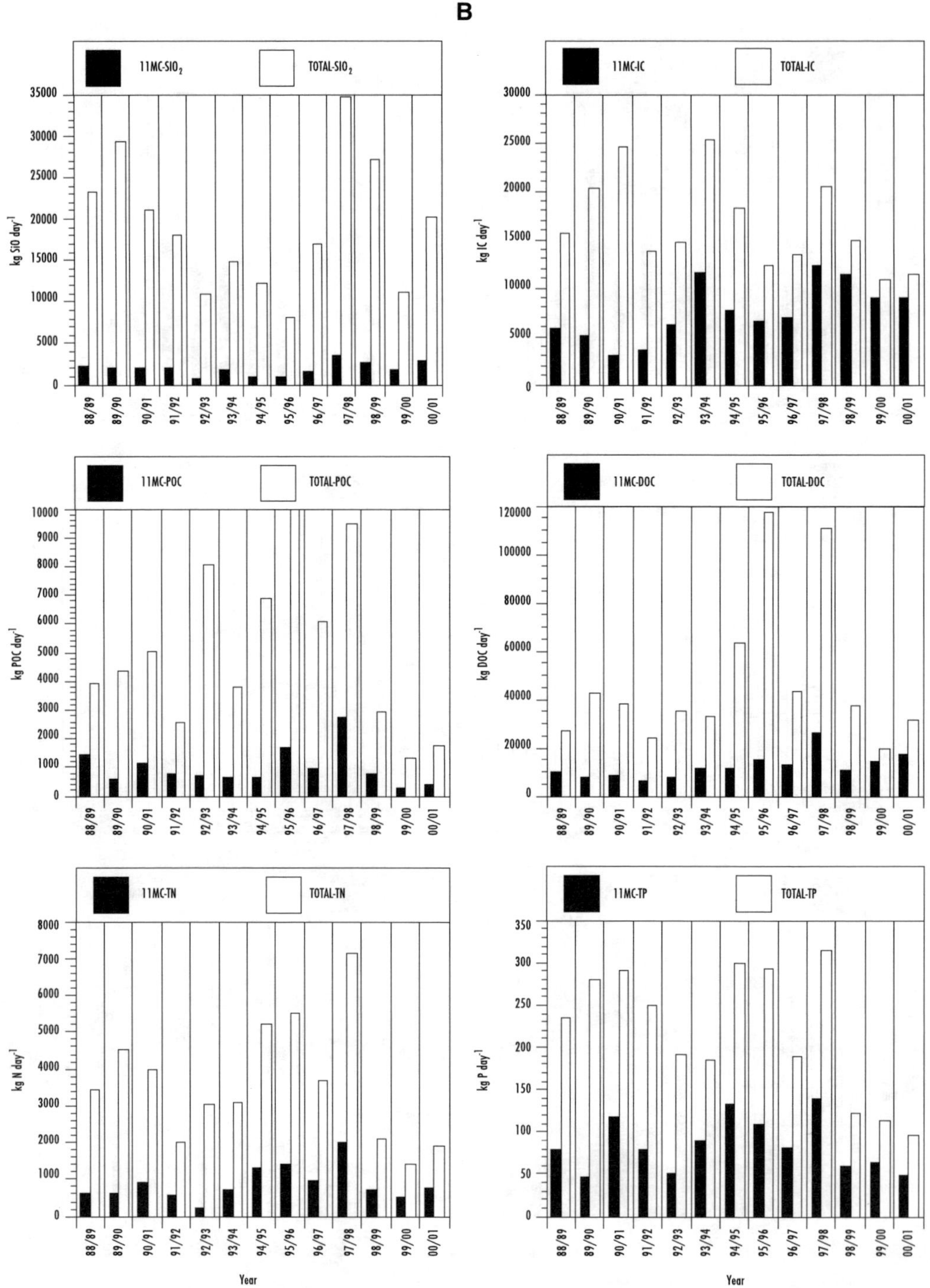

Figure 6.4 (continued) Annual loading rates (kg d^{-1}) of Elevenmile Creek vs. total upper bay loading rates (kg d^{-1}) of nitrogen, phosphorus, carbon, and silica in Perdido Bay from 1998 through 2001. (B) Silica (SiO_2), inorganic carbon (IC), particulate organic carbon (POC), dissolved organic carbon (DOC), total nitrogen (TN), and total phosphorus (TP).

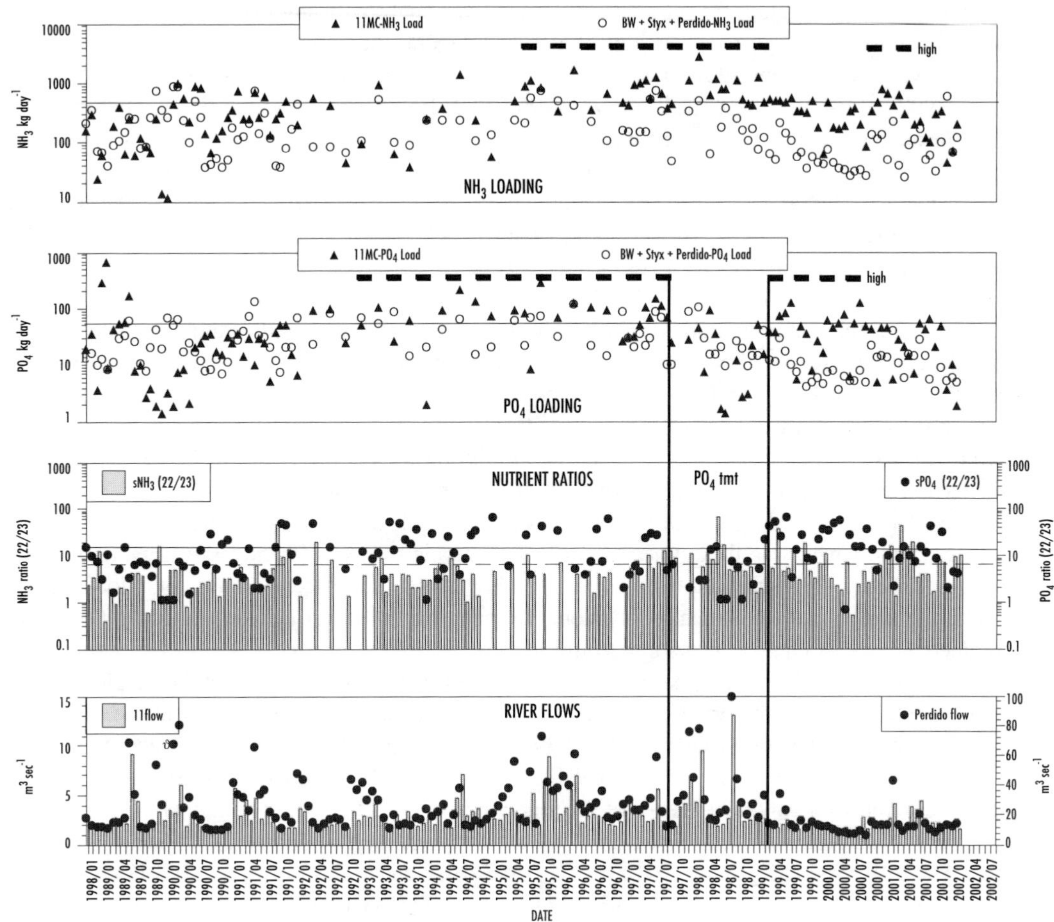

Figure 6.5 Monthly loading of orthophosphate and ammonia from the Perdido River system and Elevenmile Creek from October 1988 through April 2002. Dashed lines represent periods when orthophosphate and ammonia loading from Elevenmile Creek were significantly higher than such loading during 1989–1990. Levels where loading from Elevenmile Creek exceeded those from the Perdido River system (for ammonia, 300 kg d^{-1} and for orthophosphate, 50 kg d^{-1}) are noted by lines. Also shown are monthly Perdido River flows (m^3 s^{-1}) and nutrient ratios (stations P22/P23); dashed line represents ammonia ratio of 8.0 and solid line represents orthophosphate ratio of 14.

mill orthophosphate loading during 1992–1993. From the second quarter 1994 through the third quarter 1997, there were sustained increases of orthophosphate loading from Elevenmile Creek. The high orthophosphate loading from the mill and Elevenmile Creek occurred during the drought of 1993–1994. During late summer 1997, the pulp mill reduced orthophosphate discharges to Elevenmile Creek by treating the effluent with alum: reduced orthophosphate loading from the mill occurred from fall 1997 to winter 1999. There was a brief period of increased orthophosphate loading from the mill from late December 1998 through mid-January 1999. High orthophosphate loading from the mill commenced during early 1999 and remained periodically high through summer 2000. During spring–summer 2001, there was another period in increased orthophosphate loading from the mill. There was a general downward trend of orthophosphate loading from summer 2000 through early summer 2002. The reduction of orthophosphate loading from Elevenmile Creek from fall 1997 to winter 1999 to levels comparable to the early years (1989–1990) allowed an evaluation of the effects of reduced orthophosphate loading on the bay. These reductions occurred during a period of high river flow, whereas the later reductions were undertaken during a severe drought.

Orthophosphate loading from Elevenmile Creek was analyzed using two-sample statistical analyses (Appendix I) to determine the significance of temporal differences of such loading. The comparisons were run with the Elevenmile Creek quarterly loading data for five separate 2-year periods (1989–1990; 1991–1992; 1993–1994; 1995–1996; 1997–1998). There were no significant differences in the means and variances between the following periods: 1995–1996 and 1993–1994; 1989–1990 and 1991–1992; 1997–1998 and 1989–1990; 1997–1998 and 1991–1992; 1997–1998 and 1993–1994; 1997–1998 and 1995–1996. There were significant differences between the means of the following periods (the highest orthophosphate means are given in the first 2-year period of each pair): 1993–1994 and 1989–1990, $p = 0.0191$; 1993–1994 and 1991–1992, $p = 0.0010$; 1995–1996 and 1989–1990, $p = 0.00014$; 1995–1996 and 1991–1992, $p = 0.02$; 1995–1996 and 1997–1998, $p = 0.05$. These analyses, together with regressions showing the relationship between mill orthophosphate loadings and Elevenmile Creek orthophosphate loadings, indicate that there was a significant increase of orthophosphate loading from Elevenmile Creek (and the paper mill) to upper Perdido Bay from late 1993 through the third quarter of 1997. Based on comparisons of orthophosphate loading in the Perdido system and an evaluation of such loading during the period of 1988 to 1990 (no plankton blooms noted; Livingston, 2000), a target for mill orthophosphate loading was set at a quarterly average of 50 kg P d^{-1} during high-flow periods and 38 kg P d^{-1} during drought periods.

Elevenmile Creek was the primary source of ammonia to upper Perdido Bay (Figure 6.5). Mill ammonia loading was significantly ($p = 0.0115$, $r^2 = 0.189$) associated with ammonia loading from Elevenmile Creek, although such association was marginal based on the low r^2 value. There were sustained increases of ammonia loading from Elevenmile Creek during the 1995 to 1998 period. Quarterly ammonia loading data for Elevenmile Creek were run through a series of two-sample statistical analyses similar to that for orthophosphate. There were no significant differences in the means and variances between the following periods: 1989–1990 and 1991–1992; 1989–1990 and 1993–1994; 1993–1994 and 1991–1992; 1995–1996 and 1991–1992; 1995–1996 and 1997–1998. There were significant differences between the means of the following periods (the highest ammonia means are given in the first 2-year period of each pair): 1995–1996 and 1989–1990, $p = 0.01$; 1995–1996 and 1993–1994, $p = 0.05$; 1991–1992 and 1989–1990, $p = 0.05$; 1997–1998 and 1989–1990, $p = 0.05$; 1997–1998 and 1991–1992, $p = 0.02$; 1997–1998 and 1993–1994, $p = 0.02$. This analysis indicated that ammonia loading from Elevenmile Creek to Perdido Bay was significantly higher during the period 1995 to 1998 than that of the preceding periods. From 1999 through the first quarter of 2000, ammonia loading from the mill was relatively low. High ammonia loading from the mill occurred during winter–spring 2001. Ammonia loading from all sources was relatively low from summer 2001 through summer 2002. Both orthophosphate and ammonia loading from the Perdido River system were extremely low during the drought of 1998 to 2002. Loading rates of 300 kg N d^{-1} were established as target levels for mill ammonia loading. Creek ammonia data were distinguished from Perdido River ammonia loading of the system at quarterly average loading rates exceeding 300 kg N d^{-1}. Ammonia loading rates beneath these levels were comparable to loading rates of Elevenmile Creek and the Perdido River system during the period of relatively low ammonia and orthophosphate loading from 1988 to 1990.

C. Nutrient Concentration Gradients

Nutrient loading is an important factor in processes associated with plankton productivity in coastal systems. A postulated reduction of nutrient loading into the Apalachicola system during a drought was associated with reductions of phytoplankton activity and associated secondary production (Livingston et al., 1997). In the Perdido system, increased nutrient loading from a pulp mill combined with low creek flows during a drought led to increased nutrient concentrations that were associated with plankton blooms in areas where high nutrient gradients entered the bay (Livingston, 2000). Thus, the relationships of flow rates, natural drought–flood cycles, nutrient loading, and resultant nutrient concentrations were all related to conditions that eventually led to bloom formation in the Perdido system.

Livingston (2000) showed that nutrient loading and nutrient concentrations are related to freshwater flow rates. Even though nutrient loading may be high due to high river flow, nutrient concentrations can remain low due to dilution effects. A comparison between the relatively unpolluted Apalachicola system (Iverson et al., 1997) and the nutrient-enriched Perdido system (Livingston, 2000) indicated that nutrient loading in the Apalachicola system remained relatively high without apparent hypereutrophication, whereas relatively lower nutrient loading in the Perdido system resulted in massive phytoplankton blooms in the early to mid-1990s. The difference was the significantly higher concentrations of reactive inorganic phosphorus and nitrogen compounds in Elevenmile Creek (at the head of Perdido Bay) as a result of increased loading from a pulp mill. Thus, high nutrient concentrations, as a product of high nutrient loading into a creek with relatively low flows, could be an important factor in the initiation of plankton blooms in Perdido Bay. Nutrient loading per se was not the only factor in initiation of the bloom process. The drought–flood periodicity of river flows was another important factor in the overall pattern of eutrophication. Low flows and constant loading from the pulp mill led to higher nutrient concentrations at the entry point to the bay.

Relatively high annual average concentrations of orthophosphate in Elevenmile Creek (station P22) occurred from 1992 through 1997 (Figure 6.6). The lowest concentrations occurred during the 1997–1998 reduction of orthophosphate loading by the mill. Orthophosphate concentration averages were generally comparable during the first and last 3 years of sampling. Ammonia concentrations at station P22 were generally high from 1995 through 1998. In the bay receiving area (station P23), however, concentrations of both nutrients were relatively low, with the lowest such annual averages noted during the drought of 1998 to 2002. The reduction of orthophosphate and ammonia as water entered the bay from the creek was due to rapid dilution from Perdido River flows and uptake by phytoplankton (Livingston, 2000). By dividing the concentrations of ammonia and orthophosphate at station P22 by those noted at station P23, a nutrient gradient index for each compound was established. Based on a comparison of such ratios during low loading years (Figure 6.5), ammonia ratios >8 and orthophosphate ratios >14 were considered high relative to rates during periods of low nutrient loading. These nutrient ratios were used as a relatively simple indicator of bloom potential due to loading from Elevenmile Creek.

The long-term trends of the orthophosphate and ammonia water quality ratios generally followed the loading trends (Figure 6.5). There were periodic increases of orthophosphate ratios from 1988 through 1991. However, such ratios remained relatively low until late 1991–1992 during which time there were period increases in the nutrient ratios. From 1994 to 1997, the orthophosphate ratios consistently exceeded 14. With the application of alum from late 1997 through early 1991, the ratios remained low. There was then a resumption of high orthophosphate ratios that lasted from late 1999 through 2000. The drought of 1998 to 2002 tended to exacerbate the orthophosphate concentrations relative to loading during this period. With the exception of late 2001, orthophosphate ratios tended to be below 14 from fall 2000 through summer 2002. Ammonia concentrations systematically exceeded the target (8) from 1997 through 1999. Increased ammonia loading from late 2000 through spring 2001 was associated with increased ammonia ratios that again exceeded the target. Subsequent ammonia ratios from winter 2001 through summer 2002 were generally lower. The drought generally led to higher orthophosphate and ammonia concentrations in Elevenmile Creek relative to the loading rates for these nutrients.

D. Nutrient Limitation

Nutrient addition bioassays with phytoplankton are used to determine limiting nutrients (Maestrini et al., 1984a,b). These tests enhance understanding of the kinetics of phytoplankton growth with respect to different combinations of nutrients under specific, experimentally controlled habitat conditions. Nutrient bioassay experiments with estuarine phytoplankton communities have been carried out in various coastal systems: for example, Chesapeake Bay and tributary estuaries (D'Elia et al., 1986; Fisher et al., 1992); Narragansett Bay (Nixon et al., 1984); the Baltic Sea region

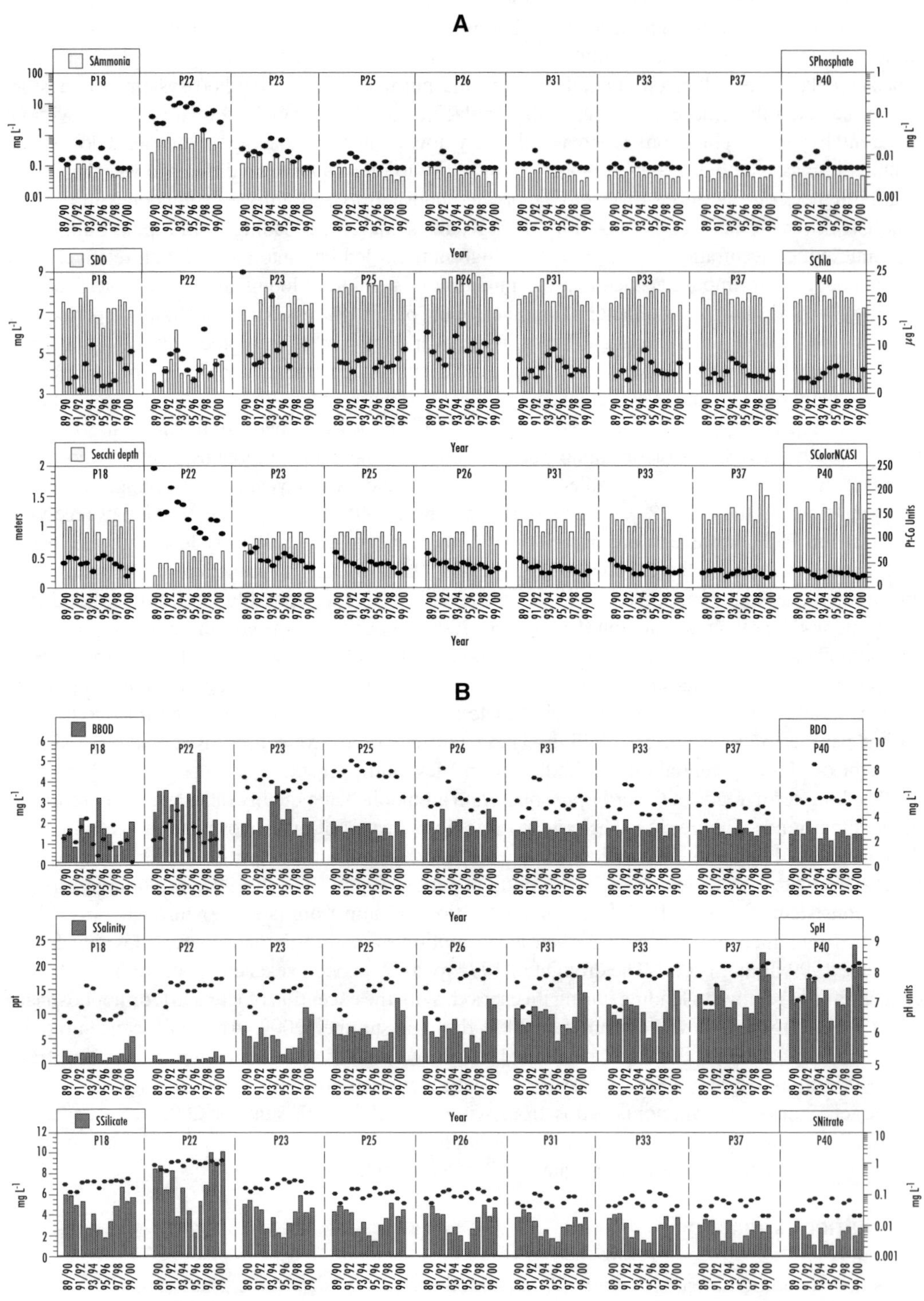

Figure 6.6 Annual averages of water quality factors at stations in the Perdido Bay system from 1988 through 2002. (A) Surface ammonia, surface orthophosphate, surface dissolved oxygen, surface chlorophyll *a*, Secchi depths, and surface color. (B) Bottom Biochemical Oxygen Demand (BOD), bottom dissolved oxygen, surface salinity, surface pH, surface silica, surface nitrate.

(Graneli, 1987; Kivi et al., 1993). Questions of nutrient limitation in the Baltic Sea are still under discussion (Hellstrom, 1996; Hecky, 1998). Phosphorus was limiting in North Sea estuaries (Postma, 1985). Oligotrophic tropical systems may be phosphorus limited (Howarth et al., 1995), but such areas may be nitrogen limited farther from shore (Corredor et al., 1999). Nutrient limitation can also switch seasonally in areas such as Chesapeake Bay (Malone et al., 1996) and parts of the Gulf of Mexico (Rabelais et al., 1999).

The effects of nitrogen and phosphorus additions on phytoplankton productivity in several Gulf coastal areas have been tested by Estabrook (1973), Myers and Iverson (1981), and Montgomery et al. (1991). In these studies, phosphorus limited phytoplankton productivity more than nitrogen. Phosphorus limitation in estuaries along the NE Gulf could be related to relatively low concentrations of phosphorus in the soils of this region. However, the relative importance of other nutrients such as nitrogen cannot be discounted. The application of a mathematical mass balance model of estuarine nutrient transport for the Ochlockonee River–estuary (Kaul and Froelich, 1984) suggested that nitrogen is more likely to be limiting than phosphorus. Similar data (Livingston, unpublished data) in the Choctawhatchee Bay system indicated mixed results that depended on the time of year and area of the bay. In Bayou Texar (Pensacola Bay system), an area noted for fish kills related to nutrient overenrichment and hypoxia, nitrate was reported to control primary productivity to a greater degree than phosphorus (Moshiri et al., 1987). Studies in Weeks Bay, off the lower Mobile Bay system (Dauphin Island Sea Laboratory, unpublished data) indicated that no single nutrient was operable in controlling estuarine primary productivity in isolation from other physical, chemical, and biological variables. Smith and Hitchcock (1994), using bottle tests, determined that surface waters on the Louisiana shelf associated with the Mississippi River plume responded to phosphorus, nitrogen, and silica as a function of season. Phytoplankton biomass in Florida Bay was shown to be deficient in phosphorus relative to nitrogen based on Redfield N:P ratios of particulate organic matter and water column ratios of dissolved inorganic nitrogen and soluble reactive phosphorus (Fourquerean et al., 1993).

Despite evidence concerning phosphorus limitation in Gulf estuaries, Howarth et al. (2000) stated that nitrogen usually controls eutrophication of coastal systems, adding that phosphorus can be important in certain situations and can become the limiting nutrient if nitrogen is managed. The ratios of inorganic nitrogen:phosphorus reflect the relative importance of these nutrients (Graneli et al., 1990; Oviatt et al., 1995), which, according to Howarth (1988) and Howarth et al. (2000), substantiates the conclusion that nitrogen limitation is the rule in most estuarine and coastal systems and that "nitrogen availability is the primary regulator of eutrophication in most coastal systems." This conclusion differs from other views that phosphorus is the primary cause of coastal eutrophication (Tyrrell, 1999). The current emphasis on nitrogen as the primary limiting nutrient to coastal systems (Howarth et al., 2000) was based, in part, on mesocosm experiments in Narragansett Bay (Oviatt et al., 1995) that showed that only nitrogen additions led to phytoplankton increases of abundance and increased primary production. In addition, long-term "natural" experiments indicate the importance of nitrogen as a limiting factor. Long-term changes of Laholm Bay in Sweden (Rosenberg et al., 1990) indicated nitrogen control.

Perdido Bay has been considered to be low to moderately enriched in nutrients compared to other Gulf estuaries (Rabalais, 1992; Macauley et al., 1995; Livingston, 2000), Previous field research indicated phosphorus as a limiting factor for phytoplankton production in upper Perdido Bay. Schropp et al. (1991) found that Elevenmile Creek, the Styx and Blackwater Rivers, and Bayou Marcus Creek were all affected by anthropogenous input of nutrients. Macauley et al. (1995) found that upper Perdido Bay had slightly increased dissolved nutrient concentrations that appeared to be related to significantly higher rates of carbon fixation and phytoplankton biomass and that the Perdido River diluted nutrients and color from Elevenmile Creek with effects restricted to an area at the mouth of Elevenmile Creek.

Flemer et al. (1997) described results of a series of nutrient limitation experiments conducted with water taken from Perdido Bay during 1991. Six experimental treatments were established in triplicate for each of three bay stations (P23, P31, P40). Treatments included three control tanks,

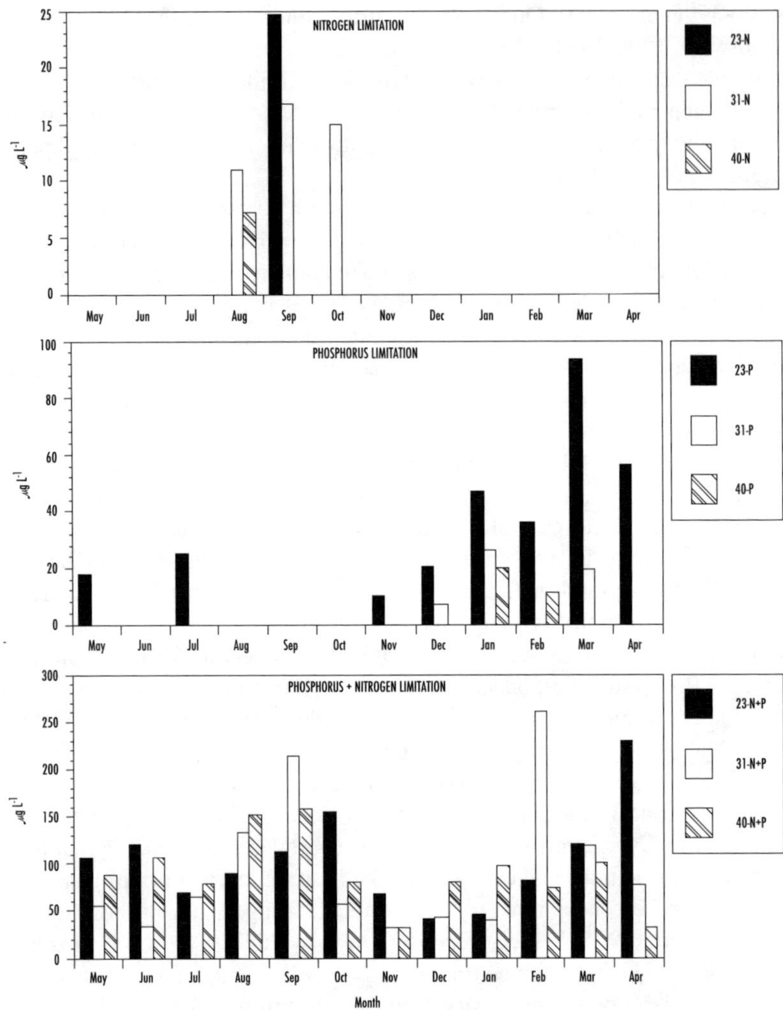

Figure 6.7 Averaged results of nutrient limitation experiments carried out with Perdido Bay water monthly from May 1990 through April 1991 (Flemer et al., 1997). Data are presented as chlorophyll *a* concentrations that were significantly higher than controls under experimental conditions of added orthophosphate, ammonia, and orthophosphate + ammonia.

three P-enriched tanks at 10 μM PO_4-P above ambient, three N-enriched tanks at 50 μM NH_3-N above ambient, and three combined NH_3 + PO_4 (referred to as N+P) above ambient as described for single additions. Flemer et al. (1997) reported that the microcosm-based phytoplankton growth responses in Perdido Bay fell into three principal categories (Figure 6.7): (1) Primary P limitation, occurring mostly during cooler months at upper (tidal brackish) and midbay (lower mesohaline) stations; (2) Primary N limitation, occurring mostly during warmer months (late summer–fall) in midbay areas and infrequently at upper and lower bay stations (upper mesohaline); and (3) Apparent N+P co-limitation, occurring throughout the year with peaks during spring and fall in the upper bay. Winter and summer–fall peaks were noted in midbay areas, with summer peaks in the lower bay. Primary orthophosphate limitation was associated with high dissolved inorganic nitrogen (DIN); DIN:DIP ratios ranged from 20 to 200. Conversely, primary N and N+P colimitation were associated with decreasing DIN:DIP ratios. Phytoplankton assemblages were not strongly nutrient limited, but, given a nutrient increase, these groups responded differentially to nitrogen and phosphorus, both seasonally and along the longitudinal salinity gradient. The combination of phosphorus and nitrogen was usually more stimulatory to phytoplankton growth in Perdido Bay than either of these nutrients alone. Overall, nutrient limitation in Perdido Bay was seasonal, with phosphorus limitation during cold months and nitrogen and/or N+P limitation during warm months.

Dissolved silicate did not limit phytoplankton production in the largely mesotrophic Apalachicola Bay (Fulmer, 1997), as was the case with the Perdido system prior to the plankton blooms (Livingston, 1992, 1997c,d; Flemer et al., 1997; Livingston et al., 1998a). The possible effects of silica limitation after the diatom blooms could have contributed to the long-term bloom succession and the replacement of many of the diatom species with raphidophytes, dinoflagellates, and cryptophytes (Livingston, 2000). In the Apalachicola system, orthophosphate availability limited phytoplankton during both low and high salinity winter periods and during the summer at stations with low salinity. Nitrogen was limiting during summer periods of moderate to high salinity in the Apalachicola estuary (Iverson et al., 1997). Light and temperature limitation was highest during winter–spring periods, thus limiting primary production during this time (Iverson et al., 1997). High chlorophyll *a* levels during winter periods were attributed to low zooplankton grazing during (Iverson et al., 1997). Nitrogen input to primary production was limited by the relatively high flushing rates in the Apalachicola system. Flow rates affected the development of nutrient limitation in the Apalachicola system; nutrient limitation was highest during low-flow summer periods. This also appears to have been the case in Perdido Bay. These conditions prevailed in the Perdido system during periods of moderate nutrient loading (1989 to 1991).

Iverson et al. (1997) noted that there had been no notable increase in chlorophyll *a* concentrations in Apalachicola Bay over the past two decades despite increases in nitrogen loading due to increased basin deposition of this nutrient. However, no phytoplankton data were taken in the Apalachicola system. The potential effects of nutrient loading from a malfunctioning sewage treatment plant were not tested. Perdido Bay, on the other hand, was subject to complex changes of both nutrient loading and nutrient concentrations due to ammonia and orthophosphate discharges from the pulp mill on Elevenmile Creek.

E. Sediment/Water Quality

1. Spatial Distribution

The Perdido drainage system is dominated physically by the Perdido River, an alluvial system of swamps, marshes, and creeks that drains into the main stem of the river. Perdido Bay has a restricted wetland system. The Elevenmile Creek system (including Eightmile Creek) is a small drainage basin that drains into the estuary about 1.5 km east of the Perdido River entry point. The relatively small Bayou Marcus Creek drains a residential area of western Pensacola with input from urban storm water runoff. Wolf Bay, to the south, receives runoff from upland agricultural sources in Alabama. The lower bay receives runoff from massive, unregulated urbanization of Perdido Key and Ono Island.

Sediment composition in Perdido Bay is directly related to depth and salinity stratification. The shallow peripheral shelf is sandy, whereas deeper parts of the bay are characterized by liquid mud or nephlos. In deeper areas (>2 m), there is a distinct halocline below which a salt wedge forms that is composed of clear, warm water during the warmer months of the year. Areas below the halocline are invariably without detectable currents even during periods of high wind (25 to 30 knots). Sediment in deeper parts of the bay has high silt-clay fractions and relatively high organic components with a surface layer of suspended particles of a flocculent nature (i.e., liquid mud). The upper, less salty layer of water is colored and turbid, and is usually characterized by water currents.

At station 18 in the Perdido River (Figure 6.1), sediments are composed of coarse sand overlain by about 30 to 50 cm of flocculent material. The floc is composed of black particles about 2 to 3 mm in diameter. These particles were very buoyant and the least disturbance set them in motion. At station 22 in Elevenmile Creek, there is relatively little liquid mud although fine floc settles in the deeper holes. In the bay at the mouth of the creek (station 23), the bottom is composed of a brown flocculent material in about 1.5 m of water. At station 25, the sediment is sandy with interspersed flocculent materials. This area is characterized by *Vallisneria americana* beds at depths <1 m and is the primary grass bed habitat in the upper bay. At station 26, the substrate is generally sandy and is subject to wind effects. There is a regular distribution of terrigenous material in the form of leaf matter and sticks in the upper bay, and particles are larger with a distinct layer of sand below a thin veneer of silty, flocculent material. The relatively shallow (<2 m) upper bay is characterized by low silt-clay fractions. In midbay (stations 31, 33) and lower bay (stations 37, 40) areas, sediments are very soft and dark with relatively well developed layers of liquid mud. There is virtually no particle movement in the liquid mud layer. Silt-clay fractions thus increase southward, with peaks of fine elements occurring in the deepest parts of the bay around station 40 off Inerarity Point. These areas are a primary depositional area for organic matter. High silt-clay fractions are also evident in Wolf Bay.

Delta sediments in Choctawhatchee Bay are composed of fine-grained brown sand. At stations farther down the bay, the bottom sediments are highly uniform with a fine flocculent layer above the sandy substrate. As in the Perdido system, Choctawhatchee Bay has a shallow, sandy perimeter that bounds deeper areas characterized by liquid mud, gray-black in color. In general appearance and consistency, the liquid mud here is almost identical to that in Perdido Bay. In both systems, a persistent salt wedge, derived from a dredged pass to the Gulf, is usually associated with the presence of liquid mud. The flocculent sediments remain undisturbed by turbulence below the well-formed halocline. Thus, in salinity-stratified areas of both bays, there is a strong development of a layer of fine particles above the sediments that remain suspended due to the lack of currents and vertical turbulence under a human-made halocline. Even during storms with high wind velocities, the liquid mud remains stable. Displacement of the fine-grained liquid mud interface occurs only during periods of high river flow, at which time the halocline is removed and vertical mixing is established. In both bays, sub-haloclinal areas are usually hypoxic during warm months.

The lower end of the Perdido River (station P18) is characterized by a number of deep holes. The same situation exists at the end of Elevenmile Creek. Water quality conditions in these deep holes are characterized by high-salinity stratification and accompanying hypoxia or anoxia below the halocline during most periods of the year. Freshwater floods break the halocline, with a return to high DO until salinity from the lower bay again migrates into the holes, thus reestablishing the previous habitat condition. Deeper stations in lower Perdido River and Elevenmile Creek (stations 9, 18, 19, 22) are highly stratified more than 80% of the time. The shallow upper bay stations (20 to 30) are not highly stratified during most of the year. Areas below the bridge (stations 31 to 40) are strongly stratified more than 70% of the time with stratification strength directly related to depth. Bay depth is associated with mean bottom salinity ($r^2 = 0.307$, $p = 0.01$) and highly stratified areas ($r^2 = 0.704$, $p = 0.0001$). Depth is inversely related to weak stratification ($r^2 = 0.479$, $p = 0.0007$) and vertical homogeneity ($r^2 = 0.406$, $p = 0.0025$). These relationships are stable from year to year regardless of interannual rainfall differences in the basin. Thus, salinity stratification is associated with depth, and is determined by stratification due to saline water entering the bay through the artificial pass. Annual surface salinity variation, as measured by the standard deviation of monthly salinity by year, is a function of position, being least variable in the river–estuaries and uniformly high in the bay. Bottom salinity variation is somewhat different, being least stable in the estuaries and the upper bay and more stable in the deeper lower bay.

Salinity stratification of Perdido Bay, as a function of depth and stability, is directly related to other water quality features. In a stepwise regression, about 44% of the variability of DO distribution in the bay can be explained by depth, salinity, and temperature. Highly stratified areas are closely associated ($r^2 = 0.443$, $p = 0.001$) with severely hypoxic conditions below the halocline during warm months. Hypoxic conditions are also associated ($r^2 = 0.169$, $p = 0.05$) with mean depth. DO is inversely related to salinity ($r^2 = 0.277$, $p = 0.0001$). Surface DO rarely goes below 4 mg l^{-1}. Thus, DO distribution in Perdido Bay is a function of temperature, depth, salinity concentration, and the level of stratification. With an increase in bottom salinity during warm periods, there is an associated reduction of DO, and highly stable stratification conditions usually are directly related to hypoxic conditions.

2. Seasonal/Interannual Trends

Long-term water quality changes in Perdido Bay have been described by Livingston (2000), and can be summarized as annual averages (Figure 6.6). Water temperature peaked during drought periods. Surface salinity in the bay varied inversely with distance from the Perdido River (Figure 6.6), and was highest during droughts. The most recent drought was generally associated with the highest salinities noted during the 14-year observation period. Water color was highest in Elevenmile Creek, peaking during the first year of sampling. The effects of the paper mill's color reduction program in 1990 were evident during subsequent years. With the exception of Elevenmile Creek, which was affected by mill discharges, water color in Perdido Bay tended to be lowest during drought periods. There was a general downward trend of color in the bay over the sampling period. Water color peaks occurred during high river flow years with reductions during droughts. Bay water color was lowest during the most recent drought. Secchi depths increased with distance from the upper bay. The effects of color reductions in Elevenmile Creek during 1990 were evident in the general increases of Secchi depths in this area of the bay through time.

Chlorophyll *a* (Figure 6.6) was inversely associated with distance from the upper bay. Peak chlorophyll *a* concentrations occurred during droughts. Although annual averages of chlorophyll *a* concentrations were generally low throughout the bay, peak values were noted at the mouth of Elevenmile Creek during 1989, 1994, and 1999 and 2001 (i.e., drought years). Overall, the most important factor concerning interannual changes of water temperature, salinity, water color, and chlorophyll *a* in Perdido Bay were the interannual drought–flood cycles. Mean surface silica concentrations (Figure 6.6) were highly variable in the upper bay, and were inversely related to distance from the Perdido River mouth. Elevenmile creek had relatively high silica concentrations during most of the study period. There were precipitous reductions of silica from 1991–1992 through 1996. Surface pH was highest during the last two drought periods. There were sustained increases of pH throughout the bay during periods of increasing numbers and biomass of the phytoplankton (Livingston, 2000). High pH appeared to be a relatively good indicator of the blooms.

Surface DO values (Figure 6.6) were usually higher than bottom values. The smallest differences of surface and bottom oxygen anomalies occurred in the upper bay sea grass beds (station P25); the greatest such differences were found at the mouth of the Perdido River and Elevenmile Creek (Livingston, 2000). Low DO in the deep hole at the mouth of the Perdido River (P18) was probably due to the effects of salinity stratification. The lowest average surface DO concentrations occurred in Elevenmile Creek (station P22) due to paper mill discharges of dissolved organic carbon and ammonia (Livingston, 2000). The highest surface and bottom DO concentrations were noted at stations P23, P26, P31, and P40 during 1992–1993. Bottom oxygen tended to be somewhat higher in the lower bay (station P40) than other deep-water areas, possibly reflecting the influx of highly oxygenated bottom water moving into the bay from the Gulf. The DO concentrations in any given part of the bay thus reflected local habitat conditions of depth, salinity stratification, pulp mill effects, the presence of sea grass beds, and the northward movement of bottom water from the Gulf. The complex, long-term changes of DO in the bay were thus determined to some degree by drought–flood conditions, vertical distribution relative to the halocline, and possibly long-term phytoplankton successions (see below).

Ammonia concentrations (Figure 6.6) were highest in Elevenmile Creek (station P22). General increases of ammonia concentrations in the creek from 1995 through 1998 coincided with loading trends. At various bay stations, average annual ammonia concentrations tended to be lowest during the last 3 years of sampling, although the trend to lower ammonia actually started in 1995. Phytoplankton blooms following increased orthophosphate loading could have accounted for reduced ammonia in the bay during periods of increased ammonia loading by the mill (Livingston, 2000). Orthophosphate concentrations in Elevenmile Creek were generally higher than anywhere in the bay with the exception of the 1997–1998 period when the mill reduced its orthophosphate loading to the bay. Creek concentrations reached maximum levels from 1992 to 1997 with a decrease in 1998 and subsequent increases in 1999 to 2001. This trend followed the long-term changes of orthophosphate loading rates by the paper mill. Peak orthophosphate concentrations in the bay occurred during 1992 with reduced concentrations thereafter, a different pattern from the creek. The lowest such concentrations were usually during the most recent drought. The overall pattern of orthophosphate in the bay was somewhat different from that in Elevenmile Creek.

The general trends of ammonia and orthophosphate in the bay indicated that, during phytoplankton blooms, there were general reductions of these nutrients. Reduced orthophosphate concentrations occurred in Perdido Bay during periods of high loading. During the 1998 period of low orthophosphate loading by the mill, concentrations of this nutrient increased at most stations, which coincided with reduced bloom activity (Livingston, 2000) Ammonia concentrations were generally reduced (except stations P22 and P23) during such periods. This is further evidence that the blooms could have affected bay nutrient concentrations in complex ways and that nutrient concentrations have been affected by the phytoplankton in varying degrees over the 14-year period of study.

Long-term water quality changes in Perdido Bay reflected drought–flood sequences and the occurrence of species-specific successions of phytoplankton blooms. In most instances, water quality trends were relatively complex so that simple and direct associations of individual factors with phytoplankton blooms and long-term, species-specific successions were not usually possible. There were both short- and long-term responses of factors such as DO and pH that indicated direct impacts due to increased phytoplankton activity associated with the blooms. According to noted habitat changes, it is likely that some blooms started as early as 1992, a period when no phytoplankton data were taken (Livingston, 2000). The droughts (especially during 1998 to 2002) had major effects on various water quality factors.

Interannual changes of water quality features were compared to nutrient loading trends in Elevenmile Creek (see Figure 6.5). Ammonia concentrations were highest at station P23 during periods of low ammonia loading. There was a general trend of lower ammonia concentrations during bloom periods; however, the lowest ammonia concentrations coincided with low flows and low loading rates during the latest drought. Orthophosphate concentrations were highest during the 1993–1994 period and were lowest following the 1997–1998 reduction of loading from the mill. Subsequent concentrations were very low during the 1999 to 2001 drought. The association of drought and low color was evident as was the increased chlorophyll *a* concentrations during these periods. Such changes were not associated with trends of nutrient loading, nutrient concentration, or phytoplankton blooms. There was no observed relationship of chlorophyll and silica concentrations with bloom occurrences. Surface and bottom DO concentrations and surface pH levels were generally higher during drought years.

F. Phytoplankton Changes in Perdido Bay

1. Introduction

a. Literature Review

Phytoplankton and benthic microalgae comprise the primary sources of organic carbon in river-dominated estuaries. Due to the high diversity, taxonomic complexity, and rapid reproduction rates of coastal plankton, community interactions of these plants are not well understood even though such organisms form the basis of many coastal food webs. Food web structure as a function of extremely complex trophodynamic processes in coastal areas is molded by the eutrophication process (Livingston, 2000). This includes long-term changes of primary producers that eventually affect the highly evolved feeding strategies of the consumers (Livingston et al., 1997). Due to myriad human activities, many temperate, river-dominated estuarine and coastal systems are quietly undergoing stress-induced changes in essential food web processes that are not detected in most coastal studies (Livingston, 2000). To correct this problem, it was necessary to understand the dynamics of phytoplankton community organization.

One of the chief complications in understanding eutrophication in coastal areas is that the processes that determine the qualitative and quantitative changes in primary productivity and the related responses of associated food webs vary considerably from one system to the next. This variation depends on factors such as the physiography, temperature, salinity, wind, tides, currents, stratification characteristics, light transmission, nutrient loading events, nutrient concentrations, sediment characteristics, primary and secondary production, and biological processes such as competition and predation. The manifestation of different combinations of these variables leads to highly variable food web responses among even closely related systems; such differences are then transformed into major differences of population dynamics, community structure, and overall secondary production through altered food web dynamics.

The diverse and poorly understood juxtapositions of the eutrophication process in coastal systems reflect varying combinations of nutrient loading and responses to seasonal and interannual alterations of nutrient loading from multiple human activities. Differentiation of the differences between natural forms of eutrophication and the hypereutrophic state is still a matter of some conjecture. The primary symptoms on which many surveys are based included decreased light availability, high chlorophyll *a* concentrations, high epiphytic/macroalgal growth rates, changes in algal dominance (diatoms to flagellates, benthic to pelagic dominance), increased decomposition of organic matter due to the high chlorophyll *a* concentrations and macroalgal growth, and associated reductions of benthic DO. However, long-term, quantitative analyses of eutrophication in a given coastal system can be confounded by how indicator factors such as nutrient concentrations and DO levels interact with phytoplankton assemblages and predator groups. This has led to various statistical approaches to the problem of distinguishing between natural and anthropogenous causes of eutrophication (Stefanou et al., 2000).

Kennish (1997) reviewed problems associated with anthropogenous loading of nutrients to coastal systems. Resulting hypereutrophication has long been correlated with increased biochemical oxygen demand (BOD), hypoxia, and anoxia in receiving estuaries. Chronic hypoxia has been a problem in coastal areas such as Chesapeake Bay, the Florida Keys, the Pamlico River estuary, Long Island Sound, and broad areas of the northern Gulf of Mexico associated with runoff from the Mississippi and Atchafalaya Rivers. Hypoxia is also considered a problem in estuaries and bays throughout Europe, the Far East, and Australia (Kennish, 1997). Based on the results of many studies, it is now generally recognized that reduced DO due to organic decomposition of increased algal biomass following increased nutrient loading is one symptom of cultural eutrophication.

Hypoxia is known to alter predator–prey relationships. Models of mortality of larval prey due to low DO have indicated significant effects on predator–prey relationships in Chesapeake Bay. Keister et al. (2000) showed how near-bottom hypoxia in the Patuxent River (Chesapeake Bay) affected depth distribution of various organisms with effects on predator–prey relationships and recruitment rates of vulnerable species. Other effects of hypoxia on fish larvae include decreased growth rates and limitation of habitat availability. The causes of reduced DO are many, and decisive cause-and-effect relationships are difficult to ascertain. Long-term reductions of bottom DO in western Long Island Sound were attributed to changes in vertical temperature stratification rather than changes in point and non-point nutrient loading (O'Shea and Brosnan, 2000). The lower DO concentrations occurred after decades of reduced BOD loading to the system from sewage treatment plants. The DO trends were not associated with changes of Secchi transparency and chlorophyll *a*; the exact causes of the DO trends remained unclear. Other studies (Livingston, 2001) indicate that bloom successions have varied effects on the sub-haloclinal DO regime on both seasonal and interannual scales, and adverse biological effects due to reduced sediment quality associated with salinity stratification can also be confused with the impacts of hypoxia on benthic organisms (Livingston, 2000).

There is considerable scientific literature concerning the relationships of limiting factors to phytoplankton populations and the eutrophication process in river–estuarine and associated coastal systems. Thompson (1998) noted complex mixtures of potentially limiting factors in space and time with annual successions of diatoms, cholorophytes, dinoflagellates, and cryptophytes in the Swan River estuary in Australia. The timing and magnitude of the physical forcing factors (mainly rainfall) were important determinants of susceptibility to summer and autumn algal blooms. It was hypothesized that midsummer blooms were supported by nitrogen that was recycled within the system. These blooms could have been facilitated by salinity stratification and deoxygenation. Smith and Hitchcock (1994) found that phosphate and silicate potentially limit phytoplankton growth in surface waters of the Louisiana Bight (especially during periods of low salinity), whereas nitrogen limitation was evident during higher salinity periods in late summer. Data from sediment analyses (Eadie et al., 1994) indicated that anthropogenous nutrient loading was responsible for water quality changes (including hypoxia; Rabelais et al., 1996, 1999).

Nutrient limitation of primary production is still considered to be a complex process. Increased nutrient loading (ammonia and phosphate) to an estuarine system in Ythan, Scotland (Balls et al., 1995) was not associated with increased chlorophyll *a*. The authors attributed this finding to the relatively short flushing time of the estuary. Pennock and Sharp (1994), working in the Delaware Estuary, indicated that there was a complex progression of seasonal limiting factors that included light (winter), phosphorus (late spring), and nitrogen (summer). However, spatial/temporal variation of phytoplankton production was also related to suspended sediments, the depth of the surface mixed layer, and nutrient inputs due to riverine inputs and *in situ* biogeochemical processes. Seasonal phosphorus limitation has been observed in areas such as Apalachicola Bay and the Econfina estuary (slightly elevated N:P ratios, unenriched conditions; Myers and Iverson, 1977, 1981). Other phosphorus-limited periods have been noted in the Delaware estuary (D'Elia et al., 1986) and Chesapeake Bay (Fisher et al., 1988, 1992): these occur during winter–spring periods when temperatures are low. The regulation of nutrient availability by physical flushing rates, mixing, geochemical equilibria reactions, and biological processes is thought to control the temporal successions of nutrient limitation of phytoplankton production (Pennock and Sharp, 1994). Guildford and Hecky (2000) contend that N or P limitation of phytoplankton is a product of the TN and TP concentration and the TN:TP ratio regardless of whether it is a freshwater or marine system. Flemer et al. (1997) found the N+P was the most important limiting combination in the Perdido Bay system.

The primary effects of nutrient loading on shallow estuaries and lagoons are generally well known (McComb, 1995). The emphasis on shallow systems highlights the importance of sediments in the nutrient loading and accumulation process (McComb and Lukatelich, 1995). Sediment resuspension as a product of water column dynamics can be a factor in nutrient storage and loading processes (de Jonge, 1995; de Jonge and van Raaphorst, 1995) although such effects are not always associated with hypereutrophication (Marcomini et al., 1995). The timed release and bioavailability of sediment nutrients, as modified by resuspension, remobilization, and regeneration through biotic activity, remain poorly understood even though such processes could have an important effect on the response of benthic microalgae and macrophytes (Thornton et al., 1995a). The lack of information regarding non-point-source nutrient loading due to urban and agricultural runoff is another consideration when reviewing the causes of hypereutrophication (King and Hodgson, 1995). de Jonge and van Raaphorst (1995) reviewed the positive and negative effects of increased nutrient loading. There can be initial increases of secondary production and fisheries output with increased nutrient loading (Thornton et al., 1995b). However, prolonged increases often end in plankton blooms and accompanying declines of coastal populations that are often not discovered until there is an advanced state of hypereutrophication.

Studies in the Barnegat Bay–Little Egg Harbor system in New Jersey (Kennish, 2001) are representative of what is happening in coastal areas throughout the United States. Uncontrolled urban development, extensive construction activities (dredging, infilling, bulkheading, lagoon construction), industrial/military activities, agricultural waste disposal, recreational factors such as boating and associated marina construction, and domestic water uses have combined to cause many changes in nutrient loading patterns and the input of various forms of toxic compounds into coastal areas. The combination of N+P loading has increased phytoplankton production. Limiting factors such as nutrients and light also include temperature, activities of benthic invertebrates, sediment resuspension, and tidal inputs. Olsen and Mahoney (2001) describe the association between increased nutrient loading and a continuous series of blooms in the Barnegat system.

Brown tides (*Nannochloris atomus* and *Stichococcus* sp.) constitute a threat to coastal fisheries in the northeastern NE United States. The impacts of such blooms in the Barnegat Bay–Little Egg Harbor system have not been defined. Similar blooms in Long Island led to losses of eelgrass beds (*Zostera marina*) and mass mortalities of bay scallops (*Argpecten irradians*) (Bricelj and Kuenstner, 1989). Brown tides also caused mass mortality of blue mussels (*Mytilus edulis*) in Narragansett Bay (Tracey, 1985). Long-term declines of hard clam resources and submerged aquatic vegetation in the Barnegat Bay–Little Egg Harbor system may reflect the impacts of such blooms. In addition, other HAB (harmful algal bloom) species such as dinoflagellates (*Prorocentrum lima, P. micans, P. minimum, P. triestinum*) have been noted in the system. Blooms of *Dinophysis acumkinata* and *D. acuta* have been noted along with the presence of other toxic species such as *Heterosigma* sp., *Scripsiella trochoidea* (= *Peridinium trochoideum*), and *Protoperidinium brevipes*. Although these bloom species may have displaced the natural phytoplankton communities, the lack of consistent funding has precluded exact estimates of the impacts of such blooms on the phytoplankton communities of the system. In addition, "many instances of adverse effects of algae, or algal blooms, go unnoticed, unreported, or uninvestigated" (Olsen and Mahoney, 2001). The persistent brown-water conditions that pose a threat to the Barnegat Bay–Little Egg Harbor system have not been addressed by funding agencies in the region. This situation is common in most of the coastal areas of the United States.

The complex processes associated with the interaction and competition of microphytes and macrophytes often preclude a uniform response of different estuaries to increased anthropogenous nutrient loading. A common aspect of cultural eutrophication in diverse estuaries includes association of nutrient loading with the development of phytoplankton blooms, associated fish kills, and the general decline of fisheries (Hodgkiss and Yim, 1995). However, in most of these studies, there is a general lack of detailed data concerning how phytoplankton blooms are initiated and how changes in the phytoplankton community structure actually affect secondary production. Despite a plethora of studies concerning hypereutrophication in estuarine and coastal systems, the processes involved in nutrient loading that ultimately lead to altered phytoplankton populations and associated food web changes remain largely undefined. The universal assumption that the 2-year funding cycle is adequate to the task continues to confound reality.

The cumulative response of coastal systems to cultural eutrophication has been reviewed and documented by Bricker et al. (1999). The various postulated water quality changes due to excessive nutrient loading include increased turbidity, reduced light penetration, and deterioration of sediment quality. Toxic agents associated with plankton blooms have been responsible for fish kills and the loss of commercially important species in addition to creating public health problems following consumption of affected sea life. Shellfish fisheries have been impaired in 46 estuaries with various forms of human influence noted (Bricker et al., 1999). The authors also asserted there was little in the way of "process-oriented data relative to the mechanisms involved in the progressive development of eutrophication." Livingston (2000) noted that thresholds of nutrient loading leading to toxic blooms remain largely undetermined, and the relationships of nitrogen and phosphorus to such blooms have not been defined. The influence of long-term climate changes on blooms remains unknown and the combination of nutrient inputs and other anthropogenous stressors remains largely undescribed. There has been virtually no information generated concerning the qualitative and quantitative changes of the phytoplankton communities relative to the generation of the blooms (Livingston, 2000). The causative factors involved in the bloom, the effects of blooms on plankton community organization, and the impact of altered plankton assemblages on secondary production through effects on coastal food webs remain largely undefined (Bricker et al., 1999).

A comprehensive understanding of the effects of cultural eutrophication on diverse coastal systems is complicated by the extreme complexity of the response of dynamic phytoplankton assemblages to background variation. Water column phytoplankton and benthic microalgae represent one of the most complex and least well understood parts of any given coastal system. Taxonomic problems with such organisms preclude detailed and quantitative determinations of ephemeral and rapidly cycling bursts of productivity, with periods of such changes measured from seconds to decades. Without such information, it is not possible to ascertain the assimilative capacity of a given coastal system for nutrient loading. Human population increases in coastal areas throughout the world have become an increasingly important source of nutrient loading to estuarine systems. The recent increase of destructive plankton blooms due to anthropogenous nutrient loading has led to renewed interest in phytoplankton systematics and natural history. However, funding for training in phytoplankton taxonomy continues to wane, and universities continue to deemphasize taxonomy as an important discipline. Consequently, information at the ecosystem level relating to the role of microalgal community structure on the eutrophication process remains poorly developed even though such processes are central to a basic understanding of the origin and succession of plankton blooms and associated effects of such blooms on the trophic organization of a given coastal system.

b. Plankton Response to Nutrient Loading

Nutrient–phytoplankton relationships in coastal areas have been well studied (Jordan et al., 1991). The generally accepted model defines freshwater areas as phosphorous limited and marine areas as nitrogen limited. This model is something of an oversimplification, however, in that various combinations of N + P may be limiting to various parts of a given coastal system during different times of the year (Flemer et al., 1997; Howarth et al., 2000). There are many processes, both allochthonous and autochthonous, that determine nutrient concentrations in estuaries and coastal systems. Competition of phytoplankton species for nutrients is complicated. For example, Jordan et al. (1991) found that regular blooms are triggered by intermittent nutrient loading from riverine inputs. Sakshaug and Olsen (1986) noted that continuous nutrient loading favored nonbloom species, whereas pulsed nutrient loading favored bloom species. Algal blooms in the York River estuary occurred in the spring with lesser blooms in the summer. Complex interactions of river discharges, as the controlling factor for flushing times, nutrient loading, light availability, tidal mixing, and residence time, explained the pattern of bloom formation. Nutrient remineralization in benthic areas of the mesohaline zone influenced phytoplankton biomass accumulation in the estuary. The authors noted that the relative importance of bottom-up control (resource limitation) relative to top-down control by zooplankton grazing. Other studies (Roegner and Shanks, 2001) noted the importance of the influence of offshore areas in nutrient/phytoplankton dynamics, thus adding upwelling and downwelling phenomena to the generally accepted model of outwelling and inwelling of materials in estuaries.

Phytoplankton blooms often have feedback control of water and sediment quality (Livingston, 2000). Light limitation in turbid areas of mixing generally gives way to phytoplankton accumulation in clearer areas downstream (Fisher et al., 1988); such areas have varying forms of nutrient limitation. Clearer areas are usually phosphorus limited. Chlorophyll maxima and consumption of nutrients by phytoplankton concentrations are influenced by water column stratification. Accompanying such distributions are increases in sub-haloclinal hypoxia. Nutrient-rich estuaries such as San Francisco Bay have indicated the importance of stratification in bloom generation and associated water quality changes (Cloern, 1979, 1996; Cloern et al., 1983). Graco et al. (2001) found that blooms were associated with a dark flocculent layer in sediments of Concepcion Bay, the so-called nepheloid benthic or bottom layer of unconsolidated clayey/organic matter that was identified in highly stratified areas of Perdido Bay and Choctawhatchee Bay in the NE Gulf of Mexico (Livingston, 2000). During summer, this layer was associated with algal mats (Graco et al., 2001) and hypoxic conditions, at which time there was inhibition of dentrification processes. Such accumulations during summer periods affected long-term eutrophication processes in Concepcion Bay.

Howarth et al. (2000) conducted a review of the effects of excess nutrients on coastal systems. This worldwide phenomenon has led to increased algal biomass, excessive concentrations of sometimes toxic algae in the form of harmful brown and red tides, reduced sea grass and coral reef habitat, altered marine biodiversity, hypoxia/anoxia, and the loss of commercial fisheries. The authors claim that 44 of the existing 139 coastal systems in the United States have some of the symptoms of excessive nutrification. The report was designed to make recommendations for the implementation of management efforts to reduce nutrient loading. These recommendations included expansion of monitoring programs, development of ways to reduce non-point sources of nitrogen and phosphorus, an increased federal role in eutrophication issues, development of a susceptibility classification scheme for management of nutrient overenrichment, improvement of comprehensive assessments of environmental quality and associated modeling efforts, and expansion of our knowledge concerning eutrophication questions.

The composition and numbers of estuarine and coastal phytoplankton depend on a delicate balance of various factors, and these aspects of phytoplankton ecology need to be viewed within the twin aspects of spatial heterogeneity and seasonal/interannual variation. Of the approximate total number of marine phytoplankton species (5000), some 300 species are known to occur at numbers high enough to discolor seawater (Sournia et al., 1991). About 40 or 50 of these species produce toxins that can affect both natural marine populations of plants and animals as well as humans (Hallegraeff et al., 1995). These HABs have become the focus of increased research over the past decade. Representative groups that have documented histories as HABs include blue-green algae, diatoms, pyrmnesiophytes, dinoflagellates, and raphidophytes. Unfortunately, there has been little unequivocal coupling established between microalgal concentrations and their harmful effects (Hallegraeff et al., 1995). Most studies of individual HABs have been undertaken after blooms are established, thus precluding the determination of cause-and-effect relationships. Smayda and Shimizu (1993) have summarized what is known in this area. However, background information concerning bloom causation and the ecological effects of HABs remains largely dependent on serial autecological studies, thus precluding realistic appraisals of how HABs are created and how such populations affect phytoplankton assemblages as a whole.

The role of nutrient loading in destabilization of plankton populations includes a broad range of adaptations of bloom species that enhance their ability to outcompete other plankton species. Such changes can be caused by altered nutrient interactions. Riegman (1998) related the occurrence of HABs to macronutrient dynamics. He noted that diatom growth during blooms associated with anthropogenous N+P loading can be inhibited by silica availability. This could be related to enhanced N:Si and P:Si supply ratios that favor nonsiliceous phytoplankton (Smayda, 1990). Shifts in N:P supply ratios can also cause changes in species composition. Nutrient enrichment with macronutrients can also be indirectly related to changes in phytoplankton community composition through induced changes in predation, resource limitation, light requirements, and biological effects on sediments. Altered nutrient speciation can also influence these changes. The ecological role of bloom-associated toxins in such interspecific interactions remains unclear (Riegman, 1998). The fact that various dinoflagellates are not usually eaten by zooplankton (compared to non-dinoflagellates) could be related to the generation of toxins. Bloom-associated toxins could also be related to protection of populations from losses due to predation. Such protection could then result in competitive advantages of bloom species leading to elimination of other species through competition for nutrients, light, or other potentially limiting factors. The production of resting spores by various plankton forms is another possible explanation for the predominance of bloom species during periods of destabilized nutrient conditions. Some dinoflagellate cysts have spines and others are toxic; the resting spores of dinoflagellates may have more toxin than the reproducing vegetative cells (Lirdwitayaprasit et al., 1990; Oshima et al., 1992; Riegman, 1998). Thus, the often-complex life history of bloom species could be associated with the mechanism of relative dominance and elimination of competing plankton species.

Despite many descriptive efforts to define plankton response to nutrient loading, the exact processes that define how bloom dominance affects phytoplankton community structure remain largely unknown. There are some data that indicate such processes work over long periods. Philippart et al. (2000) monitored a single station over a prolonged period and made identifications of various taxonomic units within the phytoplankton community of a tidal inlet in the Wadden Sea over a 21-year period. They found changes in phytoplankton biomass and species composition that coincided with shifts of a phosphorus-controlled system to a more eutrophic nitrogen-controlled system. The authors found a strong causal relationship between TN:TP ratios and phytoplankton community structure with increases of the size of the plankton.

There are many problems associated with the ingestion of seafood contaminated by toxic algal blooms. Organisms affected include fishes, seabirds, porpoises, whales, and humans (National Research Council, 1999). Documented increases of red tide incidences have been reported from Tolo Harbor, Hong Kong (Lam and Ho, 1989) and in the Inland Sea in Japan (Okaichi, 1997). Qualifications of these findings are given by Howarth et al. (2000). Increased incidence of *Chattonella* blooms has been associated with fish kills in the Seto Inland Sea (Okaichi, 1997) and the recently described dinoflagellate *Pfiesteria* also has been connected to such kills (Burkholder and Glasgow, 1997). Nutrient enrichment also adversely affects sea grass beds through changes associated with stimulation of micro- and macroalgae (Duarte, 1995; Hein et al., 1995; Valiela et al., 1997). Shumway (1990) reviewed the effects of algal blooms on shellfish and aquaculture. Depending on the species, plankton blooms can be toxic to filter-feeding organisms, can render such organisms toxic to other species through accumulation processes, can cause problems by simply clogging the gills, or can have indirect adverse effects by reducing DO in the water. Mass mortalities of shellfish are not uncommon during algal blooms. Livingston (2000) documented the direct killing of bivalves (*Rangia cuneata*) during and after plankton blooms. Shumway (1990) gives an exhaustive list of the various effects of plankton blooms on shellfish including various instances of human illness and death due to the consumption of toxic seafood. Shumway et al. (1990) also reviewed the impacts of toxic algal blooms on oysters.

Sellner et al. (1995) showed experimentally that juvenile oysters (*Crassostrea virginica*) effectively removed what was defined as "nontoxic" dinoflagellates (*Prorocentrum minimum*, *Gyrodinium uncatenum*) from suspension. The authors suggested that juvenile oysters in the field should effectively remove blooms with transference of the bloom species to sediments through deposition of feces and pseudofeces, thus shifting oxygen demand from the pelagic to the benthic environment. The dinoflagellate *P. minimum*, however, to which the authors refer as a "non-toxic" dinoflagellate, is actually quite toxic to marine organisms (Livingston, 2000). *Prorocentrum minimum* is a toxic species associated with postulated shellfish poisoning and fish kills. Nakazima (1965) indicated poisonous effects on shellfish feeding on *Prorocentrum* sp. Lassus and Berthome (1988) reported that *P. minimum* caused mortalities in old oysters. Woelke (1961) found that this species caused oyster (*Ostrea iurida*) mortalities and cessation of feeding at high densities. Wikfors and Smolowitz (1993) found that increased abundance of *P. minimum* caused shell loss. This observation was based on feeding experiments where toxins from this species caused slow growth and mortality in shellfish.

The question of the interaction between bivalve mollusks and phytoplankton blooms is complex. Jackson et al. (2001) state that oysters in Chesapeake Bay limited phytoplankton blooms until overfishing of oysters in the early 20th century limited their numbers. According to this point of view, only after overfishing occurred did eutrophication and accompanying water quality deterioration occur. This top-down explanation was not supported by actual evidence, however, and experimental determination of which comes first, oyster mortality or bloom decimation, would help to decide the issue. Although it is true that human overfishing and associated aberrations involving predator–prey interactions can have important impacts on marine food webs, it is also true that anthropogenous nutrient loading is a cause of toxic bloom formation. Various coastal areas that historically have not supported filter-feeding populations have been adversely affected by hypereutrophication and plankton blooms. Other areas with healthy populations of bivalve mollusks have lost such populations due to such loading (Livingston, 2000). In the absence of adequate, long-term ecosystem-level data, the question of bottom-up or top-down control of plankton blooms becomes academic where, in many instances, it is likely that both processes occur somewhere in the food web response to both natural and anthropogenous eutrophication processes.

Burkholder et al. (1992) illustrated the complexity of attempting to relate the natural history of plankton to nutrient loading on the one hand, and effects on food webs on the other. Relatively little is known about the life history of many bloom species. Indeed, taxonomic confusion has led to embarrassing assumptions by plankton specialists. The effects of toxic plankton blooms may be associated with a number of complex life history stages. Dinoflagellates, for example, can have toxic vegetative cells and cysts that produce neurotoxins. However, relatively few species of cyst-forming dinoflagellates are toxic, and very few cells actively form cysts late in their brief lives. Recent studies with the dinoflagellate *Pfiesteria piscicida* (Steidinger et al., 1998a,b) indicate that there can be direct attacks by specific life stages with attachment to a dying fish and subsequent digestion of tissue debris. Lethal flagellated vegetative cells can move from the sediment surface to the water column in response to finfishes that have been adversely affected by the release of neurotoxins from this dinoflagellate. The organism then kills the fish and rapidly descends once again so that the entire process is relatively ephemeral and cannot be detected using routine field sampling methods.

Regardless of the mechanism of toxicity, there is strong evidence that there is a direct correlation of toxic phytoplankton blooms with anthropogenous loading of nitrogen and phosphorus compounds. High nutrient loading makes the estuarine habitat more favorable for algal blooms and toxic effects (Burkholder et al., 1992). In addition, there is some evidence that dinoflagellates may have symbiotic relationships with estuarine/marine bacteria that could also be involved in the ecological consequences of the blooms. These general observations are consistent with what we know of other forms of HABs. Gilbert et al. (2001), in a description of HAB occurrences in Chesapeake Bay, noted that nutrient input to the bay in the organic form has been increasing in the past decade, and that the availability of DOC and DOP may provide a substrate for some bloom species. The authors noted that the timing of the nutrient delivery may also be important in the success of some bloom species. Lucas et al. (1999a,b), using models of the development and distribution of phytoplankton blooms in San Francisco Bay, found that local growth conditions that depend on factors such as water column depth, tidal amplitude, light availability, and grazing rates determine if a bloom can occur. However, large-scale transport plus spatiotemporal variation in the local processes determine if and where a bloom will occur.

The basic lack of understanding of phytoplankton community processes should lead to caution in the attribution of specific water quality impacts, on the one hand, and the population-specific responses to actual phytoplankton communities, on the other. Few studies are designed to measure long-term changes of both the primary and secondary producers as a response to well-documented patterns of nutrient loading. As for the impacts on secondary producers, very little has been shown in the way of adequate scientific documentation. Since many of the HAB species have relatively complex and poorly understood life histories, the relationships of bloom outbreaks and responses of co-occurring species remain vague and poorly documented.

It is with the above in mind that I will proceed to outline the results of a 14-year study that was designed to relate changes of nutrient loading in the Perdido Bay system to phytoplankton community response and the impacts of seasonal and interannual bloom successions on associated food webs.

2. *Perdido Bay Phytoplankton Studies*

a. Background

The long-term (14-year) studies in Perdido Bay were carried out through monthly to quarterly collections of water and sediment samples along with comprehensive biological analyses that included synoptic field sampling of phytoplankton, zooplankton, benthic microphytes, infaunal macroinvertebrates, epibenthic macroinvertebrates, and fishes (see Figure 3.2). The descriptive field data were supplemented with laboratory and field experiments to determine the nature of important processes such as nutrient limitation, the origin and distribution of nitrogen, phosphorus, and carbon particulate matter, and the response of the food web to changes in the phytoplankton community. Trends could be evaluated before and after the initiation of phytoplankton blooms that started sometime between 1992 and 1993. Phytoplankton were taken with both net and whole-water samples (see Appendix I) and were identified to species where possible. The monthly sampling pattern was inadequate as a quantitation of short-term bloom sequences, but the long-term sampling effort was coupled with complete phytoplankton community data from which various indices could be used as indicators of changes in phytoplankton composition. These indices, were more stable in time than the relatively evanescent plankton blooms: these indices (cumulative and average species richness, Shannon species diversity, Shannon evenness, total biomass and total numbers of cells per liter, biomass and numbers of cells per liter of bloom species) served as indicators of the effects of bloom activity on the phytoplankton assemblages in Perdido Bay (Livingston, 2000). Phytoplankton species richness, diversity, and evenness, as monthly or cumulative (annual) indices, were used to indicate the response of the complex phytoplankton assemblages to patterns of bloom species abundance. Relative dominance of bloom species (numbers and biomass) was also used as an index of phytoplankton community structure in addition to the usual count identification of bloom incidence (i.e., $>10^6$ cells l^{-1}). The long-term changes of the community structure of plankton assemblages were thus analyzed within the context of trends of nutrient loading, nutrient gradients, water and sediment quality, and changes in the primary food webs of the Perdido Bay system.

b. Bloom Definition

The designation of bloom status in phytoplankton assemblages is dependent to a considerable degree on the natural history characteristics of individual species. A primary definition involves the concept of high numbers of cells and low species diversity (Fryxell and Villac, 1999). Anderson (1996) included in the definition of blooms the concentration(s) of one or more species that cause harm to other species or that cause accumulations of toxins in such a way as to harm those who may eat toxic species. Many algal specialists consider a bloom to be defined as a population density in excess of 1×10^6 cells l^{-1}. This definition leads to other complications regarding which measure to use in the bloom definition; numbers, biomass, cell volume, or indices calculated on the basis of chlorophyll *a* concentrations (Fryxell and Villac, 1999). However, Livingston (2000) noted that there are subdominant plankton populations that occur over a range from 1×10^4 to 1×10^6 cells l^{-1} that can be important indicators of bloom dynamics. These populations were found to be significant in terms of the overall composition of the phytoplankton community. Determination of relative dominance requires a community analysis to the species level. The methods and timing of the plankton collections (whole-water and net samples taken at monthly intervals) can preclude a fully representative determination of bloom dynamics because of the scale of bloom variation relative to the intervals of sampling frequency.

Another research question concerns how community characteristics of phytoplankton assemblages as a whole are affected by periodic outbreaks of bloom species. This aspect can ultimately be related to changes in associated food webs. The size, shape, and biomass of any given microalgal species may not be related to the relative impact on water quality and/or toxicity. However, the relative dominance (numbers or biomass) of a set of bloom species can comprise an important indicator of the effects of such blooms on overall phytoplankton community structure. In this book, I will use the conventional definition of a bloom as a population density that exceeds 1×10^6 cells l^{-1}. However, associated indices of the phytoplankton assemblages will be used as determinants of the influence of plankton blooms on the trophic response of the system at various levels of biological organization (Livingston, 2000). The emphasis of the research is therefore based on (1) qualitative and quantitative aspects of the species-specific composition of the phytoplankton community relative to long-term changes of nutrient loading and (2) associated effects on the trophic organization of the system.

c. Plankton Species Composition

The various bloom species documented in the Perdido Bay system since 1994 (Table 6.1) represent a taxonomically diverse group dominated by diatoms, raphidophytes, dinoflagellates, and blue-green algae (Livingston, 2001). The class Bacillariophyceae (diatoms) are usually more diverse in areas characterized by relatively high nutrient levels. Their silica-based cell wall is unique. The class Chrysophyceae (the golden brown algae) has relatively few marine species, and individuals appear to prefer oligotrophic waters where they are good competitors for phosphorus (van den Hoek et al., 1995). The cryptophytes (division Cryptophyta) are flagellated microalgae that are known to be bloom forming, but are generally nontoxic (A. K. S. K. Prasad, personal communication). This group is characterized by developed chloroplasts, although cryptophytes are known to live on organic compounds, as well as by asymmetrical flagella, a gullet lined with two or more rows of trichocysts, one or two plastids, and a starch storage product. Nutrition is photomictic or heterotrophic (Throndsen, 1993). These microalgae can act as facultative phagotrophs.

The division Dinophyta (dinoflagellates) is primarily marine, although freshwater forms occur. Pigments include chlorophyll *a*, chlorophyll *c*, dinoxanthrin, diadinoexanthrin, diatoxanthirin, and β-carotene (Latasa and Bidigare, 1998). The division Chlorophyta (green algae) is characterized by carotenoids as well as chlorophyll *a* and *b*. The division Cyanophyta (cyanobacteria or blue-green algae) is represented by some marine forms that are often members of the picoplankton (0.2 to 2.0 μm). Many of these forms can grow well at low light levels. Dinoflagellates are flagellated, with extensive horns, ridges, and wings. About half of the dinoflagellate species lack chlorophyll, and are obligate phagotrophs (van den Hoek et al., 1995). Some dinoflagellates are facultative heterotrophs, and are able to switch from autotrophy to heterotrophy in response to ambient environmental conditions. Some dinoflagellates are considered bloom species with various toxic forms. Species of the class Raphidophyceae (the raphidophytes) are often quite large in size (50 to 100 μm), and are represented by a variety of toxic bloom species that occur in coastal waters. The euglenophytes and pyaminesiphytes were not well represented in Perdido Bay (Livingston, 2000).

Studies in the Perdido Bay system over the past 14 years indicate the occurrence of around ten bloom species. The numerically dominant Perdido Bay phytoplankton bloom species have highly variable structural components. The chain-forming diatom *Leptocylindrus danicus* has relatively large cells (29.0 to 47.3 μm in length; biomass of 404 pg ash-free dry weight [AFDW] cell^{-1}). This species has numerous plastids. The diatom *Cyclotella choctawhatcheeana* has relatively small cells (3.0 to 5.0 μm in length; 40 pg AFDW cell^{-1}) with few plastids and little chlorophyll. The raphidophyte *Heterosigma akashiwo* has numerous chloroplasts. This species is relatively large (length, 8.8 to 12.5 μm, 207 pg AFDW cell^{-1}). The diatom *Miraltia throndsenii* has very small plastids and is itself very small (length, 5.4 μm; 19 pg AFDW cell^{-1}). The diatom *Synedropsis* sp. has only two plastids, whereas *C. choctawhatcheeana* has more but not as many as *H. akashiwo* or *L. danicus*. *Synedropsis* sp. is somewhat larger (length, 20.0 to 23.1 μm; 26 pg AFDW cell^{-1}) than *C. choctawhatcheeana*. The dinoflagellate *Prorocentrum minimum* is photosynthetic with relatively small chloroplasts. This species is relatively large (length, 17.5 to 22.0 μm; 518 pg AFDW cell^{-1}). The raphidophyte *Chattonella subsalsa* is also relatively large (length, 23 to 43 μm; 1933 pg AFDW cell^{-1}). The colony-forming blue-green alga *Merismopedia tenuissima* is common in stagnant eutrophic fresh waters, mainly

Table 6.1 Occurrences of Blooms (>1 × 10^6 cells l^{-1}) in the Perdido Bay System from October 1988 through February 2002

Station	Date	Depth	Species	No. Cells/Liter	Chl *a*	Chl *b*	Chl *c*
26	2/22/94	S	*Leptocylindrus danicus* Cleve	1.95 × 10^6	10.7	<0.2	2.7
29	2/22/94	S	*Leptocylindrus danicus* Cleve	1.64 × 10^6	4.0	<0.2	0.5
23	3/22/94	S	*Miraltia throndsenii* Marino et al.	1.25 × 10^6	7.9	<0.2	1.5
26	3/22/94	S	*Miraltia throndsenii* Marino et al.	4.78 × 10^6	10.1	<0.2	2.1
29	3/22/94	S	*Miraltia throndsenii* Marino et al.	5.62 × 10^6	11.3	<0.2	2.4
23	4/12/94	S	*Cyclotella choctawhatcheeana* Prasad	2.43 × 10^6	124.4	<0.2	17.9
26	4/12/94	S	*Cyclotella choctawhatcheeana* Prasad	8.62 × 10^6	12.1	0.6	2.2
29	4/12/94	S	*Cyclotella choctawhatcheeana* Prasad	8.53 × 10^6	11.1	<0.2	2.1
33	4/12/94	S	*Cyclotella choctawhatcheeana* Prasad	7.65 × 10^6	6.5	<0.2	<0.2
23	6/24/96	S	*Synedropsis* sp.	5.75 × 10^6	14.8	0.2	1.3
SC1	7/17/96	S	*Synedropsis* sp.	2.28 × 10^6	6.9	<0.2	0.6
23	7/18/96	S	*Heterosigma akashiwo* (Hada) Hada	23.93 × 10^6	4.8	2.2	1.2
23	7/18/96	S	*Synedropsis* sp.	2.66 × 10^6	4.8	2.2	1.2
37	7/18/96	S	*Synedropsis* sp.	20.68 × 10^6	7.3	<0.2	0.3
23	8/12/96	S	*Synedropsis* sp.	3.34 × 10^6	15.3	<0.2	1.5
37	8/13/96	S	*Synedropsis* sp.	2.97 × 10^6	4.7	2.5	<0.2
31	3/24/97	S	*Miraltia throndsenii* Marino et al.	11.19 × 10^6	6.2	0.3	1.7
31	3/24/97	B	*Miraltia throndsenii* Marino et al.	26.18 × 10^6	4.4	<0.2	0.8
37	3/26/97	S	*Miraltia throndsenii* Marino et al.	67.81 × 10^6	7.0	<0.2	1.6
40	3/26/97	S	*Miraltia throndsenii* Marino et al.	18.59 × 10^6	6.9	<0.2	0.8
25	3/27/97	S	*Heterosigma akashiwo* (Hada) Hada	11.80 × 10^6	13.7	<0.2	0.6
25	3/27/97	B	*Heterosigma akashiwo* (Hada) Hada	9.23 × 10^6	8.8	<0.2	0.5
26	3/27/97	S	*Heterosigma akashiwo* (Hada) Hada	1.37 × 10^6	17.2	<0.2	2.5
25	3/27/97	S	*Miraltia throndsenii* Marino et al.	12.74 × 10^6	13.7	<0.2	0.6
25	3/27/97	B	*Miraltia throndsenii* Marino et al.	13.95 × 10^6	8.8	<0.2	0.5
26	3/27/97	S	*Miraltia throndsenii* Marino et al.	10.83 × 10^6	17.2	<0.2	2.5
26	3/27/97	B	*Miraltia throndsenii* Marino et al.	12.33 × 10^6	2.8	0.2	<0.2
29	3/28/97	B	*Heterosigma akashiwo* (Hada) Hada	8.76 × 10^6	17.9	<0.2	2.8
29	3/28/97	S	*Miraltia throndsenii* Marino et al.	21.92 × 10^6	22.6	<0.2	4.0
29	3/28/97	B	*Miraltia throndsenii* Marino et al.	36.14 × 10^6	17.9	<0.2	2.8
29	3/28/97	S	*Synedropsis* sp.	1.05 × 10^6	22.6	<0.2	4.0
23	5/13/97	S	*Heterosigma akashiwo* (Hada) Hada	7.67 × 10^6	10.1	<0.2	0.6
23A	5/13/97	S	*Heterosigma akashiwo* (Hada) Hada	4.69 × 10^6	6.9	<0.2	1.0
25	7/15/97	S	*Heterosigma akashiwo* (Hada) Hada	1.39 × 10^6	13.9	<0.2	1.3
26	7/15/97	S	*Synedropsis* sp.	1.32 × 10^6	8.4	<0.2	0.9
26	9/9/97	S	*Heterosigma akashiwo* (Hada) Hada	1.54 × 10^6	12.0	<0.2	1.4

24A	9/9/97	S	*Heterosigma akashiwo* (Hada) Hada	45.28 $\times 10^6$	132.9	<0.2	16.1
22	11/19/97	S	*Merismopedia tenuissima* Lemmermann	1.07 $\times 10^6$	—	—	—
31	4/14/98	S	*Cyclotella choctawhatcheeana* Prasad	1.65 $\times 10^6$	5.1	<0.2	0.9
37	4/14/98	S	*Cyclotella choctawhatcheeana* Prasad	2.27 $\times 10^6$	3.7	<0.2	0.7
40	4/14/98	S	*Cyclotella choctawhatcheeana* Prasad	1.30 $\times 10^6$	5.1	<0.2	0.8
23	5/12/98	S	*Cryptomonas* sp.	1.98 $\times 10^6$	14.1	0.6	1.5
25	5/12/98	S	*Cyclotella choctawhatcheeana* Prasad	4.59 $\times 10^6$	3.7	<0.2	0.4
26	5/12/98	S	*Cyclotella choctawhatcheeana* Prasad	2.14 $\times 10^6$	16.8	<0.2	1.5
29	5/12/98	S	*Cyclotella choctawhatcheeana* Prasad	3.80 $\times 10^6$	7.7	<0.2	0.5
31	5/12/98	S	*Cyclotella choctawhatcheeana* Prasad	6.21 $\times 10^6$	4.6	<0.2	0.3
37	5/12/98	S	*Cyclotella choctawhatcheeana* Prasad	5.14 $\times 10^6$	4.5	<0.2	0.4
40	5/12/98	S	*Cyclotella choctawhatcheeana* Prasad	1.67 $\times 10^6$	2.4	<0.2	0.3
26	5/12/98	S	*Heterosigma akashiwo* (Hada) Hada	2.00 $\times 10^6$	16.8	<0.2	1.5
22	5/12/98	S	*Pedinophora* sp. (4 fl. in mucus)	1.02 $\times 10^6$	44.2	5.3	4.8
23	6/9/98	S	*Cryptomonas* sp. (Teleaulax-like)	1.68 $\times 10^6$	17.1	<0.2	1.3
23	6/9/98	S	*Heterosigma akashiwo* (Hada) Hada	7.47 $\times 10^6$	17.1	<0.2	1.3
25	6/9/98	S	*Synedropsis* sp.	2.68 $\times 10^6$	7.8	<0.2	0.6
26	6/9/98	S	*Synedropsis* sp.	1.66 $\times 10^6$	12.1	<0.2	1.2
29	6/9/98	S	*Synedropsis* sp.	1.51 $\times 10^6$	5.3	<0.2	0.6
31	6/9/98	S	*Synedropsis* sp.	1.47 $\times 10^6$	3.6	<0.2	0.5
25	7/21/98	S	*Cyclotella choctawhatcheeana* Prasad	1.07 $\times 10^6$	9.4	<0.2	0.3
25	7/21/98	S	*Synedropsis* sp.	1.24 $\times 10^6$	9.4	<0.2	0.3
29	7/21/98	S	*Synedropsis* sp.	1.37 $\times 10^6$	10.0	<0.2	0.5
26	8/25/98	S	*Heterosigma akashiwo* (Hada) Hada	1.26 $\times 10^6$	10.9	<0.2	0.7
25	10/20/98	S	*Urosolenia eriensis* (Smith) Round	2.97 $\times 10^6$	6.1	0.5	1.1
26	10/20/98	S	*Urosolenia eriensis* (Smith) Round	1.22 $\times 10^6$	2.3	<0.2	<0.2
29	10/20/98	S	*Urosolenia eriensis* (Smith) Round	2.44 $\times 10^6$	0.4	<0.2	<0.2
31	10/20/98	S	*Urosolenia eriensis* (Smith) Round	1.55 $\times 10^6$	7.8	<0.2	1.1
22	12/15/98	S	*Merismopedia tenuissima* Lemmermann	5.10 $\times 10^6$	0.7	0.2	<0.2
23	12/15/98	S	*Merismopedia tenuissima* Lemmermann	2.99 $\times 10^6$	2.9	<0.2	0.7
22	1/20/99	S	*Merismopedia tenuissima* Lemmermann	1.22 $\times 10^6$	0.9	<0.2	0.3
23	1/20/99	S	*Prorocentrum minimum* (Pavillard) Schiller	1.06 $\times 10^6$	4.6	<0.2	1.5
18	4/20/99	S	*Cyclotella choctawhatcheeana* Prasad	3.05 $\times 10^6$	5.8	<0.2	0.6
23	4/20/99	S	*Cyclotella choctawhatcheeana* Prasad	20.67 $\times 10^6$	3.7	<0.2	0.5
25	4/20/99	S	*Cyclotella choctawhatcheeana* Prasad	8.35 $\times 10^6$	2.2	<0.2	0.4
26	4/20/99	S	*Cyclotella choctawhatcheeana* Prasad	36.38 $\times 10^6$	2.1	<0.2	0.4
29	4/20/99	S	*Cyclotella choctawhatcheeana* Prasad	22.63 $\times 10^6$	10.6	<0.2	1.3
31	4/20/99	S	*Cyclotella choctawhatcheeana* Prasad	12.47 $\times 10^6$	6.4	<0.2	0.9

(continued)

Table 6.1 (continued) Occurrences of Blooms (>1 × 10^6 cells l^{-1}) in the Perdido Bay System from October 1988 through February 2002

Station	Date	Depth	Species	No. Cells/Liter	Chl *a*	Chl *b*	Chl *c*
37	4/20/99	S	*Cyclotella choctawhatcheeana* Prasad	4.11 × 10^6	4.7	<0.2	0.7
40	4/20/99	S	*Cyclotella choctawhatcheeana* Prasad	3.55 × 10^6	3.4	<0.2	0.4
23	4/20/99	S	*Miraltia* sp.	1.69 × 10^6	3.7	<0.2	0.5
26	4/20/99	S	*Miraltia* sp.	5.40 × 10^6	2.1	<0.2	0.4
29	4/20/99	S	*Miraltia* sp.	2.27 × 10^6	10.6	<0.2	1.3
23	4/20/99	S	*Miraltia throndsenii* Marino et al.	2.33 × 10^6	3.7	<0.2	0.5
26	4/20/99	S	*Miraltia throndsenii* Marino et al.	6.40 × 10^6	2.1	<0.2	0.4
29	4/20/99	S	*Miraltia throndsenii* Marino et al.	2.27 × 10^6	10.6	<0.2	1.3
26	4/20/99	S	*Nitzschia reversa* Wm. Sm.	2.10 × 10^6	2.1	<0.2	0.4
25	5/12/99	S	*Cyclotella choctawhatcheeana* Prasad	2.19 × 10^6	4.5	<0.2	0.3
26	5/12/99	S	*Cyclotella choctawhatcheeana* Prasad	7.70 × 10^6	12.7	<0.2	1.3
29	5/12/99	S	*Cyclotella choctawhatcheeana* Prasad	5.10 × 10^6	10.3	<0.2	1.0
31	5/12/99	S	*Cyclotella choctawhatcheeana* Prasad	6.10 × 10^6	2.9	<0.2	0.3
37	5/12/99	S	*Cyclotella choctawhatcheeana* Prasad	1.02 × 10^6	2.5	<0.2	0.4
23	6/23/99	S	*Chattonella c.f. subsalsa* Biecheler	2.39 × 10^6	56.3	<0.2	6.8
18	6/23/99	S	*Heterosigma akashiwo* (Hada) Hada	1.26 × 10^6	32.7	<0.2	4.1
22	8/24/99	S	*Merismopedia tenuissima* Lemmermann	2.90 × 10^6	5.9	0.2	<0.2
23	8/24/99	S	*Merismopedia tenuissima* Lemmermann	3.18 × 10^6	11.6	<0.2	0.7
18	9/8/99	S	*Heterosigma akashiwo* (Hada) Hada	1.61 × 10^6	24.2	<0.2	2.5
22	9/8/99	S	*Merismopedia tenuissima* Lemmermann	10.01 × 10^6	7.4	0.7	0.3
22	9/8/99	B	*Merismopedia tenuissima* Lemmermann	2.71 × 10^6	87.0	254.1	<0.2
22	10/5/99	S	*Merismopedia tenuissima* Lemmermann	1.52 × 10^6	4.8	4.5	<0.2
22	12/7/99	S	*Merismopedia tenuissima* Lemmermann	34.81 × 10^6	1.8	0.9	0.2
23	12/7/99	S	*Merismopedia tenuissima* Lemmermann	17.27 × 10^6	7.1	<0.2	1.2
25	1/11/00	S	*Cyclotella choctawhatcheeana* Prasad	1.02 × 10^6	4.5	<0.2	0.5
26	1/11/00	S	*Cyclotella choctawhatcheeana* Prasad	1.14 × 10^6	4.6	<0.2	0.7
22	1/11/00	S	*Merismopedia tenuissima* Lemmermann	5.87 × 10^6	1.8	<0.2	<0.2
25	2/8/00	S	*Cyclotella choctawhatcheeana* Prasad	1.18 × 10^6	3.3	0.3	0.6
22	2/8/00	S	*Merismopedia tenuissima* Lemmermann	43.27 × 10^6	1.6	<0.2	0.4
23	2/8/00	S	*Merismopedia tenuissima* Lemmermann	36.14 × 10^6	4.9	0.3	0.9
25	2/8/00	S	*Merismopedia tenuissima* Lemmermann	1.30 × 10^6	3.3	0.3	0.6
26	2/8/00	S	*Merismopedia tenuissima* Lemmermann	1.18 × 10^6	3.4	0.3	0.7
29	2/8/00	S	*Merismopedia tenuissima* Lemmermann	2.43 × 10^6	3.5	0.4	0.7
22	3/7/00	S	*Merismopedia tenuissima* Lemmermann	50.59 × 10^6	1.3	<0.2	<0.2
22	4/11/00	S	*Merismopedia tenuissima* Lemmermann	7.46 × 10^6	6.8	0.4	0.4
23	5/9/00	S	*Heterosigma akashiwo* (Hada) Hada	2.96 × 10^6	18.3	0.4	1.1

22	5/9/00	S	*Skeletonema potamus* (Weber) Hasle	1.89	$\times 10^6$	12.0	0.2	1.3
22	7/11/00	S	*Merismopedia tenuissima* Lemmermann	1.11	$\times 10^6$	5.6	0.8	0.4
22	7/11/00	S	*Spermatozopsis exsultans*	1.16	$\times 10^6$	5.6	0.8	0.4
42B	8/9/00	S	*Cyclotella cf. atomus* Hustedt	30.79	$\times 10^6$	—	—	—
22	8/9/00	S	*Merismopedia tenuissima* Lemmermann	2.21	$\times 10^6$	7.5	0.8	0.5
42B	8/9/00	S	*Prorocentrum minimum* (Pavillard) Schiller	1.52	$\times 10^6$	—	—	—
22	9/6/00	S	*Merismopedia tenuissima* Lemmermann	2.29	$\times 10^6$	12.3	10.3	<0.2
22	10/3/00	S	*Merismopedia tenuissima* Lemmermann	1.34	$\times 10^6$	7.2	2.8	<0.2
22	11/7/00	S	*Merismopedia tenuissima* Lemmermann	15.65	$\times 10^6$	2.5	0.9	<0.2
22	12/5/00	S	*Merismopedia tenuissima* Lemmermann	28.86	$\times 10^6$	0.8	0.3	0.7
23	12/5/00	S	*Merismopedia tenuissima* Lemmermann	7.01	$\times 10^6$	3.3	<0.2	0.2
25	1/9/01	S	*Cyclotella choctawhatcheeana* Prasad	1.67	$\times 10^6$	1.7	<0.2	0.5
26	1/9/01	S	*Cyclotella choctawhatcheeana* Prasad	1.87	$\times 10^6$	4.9	<0.2	0.8
29	1/9/01	S	*Cyclotella choctawhatcheeana* Prasad	1.60	$\times 10^6$	5.4	<0.2	0.9
31	1/9/01	S	*Cyclotella choctawhatcheeana* Prasad	1.96	$\times 10^6$	5.5	<0.2	1.3
37	1/9/01	S	*Cyclotella choctawhatcheeana* Prasad	2.37	$\times 10^6$	3.9	<0.2	1.0
40	1/9/01	S	*Cyclotella choctawhatcheeana* Prasad	2.12	$\times 10^6$	7.2	<0.2	1.7
42B	1/9/01	S	*Cyclotella choctawhatcheeana* Prasad	1.94	$\times 10^6$	50.1	<0.2	15.5
22	1/9/01	S	*Merismopedia tenuissima* Lemmermann	7.25	$\times 10^6$	3.4	<0.2	0.5
42B	1/9/01	S	*Prorocentrum minimum* (Pavillard) Schiller	21.28	$\times 10^6$	50.1	<0.2	15.5
42B	1/9/01	S	*Skeletonema costatum* (Grev.) Cleve	2.59	$\times 10^6$	50.1	<0.2	15.5
25	2/28/01	S	*Chaetoceros subtilis* Cleve	1.36	$\times 10^6$	3.5	0.2	1.1
26	2/28/01	S	*Chaetoceros subtilis* Cleve	1.65	$\times 10^6$	5.4	<0.2	0.9
29	2/28/01	S	*Chaetoceros subtilis* Cleve	1.42	$\times 10^6$	3.7	<0.2	0.7
31	2/28/01	S	*Chaetoceros subtilis* Cleve	1.58	$\times 10^6$	7.2	<0.2	1.5
37	2/28/01	S	*Chaetoceros subtilis* Cleve	1.16	$\times 10^6$	3.6	<0.2	0.9
40	2/28/01	S	*Chaetoceros subtilis* Cleve	1.11	$\times 10^6$	4.0	<0.2	0.7
42B	2/28/01	S	*Cyclotella choctawhatcheeana* Prasad	1.63	$\times 10^6$	—	—	—
22	2/28/01	S	*Merismopedia tenuissima* Lemmermann	1.70	$\times 10^6$	1.3	<0.2	0.2
23	2/28/01	S	*Merismopedia tenuissima* Lemmermann	1.09	$\times 10^6$	8.7	<0.2	1.6
42B	2/28/01	S	*Skeletonema costatum* (Grev.) Cleve	1.53	$\times 10^6$	—	—	—
42B	3/20/01	S	*Cyclotella choctawhatcheeana* Prasad	1.06	$\times 10^6$	—	—	—
22	3/20/01	S	*Merismopedia tenuissima* Lemmermann	5.98	$\times 10^6$	0.7	0.2	0.2
23	3/20/01	S	*Merismopedia tenuissima* Lemmermann	2.64	$\times 10^6$	1.3	0.2	0.2
42B	3/20/01	S	*Skeletonema costatum* (Grev.) Cleve	4.93	$\times 10^6$	—	—	—
18	4/10/01	S	*Cyclotella choctawhatcheeana* Prasad	2.08	$\times 10^6$	30.3	<0.2	2.4
23	4/10/01	S	*Cyclotella choctawhatcheeana* Prasad	15.55	$\times 10^6$	22.9	<0.2	2.5
25	4/10/01	S	*Cyclotella choctawhatcheeana* Prasad	17.50	$\times 10^6$	13.2	<0.2	1.7

(continued)

Table 6.1 (continued) Occurrences of Blooms (>1 × 10^6 cells l^{-1}) in the Perdido Bay System from October 1988 through February 2002

Station	Date	Depth	Species	No. Cells/Liter	Chl *a*	Chl *b*	Chl *c*
26	4/10/01	S	*Cyclotella choctawhatcheeana* Prasad	16.82 × 10^6	14.2	0.3	2.1
29	4/10/01	S	*Cyclotella choctawhatcheeana* Prasad	15.52 × 10^6	12.0	<0.2	1.5
31	4/10/01	S	*Cyclotella choctawhatcheeana* Prasad	17.34 × 10^6	9.3	<0.2	1.2
37	4/10/01	S	*Cyclotella choctawhatcheeana* Prasad	4.45 × 10^6	3.0	<0.2	0.4
40	4/10/01	S	*Cyclotella choctawhatcheeana* Prasad	3.38 × 10^6	2.6	<0.2	0.2
42B	4/10/01	S	*Cyclotella choctawhatcheeana* Prasad	1.25 × 10^6	11.4	<0.2	1.6
18	4/10/01	S	*Heterosigma akashiwo* (Hada) Hada	4.93 × 10^6	30.3	<0.2	2.4
23	4/10/01	S	*Heterosigma akashiwo* (Hada) Hada	3.04 × 10^6	22.9	<0.2	2.5
22	4/10/01	S	*Merismopedia tenuissima* Lemmermann	6.15 × 10^6	30.4	0.3	3.1
25	4/10/01	S	*Miraltia* sp.	1.07 × 10^6	13.2	<0.2	1.7
26	4/10/01	S	*Miraltia* sp.	1.67 × 10^6	14.2	0.3	2.1
31	4/10/01	S	*Miraltia* sp.	1.43 × 10^6	9.3	<0.2	1.2
23	4/10/01	S	*Miraltia throndsenii* Marino et al.	2.27 × 10^6	22.9	<0.2	2.5
25	4/10/01	S	*Miraltia throndsenii* Marino et al.	1.51 × 10^6	13.2	<0.2	1.7
26	4/10/01	S	*Miraltia throndsenii* Marino et al.	3.20 × 10^6	14.2	0.3	2.1
29	4/10/01	S	*Miraltia throndsenii* Marino et al.	1.82 × 10^6	12.0	<0.2	1.5
31	4/10/01	S	*Miraltia throndsenii* Marino et al.	1.73 × 10^6	9.3	<0.2	1.2
23	5/8/01	S	*Cyclotella choctawhatcheeana* Prasad	3.57 × 10^6	8.2	<0.2	0.7
25	5/8/01	S	*Cyclotella choctawhatcheeana* Prasad	3.13 × 10^6	5.5	<0.2	0.4
26	5/8/01	S	*Cyclotella choctawhatcheeana* Prasad	1.23 × 10^6	4.6	<0.2	0.6
29	5/8/01	S	*Cyclotella choctawhatcheeana* Prasad	2.39 × 10^6	3.1	<0.2	0.2
31	5/8/01	S	*Cyclotella choctawhatcheeana* Prasad	1.75 × 10^6	2.6	<0.2	0.2
37	5/8/01	S	*Cyclotella choctawhatcheeana* Prasad	1.07 × 10^6	1.9	<0.2	0.2
22	5/8/01	S	*Merismopedia tenuissima* Lemmermann	17.81 × 10^6	14.5	2.2	1.1
22	6/5/01	S	*Merismopedia tenuissima* Lemmermann	6.72 × 10^6	10.5	2.2	1.1
42B	7/11/01	S	*Gymnodinium* spp.	1.35 × 10^6	23.0	0.3	3.4
22	7/11/01	S	*Merismopedia tenuissima* Lemmermann	23.94 × 10^6	7.9	1.4	0.3
23	7/11/01	S	*Merismopedia tenuissima* Lemmermann	1.66 × 10^6	22.4	<0.2	2.1
42B	8/8/01	S	*Heterocapsa rotundata* (Lohmann) Hansen	7.59 × 10^6	—	—	—
23	8/8/01	S	*Heterosigma akashiwo* (Hada) Hada	1.52 × 10^6	27.2	<0.2	3.5
25	9/5/01	S	*Heterosigma akashiwo* (Hada) Hada	1.19 × 10^6	15.8	<0.2	0.9

Note: Also shown are chlorophyll *a*, *b*, and *c* concentrations (µg l^{-1}) in water quality samples taken at the same time as the whole-water phytoplankton sampling. Data are shown by month and year in various areas of the bay.

fertilized fish ponds. This species also occurs in brackish waters with some capacity to tolerate low salinities, particularly during warm periods (Komarek and Anagnostidis, 1998). However, the small size of the cells (1 to 2 μm) makes it difficult to determine viability; therefore, *M. tenuissima* was reported primarily in the freshwater entry points to upper Perdido Bay. The considerable variability among the various bloom species with respect to size, biomass, and chloroplast development could be related to the lack of a statistical relationship between the chlorophyll observations and the incidence and intensity of the blooms in Perdido Bay (Table 6.1; Livingston, 2000).

The diatom *L. danicus* has a worldwide distribution in a variety of tropical coastal environments (Hargraves, 1990), and is often a constituent of spring phytoplankton outbursts as well as fall growing periods (Marshall, 1988). This species has been associated with red tide events in Japan (Fukuyo et al., 1990). According to Fryxell and Villac (1999), *L. danicus* is believed to be harmful to caged sea trout (*Cynoscion regalis*), Atlantic salmon (*Salmo salar*), and smolt of coho salmon (*Oncothuynchus kisutch*). Another diatom, *Miraltia throndsenii* Marino, Montresor & Zingone has been associated with blooms stimulated by effluents from sewage and industrial outfalls (Marino et al., 1987). *Miraltia throndsenii* is congeneric with *Chaetoceros ehrenberg* on the basis of fine structure of vegetative cell and resting spore morphology and the resulting combination was *C. throndsenii* (Marino, Montresor & Zingone) (Marino et al., 1992). The diatom *Cyclotella choctawhatcheeana* was first described from samples taken in the Choctawhatchee Bay system (Prasad et al., 1990), where short chains were found during each month of the year. Small blooms were observed in some areas. According to Livingston and Prasad (unpublished data), *C. choctawhatcheeana* responds to combinations of environmental factors that range along inshore–offshore gradients of light and nutrients with salinity as a possible modifying factor. Cooper (1995a,b) found that *C. choctawhatcheeana* in Chesapeake Bay had increased abundances that coincided with increased sedimentation, turbidity, and eutrophication. Overall, this species has a high tolerance for wide ranges of salinity and is stimulated by increased nutrient loading.

During the last two decades, there have been an increasing number of red tide blooms along various coasts, especially in gulfs and inland seas. Among the most notable of the many kinds of bloom-forming flagellates is *H. akashiwo,* which frequently causes extensive red tide events (Hara and Chihara, 1987). Recently, considerable attention has been focused on the genus *Heterosigma* as the cause of fish kills in New Zealand, Chile, and British Columbia (Chang et al., 1990). Smayda (1997c) detailed the ecophysiology and bloom dynamics of *H. akashiwo*, a species well known for ichthyotoxic blooms (Honjo, 1993). Honjo (1993) found that *H. akashiwo* blooms were associated with river runoff, low bottom oxygen, and wind-induced turbulence of bottom sediments. Growth occurs from overwintering motile forms and/or germinating resting spores. This species has a high growth potential and can form a bloom in a short time. It excretes allelopathic substances that can suppress diatom growth and may dramatically reduce cell numbers in other phytoplankton species. In addition to making vertical migrations, *H. akashiwo* can explore diverse spatial habitats in search of favorable growth conditions. Nutrient limitation stimulates migration tactics. There are relatively high nutrient requirements for *H. akashiwo*. Bactivorous feeding is stimulated by phosphorus limitation with iron and manganese needed for bloom continuance. Nutrient enrichment supports blooms of this species with dependence on river flows in various regions. Temperature, nutrients, and competition appear to be limiting for bloom generation by *H. akashiwo*. Smayda (1997) indicated multifactorial control of blooms that, once released, have a "remarkably high degree of broad spectrum allelochemical allelopathic antagonisms" that are used against competing species and potential grazers.

The fish-killing mechanism of raphidophyte blooms is still poorly understood. Both physical clogging of gills by mucus excretion as well as gill damage by hemolytic substances such as polyunsaturated fatty acids may be involved (Shimoda et al., 1983; Chang et al., 1990). There is accumulating evidence that production of superoxide radicals represents the primary mechanism of fish mortality. Imae et al. (1997) reported details concerning life cycle and bloom dynamics of *Chattonella*, a genus with two known fish-killing species (*C. antiqua* and *C. marina*). Nutrients and competitors (mainly diatoms) appear to affect the development of *Chattonella* populations. Anything interfering with diatom proliferation could give *Chattonella* an advantage. During a severe phytoplankton outbreak in the Seta Sea in 1972, a raphidophyte red tide killed 14 million cultured yellow tail fish. Effluent controls were then initiated to reduce the organic carbon loading and the discharge of phosphates from household detergents. Following a time lag of 4 years, the frequency of red tide events in the Seta Sea then decreased by about twofold to a more stable level (Hallegreaff, 1995). A similar pattern of long-term loading of coastal waters was evident for the North Sea in Europe (Smayda, 1990). Since 1955, the phosphorus loading of the River Rhine has increased 7.5-fold, and nitrogen levels have also increased. This has resulted in a significant sixfold decline in Si:P ratios because long-term reactive silicate concentrations (a nutrient derived from natural land weathering) have remained constant. There is considerable concern that altered nutrient ratios in coastal waters may favor blooms of nuisance flagellate species, which replace the normal spring and autumn blooms of siliceous diatoms (Smayda, 1990).

The dinoflagellate *Prorocentrum minimum* is a toxic bloom species associated with postulated shellfish poisoning and fish kills. This species is mostly estuarine and cosmopolitan in cold temperate to tropical waters. Nakazima (1965) indicated poisonous effects on shellfish feeding on *Prorocentrum* sp. The toxic substance venerupin has been associated with *P. minimum* although there is some question concerning this association. Hansen (1997) reported that *P. minimum* can ingest certain cryptophytes (*Cryptomonas* sp.) and ciliates. Mixotrophy (being photosynthetic and phagotrophic) is widespread among dinoflagellates with a variety of feeding mechanisms found in this group. The mechanisms of prey selectivity are not well understood. Control of plankton organisms by phagotrophy is considered to be highly variable (Granelli and Carlsson, 1997). However, losses of plankton to pigmented flagellates must be considered according to these authors. Sellner et al. (1995) indicated that *P. minimum* did not adversely affect oysters (*Crassostrea virginica*) in Chesapeake Bay; oysters effectively reduced these bloom-forming dinoflagellates. However, Lassus and Berthome (1988) reported that *P. minimum* caused mortalities in old oysters. Woelke (1961) found that this species caused oyster (*Ostrea iurida*) mortalities and cessation of oyster feeding at high densities. Wikfors and Smolowitz (1993) found that increased abundance of *P. minimum* could cause shell losses. Cardwell et al. (1979) found that *Gymnodinium splendens*, another dinoflagellate, was acutely toxic to oyster larval stages.

There is relatively little detailed scientific literature concerning the long-term effects of nutrient loading on phytoplankton communities, particularly with respect to the origin of plankton blooms (Livingston, 2000). Howarth et al. (2000) point out that increased loading of nitrogen and phosphorus may be associated with silica reductions that, in turn, limit diatom growth leading to changes in phytoplankton community composition (Jorgensen and Richardson, 1996). This can affect the size distribution as well as the species-specific content of the plankton assemblages. Changes from diatoms to flagellates can lead to food web changes that can then lead to hypothetical alterations of fish populations (Greve and Parsons, 1977). Radach et al. (1990) found that, with increases of the N:Si and P:Si ratios over a 23-year period, diatoms decreased and flagellates increased along the German coast. Shifts of N:P ratios increased summer bloom incidence of the haptophyte *Phaeocystis* (Riegman et al., 1992), which was associated with phosphorus limitation. Riegman (1998) indicated that phytoplankton species composition may be changed by the dynamic interactions among macronutrients. However, as in the relatively few studies listed above, detailed, quantitative, long-term phytoplankton community changes due to altered nutrient loading have not been studied in a systematic fashion.

d. Phytoplankton Changes in Perdido Bay

Analyses of the origin and succession of phytoplankton blooms in response to nutrient loading by a pulp mill to the Perdido Bay system have been described by Livingston (2000). Perdido Bay in the NE Gulf is a relatively small system with only one primary anthropogenous source of nutrification to the upper bay, a pulp mill. Kraft mill effluents tend to have relatively high concentrations of orthophosphate and ammonia. Most of the land in the upper Perdido drainage basin is lightly populated. The most concentrated agricultural and urban runoff is located in the lower bay.

This combination of factors established upper Perdido Bay as an appropriate subject for research concerning nutrient loading and eutrophication. The Perdido River, an alluvial stream, also loads nutrients into upper Perdido Bay. A detailed analysis of the nutrient loading characteristics of the Perdido drainage system indicated that, during early years of analysis, orthophosphate and ammonia loading from the pulp mill enhanced secondary production in the immediate receiving area of the upper estuary. Plankton blooms were not present during this period. The chrysophytes, and to a lesser degree, the chlorophytes, tended to be abundant and were associated with a balanced food web and relatively high secondary production in the upper bay.

During a winter–spring drought in 1993–1994, the pulp mill increased orthophosphate loading to the bay. The combination of drought conditions and increased orthophosphate loading was associated with the initiation of a series of phytoplankton blooms dominated by diatom species. There was an orderly seasonal succession of these bloom species in the bay. Continued high orthophosphate loading over the next 3 years led to increases of phytoplankton bloom frequency and intensity throughout the bay. From 1996 through 1998 there was increased ammonia loading to the upper bay by the pulp mill. Plankton response included a pattern of the individual bloom species that were attributed to seasonal differences in nutrient requirements of the bloom species. Interannual qualitative and quantitative phytoplankton dominance shifts occurred from 1993 through 1999 whereby larger-celled raphidophyte and dinoflagellate species replaced diatom species that had been predominant during the initial blooms. Increased dominance of bloom species was usually accompanied by reductions of phytoplankton species richness. There was a long-term increase of plankton numbers and biomass during the years of increased nutrient loading.

From late summer 1997 through spring 1999, the pulp mill reduced its orthophosphate loading and there were concurrent reductions of the high relative dominance of bloom species, especially during winter–spring periods. There was a partial recovery of the phytoplankton associations during the period of low orthophosphate loading to upper Perdido Bay. However, by spring 1999, the mill again resumed high loading of orthophosphate to the bay. This loading was again accompanied by baywide spring and early summer blooms, increased dominance of bloom species, and associated reductions of phytoplankton species richness. There was considerable background noise in the phytoplankton response, and it is possible that other nutrients could have played a role in the noted changes. However, the postulated effects of increased orthophosphate and ammonia loading were usually correlated with general sequences of bloom species and associated changes in the phytoplankton community structure that were consistent with observed natural history characteristics of the diatoms, raphidophytes, and dinoflagellates that comprised the bloom types. There were distinct concentration gradients of the orthophosphate and ammonia as water from Elevenmile Creek entered the bay; such areas were noted as primary sites of bloom origin. The concentration gradients appeared to provide the spark that ignited at least some of the plankton blooms.

Seasonal and interannual successions of plankton blooms could have been affected by interspecific competition and predation within the plankton community. High nutrient loading tended to favor cryptophytes, cyanophytes, dinoflagellates, and raphidophytes. During periods of high winter orthophosphate loading, the dinoflagellates were numerically dominant, whereas summer ammonia loading was associated with increased dominance by the raphidophytes. The chrysophytes, and to a lesser degree, the chlorophytes, tended to be more abundant during periods of low nutrient loading. Diatom blooms were noted during winter–spring periods and could have been affected by the predominant raphidophytes during periods of high raphidophyte biomass. The responses of the different groups of plankton to changes in nutrient loading were consistent with the results of nutrient limitation experiments that indicated phosphorus limitation during cool months and P+N limitation during warm months.

Detailed characterization of the long-term changes of plankton assemblages in Perdido Bay led to an examination of water quality factors as indicators of phytoplankton activity. For example, chlorophyll *a*, widely used as an indicator of phytoplankton assemblages in both descriptive analyses and modeling studies, was not representative of the incidence and succession of phytoplankton blooms.

Chlorophyll *a* was significantly associated with phytoplankton numbers and biomass, but accounted for relatively little of the variation of such variables. Although DO is commonly used as an indicator of hypereutrophication, this factor went through complex changes during the long-term plankton successions. Hypoxia was seasonal and interannual trends of DO were not always consistent with hypoxic conditions during plankton bloom activity. Nutrient concentrations in the bay also underwent complex interannual responses to such blooms. Benthic microalgal community structure proved to be a more reliable indicator of the blooms than the usual water quality indices that are in use today.

The data indicated that the response of coastal systems to anthropogenous nutrient loading should be placed within the context of regional habitat conditions, regular interannual sequences of drought and flood conditions, and anthropogenous nutrient loading. Seasonal temperature and salinity changes, interannual cycles of river flows and associated estuarine habitat variables, long-term nutrient loading and associated nutrient concentration gradients, and anthropogenous changes in the physiography of Perdido Bay due to dredged openings to the Gulf of Mexico affected the initiation and temporal successions of plankton blooms. A comparison of the Perdido results with other bay systems in the NE Gulf of Mexico confirmed the probable association of bloom generation and altered phytoplankton community structure. In Wolf Bay there was increased nutrient loading due to agricultural runoff. This loading was associated with increased relative abundance of plankton bloom species and reduced phytoplankton species richness. The general absence of well-defined nutrient gradients in Wolf Bay indicated that loading of phosphorus and nitrogen compounds per se could lead to plankton blooms. The relationship of reduced phytoplankton species richness and increased dominance by bloom species was also noted in Escambia Bay, which had been subjected to historically high anthropogenous nutrient loading and phytoplankton blooms that occurred during the late 1960s and 1970s. Upper Escambia Bay was still characterized by increased dominance by phytoplankton bloom species during a recent survey; it was noted that point-source pollution and urban storm water runoff were key sources of anthropogenous nutrient loading in the Pensacola Bay system. Non-point sources of nutrient loading were postulated as the cause of recent blooms in Choctawhatchee Bay that were associated with fish kills and the deaths of hundreds of dolphins.

A restoration program of proposed reductions of orthophosphate and ammonia loading by the pulp mill to Elevenmile Creek was undertaken in 1999. The restoration program was based on the long-term Perdido Bay database, and, for the first time in Florida, the Florida Department of Environmental Protection issued a permit to the paper mill that uses the proposed orthophosphate and ammonia loading rates based on the long-term research in the Perdido system. Orthophosphate and ammonia loading by the pulp mill was targeted for reductions to levels that approximated loadings during periods when the bay was free of blooms. Another 3 years of data were taken to determine the response of the Perdido Bay system to the proposed reductions of nutrient loading from the pulp mill. The following is an examination of the response of the Perdido system to changes in nutrient loading.

3. Seasonal Bloom Occurrences

Seasonal changes of monthly averages of nutrient loading in the Perdido system (Figure 6.8) indicate that the highest ammonia loading in Elevenmile Creek occurred during early summer months (June–July). Ammonia loading from the Perdido River followed the same seasonal pattern, peaking from May through July. Orthophosphate loading in the creek was highest during summer months. In the Perdido River, such loading again followed a similar pattern, peaking in May. Relative abundance (% total numbers) of diatoms (BACIL) occurred during late winter–spring months. There was no statistical relationship between diatom abundance and silica loading. Raphidophyte (RAPHID) relative abundance peaked in June and July. The Raphidophytes were virtually absent from the bay during cold months of the year. Dinoflagellate (DINO) relative abundance peaked in January. The lowest dinoflagellate relative abundance occurred during April when diatoms were at peak levels of relative abundance. There was little overlap in the relative abundances of the numerically dominant phytoplankton groups. Total phytoplankton biomass peaked in June,

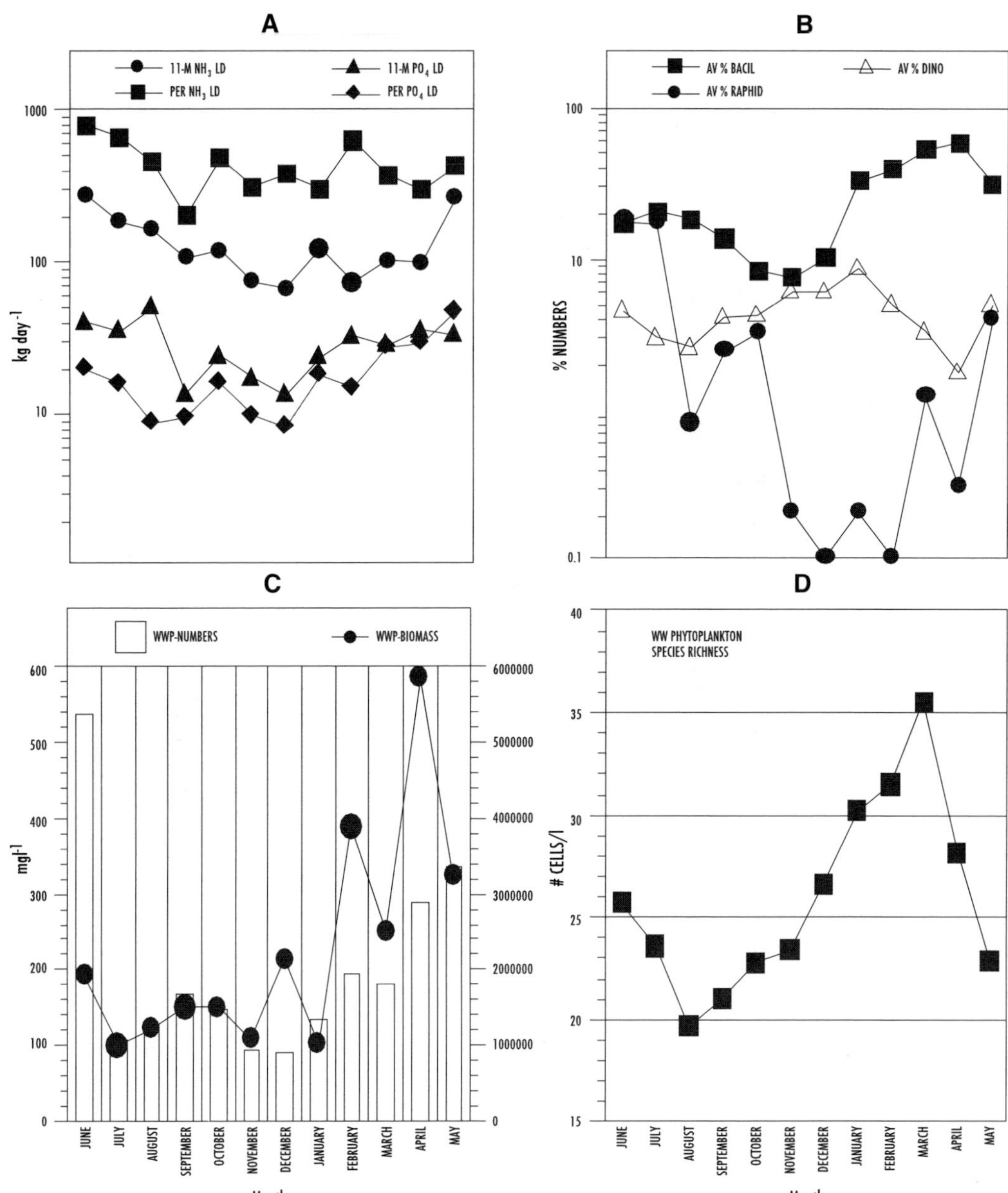

Figure 6.8 Monthly averages of nutrient loading and phytoplankton community changes in Perdido Bay from 1988 through 2002. (A) Ammonia and orthophosphate loading from Elevenmile Creek and the Perdido River system. (B) Average percent of numbers of diatoms (class Bacillariophyceae), raphidophytes (class Raphidophyceae), and dinoflagellates (division Dinophyta). (C) Phytoplankton numbers and biomass. (D) Phytoplankton species richness.

reflecting the greater size of the raphidophytes, whereas phytoplankton numbers peaked during April, a reflection of the generally smaller diatoms (i.e., *Cyclotella choctawhatcheeana*). Phytoplankton species richness was highest during March with the least numbers of phytoplankton species noted from August through October.

The spatiotemporal distribution of the plankton blooms in the Perdido system are shown in Table 6.2. There was a distinct species-specific pattern to the seasonal occurrence of the bloom

Table 6.2 Occurrences of Blooms (>1 × 10^6 cells l^{-1}) in the Perdido Bay System from October 1988 through June 2002

Month	Bloom Area	Bloom Species	Bloom Year	Bloom Area	Bloom Species	Bloom Year
December						
January	Baywide	*C. choctawhatcheeana*	2000, 2001			
February						
March				Upper	*M. throndsenii*	1994, 1997
April	Baywide	*C. choctawhatcheeana*	1994, 1998,1999, 2001	Upper	*M. throndsenii*	1999, 2001
May	Baywide	*C. choctawhatcheeana*	1998, 1999			
June						
July						
August						
September						
October						
November						
December						
January				Upper	*P. minimum*	1999
February						
March	Upper	*H. akashiwo*	1997			
April	Upper	*H. akashiwo*	2001			
May	Upper	*H. akashiwo*	1997, 1998, 2000			
June	Upper	*H. akashiwo*	1998, 1999			
July	Upper	*H. akashiwo*	1996			
August	Upper	*H. akashiwo*	1998	Lower	*P. minimum*	2000
September	Upper	*H. akashiwo*	1997, 1999			
October						
November						
December						

January						
February	Upper	*L. danicus*	1994			
March						
April						
May						
June				Upper	*Synedropsis* sp.	1996, 1998
July				Upper	*Synedropsis* sp.	1996, 1997, 1998, 1996
August				Upper	*Synedropsis* sp.	
September						
October						
November						
December	Elevenmile Creek	*M. tenuissima*	1998, 1999			
January	Elevenmile Creek	*M. tenuissima*	1999, 2000, 2001			
February	Elevenmile Creek	*M. tenuissima*	2000			
March	Elevenmile Creek	*M. tenuissima*	2000			
April	Elevenmile Creek	*M. tenuissima*	2000, 2001			
May						
June						
July	Elevenmile Creek	*M. tenuissima*	2000			
August	Elevenmile Creek	*M. tenuissima*	1999, 2000			
September	Elevenmile Creek	*M. tenuissima*	1999, 2000			
October	Elevenmile Creek	*M. tenuissima*	1999, 2000	Upper	*U. eriensis*	1998
November	Elevenmile Creek	*M. tenuissima*	1997			

Note: Data are shown by month and year in various areas of the bay.

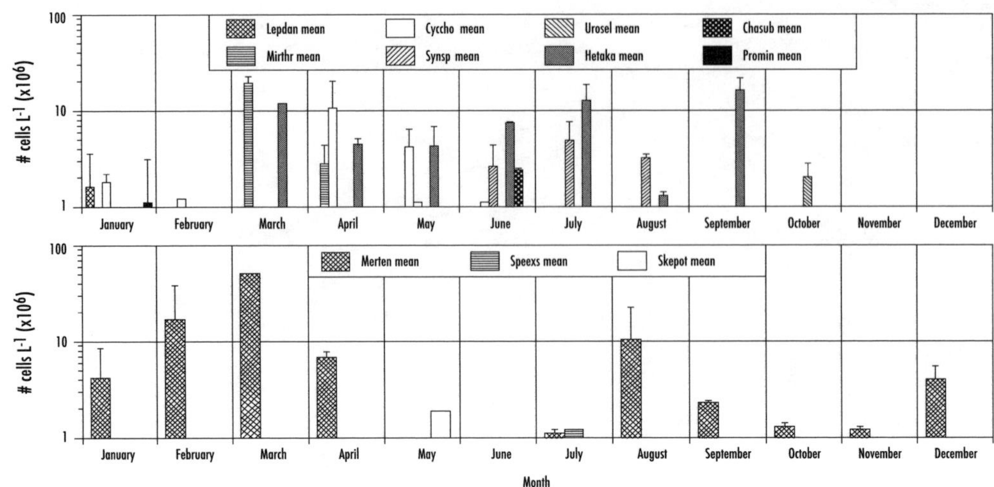

Figure 6.9 Monthly averages of numerical abundance of phytoplankton bloom species in Perdido Bay from 1988 through 2002. Species include *Leptocylindrus danicus* (Lepdan), *Miraltia throndsenii* (Mirthr), *Cyclotella choctawhatcheeana* (Cyccho), *Synedropsis* sp. (Synsp), *Urosolenia eriensis* (Urosel), *Heterosigma akashiwo* (Hetaka), *Chattonella c.f. subsalsa* (Chasub), *Prorocentrum minimum* (Promin), and *Merismopedia tenuissima* (Merten).

species in Perdido Bay. In the upper bay, *Prorocentrum minimum* and *Leptocylindrus danicus* were restricted to winter peaks, whereas the raphidophyte blooms (*Heterosigma akashiwo* and *Chattonella subsalsa*) occurred during warm months of the year. The diatoms *Cyclotella choctawhatcheeana* and *Miraltia throndsenii* bloomed mainly during spring months whereas *Synedropsis* sp. bloomed during the summer. The raphidophyte and blue-green algae blooms occurred later in the interannual bloom succession, and it is possible that *H. akashiwo* displaced *C. choctawhatcheeana* when it reached bloom numbers. There were also differences in percent dominance between numerical abundance and biomass transformations of the data. The larger species such as *H. akashiwo* and *P. minimum* assumed higher dominance levels when defined as biomass percentages.

Species-specific monthly averages of numerical abundance of the primary bloom species in Perdido Bay (Figure 6.9) indicated January peaks of *P. minimum* and the diatom *L. danicus*. The diatom *M. throndsenii* peaked in March, whereas *C. choctawhatcheeana* was prevalent during April and May. The raphidophyte *H. akashiwo* was most abundant from March through July with a peak in September. The diatom *Synedropsis* sp. peaked from June through August and *Urosolenia eriensis* reached high numbers during October. The primary dominant of the blue-green algae (*Merismopedia tenuissima*) in Elevenmile Creek was most prevalent from January through April and from August through December. Overall, there were distinct, species-specific seasonal patterns of bloom abundance in the Perdido system that generally followed the seasonal bloom sequences (see Table 6.2).

4. Long-Term Bloom Responses to Nutrient Loading

a. Net Phytoplankton and Zooplankton

Net phytoplankton collections were used to determine the effects of the first blooms noted during 1993–1994. A comparison was made of the net phytoplankton taken during the drought years of relatively low nutrient loading (1988 through 1991) and the drought period characterized by the initiation of consistently high orthophosphate loading (1993–1994) (Figure 6.10). During all periods of comparison, diatoms were dominant. Phytoplankton numbers were higher at all stations during 1993–1994 (Figure 6.10). Numerical abundance of net phytoplankton

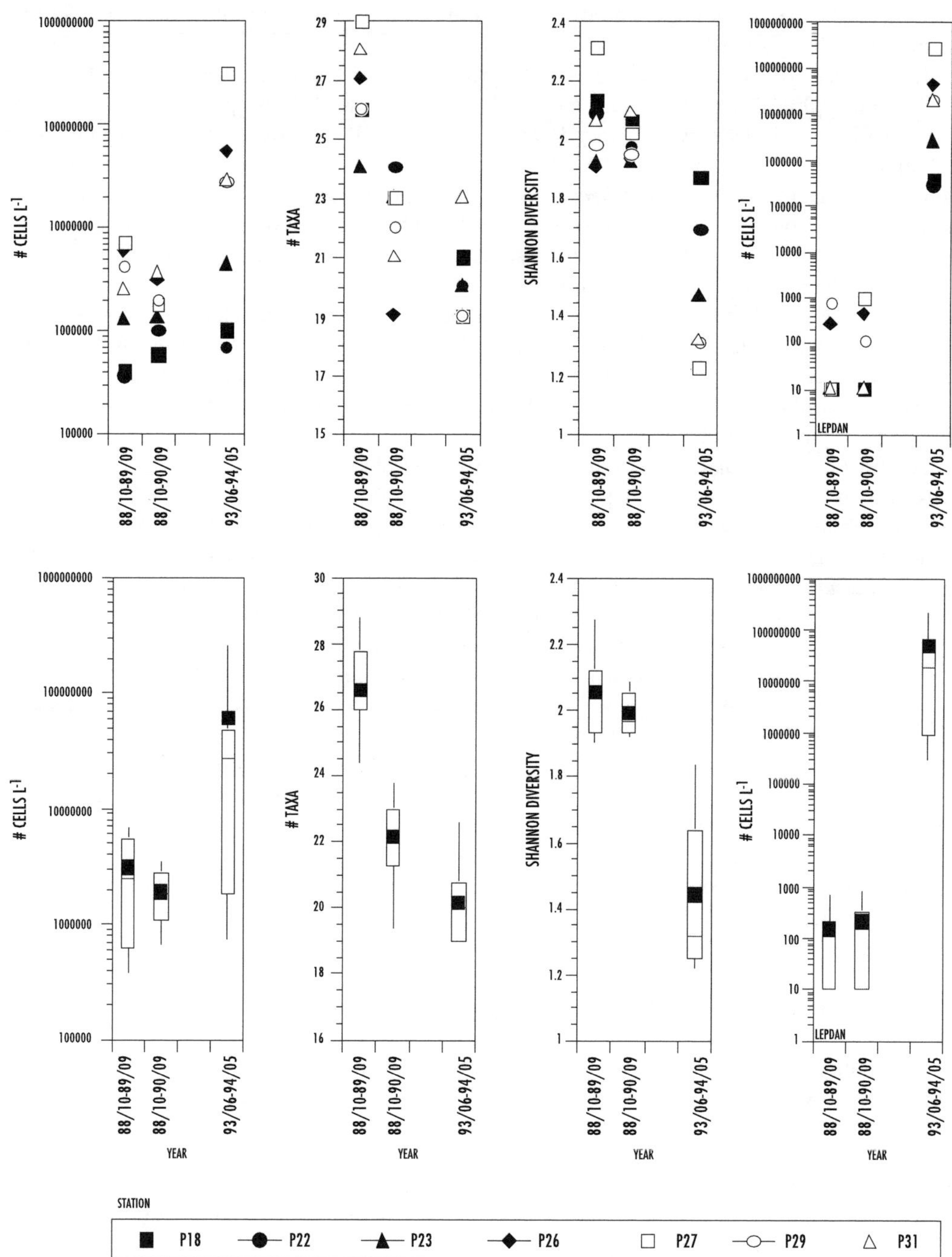

Figure 6.10 Net phytoplankton numbers per liter, species richness and diversity, and numbers of *Leptocylindrus danicus* at stations in Perdido Bay. The indices were averaged annually during the relatively low nutrient loading years (1988–1989; 1989–1990) and during the period of consistently high orthophosphate loading period (1993–1994).

was significantly (p = 0.05) higher during all months in 1993–1994 (Livingston, 2000). Phytoplankton numbers were lowest in the upper bay during all three periods of sampling. Phytoplankton species richness was highest during 1988–1989, although species diversity was comparable and uniformly higher during the early years when compared to 1993–1994.

Phytoplankton species richness and diversity were significantly ($p = 0.05$) lower in 1994 in 7 of the 9 months analyzed.

During 1993–1994, there was an orderly progression of bloom species relative abundance (Livingston, 2000). During fall 1993, *Falcula hyalina* dominated. This species was followed by *L. danicus* blooms in January–February 1993 (Figure 6.10). Blooms were noted throughout upper and midbay areas. During March 1994, *Miraltia throndsenii* was dominant. There were extensive, baywide blooms of *C. choctawhatcheeana* during April 1994. With the exception of April 1994, orthophosphate concentration increases at station P23 (receiving runoff from Elevenmile Creek) and associated blooms were not followed by like increases in chlorophyll *a* concentrations. The relatively high chlorophyll *a* concentration at station P23 in April 1994 was the only sign of bloom activity throughout the study period. The pH of the water was higher during 1993–1994 than 1988–1989 (Livingston, 2000). These data suggest that an orderly succession of phytoplankton blooms occurred with the 1993–1994 drought conditions and these blooms were associated with increased mill orthophosphate loading. Bloom activity was also associated with reduced phytoplankton species richness and diversity.

Analysis of the zooplankton (Livingston, 2000) indicated that, with the exception of February 1994, the above-noted changes of phytoplankton composition during the sustained increases of orthophosphate loading were accompanied by increased zooplankton abundance. High zooplankton numbers at stations P23, P26, and P31 were associated with increased numbers of *C. choctawhatcheeana* and *M. throndsenii* during spring 1994. The zooplankton dominant, *Acartia tonsa*, tended to have comparable numbers between 1988–1989 and 1993–1994. Increased zooplankton numbers were associated with high numbers of phytoplankton from January 1994 through March 1994. Significant increases of zooplankton numbers occurred during April and May 1994 in Perdido Bay. Zooplankton species richness was significantly higher from September 1993 through January 1994 than that observed during the nonbloom period (1988–1989). However, relatively lower numbers of zooplankton species were noted during succeeding bloom months, which could have been associated with the high dominance of the leading species *A. tonsa* (Livingston, 2000). The inverse relationship of zooplankton numbers and zooplankton species richness could have been related to possible interspecific competition. Thus, the diatom blooms of 1993–1994 appeared to have little adverse effect on the zooplankton. These numbers actually increased during this initial bloom period and could have been associated with relatively steep declines of phytoplankton during late spring–summer months.

b. *Whole-Water Phytoplankton*

The initial response of the whole-water phytoplankton during 1993–1994 was a seasonal succession of winter–spring diatom blooms (*L. danicus*, *C. choctawhatcheeana*, *M. throndsenii*) that was similar to that described with the net plankton succession (Figure 6.11). Species such as *C. choctawhatcheeana*, *M. throndsenii*, *L. danicus,* and *Prorocentrum minimum* were present in the bay before the 1993–1994 increases in nutrient loading, whereas the diatoms *Synedropsis* sp. and *Urosolenia eriensis*, the blue-green algae *Merismopedia tenuissima*, and the raphidophytes (*Heterosigma akashiwo*, *Chattonella subsalsa*) were generally absent from the bay prior to such loading. The diatom *L. danicus* appeared to be a pioneer bloom species, occurring only once as a bloom during winter 1994. The diatom *Cyclotella choctawhatcheeana*, the raphidophyte *H. akashiwo*, and the dinoflagellates *P. minimum* and *Gymnodinium* spp. appeared to occur baywide, whereas other species were generally restricted to the upper bay. The blue-green algae, primarily *M. tenuissima*, were found largely in Elevenmile Creek and bay areas at the mouth of the creek (P23).

During 1995–1996, both ammonia and orthophosphate were being loading into the bay by the mill at unprecedented rates (see Figure 6.5). During this period and running into 1996–1997, the raphidophyte *H. akashiwo* was at peak bloom activity in the upper bay (Figure 6.11). Over the

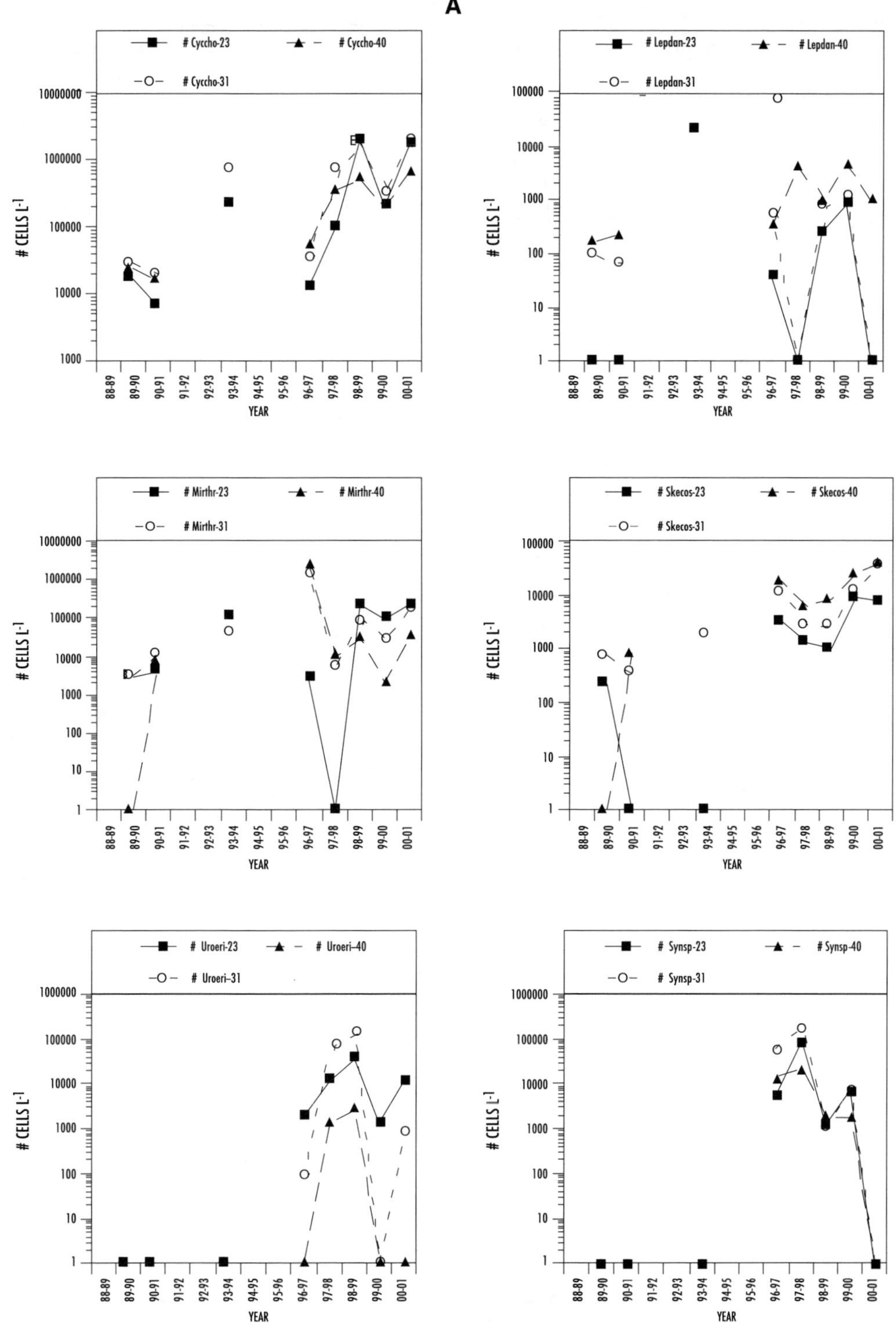

Figure 6.11 Annual averages of whole-water phytoplankton bloom species numbers per liter at stations P23, P31, and P40 in Perdido Bay taken over a 13-year period (1988 to 2001). (A) *Cyclotella choctawhatcheeana* (Cyccho), *Leptocylindrus danicus* (Lepdan), *Miraltia throndsenii* (Mirthr), *Skeletonemia costatum* (Skecos), *Urosolenia eriensis* (Uroeri), *Synedropsis* sp. (Synsp). (continued)

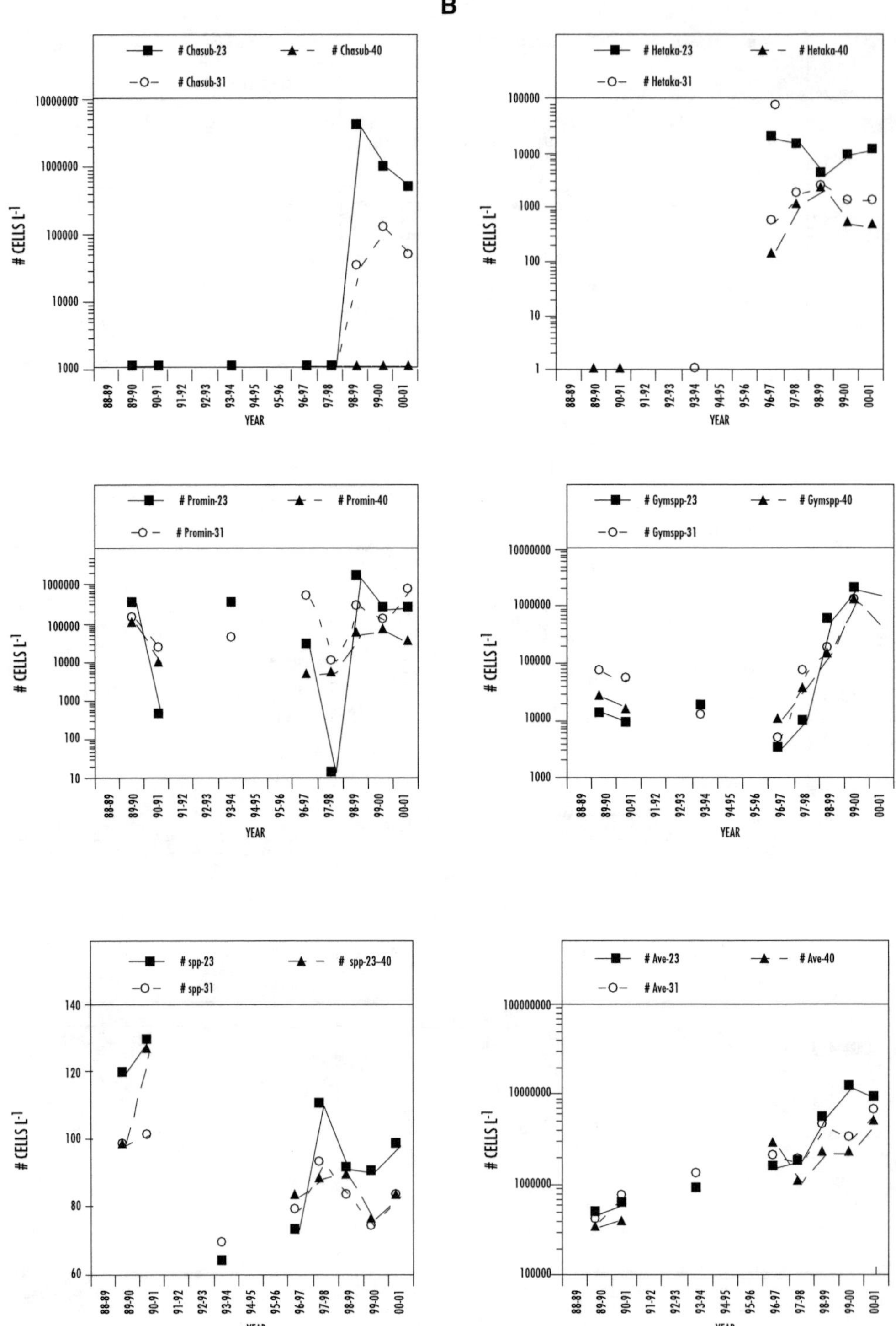

Figure 6.11 (continued) Annual averages of whole-water phytoplankton bloom species numbers per liter at stations P23, P31, and P40 in Perdido Bay taken over a 13-year period (1988 to 2001). (B) *Chattonella c.f. subsalsa* (Chasub), *Heterosigma akashiwo* (Hetaka), *Prorocentrum minimum* (Promin), *Gymnodinium* spp. (Gymspp), cumulative number of species (annual total), average numbers of cells per liter (12-month average).

following 2 years, there were reductions of this species in the upper bay, whereas its numbers increased in midbay to lower bay areas. From 1999 to 2001, there were reductions in these areas while *H. akashiwo* in the upper bay increased. During this period, there were also blooms of *Synedropsis* sp. (1997–1998), *Miraltia throndsenii* (1996–1997), *C. choctawhatcheeana* (1998–1999), *Chattonella c.f. subsalsa* (1998–1999), and *P. minimum* (1998–199). During the low orthophosphate loading (late 1997–1998), the cell numbers of various bloom species was down. No winter blooms were noted during this period, and *P. minimum* almost disappeared from the upper bay. This was the result of low phosphorus loading from the mill and high river flows during winter 1999. The resumption of orthophosphate loading during winter 1999 coincided with winter blooms of *P. minimum* in addition to increased bloom activity by *Synedropsis* sp., *M. throndsenii*, and *Cyclotella choctawhatcheeana*. The reduced levels of ammonia loading during 1998–1999 coincided with reductions of the bloom activity of *H. akashiwo* in the upper bay; *H. akashiwo* is a spring–summer bloom species. Continued orthophosphate loading during 2000 was associated with increased concentrations of *H. akashiwo*; resumption of ammonia loading during winter–spring of 2001 was accompanied by increased bloom activity by the spring–summer bloom species *C. choctawhatcheeana*, *M. throndsenii*, and *H. akashiwo*. During the last year of reduced mill nutrient loading (2001–2002), bloom formation ceased, an indication of the efficacy of the nutrient loading guidelines as part of the bay restoration program.

Phytoplankton community indices (Figure 6.11) indicated relatively low numbers of phytoplankton taxa during 1993–1994. By 1996–1997, low cumulative phytoplankton species richness was evident throughout the bay. There was a peak of post-bloom species richness in the upper bay during the period of low orthophosphate loading in 1997–1998. There were minor increases of this index from 1999 to 2001, although such numbers remained relatively low, especially in midbay and lower bay areas. Over the period of study, there was a general logarithmic increases in phytoplankton numbers in all parts of Perdido Bay from 1989–1990 through 2000–2001. Shannon diversity and evenness indices were lowest during periods of peak numbers of blooms and phytoplankton biomass (1995–1996), a period when peaks of orthophosphate and ammonia loading overlapped (Livingston, 2000). The general distribution of phytoplankton community indices were related to nutrient loading trends, distance from upper bay areas of high nutrient loading, and the succession of species-specific blooms.

When viewed as percentages of numerical abundance of the major bloom species through time (Figure 6.12), most of the primary populations showed declining percentages despite that there were major increases of phytoplankton numbers in Perdido Bay over the past 5 years. These changes occurred as there was a general decline in the overall species richness of the phytoplankton communities in the bay. The data thus indicated that there was another source of plankton increases that were not related to the individual bloom species noted in the previous analyses. The answer was found in changes of the nanoplankton in the system.

It has been established that nanoplankton (2 to 20 μm) increase in numbers with nutrification (Bell and Kalff, 2001) with the importance of the small phytoplankton relative to the larger plankton hypothesized to decline with increasing nutrient load. This has been explained in two ways (Bell and Kalff, 2001). First, it is possible that small plankton biomass is constrained by high loss rates due to predation. Second, it is hypothesized that the larger microalgae outcompete the smaller types for nutrients and are not preyed upon at the same level as the nanoplankton. The higher rates of sedimentation of the larger microalgae also may be a factor with system depth tending to complicate the relationship of the various size classes relative to the nutrient loading status of the system in question.

The overall importance of the nanoplankton is generally accepted (Van Valkenburg and Flemer, 1974). In the present study, we have five primary groups of nanoplankton that were numerically dominant in Perdido Bay but were not identified to species: nanococcoids (Nanoc), cryptophytes (Undcry), the pinnate diatoms (Undpendia), *Cryptomonas* spp. (Crysp), and unidentified nanoflagellates (Undnam). Nanococcoids represent a catch basket of species that cross taxonomic boundaries,

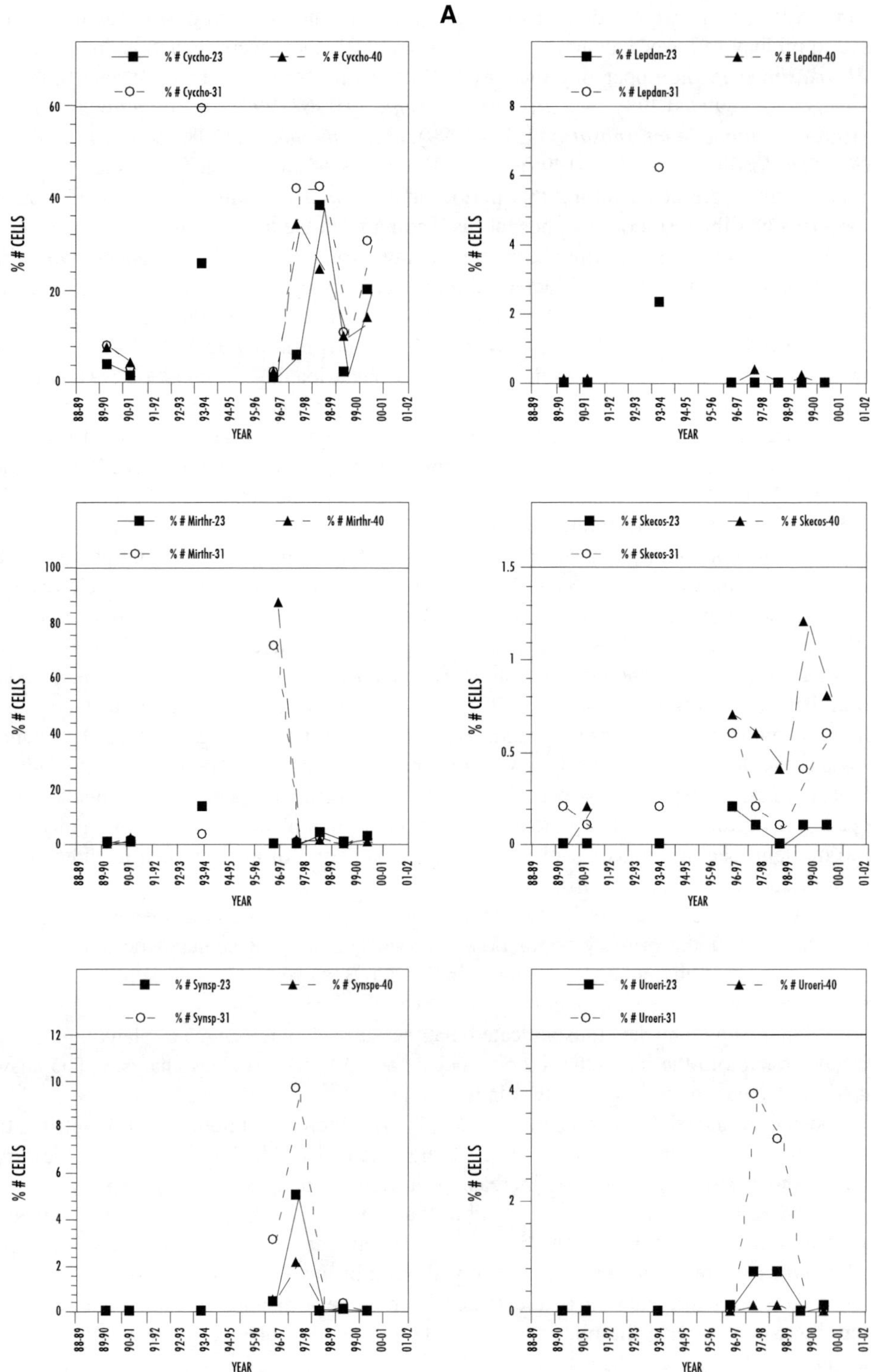

Figure 6.12 Annual averages of whole-water phytoplankton bloom species as percentage of numbers per liter at stations P23, P31, and P40 in Perdido Bay taken over a 13-year period (1988 to 2001). (A) *Cyclotella choctawhatcheeana* (Cyccho), *Leptocylindrus danicus* (Lepdan), *Miraltia throndsenii* (Mirthr), *Skeletonemia costatum* (Skecos), *Synedropsis* sp. (Synsp), *Urosolenia eriensis* (Uroeri). (continued)

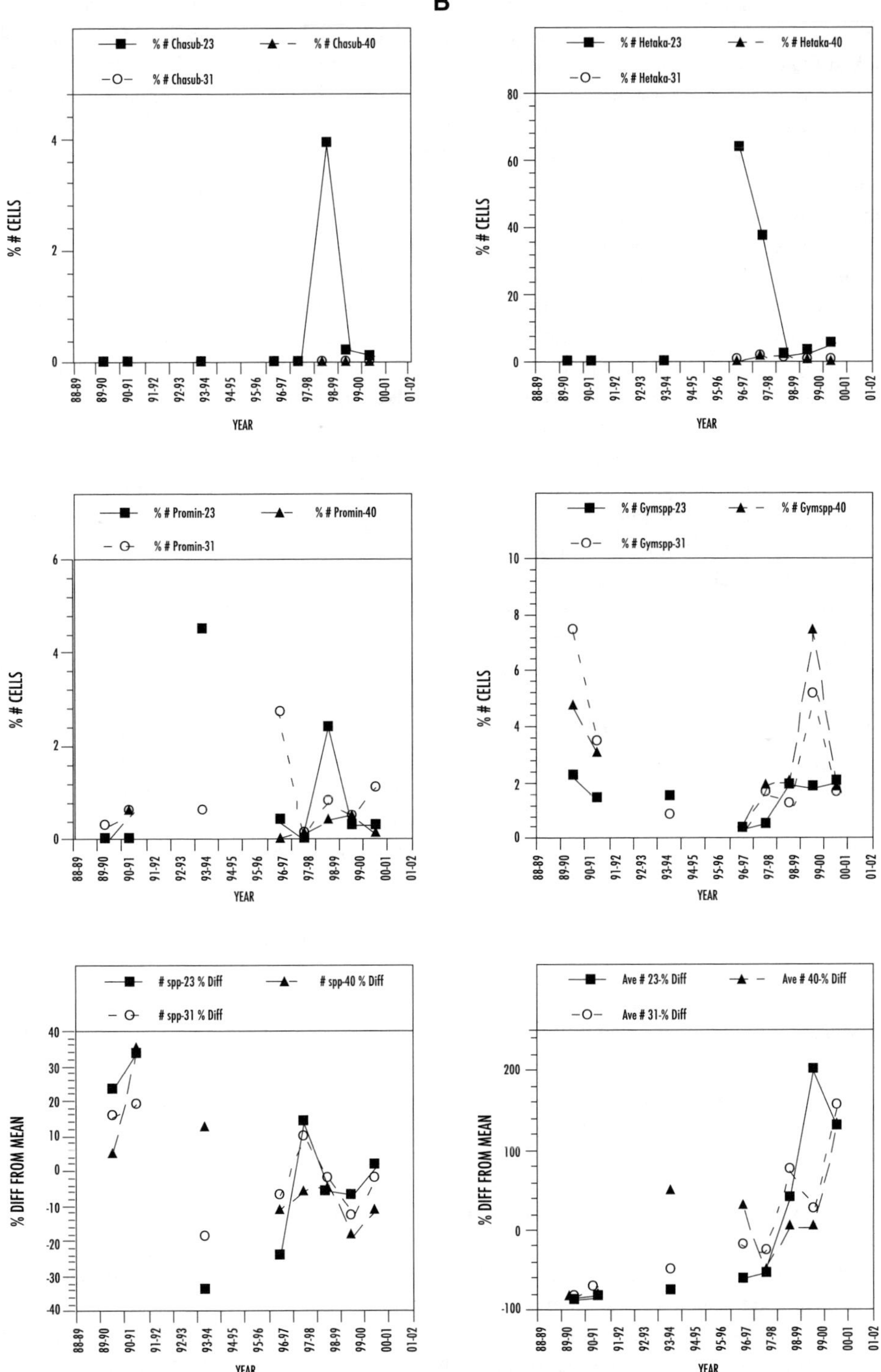

Figure 6.12 (continued) Annual averages of whole-water phytoplankton bloom species as percentage of numbers per liter at stations P23, P31, and P40 in Perdido Bay taken over a 13-year period (1988 to 2001). (B) *Chattonella c.f. subsalsa* (Chasub), *Heterosigma akashiwo* (Hetaka), *Prorocentrum minimum* (Promin), *Gymnodinium* spp. (Gymspp). Differences of the means of cumulative number of species (annual total) and differences of the means of average numbers of cells per liter (12-month average) are shown.

and are typically nonflagellated bloom types with cell walls that are spherical to elliptical in shape and often include blue-green algae (A. K. S. K. Prasad, personal communication). The cryptophytes are naked flagellates that are known to have high growth rates. This group is typified by a pellicle (covering) and trichocysts or organelles on the plasma membrane. The cryptophytes are not considered to be toxic. Pinnate diatoms are generally difficult to identify due to their size, and are also not generally considered as toxic algae. The nanoflagellates have no cell wall, and have well-developed chloroplasts. This group is also composed of various species that cross taxonomic lines, and are thought to be bloom forming.

The long-term distribution in space and time (Figure 6.13) indicates that all five groups had relatively substantial increases in numerical abundance during the period 1996 to 2001. The most abundant of the nanoplankton included the cryptophytes and the nanoflagellates, which were particularly abundant during the period 1999 to 2001. When viewed as percentages of the total numbers (Figure 6.14), the cryptophytes and the nanoflagellates reached relatively high percentages of the overall phytoplankton populations in Perdido Bay. However, only the cryptophytes reached percentages comparable to those taken during the earlier periods of lower nutrient loading. The other groups generally had lower percentages of the overall numerical abundances of phytoplankton in the system during later years of increased population growth. With the exception of *Cryptomonas* spp., the various nanoplankton groups appeared to be on the upswing in Perdido Bay, and these groups were generally distributed throughout the bay.

Elevenmile Creek is the primary source of increased ammonia and orthophosphate loading to the bay. A review of annual averages of various key elements of nutrient loading, habitat changes, and the response of phytoplankton assemblages (Figure 6.15) indicates that the orthophosphate and ammonia loading peak differencing and the orthophosphate and ammonia ratio peak differencing (1994 to 1997) were somewhat different in that loading patterns did not coincide with the nutrient concentration peaks during the 1998 to 2001 drought. The drought exacerbated the nutrient concentrations even though loading was lower than the previous years. Because of differences in the ammonia and orthophosphate loading and nutrient ratio scales, the loading and ratio data were standardized by adding the differenced data. This allowed an even greater differentiation of these indices due to the drought conditions. During flooding events, increased loading was associated with lower concentrations due to dilution effects. The constructed N+P indices standardized the differences. Chlorophyll *a* concentrations in Elevenmile Creek generally were highest during the three drought periods. These peaks were not associated with peak annual averages of phytoplankton numbers, which were generally highest during the bloom years (1996 to 2001). The pattern of blue-green algae blooms in Elevenmile Creek (dominated by *Merismopedia tenuissima*) were followed by the TOC trends in the creek. Phytoplankton species richness and diversity were inversely related to biomass/numerical abundance and bloom occurrence. Although the annual averages of these factors tended to obscure specific seasonal trends, these figures indicate that interannual trends are consistent with relationships reported previously (Livingston, 2000).

A series of simple regressions were run with the top three bloom species as dependent variables over the 13-year period of annual averaging (Table 6.3). To evaluate the possible interactions of ammonia and orthophosphate loading rates and concentration gradients on blooms of *Prorocentrum minimum*, the winter loading and concentration ratio data were run through a differencing program that was designed to calculate (for each month in the time series) the difference between the monthly value and the long-term average for that month of the year. By using averages of the differenced data for orthophosphate and ammonia loadings and concentration ratios, the data were standardized so that the combined effects of these two nutrients could be evaluated.

The dinoflagellate *P. minimum* was significantly associated (positively) with orthophosphate ratios, which is consistent with this winter dominant that usually blooms during periods of orthophosphate limitation. There were no close associations of the diatom *Cyclotella choctawhatcheeana*. This may be explained by the fact that blooms of this species usually occurred in the bay and not

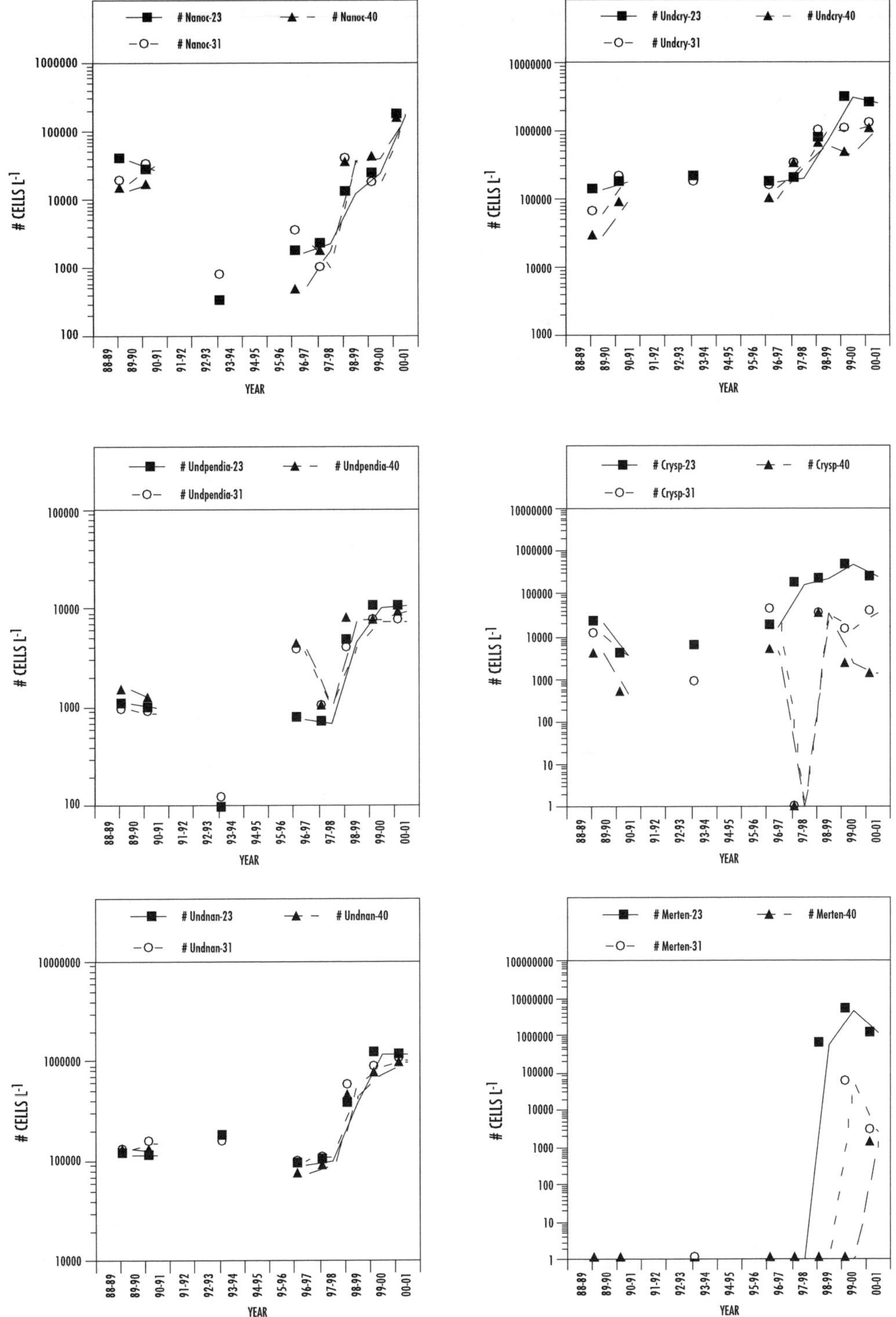

Figure 6.13 Annual averages of nanoplankton numbers per liter at stations P23, P31, and P40 in Perdido Bay taken over a 13-year period (1988 to 2001). Nanoplankton were ordered as nanococcoids (Nanoc), cryptophytes (Undcry), the pinnate diatoms (Undpendia), *Cryptomonas* spp. (Crysp), and unidentified nanoflagellates (Undnan). Also shown are the distributions of blue-green alga *Merismopedia tenuissima* (Marten).

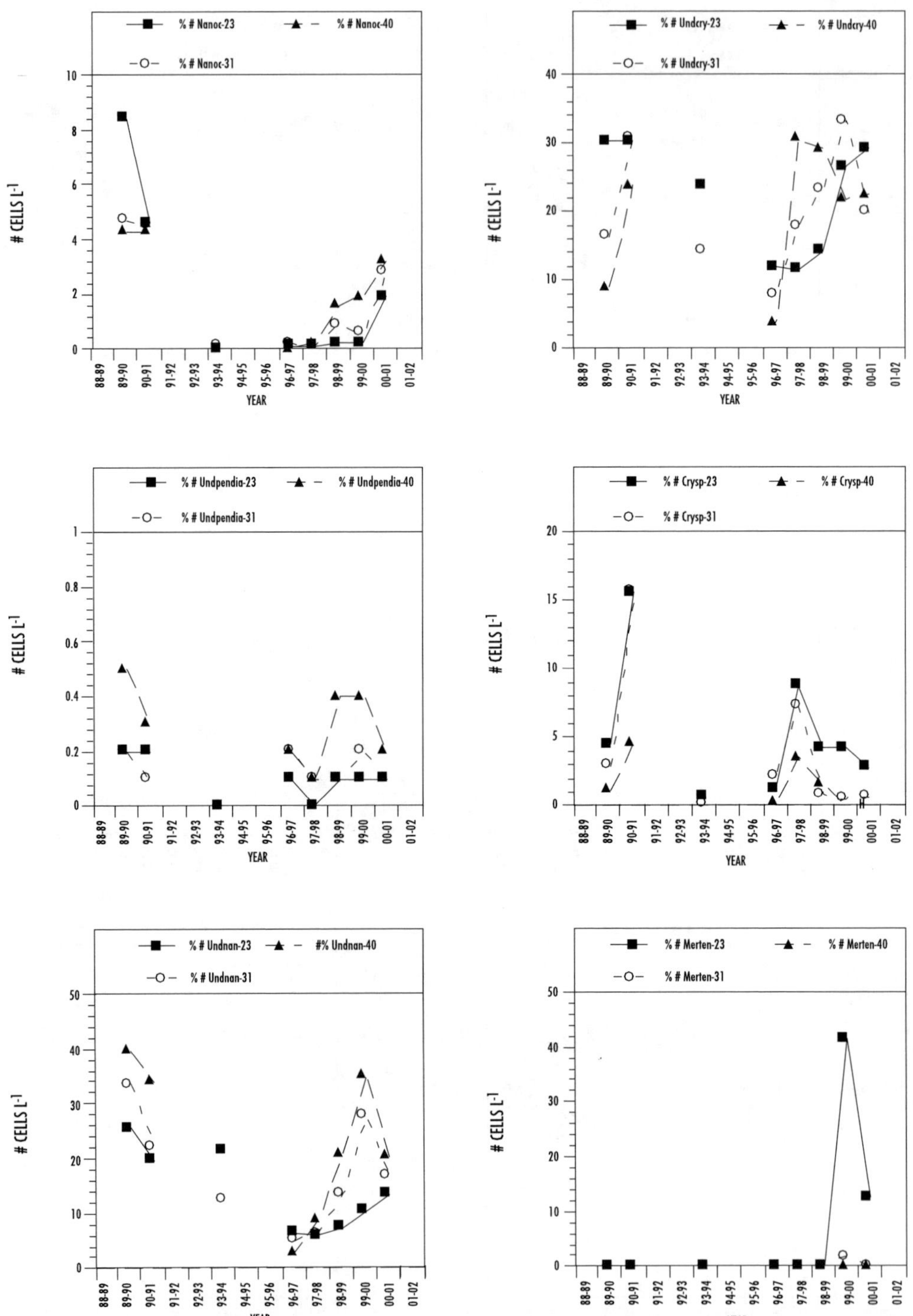

Figure 6.14 Annual averages as percentage of nanoplankton numbers per liter at stations P23, P31, and P40 in Perdido Bay taken over a 13-year period (1988 to 2001). Nanoplankton were ordered as nanococcoids (Nanoc), cryptophytes (Undcry), the pinnate diatoms (Undpendia), *Cryptomonas* spp. (Crysp), and unidentified nanoflagellates (Undnan). Also shown are percent abundance of the blue-green alga *Merismopedia tenuissima* (Merten).

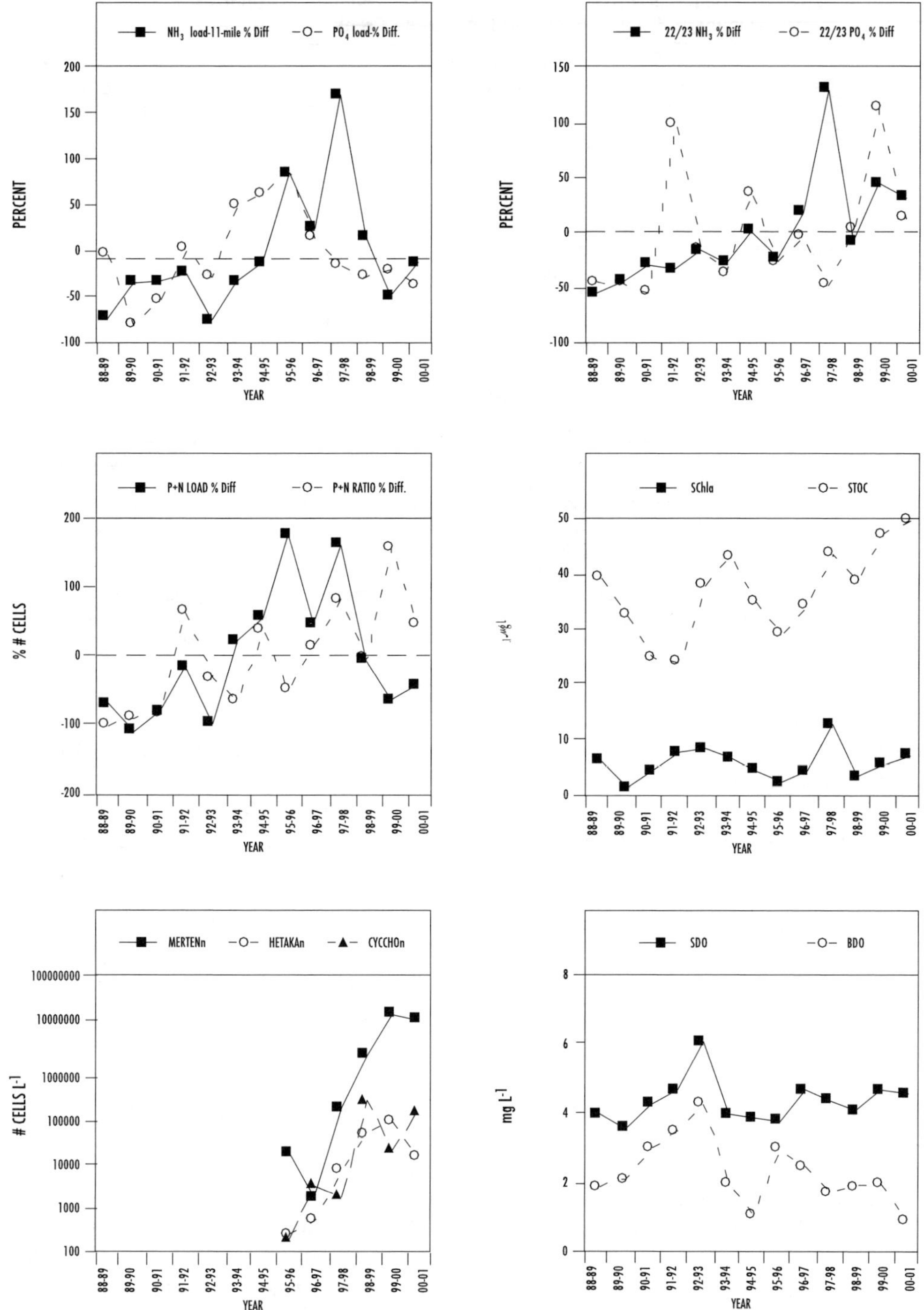

Figure 6.15 Data taken at station P22 (Elevenmile Creek) as annual averages from 1988 to 2001. Ammonia and orthophosphate loading, as percent differences from the grand mean; ammonia and orthophosphate ratios (P22/P23) as percent differences from the grand mean; ammonia + orthophosphate load percent differences (added) from the grand mean and ammonia + orthophosphate ratios (P22/P23) as percent differences (added) from the grand mean; surface chlorophyll *a* and surface total organic carbon (TOC); numbers per liter of the top bloom species, *Merismopedia tenuissima* (Merten), *Heterosigma akashiwo* (Hetaka), and *Cyclotella choctawhatcheeana* (Cyccho); surface and bottom dissolved oxygen.

Table 6.3 Regressions Run with Annual Averages of Numerical Abundance of the Three Dominant Bloom Species in Elevenmile Creek

Dependent Variable	Independent Variable	r^2	Significance	Sign
P. minimum				
	P22/P23 PO_4 ratios	0.81	0.03	+
C. choctawhatcheeana				
M. tenuissima				
	ssalinity	0.81	0.002	+
	sTOC	0.6	0.05	+
	pH	0.74	0.02	+
	Perdido NH_3 load	0.75	0.02	–
	11 Mile NH_3 load	0.55	0.05	–
	Diff. PO_4/NH_3 load	0.62	0.05	–
	Phytoplankton div.	0.61	0.05	–

Notes: Regressions are run on the three dominant species (*Cyclotella choctawhatcheeana, Prorocentrum minimum, Merismopedia tenuissima*) as dependent variables against annual averages of independent variables (salinity [s], dissolved oxygen [s,b], total organic carbon [sTOC], ammonia [s], orthophosphate [s], P22/P23 orthophosphate ratio, P22/P23 ammonia ratio, Elevenmile Creek orthophosphate loading, Elevenmile Creek ammonia loading, Perdido River orthophosphate loading, Perdido River ammonia loading rates, differenced ammonia + orthophosphate loading rates, differenced ammonia + orthophosphate ratios, silica [s], pH [s], Secchi depth, cumulative number of phytoplankton taxa, cumulative phytoplankton Shannon diversity indices). Only significant ($p \leq 0.05$) regressions are given.

the creek, and it is likely that diatoms in the creek originated in the bay. The primary dominant in Elevenmile Creek was *M. tenuissima*; blooms were noted during the drought of 1998 to 2002. Numerical abundance of this blue-green alga was closely associated positively with salinity. Other positive associations included TOC and pH, which appear to be reliable indicators of bloom activity in the creek. The negative associations with ammonia loading from the Perdido River and Elevenmile Creek are consistent with loading patterns. Although nutrient ratios were not significantly associated with *M. tenuissima* abundance, there were relatively close positive associations with orthophosphate ratios ($r^2 = 0.45$, $p = 0.07$) and the N+P differenced data ($r^2 = 0.40$, $p = 0.09$), which would indicate that the high orthophosphate concentrations during the drought were associated with the increases of this bloom species. The significant negative association with phytoplankton diversity indices reflects the impact of *M. tenuissima* dominance on this index. The lack of a significant association with phytoplankton species richness indices indicates no serious impact of this species on other plankton populations.

Interannual habitat changes in Perdido Bay (Figure 6.16) were indicative of basic changes in the Perdido Bay habitat during the period of observation. Ammonia concentrations were generally low in the midbay and lower bay areas with such concentrations peaking during the period 1989 to 1992 at station P23. There were generally low ammonia levels at this station during the three drought periods. There was a general downward trend of ammonia over the study period that ran contrary to the ammonia loading pattern. Orthophosphate was also lower in the midbay to lower bay areas with peaks in the upper bay occurring during 1993–1994 and 1995–1996 (peak loading years). The lowest such concentrations occurred during the last drought period. Chlorophyll *a* concentrations peaked during the three drought periods and were highest during 1988–1989 at which time there were all-time lows of phytoplankton cell numbers (and no observed plankton blooms). These data reflect the strong influence of drought on interannual changes of bay habitat conditions. At the same time, factors such as surface pH and DO tended to follow interannual trends of drought–flood cycles and phytoplankton abundance. The pH levels were generally highest during periods of high phytoplankton numbers. Bottom DO in the bay tended to decrease over the

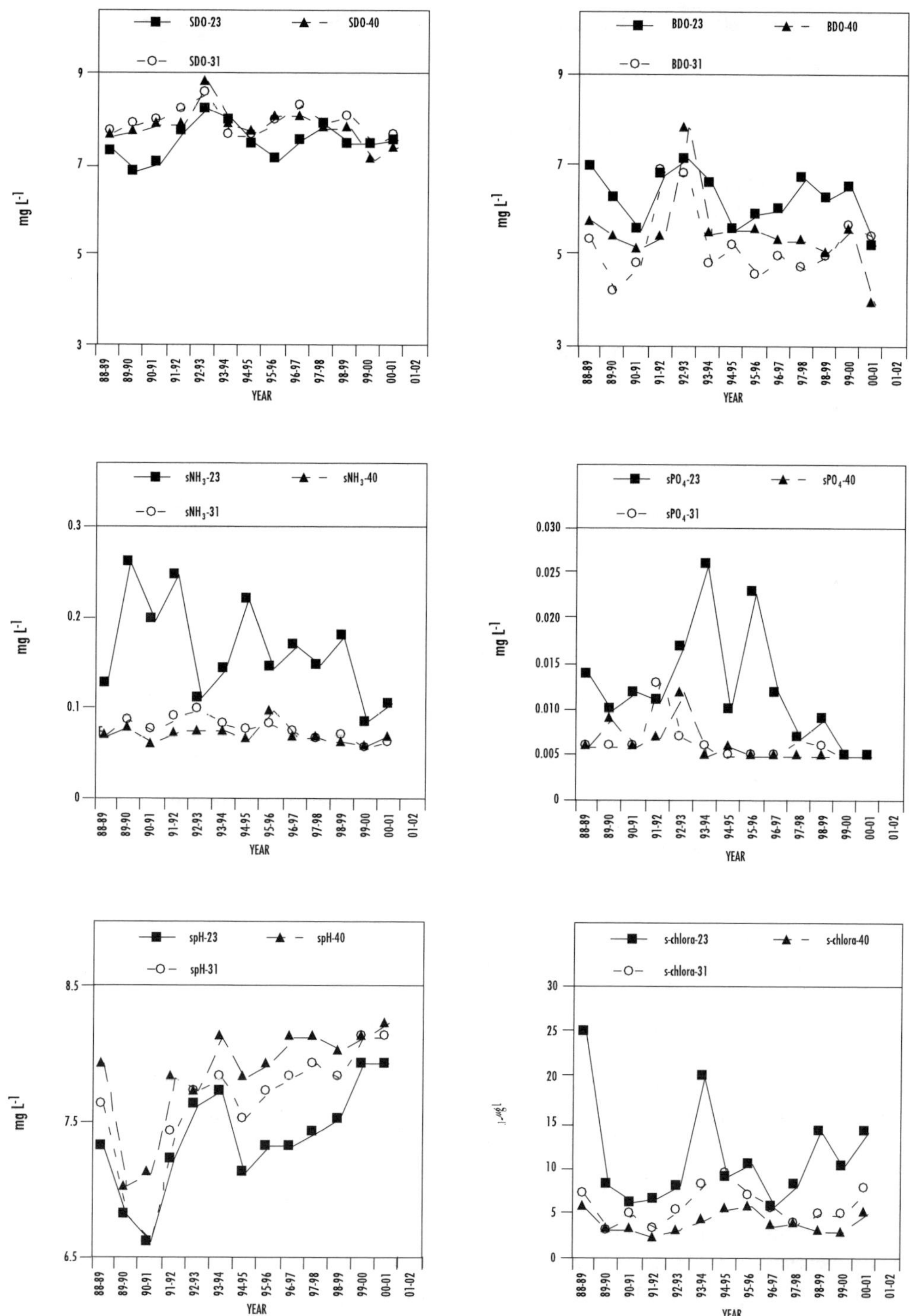

Figure 6.16 Data taken at stations P23, P31, and P40 as annual averages from 1988 to 2001. Surface and bottom dissolved oxygen, surface ammonia and orthophosphate concentrations, surface pH, and surface chlorophyll *a*.

Table 6.4 Regressions Run with Annual Averages of Numerical Abundance of Dominant Bloom Species in Perdido Bay as Dependent Variables

Dependent Variable	Independent Variable	r^2	Significance	Sign
P. minimum[a]				
	P22/P23 PO_4 ratios	0.44	0.05	+
	Phytoplankton no. of taxa	0.53	0.04	–
C. choctawhatcheeana[b]				
	P22/P23 NH_3 ratios	0.62	0.02	+
	Diff. PO_4/NH_3 ratios	0.79	0.003	+
M. throndsenii[b]				
	Diff. PO_4/NH_3 ratios	0.73	0.01	+
H. akashiwo[c]				
	NH_3 loading	0.4	0.05	+
	Phytoplankton no. of taxa	0.83	0.01	–
	Diff. PO_4/NH_3 ratios	0.89	0.005	+
	Diff. PO_4/NH_3 loading	0.66	0.05	+
Synedropsis sp.[c]				
	P22/P23 PO_4 ratios	0.76	0.05	+
	Phytoplankton no. of taxa	0.75	0.05	–
	Diff. PO_4/NH_3 ratios	0.64	0.05	+
	Diff. PO_4/NH_3 loading	0.87	0.02	+

Notes: (a) Winter: *Prorocentrum minimum* (Promin); (b) spring: *C. choctawhatcheeana* and *Miraltia throndsenii*; (c) summer: *Heterosigma akashiwo* and *Synedropsis* sp. Independent variables included annual averages of P22/P23 orthophosphate ratio, P22/P23 ammonia ratio, Elevenmile Creek orthophosphate loading, Elevenmile Creek ammonia loading, rates, differenced ammonia + orthophosphate loading rates, differenced ammonia + orthophosphate ratios, cumulative number of phytoplankton taxa, cumulative phytoplankton Shannon diversity indices). Only significant ($p \leq 0.05$) regressions are given.

study period with lows occurring during the increases in phytoplankton biomass and TOC during the drought of 1998 to 2001. The lowest such concentrations were in the depositional areas of the lower bay. These data indicate that phytoplankton abundance was strongly related to key habitat variables such as pH, bottom DO, and nutrient concentrations. Chlorophyll *a*, however, was not a reliable indictor of bloom incidence.

Based on the seasonal occurrences of the various bloom species (see Table 6.1), the long-term database was reorganized by winter (December–February), spring (March–May), and summer (April–August) sequences (Table 6.4; Figure 6.17). Some overlapping months between seasons were used based on the monthly patterns of species-specific bloom sequences.

Winter data (Table 6.4; Figure 6.17) indicated increased orthophosphate loading from Elevenmile Creek from 1992 through 1996 with increased orthophosphate loading in February 1999 and January–February 2002. Increased ammonia loading from Elevenmile Creek was indicated from 1997 to 1999 with smaller increases during winter 2000–2001. The diatom *Cyclotella choctawhatcheeana,* a species that occasionally bloomed during winter months, showed baywide increases during winters from 1996 to 2001. The dinoflagellate *Prorocentrum minimum* reached winter peaks in the upper bay during the winters of 1994 and 1999. There were general increases of this species during winter 2001. Regressions of the distribution of *P. minimum* indicated significant positive associations with orthophosphate concentration ratios. The association of this species

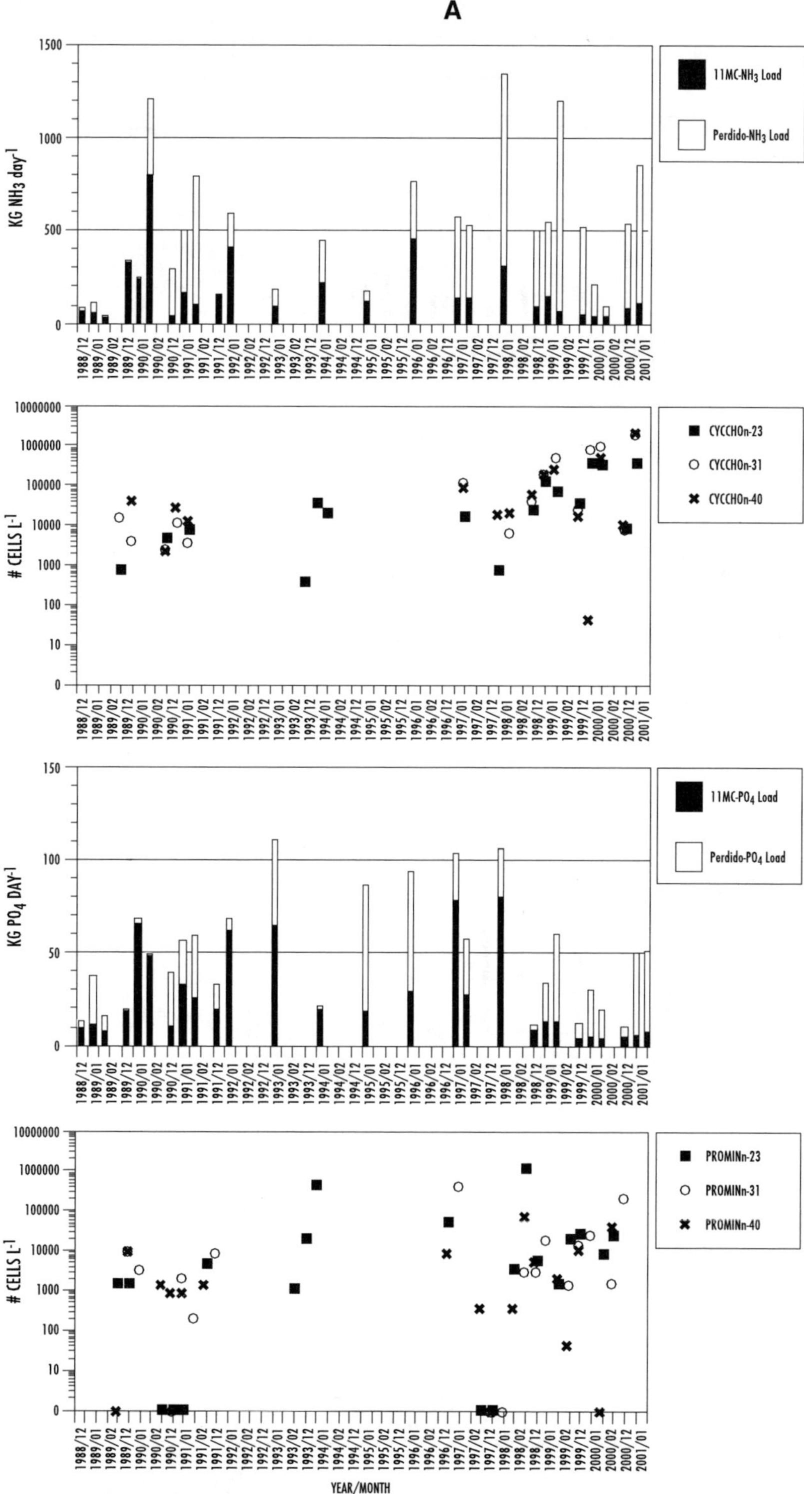

Figure 6.17 Seasonal monthly ammonia and orthophosphate loading data from the Perdido River and Elevenmile Creek along with numerical abundance of the top bloom species at stations P23, P31, and P40 from 1988 to 2002. (A) Winter: *Cyclotella choctawhatcheeana* (Cyccho) and *Prorocentrum minimum* (Promin). (continued)

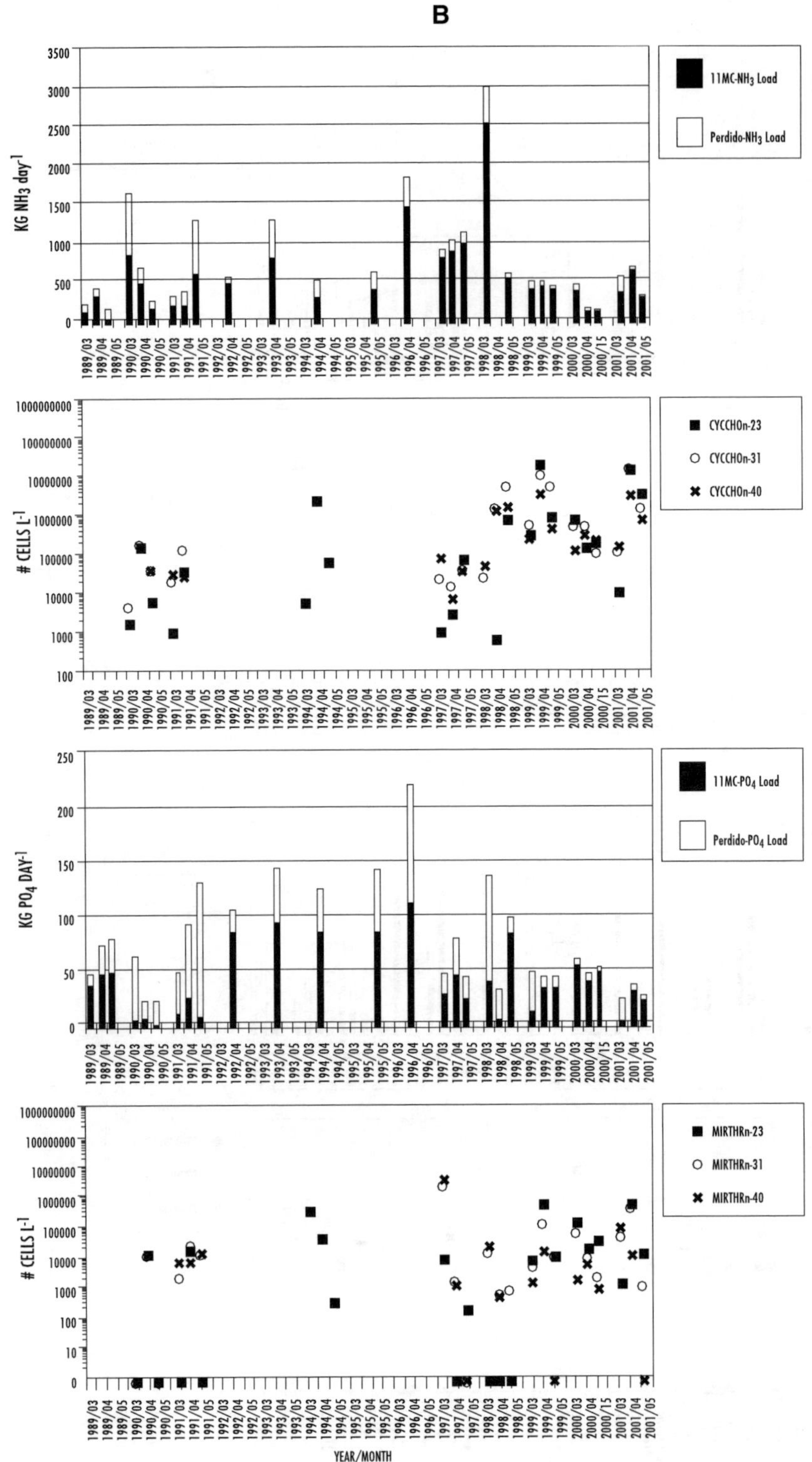

Figure 6.17 (continued) Seasonal monthly ammonia and orthophosphate loading data from the Perdido River and Elevenmile Creek along with numerical abundance of the top bloom species at stations P23, P31, and P40 from 1988 to 2002. (B) Spring: *Cyclotella choctawhatcheeana* (Cyccho) and *Miraltia throndsenii* (Mirthr). (continued)

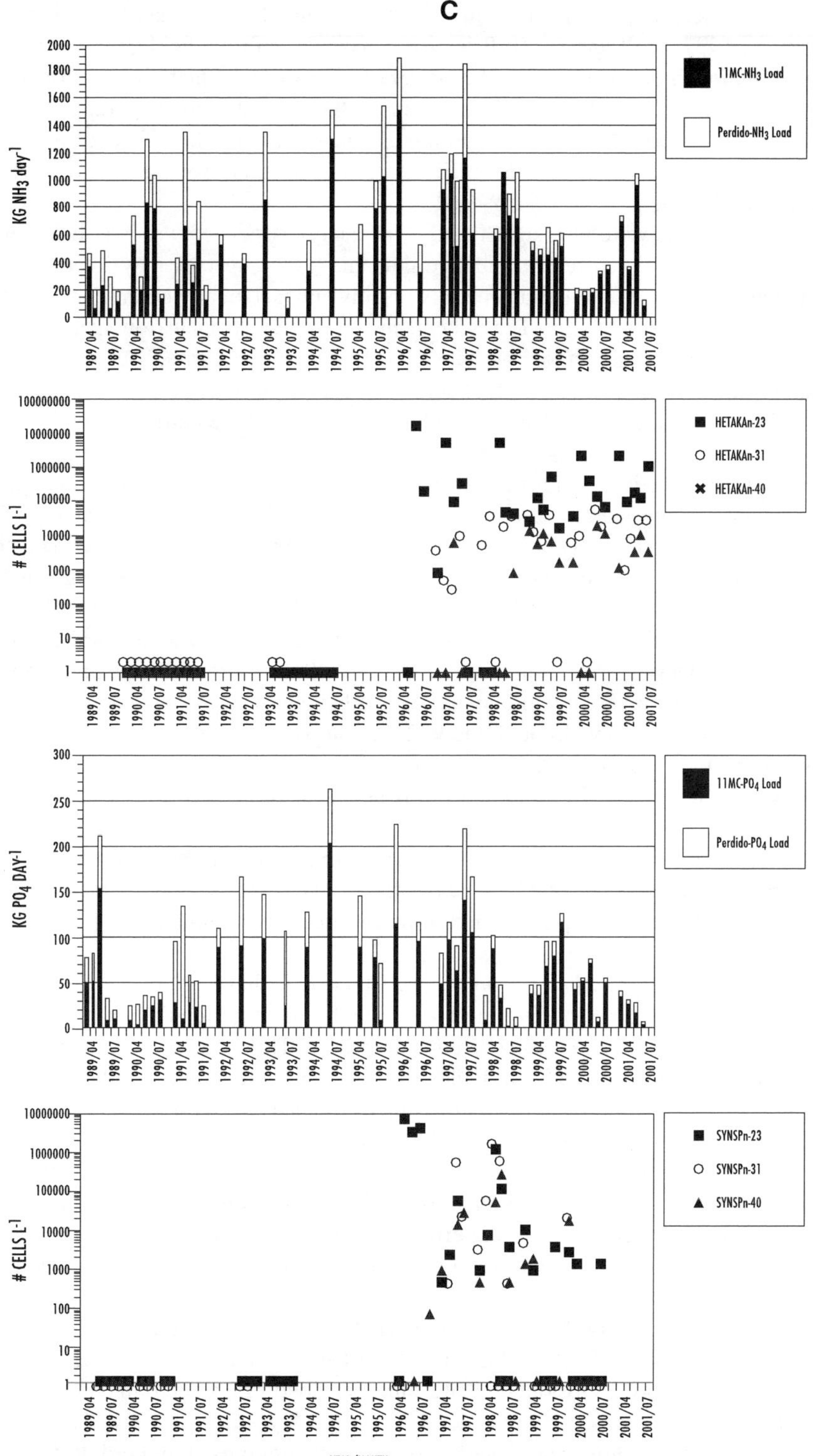

Figure 6.17 (continued) Seasonal monthly ammonia and orthophosphate loading data from the Perdido River and Elevenmile Creek along with numerical abundance of the top bloom species at stations P23, P31, and P40 from 1988 to 2002. (C) Summer: *Heterosigma akashiwo* (Hetaka) and *Synedropsis* sp. (Synsp).

with orthophosphate concentrations is consistent with the results of the nutrient limitation experiments (Flemer et al., 1997; see Figure 6.7). It is also consistent with the field data whereby very low numbers of *P. minimum* occurred during the winter of 1998 during which time orthophosphate loading from the mill had been substantially reduced. Orthophosphate loading rates during the drought of 1988 to 2002 were relatively low, and this is consistent with the effects of the drought whereby high orthophosphate concentrations in Elevenmile Creek were somehow related to stimulation of *P. minimum* blooms.

The *P. minimum* numbers were also significantly (inversely) associated with whole-water phytoplankton species richness. These results indicate that this bloom species may interfere with a number of phytoplankton species and has a generally negative impact on phytoplankton community structure in many areas of Perdido Bay. The general increases of phytoplankton species richness in upper Perdido Bay during the orthophosphate reduction program carried out by the mill in 1997–1998 along with general increases of Shannon diversity and evenness during this period present further evidence of the relationship of this bloom species with the Perdido Bay phytoplankton community.

Analyses of the long-term distributions of spring information (Table 6.4; Figure 6.17) indicated that increased orthophosphate loading started during spring 1992 and continued at high levels through spring 1996. Thereafter, spring orthophosphate loading remained relatively low. The diatom *C. choctawhatcheeana* had increased numbers (relative to 1990–1991) during spring 1994; this coincided with increased orthophosphate loading from the mill (Elevenmile Creek). There were subsequent low numbers of this species during spring 1997 that coincided with low orthophosphate loading from the mill. During the springs of 1999 and 2001, there were peak numbers of *C. choctawhatcheeana* in the upper and middle parts of the bay. Both ammonia and orthophosphate loading from the mill remained low although the P22/P23 ratios were higher due to the prolonged drought during this period (see Figure 6.5). Numerical abundance of *C. choctawhatcheeana* was significantly associated with P22/P23 ammonia ratios and with combined (averaged) differences of orthophosphate and ammonia P22/P23 ratios taken during spring periods from 1989 through 2001. This result is further evidence that both ammonia and orthophosphate concentrations combine in the stimulation of spring blooms of *C. choctawhatcheeana*. There was no evidence that this species had an adverse effect on other phytoplankton species in terms of cumulative species richness and diversity indices.

The spring-blooming diatom *Miraltia throndsenii* tended to have relatively high numbers during 1994 (high orthophosphate loading) and 1997 (high ammonia loading) with subsequent decreasing numbers as the loading of these nutrients decreased. This species was not significantly associated with any of the loading ratio variables or with the phytoplankton indices. However, there was a significant association of numbers of cells of this species with the mean differences of the nutrient ratio means, which would explain the increased numbers during the latest drought. Thus, both of the spring dominant bloom species appeared to be most associated with nutrient gradients (orthophosphate + ammonia), which is further indication that N+P concentrations stimulate the spring bloom species. The strong effects of orthophosphate and N+P concentrations during the nutrient limitation experiments (see Figure 6.7) support the field associations of these species with these nutrients.

The long-term summer data (Table 6.4; Figure 6.17) indicated that ammonia loading from Elevenmile Creek peaked during the period from 1994 to 1998 with decreases during 1999–2000 and another increase in 2001. Orthophosphate summer loading peaked in summers from 1994–1995 after which there was a general decrease to the low level in 2001. The dominant summer bloom species, *H. akashiwo* and *Synedropsis* sp., were not present in Perdido Bay during the early sampling years (1990–1991; 1993–1994). Peak numbers of *H. akashiwo* in the upper bay were noted during summer 1996, a period of high ammonia and orthophosphate loading. There was a general decline of this species in the upper bay during succeeding summers of reduced ammonia and orthophosphate loading. However, *H. akashiwo* numbers in mid to lower parts of the bay tended to rise during this period. Although these numbers were substantially lower than those in the upper bay, there appeared to be differences in the distribution of *H. akashiwo* from 1996 to 2001. This same temporal pattern was noted in the distribution of *Synedropsis* sp. This species also had peaks during the high nutrient loading year of 1996. However, unlike *H. akashiwo*, this species had much steeper decreases in numerical abundance, and, by 2000–2001, had virtually disappeared from the Perdido system.

The relationships of the *H. akashiwo* numbers relative to the key loading and phytoplankton community indices (Table 6.4) indicated significant positive correlations with ammonia loading rates. *Heterosigma akashiwo* numbers were significantly (positively) associated with combined differences of the orthophosphate and ammonia concentration ratios and with the differenced N+P loading data. This result is consistent with the nutrient limitation results (see Figure 6.7), which showed that N+P concentrations were limiting during summer months. The *Heterosigma* blooms were negatively associated with summer phytoplankton species richness indices; this is consistent with background literature concerning this species. Regression analyses of the *Synedropsis* sp. data indicated significant relationships with the P22/P23 orthophosphate ratios, and the N+P differenced data for both concentration ratios and loading. There was also a significant negative correlation of *Synedropsis* sp. with phytoplankton species richness.

The spatial differences of the temporal patterns of *Heterosigma* and *Synedropsis* abundance may be indicative of the complicated patterns of bloom occurrence in the Perdido Bay system. The population increases of both species in midbay to lower bay areas as such numbers decreased in the upper bay could be due to changes related to sedimentation (i.e., nutrient deposition) rates in the depositional areas of the lower bay. According to Niedoroda (1992), comparative analyses of two surveys taken in 1937 and 1991 indicated that sediment provided by the drainages into upper Perdido Bay accumulated in the lower bay. Livingston (2000) showed that phosphorus deposition in sediments followed nutrient loading patterns with immediate deposition in upper bay areas, and delayed deposition in midbay to lower bay areas. The highest sediment phosphorus concentrations in Perdido Bay were found at stations P26, P29, P31, P33, and P37 and in parts of Wolf Bay (Livingston, 2000). These areas coincided generally with depositional areas in deeper sections of the bay (i.e., station P40). Sediments in upper parts of Perdido Bay had relatively lower phosphorus concentrations. Coffin and Cifuentes (1992, 1999), in a series of stable isotope analysis of the nutrient cycling in Perdido Bay, found that primary production was linked to an important fraction of the ^{13}C-depleted suspended particulate matter (SPM), and that phytoplankton activity was responsible for such production. Phytoplankton appeared to be an important source of the organic carbon pool in Perdido Bay sediments. These studies thus made the connection between phytoplankton in the water column and benthic processes that were important to the overall food web of the bay (Livingston, 2000). It is possible that the nutrient loading to the upper bay, along with associated bloom activity, could have had a delayed effect on midbay and lower bay areas so that delayed nutrient deposition, and associated nutrient mobilization from the sediments into the water column, could have accounted for the delayed response of *Heterosigma* and *Synedropsis* abundances in the lower bay.

The data (Table 6.4) indicate that the primary bloom species in Perdido Bay are closely associated with both loading and concentrations of ammonia and orthophosphate on a seasonal basis. Ambient nutrient concentrations in Perdido Bay depend on complex interactions with the phytoplankton community. The questions of nutrient colimitation (Fuhs et al., 1972; Gerhart and Likens, 1975; Norin, 1977; Caraco et al., 1987; Suttle and Harrison, 1988; Paasche and Erga, 1988; Pennock and Sharp, 1994) and incipient colimitation (Powers et al., 1972; Graneli, 1987; Kivi et al., 1993) are directly related to such nutrient concentrations. Nutrient N+P co-limitation of phytoplankton growth rates (Flemer et al., 1997) in Perdido Bay may at times involve an escape of these phytoplankton assemblages from grazing pressure. The complexity of the interaction of these phytoplankton assemblages and ambient nutrient concentrations confounds a clear determination of exact cause-and-effect relationships. However, the relative stability of the seasonal succession

Table 6.5 Regressions Run with Annual Averages (Stations P23, P31, P40) of Numerical Abundance of Nanoplankton Groups in Perdido Bay

Dependent Variable	Independent Variable	r^2	Significance	Sign
Undetermined cryptophytes				
	Perdido-PO_4 load	0.79	0.02	–
	Perdido-NH_3 load	0.56	0.03	–
	sNH3 (22/23)	0.44	0.05	+
	Diff. PO_4/NH_3 ratios	0.82	0.001	+
	Gymspp.	0.91	0.0003	+
	Merten	0.57	0.03	+
Undetermined nanoflagellates				
	av s DO	0.46	0.05	–
	Perdido-PO_4 load	0.69	0.01	–
	Perdido-NH_3 load	0.53	0.04	–
	sNH_3 (22/23)	0.41	0.05	+
	Diff. PO_4/NH_3 ratios	0.78	0.003	+
	Gymspp.	0.90	0.0003	+
	Merten	0.55	0.04	+
Undetermined pennate diatoms				
	Perdido-PO_4 load	0.60	0.02	–
	Perdido-NH_3 load	0.49	0.05	–
	Diff. PO_4/NH_3 ratios	0.72	0.008	+
	Gymspp.	0.84	0.003	+
	Merten	0.53		+
Cryptomonas spp.				
	Perdido-PO_4 load	0.50	0.05	–
	sPO_4 (22/23)	0.49	0.05	+
	Diff. PO_4/NH_3 ratios	0.74	0.006	+
	Gymspp.	0.87	0.0007	+
	Merten	0.75	0.006	+
Undetermined nanococcoids				
	av b DO	0.46	0.05	–
	sNH_3 (22/23)	0.67	0.01	+

Note: Regressions run with dependent variables against annual averages of independent variables (salinity [s], dissolved oxygen [s,b], total organic carbon [sTOC], ammonia [s], orthophosphate [s], P22/P23 orthophosphate ratio, P22/P23 ammonia ratio, Elevenmile Creek orthophosphate loading, Elevenmile Creek ammonia loading, Perdido River orthophosphate loading, Perdido River ammonia loading rates, differenced ammonia + orthophosphate loading rates, differenced ammonia + orthophosphate ratios, silica [s], pH [s], Secchi depth, cumulative number of phytoplankton taxa, cumulative phytoplankton Shannon diversity indices, and numbers per liter of *Cyclotella choctawhatcheeana*, *Prorocentrum minimum*, *Merismopedia tenuissima*, *Miraltia throndsenii*, *Heterosigma akashiwo*, *Synedropsis* sp., *Chattonella c.f. subsalsa*, and *Gymnodinium* spp. Only significant ($p \leq 0.05$) regressions are given.

of bloom species in the Perdido system indicates that regular seasonal changes of bloom successions interact with nutrient loading/nutrient concentration phenomena to produce the complicated interannual bloom successions that have been documented in the Perdido system. In addition, bloom species such as *P. minimum, H. akashiwo,* and *Synedropsis* sp. appear to have serious, adverse effects on the phytoplankton assemblages of Perdido Bay as measured by reduced species richness and diversity indices.

To determine the relationships of the nanoplankton groups to nutrient conditions and bloom distributions in space and time, a series of regressions were run (Table 6.5). The cryptophytes were generally associated with the differenced orthophosphate/ammonia concentration ratios, *Gymnodinium* spp., and the blue-green alga *Merismopedia tenuissima*. These relationships were also evident for the nanoflagellates, the pennate diatoms, and *Cryptomonas* spp. Also noted as a negative association with these groups was the orthophosphate and ammonia loading for the Perdido River system (except ammonia loads for *Cryptomonas* spp.). These data indicate that the drought conditions during the last few years were implicated in the massive increases of these nanoplankton groups. Surface ammonia concentrations were positively associated with the nanococcoids, which is consistent with the drought hypothesis. The nanococcoids were negatively associated with bottom DO and the nanoflagellates were negatively associated with surface DO, which would explain the reductions of DO in Perdido Bay during the latest drought. These trends also show that the pattern of recovery of the bay as nutrient loading from the mill is reduced will not be the mirror image of the noted impacts. Recovery will undoubtedly be related to continued low nutrient loading and the climactic conditions during this phase of the interactions of nutrients with the phytoplankton assemblages in Perdido Bay. The recent reduction of blooms with reduced mill nutrient loading (see Table 6.1) is evidence that increased orthophosphate and ammonia loading was responsible for the formation of phytoplankton blooms in Perdido Bay.

G. Phytoplankton Changes in Wolf Bay

Wolf Bay (see Figure 2.2) has been subjected to unregulated agricultural runoff for years (Livingston et al., 1998a; Livingston, 2000). The Wolf Bay drainage basin has been removed from federal land use plans. Agricultural activities have increased in recent times (Livingston, 2000). The Wolf Bay system has been sampled for water quality since October 1993. Whole-water phytoplankton samples have been taken monthly in Wolf Bay from August 1998 through the present time. Sediment quality in Wolf Bay is similar to that in mid to lower Perdido Bay (Livingston, 2000). Agricultural activities in the Wolf Bay drainage basin (Livingston et al., 1998a) are responsible for extensive nutrient loading to the loading Perdido Bay system (Livingston, 2000), and this non-point, anthropogenous source of nutrients thus presents an interesting comparison with the point-source nutrient loading in the upper bay.

The phytoplankton biota of Wolf Bay has been dominated by specific bloom indicators such as Cryptophytes along with the primary bloom species noted in upper Perdido Bay (*Cyclotella choctawhatcheeana, Skeletonemia costatum, Gymnodinium* spp., *Prorocentrum minimum, Heterosigma akashiwo*, *Chattonella c.f. subsalsa*) (Figure 6.18). There was a seasonal pattern of bloom dominance as noted in the upper bay. The near absence of the chrysophytes was further indication that Wolf Bay was dominated by similar bloom species progressions and alterations in the plankton community structure as were found in upper Perdido Bay during increased nutrient loading from Elevenmile Creek. There was a general increase of bloom species occurrence over the study period with peaks of most such species noted from late 2000 through 2001. The dinoflagellate *P. minimum* was particularly dominant during this period along with blooms of *S. costatum,* and *Cyclotella choctawhatcheeana*. There were peaks of *Gymnodinium* spp., *H. akashiwo*, and *Chattonella c.f. subsalsa* during this period. There were general declines of water quality features such as bottom oxygen anomalies and Secchi depths that paralleled bloom occurrence (Figure 6.19). The low points of these indicators generally coincided with the bloom peaks in late 2000–2001.

The Wolf Bay data indicate that nutrient loading from non-point sources (i.e., agricultural activities) has led to increased bloom activity. This is in contrast to the improving conditions in the upper bay following reductions of the point-source nutrient loading during this period. However, it should be pointed out that there are some differences between point- and non-point-source nutrient loading with respect to initiating plankton blooms. Most point sources are continuous, whereas many non-point sources load nutrients mainly when it rains, and are therefore intermittent nutrient sources. In the case of the pulp mill, this continuous source of nutrients occurred during drought as well as flood periods whereas non-point (agricultural) sources do not load as regularly during droughts. Also, point sources are more easily controlled than non-point sources, which, in many urban and agricultural situations, are usually ignored by regulatory agencies. This could mean that non-point nutrient sources may have different effects than point sources during drought periods.

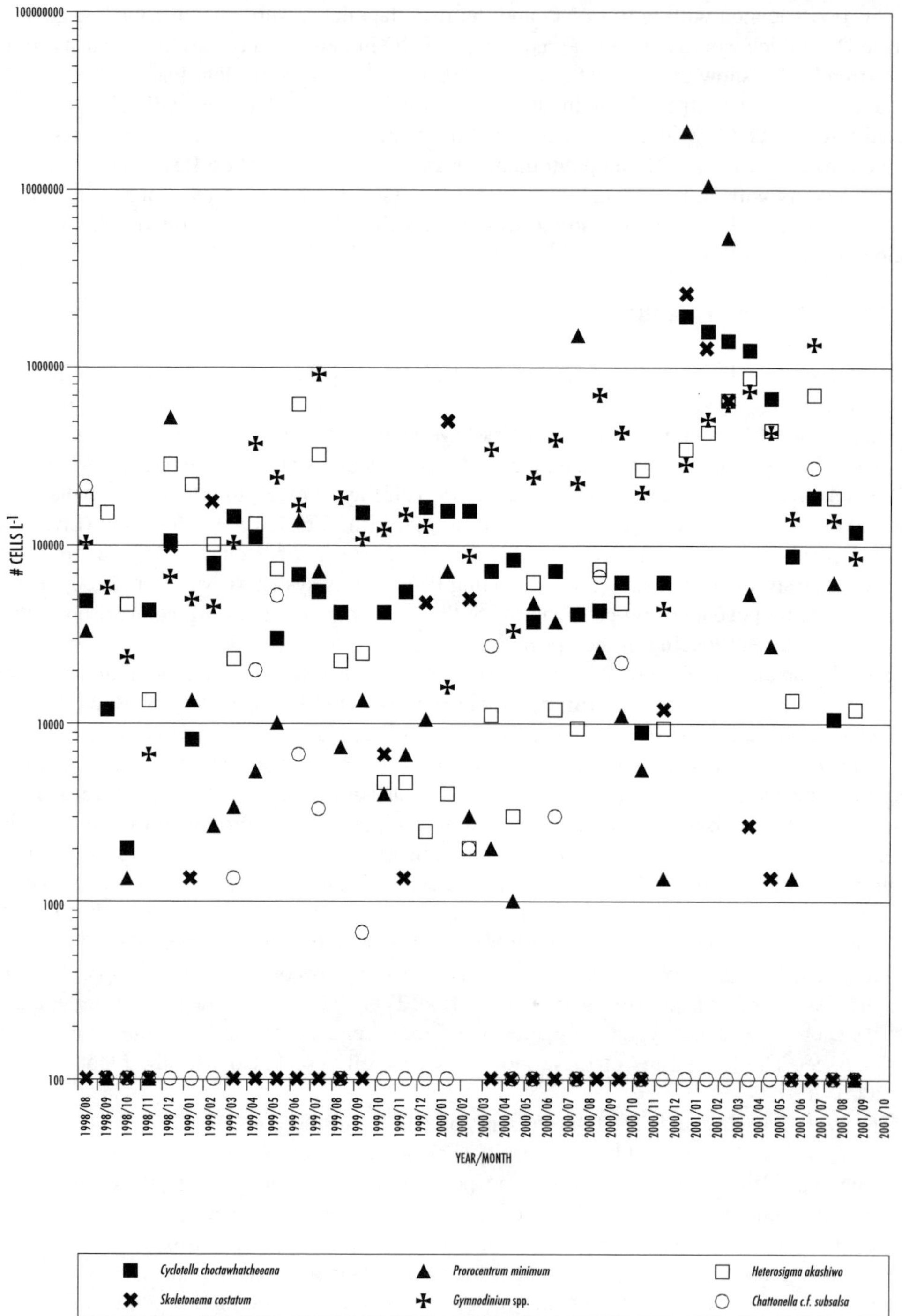

Figure 6.18 Numerical abundance of bloom populations taken at station 42b in Wolf Bay monthly from August 1998 through October 2001.

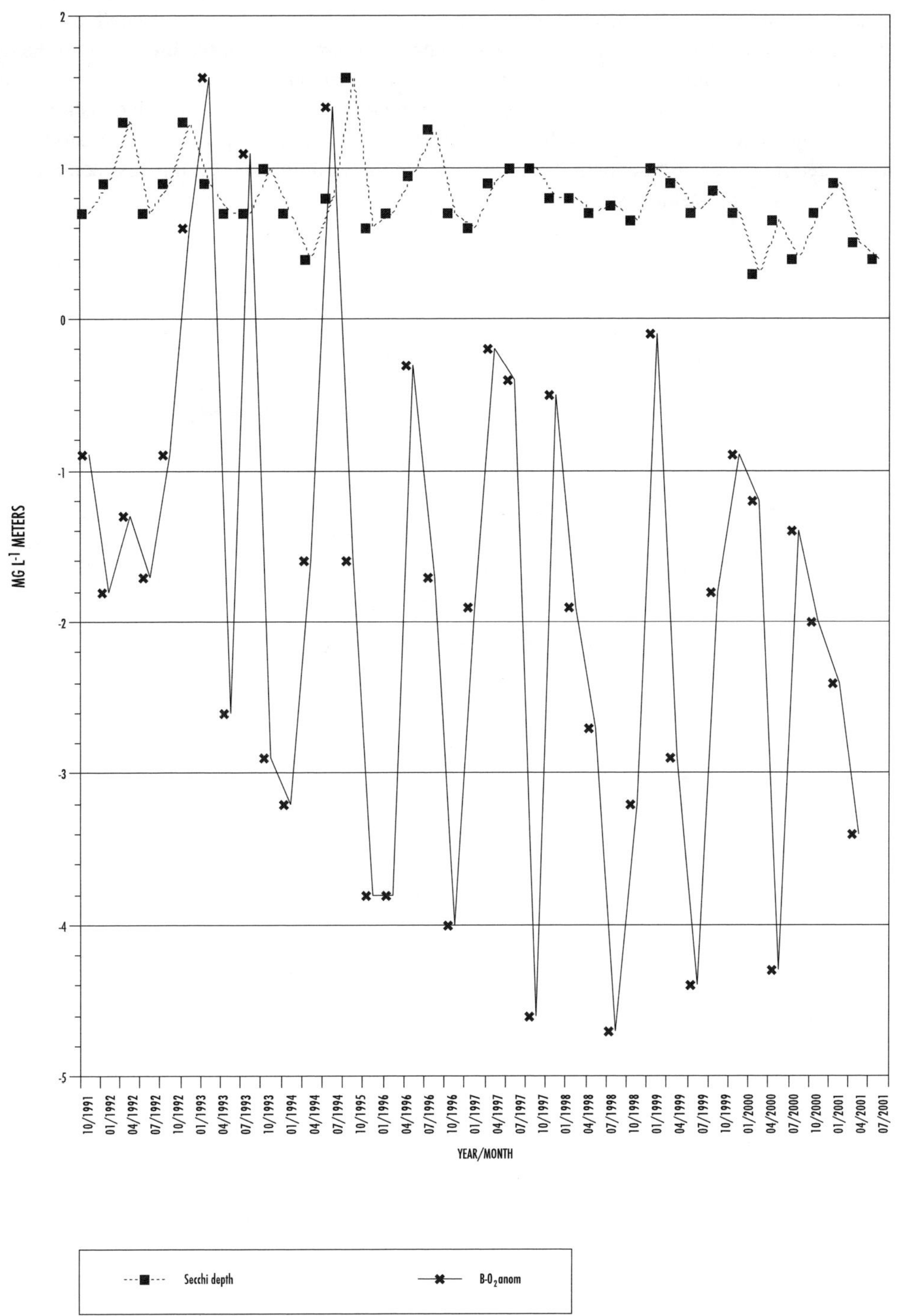

Figure 6.19 Water quality data taken at station 42 in Wolf Bay quarterly from October 1991 through July 2001.

Livingston (2000) noted that regression analyses of the data indicated significant associations of high nutrient concentrations, altered phytoplankton community indices through increased representation by bloom species, deterioration of the benthic habitat, and losses of benthic secondary production in the form of infaunal macroinvertebrates. The proximity of the lower bay to the open Gulf of Mexico adds another dimension to the generation of potentially toxic species such as *P. minimum*, *Gymnodinium* spp., *H. akashiwo*, and *Chattonella c.f. subsalsa*. The relationship of these blooms to so-called red tide outbreaks has not been evaluated, but the impact of the plankton blooms on associated food web components represents a major question in the overall assessment of bloom impacts on offshore coastal systems.

CHAPTER 7

Food Web Response to Plankton Blooms

I. INTRODUCTION

Phytoplankton blooms adversely affect invertebrate and fish populations (invertebrates: Buskey and Stockwell, 1993; Buskey et al., 1997; Steidinger et al., 1998; fishes: Shimada et al., 1983, 1996; Chang et al., 1990; Burkholder et al., 1992; Burkholder and Glasgow, 1997). Blooms also harm coastal fisheries (Smayda and Shimizu, 1993; Hallegreaff, 1995; Hallegreaff et al., 1995; Anderson and Garrison, 1997). It is not clear how specific changes in phytoplankton assemblages alter food webs, however. Riegman (1998) reviewed the effects of altered nutrients on phytoplankton species composition although little is known about long-term changes of these coastal assemblages. Fryxell and Villac (1999) noted that data are rarely available for events leading up to blooms, and even less is known about the effects of blooms on associated food webs. Sorokin et al. (1996), working in the Comacchio Lagoons (Ferrar, Italy), noted reduced zooplankton and losses of benthic organisms and fishes due to increased nutrient loading and dominance of "inedible cyanobacteria." However, the effects of plankton blooms are often cumulative, and occur in complex ways (Fryxell and Villac, 1999). Toxic diatoms may have direct or indirect effects on grazers, but it is more common that the primary adverse effect is on the consumer of the affected grazer. Toxic diatoms can cause mechanical injuries to fish gills. Bloom secretions can clog gills. Blooms also cause hypoxia or anoxia. Ryther (1954) noted the association of nitrogen fertilization of Long Island Sound with proliferation of chlorophytes, with impacts on associated food webs. However, these early studies have not been followed up by long-term, comprehensive analyses, and published statements claiming bloom-caused ecological disruption are rarely based on ecological data (Smayda, 1997a).

II. HABITAT TRENDS IN PERDIDO BAY

The distribution of animal associations of shallow estuaries and coastal systems in the NE Gulf of Mexico are relatively well known (Sheridan, 1978, 1979; Laughlin, 1979; Sheridan and Livingston, 1979, 1983a; Livingston, 1984a, 1991c). Infaunal macroinvertebrates comprise the base of the macrotrophic food web (Livingston et al., 1997). Most invertebrate and fish populations in shallow coastal systems are benthic feeders (Livingston, 1984a). The meio- and macroinfaunal populations, living in or on the sediments, provide a direct connection between primary production and the upper levels of the benthic food webs (Livingston et al., 1997). Interactions of these groups are mediated by competition and predator–prey relationships (Virnstein, 1977; Peterson, 1982; Whitlach and Zajac, 1985).

The trophic organization of Perdido Bay should be placed within the context of long-term drought–flood cycles and the overall distribution of habitats through the 14-year period of observation. There is no intensive commercial fishing pressure on the upper Perdido system. The opening of Perdido Pass led to widespread salinity stratification in deeper areas with associated sub-haloclinal hypoxia during warm periods and the creation of habitat-damaging liquid mud at depth (Livingston, 1997a). This has resulted in relatively low secondary production in affected areas (Livingston, 2000). These effects, together with seasonal and interannual changes in Perdido River flow, provided the background for food web response to changes in the phytoplankton assemblages due to altered nutrient loading.

Reviews of sediment quality of Perdido Bay (Livingston, 2000) show that sandy conditions predominate in the river and creek approaches, the shallow (<1 m) periphery of the bay, and in western sections of the upper bay. Station P18, at the mouth of the Perdido River, is located in a deep hole, and salinity stratification causes hypoxic conditions, proliferation of liquid mud, and generally deteriorated water quality. The most pronounced hypoxic conditions occur in the deep holes of the lower Perdido River where there is maximal salinity stratification. Elevenmile Creek has been adversely affected by pulp mill effluents, with hypoxic conditions due to high dissolved organic carbon (DOC), relatively high ammonia concentrations, high conductivity (in fresh water areas), and increased nutrient concentrations (Livingston, 1991). In areas where there was less difference between the surface and bottom salinities (shallow shelf areas of upper Perdido Bay), there was less pronounced hypoxia. Sub-haloclinal hypoxia was not prevalent in shallow (upper bay) stations (P23) and in the grass beds in western sections of upper Perdido Bay (P25). Although salinity stratification is an important factor in the reduction of sub-haloclinal dissolved oxygen (DO), the long-term trends of DO have indicated that other factors are also important determinants of benthic hypoxia in Perdido Bay (Livingston, 2000).

The shallow parts of upper Perdido Bay have relatively low silt concentrations. Station P23, at the mouth of Elevenmile Creek, had generally good water quality prior to the outbreak of plankton blooms (Livingston, 1991). During the early years of sampling, secondary production here was comparable to that in similar areas of unpolluted Apalachicola Bay (Livingston, 2000). Station P25 is characterized by *Vallisneria americana* beds that were, during the early years, outside of the influence of the pulp mill due to Perdido River flow influences and prevailing currents in the upper bay (Livingston, 1991). The grassy areas in the western upper bay are sandy with interspersed flocculent materials. At station P26, the substrate is sandy and is subject to wind effects and direct influence from the Perdido River and Elevenmile Creek (Livingston, 2000). Midbay areas (P31, P33) are still shallow, and are characterized by increased primary production (Livingston, 1991). In deeper parts of the lower bay (stations P37, and P40), salinity stratification is maximal. The sediments are soft and dark with heavy infiltration of liquid mud. Depths of the liquid mud in the lower bay vary but are often greater than 1 m. Liquid mud characterizes most of Wolf Bay. Sediments in lower Perdido Bay are sandy. Seasonal (warm weather) hypoxia prevails below the halocline, although DO is periodically higher at station 40 due to influxes of Gulf sub-haloclinal water (Livingston, 1991, 2000).

III. BIOMASS DISTRIBUTION AND SPECIES RICHNESS

The distribution of infaunal (Inf), epibenthic macroinvertebrate (Inv), and fish biomass in space and time (Figure 7.1) shows increased secondary production at all levels in upper bay areas (P23 and P25). However, infaunal and epibenthic invertebrate and fish biomass was uniformly low at the mouth of the Perdido River (P18), where infaunal biomass increased somewhat during later years. This trend was reversed in Elevenmile Creek where infaunal and fish biomass decreased to a low in 2001–2002. The very low biomass during the recent period coincided with increased blue-green algae bloom frequency. Average infaunal biomass was uniformly high in the upper bay sea grass beds (station P25) over the entire 14-year period. Invertebrate biomass was relatively low here, and the generally high fish biomass dropped during the 2001–2002 period. The general trends at the mouth of Elevenmile Creek (P23) were similar to but more pronounced than the grass bed

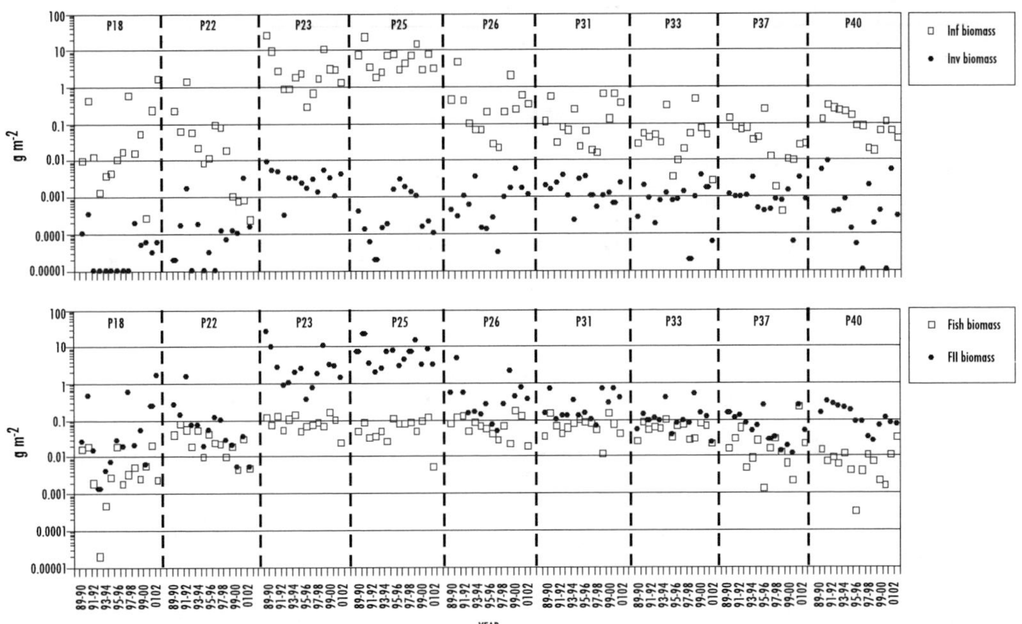

Figure 7.1 Distribution of infaunal macroinvertebrate (Inf), epibenthic macroinvertebrate (Inv), and fish dry weight biomass m^{-2} in Perdido Bay as annual averages taken from 1989–1990 through 2001–2002. The total (fish, invertebrate, infauna [F11] biomass m^{-2}) is also shown.

areas with high infaunal biomass during the early years and decreases to a low point in 1995–1996. This was followed by increases to 1999 after which infaunal biomass again decreased. Invertebrate biomass declined over the study period whereas fish biomass remained relatively stable over the 14-year observation period. Temporal changes of infaunal and total biomass in midbay and lower bay areas tended to follow that in the upper bay stations with steep declines during the middle years, followed by increases up to 1999, followed by subsequent decreases. Infaunal and fish biomass tended to decline along the north–south axis of the bay. Fish and invertebrate trends were similar to the infaunal distributions. In lower parts of the bay (P37, P40), there was a steady decrease of biomass to a low in 1997–1998. This pattern was somewhat different from that in the upper bay (P23, P26), where lows were reached during early periods (1995–1996). Overall, the biomass levels reflected specific habitat distributions in the bay, whereas the temporal trends reflected somewhat different processes in upper and lower parts of the bay.

Species richness peaked in all three groups during the first 2 years of sampling, with general declines that reached low points in 1994–1995 (infauna), 1993–1996 (invertebrates), and 1993–1994 and 1998–1999 (fishes) (Figure 7.2). With the exception of 1999–2000 for the invertebrates, the high levels noted during the early years were not reached during subsequent samplings. The range maxima showed similar patterns to the means in most cases.

IV. TROPHIC ORGANIZATION

The distribution of trophic unit numbers in the Perdido system (Figure 7.3) indicates several different trophic patterns in the Perdido system. The upper bay stations (P23, P25) were clearly the most diverse in terms of numbers of trophic units. There was a recurring pattern of initial high numbers of trophic units followed by declines with some recovery during the period 1997–1999, and subsequent reductions of the trophic indices. This pattern roughly approximates the pattern of bloom formation in the Perdido system. The same pattern, at a lower level, was evident at the

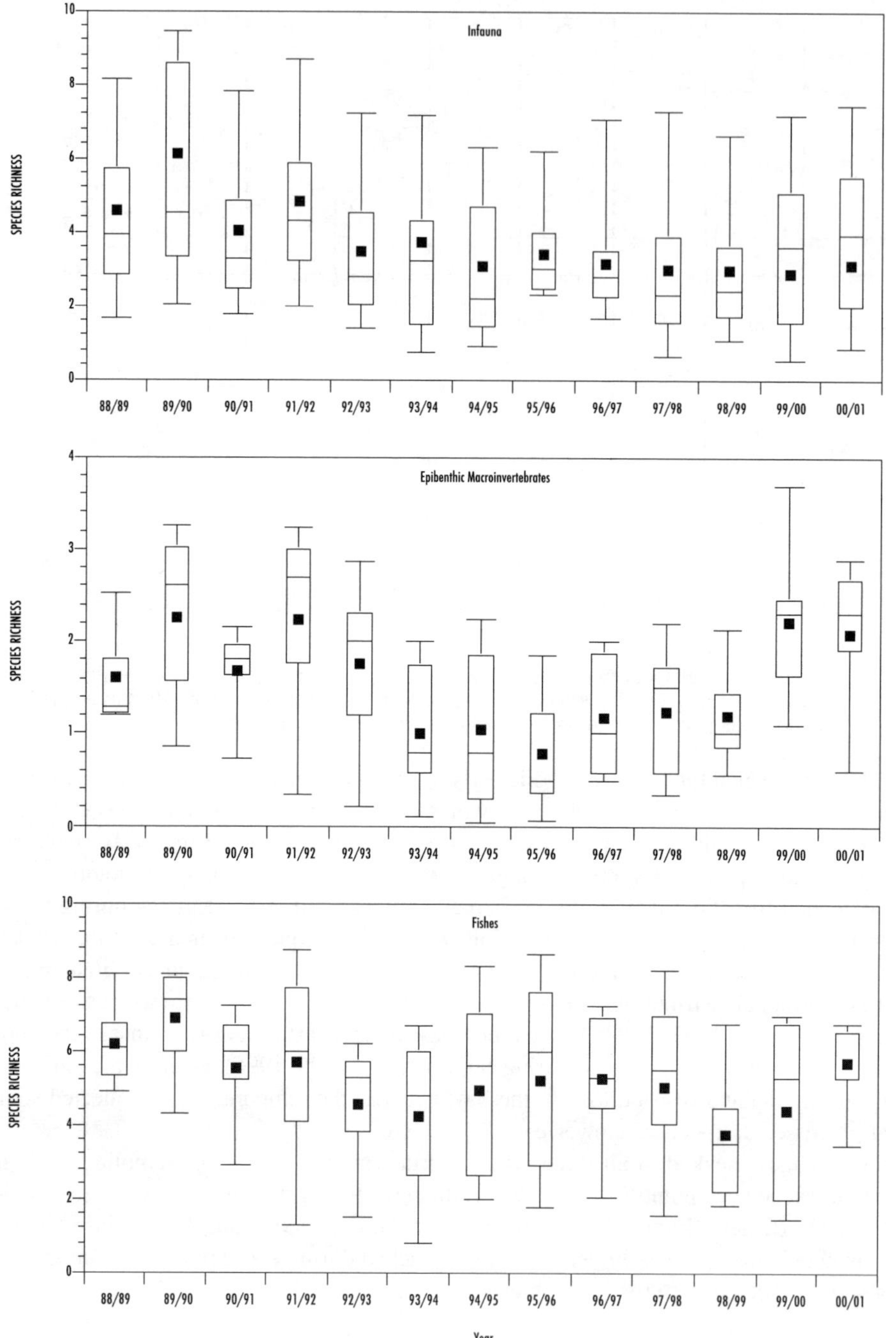

Figure 7.2 Trends of infaunal macroinvertebrate, epibenthic macroinvertebrate, and fish species richness in Perdido Bay from 1988–1989 through 2000–2001. Data are shown as means and the 10th, 25th, 75th, and 90th percentiles.

mouth of the Perdido River (station P18). In Elevenmile Creek, there was a general pattern of decline with low levels reached during the early bloom years and the later blue-green algae outbreaks. In midbay to lower bay areas, there was a general decline in the number of trophic units with time; in the lower bay, the invertebrates just about disappeared with time.

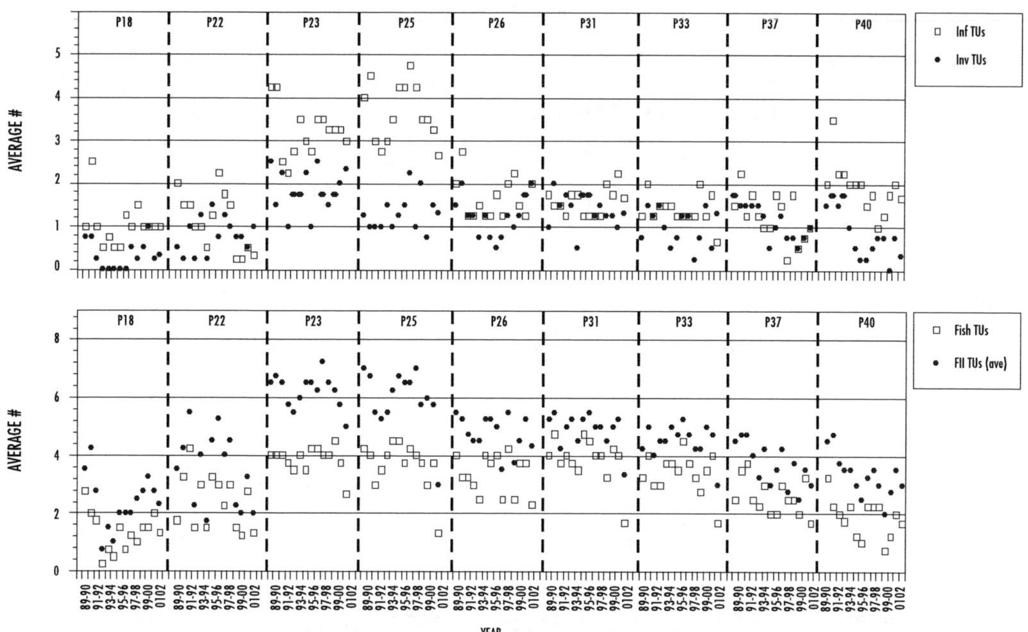

Figure 7.3 Trends of infaunal macroinvertebrate (Inf.), epibenthic macroinvertebrate (Inv.), and fish numbers of trophic units (TUs) in Perdido Bay (1988–1989 through 2001–2002).

The overall trophic organization changes in the Perdido system are given in Figure 7.4. For orientation purposes, the period of maximal N+P loading and bloom activity is noted by the down arrow, and the period of low orthophosphate loading (1997–1998) is noted by the up arrow at station 23. Changes in herbivore and omnivore groups tended to follow the loading patterns with herbivores showing declines during the bloom periods and recovery during the low nutrient loading period and omnivores with an inverse relationship to the change in the herbivore groups. Recovery patterns of the C1 and C2 carnivores appeared to be delayed, which could fit the models noted by Livingston et al. (1997). There were virtually no herbivores in Elevenmile Creek and relatively few at the mouth of the Perdido River. However, the omnivores deteriorated during the last 4 years in the creek, whereas this group showed no such pattern in the Perdido River. Both primary and secondary carnivores deteriorated in both areas over the study period with low levels again reached during blue-green algae bloom years in Elevenmile Creek.

In midbay to lower bay areas, herbivores were just about non-existent. Omnivore patterns tended to deteriorate during the bloom periods with some recovery during the last 3 to 4 years of the study period. This pattern was somewhat different from that of the upper bay. A more detailed analysis of the general trophic organization of planktivorous herbivores (the most important of the herbivore groups) (HP), detritivores (OD), general omnivores (OR), and carnivorous omnivores (OC) (Appendix II) is shown in Figure 7.4. Both the planktivorous herbivores and detritivores were largely restricted to the upper bay. Temporal patterns of the distribution of these groups tended to follow those described above. There were virtually no detritivores in the river or creek areas and in midbay to lower bay areas. The general omnivores were also largely restricted to the upper bay, and followed a similar pattern to that of the planktivorous herbivores and detritivores.

The distribution of the FII feeding mode indices is given in Figure 7.5. Filter feeders, suspension feeders, and subsurface deposit feeders were biomass dominants among the infaunal macroinvertebrates, whereas epibenthic macroinvertebrates and fishes were mainly browser-grazers. Filter feeders were located mainly in the upper bay (stations P23 and P25). This group was generally absent in Elevenmile Creek and midbay to lower bay areas. Temporal distributions of the filter feeders in the upper bay followed those described for the herbivores. Filter feeders were more

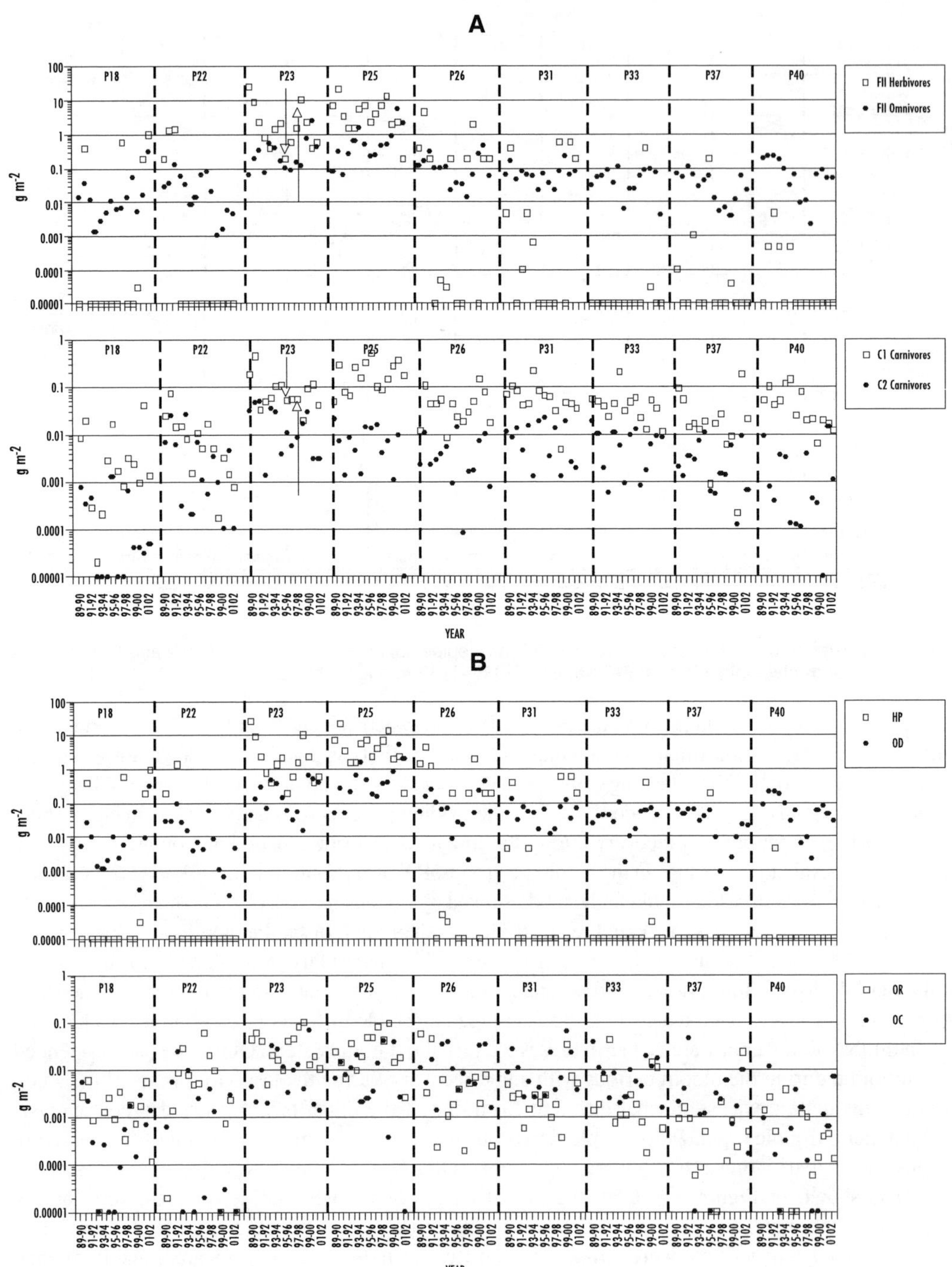

Figure 7.4 Trends of FII (infaunal macroinvertebrates, epibenthic macroinvertebrates, fishes) trophic organization in Perdido Bay from 1988–1989 through 2001–2002: (A) herbivores, omnivores, C1 carnivores, C2 carnivores. (B) Phytoplanktivorous herbivores (HP), detritivores (OD), general omnivores (OR), carnivorous omnivores (OC).

evenly distributed, but still predominated in upper bay areas. They were largely absent or rare in Elevenmile Creek and at the mouth of the Perdido River. There were reductions of filter feeders at station P23 during the mid-study period of prevalent blooms. In the lower bay, the filter feeders underwent reductions over the study period.

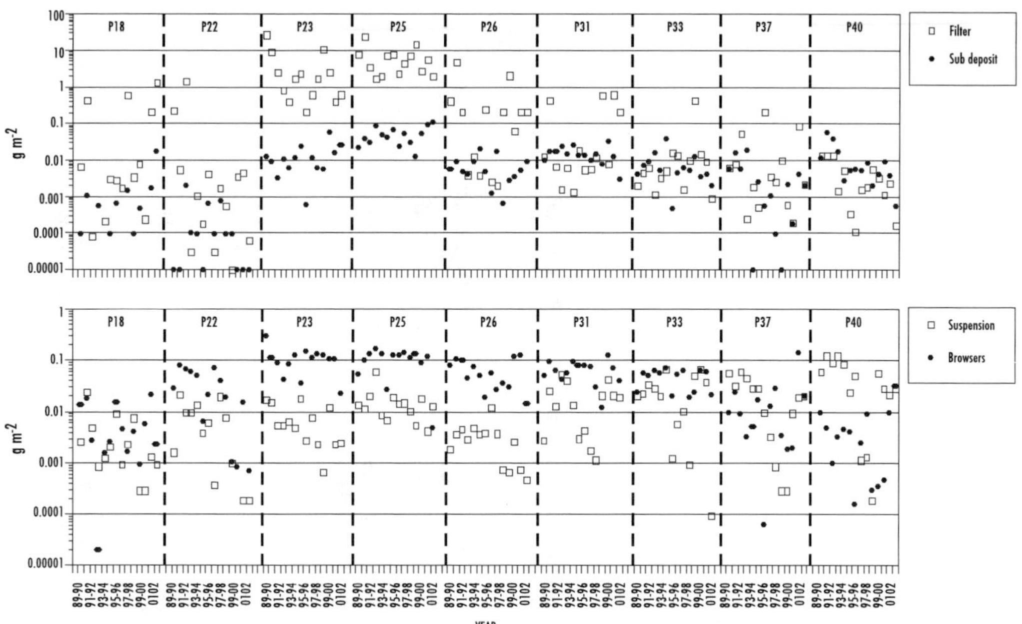

Figure 7.5 Trends of the distribution of FII (infaunal macroinvertebrates, epibenthic macroinvertebrates, fishes) feeding modes (filter-feeding, submergent deposit feeders, suspension-feeders, browser/grazer groups) in Perdido Bay from 1988–1989 through 2001–2002.

Results of regressions of the top bloom species and the primary trophic (FII) indices run against independent variables in Elevenmile Creek over the study period are shown in Table 7.1. The primary bloom species, *Merismopedia tenuissima*, was significantly associated with drought periods, orthophosphate ratios, and with *Prorocentrum minimum* biomass. *Merismopedia tenuissima* was negatively associated with phytoplankton diversity and evenness, thus indicating that this species had an adverse effect on phytoplankton associations in Elevenmile Creek during drought periods. The bloom species *Heterosigma akashiwo* showed similar significant associations. FII total biomass was negatively associated with pH, orthophosphate ratios, phytoplankton biomass, and *M. tenuissima* biomass. The FII total numbers of trophic units and FII species richness were also negatively associated with salinity, phytoplankton biomass, and the biomass of three bloom species. The other trophic indices were also negatively associated with various phytoplankton indices such as pH, phytoplankton species richness, total organic carbon, and bottom DO. The FII C2 biomass was also associated with FII herbivore biomass. These data show that the blooms occurred during periods of high orthophosphate ratios under drought conditions, and that these blooms were negatively associated with important trophic indices in Elevenmile Creek.

Major temporal trends of the trophic indices at upper, midbay and lower bay stations are shown in Figure 7.6; associated regression data are given in Table 7.2. Herbivore biomass was generally lowest from 1992 to 1997. Recovery was noted during the winter of 1998–1999 with subsequent reductions through 2001. Upper bay herbivores were negatively associated with *P. minimum* biomass during winter periods, but were positively associated with orthophosphate concentrations and negatively associated with DO anomalies during fall. In midbay areas, herbivores were positively associated with nutrient conditions and phytoplankton/diatom biomass during winter and summer periods. In the lower bay, herbivore biomass was again positively associated with phytoplankton/diatom indices during summer and fall periods. Since herbivores are located largely in the upper bay (Figure 7.6), the negative associations of the group with a winter bloom species constitute direct evidence of the adverse effects of blooms on the trophic order of the bay.

Table 7.1 Regressions Run with Annual Averages of Biomass of *Merismopedia tenuissima* (Merten) and *Heterosigma akashiwo* (Hetaka)

Station P22 Dependent Variable	Independent Variable	r^2	Significance	Sign	Dependent Variable	Independent Variable	r^2	Significance	Sign
Merten biomass					FII no. of taxa	ssal	0.57	0.002	–
	Perdido River flow	0.72	0.03	–		Phyto biomass	0.7	0.04	–
	ssal	0.85	0.008	+		Hetaka biomass	0.66	0.04	–
	22/23 PO_4 ratio	0.73	0.03	+		Merten biomass	0.6	0.05	–
	PO_4% diff	0.73	0.03	+		Gymspp biomass	0.67	0.04	–
	sNH_3	0.81	0.01	–					
	Phyto diversity	0.62	0.05	–					
	Phyto evenness	0.73	0.03	–	FII herb biomass	spH	0.63	0.001	–
	Promin biomass	0.73	0.03	+					
Hetaka biomass					FII omn biomass	ssal	0.46	0.01	–
	ssal	0.74	0.03	+		STOC	0.4	0.01	–
	22/23 PO_4 ratio	0.79	0.01	+		Merten biomass	0.66	0.04	–
	PO_4% diff	0.76	0.01	+		Phyto no. of spp	0.83	0.01	–
	Gymspp biomass	0.86	0.007	+		bDO	0.4	0.02	+
	Phyto biomass	0.87	0.04	+					
					FII C1 biomass				

FII tot. biomass				
	22/23 PO_4 ratio	0.39	0.05	–
	spH	0.75	0.02	–
	Phyto biomass	0.52	0.05	–
	Merten biomass	0.66	0.05	–
FII tot. troph units				
	Perdido River flow	0.37	0.02	+
	Total phyto no.	0.69	0.04	–
	Phyto biomass	0.75	0.02	–
	Hetaka biomass	0.86	0.007	–
	Merten biomass	0.65	0.05	–
	Gymspp biomass	0.73	0.03	–
	Gymspp biomass	0.56	0.05	–
FII C2 biomass				
	spH	0.58	0.002	–
	Phyto no. of spp	0.76	0.02	+
	Inf. biomass	0.99	0.0001	+
	FII herb biomass	0.99	0.0001	+

Note: FII total biomass, FII total trophic units, FII no. of taxa, FII herbivore, omnivore, C1 carnivore, and C2 carnivore biomass are dependent variables taken in Elevenmile Creek (station 22). These indices were run against annual averages of independent variables (salinity [s], dissolved oxygen [s, b], total organic carbon [sTOC], ammonia[s], orthophosphate [s], P22/P23 orthophosphate ratio, P22/P23 ammonia ratio, Elevenmile Creek orthophosphate loading, Elevenmile Creek ammonia loading, Perdido River orthophosphate loading, Perdido River ammonia loading rates, differenced ammonia + orthophosphate loading rates, differenced ammonia + orthophosphate ratios, silica [s], pH [s], Secchi depth, cumulative number of phytoplankton taxa, cumulative phytoplankton Shannon diversity indices, and numbers per liter of *Cyclotella choctawhatcheeana* (Cyccho), *M. tenuissima* (Merten), *Prorocentrum minimum* (Promin), *Miraltia throndsenii* (Mirthr), *H. akashiwo* (Hetaka), *Synedropsis* sp. (Synsp), *Chattonella c.f. subsalsa* (Chasub), and *Gymnodinium* spp. (Gymspp). Only significant ($p \leq 0.05$) regressions are given.

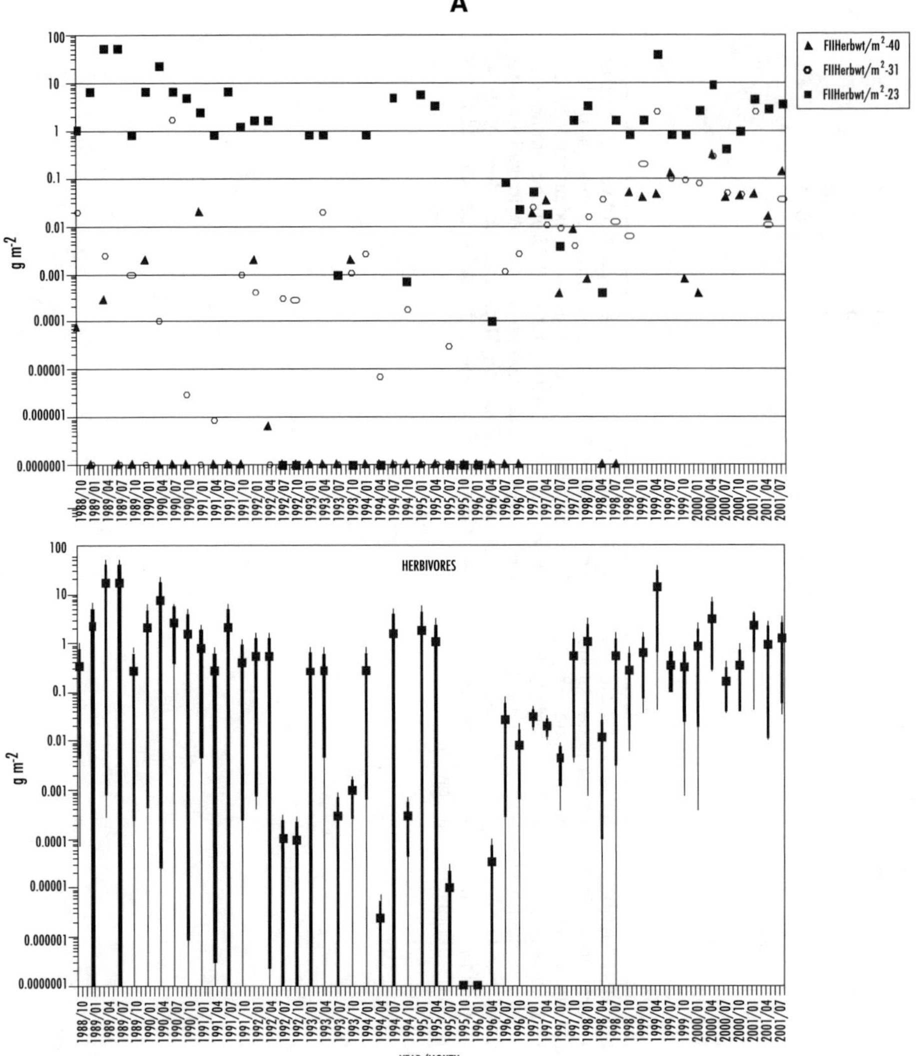

Figure 7.6 Trends of FII (infaunal macroinvertebrates, epibenthic macroinvertebrates, fishes) biomass of trophic variables. (A) Herbivores. (Data were taken in Perdido Bay from 1988 through 2002, and are shown by station and as means and the 10th, 25th, 75th, and 90th percentiles of the three stations.) (continued)

Omnivore distribution (Figure 7.6; Table 7.2) was variously distributed through time with some decreases noted during 1992 and 1994–1995. Omnivore biomass in the upper bay was positively associated with *P. minimum* during the winter, surface ammonia during the spring, and bloom numbers in the summer. This indicates that the main omnivore groups prevail during bloom periods as omnivores take over areas once dominated by herbivores, In the lower bay, omnivores were positively associated with the numbers of winter plankton blooms and with summer N+P loading differences; this constituted further proof that omnivores were favored during certain seasons by nutrient blooms and nutrient N+P loading.

Primary carnivores (Figure 7.6; Table 7.2) underwent reductions during the period from 1994 through 1999. In the upper bay, primary carnivores were negatively associated with orthophosphate ratios, phytoplankton biomass, and bloom species biomass. During summer, this group was negatively associated with numbers of blooms, and during the fall, primary carnivores were negatively associated with *P. minimum*. In the lower bay, primary carnivore biomass was negatively associated with winter

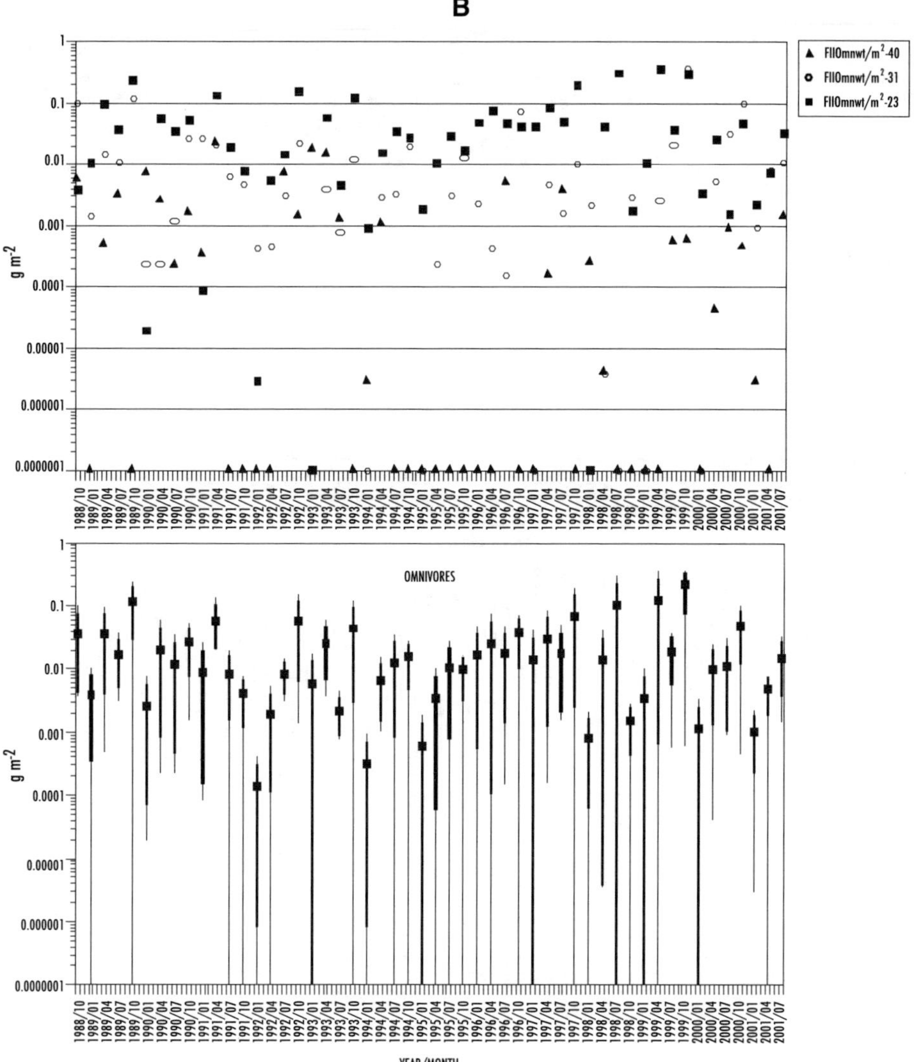

Figure 7.6 (continued) Trends of FII (infaunal macroinvertebrates, epibenthic macroinvertebrates, fishes) biomass of trophic variables. (B) Omnivores. (Data were taken in Perdido Bay from 1988 through 2002, and are shown by station and as means and the 10th, 25th, 75th, and 90th percentiles of the three stations.) (continued)

bloom numbers and raphidophyte biomass; during the summer, there were negative associations with N+P ratio differences, and raphidophyte/*Heterosigma akashiwo* biomass. During fall, C1 carnivores were negatively associated with ammonia loading, and *H. akashiwo*/*Cyclotella choctawhatcheeana* biomass. The data thus show that the C1 carnivores were negatively affected by seasonally changing bloom species, and by nutrient loading/concentration factors that caused the blooms.

Secondary carnivores (Figure 7.6; Table 7.2) had a complex, long-term pattern of occurrence. In the upper bay, the primary seasonally directed significant factors included food items (*Rangia cuneata*, FII omnivores). In midbay areas, however, C2 carnivores were negatively associated with winter chlorophyll *a* concentrations and orthophosphate concentrations. During spring in midbay areas, this group was negatively associated with *C. choctawhatcheeana* biomass, and, during summer, C2 carnivores were negatively associated with surface orthophosphate and chlorophyll *a* concentrations. In the lower bay, C2 carnivores were negatively associated with winter bloom numbers, and raphidophyte/*H. akashiwo* biomass. During spring in the lower bay, this group was negatively

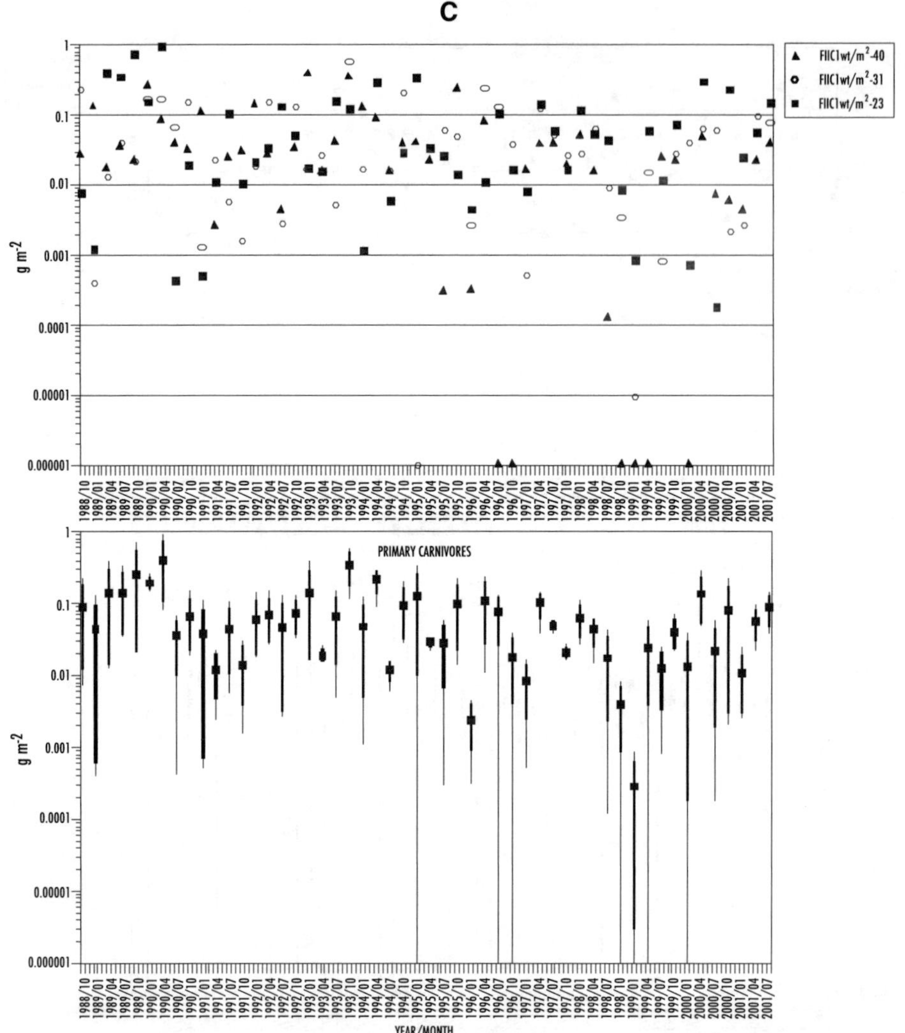

Figure 7.6 (continued) Trends of FII (infaunal macroinvertebrates, epibenthic macroinvertebrates, fishes) biomass of trophic variables. (C) C1 carnivores. (Data were taken in Perdido Bay from 1988 through 2002, and are shown by station and as means and the 10th, 25th, 75th, and 90th percentiles of the three stations.) (continued)

associated with various nutrient concentration indices, total phytoplankton biomass, and bloom biomass. During summer in the lower bay, C2 biomass was negatively associated with nutrient concentration indices and raphidophyte/*H. akashiwo* biomass. In the fall in the lower bay, this group was negatively associated with ammonia loading, and *C. choctawhatcheeana*/*H. akashiwo* biomass.

Statistical analyses of the relationships of feeding modes with various physical, chemical, and biological variables (Table 7.3) indicated that deposit feeders and filter feeders in the upper bay and midbay had positive associations with chlorophyll *a* and some bloom species. In the lower bay, both feeding groups were negatively associated with various nutrient loading indices and bloom numbers. This dichotomy of response by region of these important feeding groups could be related to the fact that the upper bay is the center of both autochthonous and allochthonous forms of primary production. More adverse habitat conditions in the lower bay could be exacerbated by bloom generation as the plankton fall through the halocline into the hypoxic conditions at the bottom.

The statistical analyses are consistent with the direct association (negative for herbivores, C1 and C2 carnivores; positive for omnivores) of phytoplankton blooms and the long-term trends of

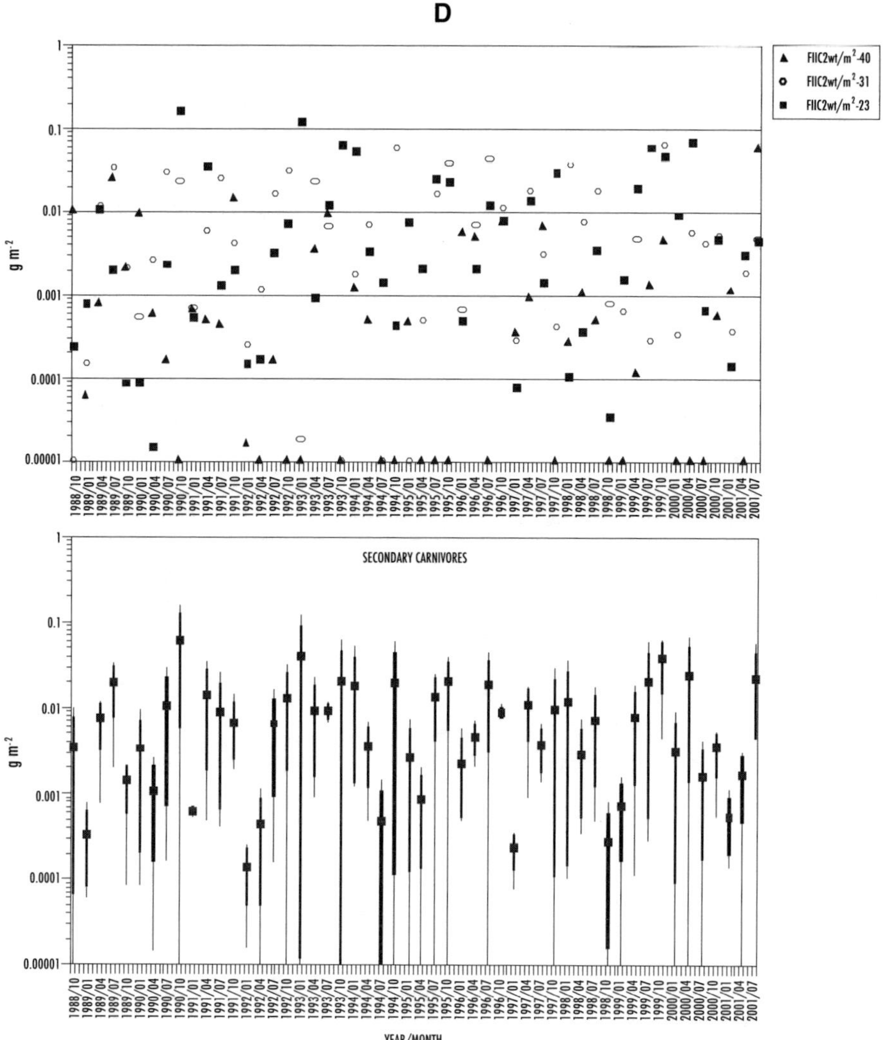

Figure 7.6 (continued) Trends of FII (infaunal macroinvertebrates, epibenthic macroinvertebrates, fishes) biomass of trophic variables. (D) C2 carnivores. (Data were taken in Perdido Bay from 1988 through 2002, and are shown by station and as means and the 10th, 25th, 75th, and 90th percentiles of the three stations.)

key food web indices in Perdido Bay. The spatial and temporal occurrences of such impacts are complex, yet they are consistent with the species-specific responses of the phytoplankton blooms to nutrient loading from the pulp mill into Elevenmile Creek. Periodic drought occurrences exacerbate the bloom incidences and associated effects. The seasonal and interannual trends of nutrient loading superimposed over these climatological features thus determine not only the levels of secondary production, but also specific changes in associated food webs in different parts of the bay. Auxiliary habitat features, possibly unrelated to the blooms, also affect the specific cause-and-effect relationships of nutrient loading, qualitative and quantitative aspects of bloom formation, and associated changes in food web processes. The recovery of the bay in response to reduced nutrient loading appears to have been altered by the extended drought of 1998 to 2002. Another explanation for the lack of recovery of bay food webs after bloom cessation could be natural delays of response as noted by Livingston et al. (1998) in the Apalachicola system. In any case, recovery as a response to restoration activities is not a mirror image of the impact history, and will probably be complicated by changes of the nannoplankton populations.

Table 7.2 Regressions Run with Annual Averages of FII Total Biomass, FII Total Trophic Units, FII Herbivore, Omnivore, C1 Carnivore, and C2 Carnivore Biomass as Dependent Variables taken in Perdido Bay (Stations 23, 31, 40)

Dependent Variable	Independent Variable	r^2	Significance	Sign	Independent Variable	r^2	Significance	Sign
Station P23								
	Winter				**Spring**			
FII tot. biomass	Gymspp biomass	0.57	0.03	+				
FII tot. troph units	Promin biomass	0.56	0.05	–				
FII herb biomass	Promin biomass	0.46	0.05	–				
FII omn biomass	Promin biomass	0.79	0.04	+	sNH_3	0.68	0.05	+
FII C1 biomass	22/23 PO_4 ratio	0.48	0.05	–	*R. cuneata*	0.79	0.003	+
	sanom O^2	0.72	0.01	–				
	Secchi	0.56	0.05	–				
	Total phyto biomass	0.57	0.05	–				
	Cyccho biomass	0.48	0.05	–				
	Gymspp biomass	0.49	0.05	–				
FII C2 biomass	No. of *R. cuneata*	0.92	0.0007	+				
	Summer				**Fall**			
FII tot. biomass					anom O^2	0.41	0.05	–
FII tot. troph units	22/23 PO_4 ratio	0.36	0.05	–	Cyccho biomass	0.49	0.03	+
FII herb biomass					sPO_4	0.42	0.05	+
					sanom O^2	0.41	0.05	–
FII omn biomass	No. of blooms	0.56	0.01	+				
FII C1 biomass	sChlor *a*	0.45	0.03	–	Gymspp biomass	0.42	0.05	+
					Promin biomass	0.79	0.001	–

					C. subsalsa biomass	0.93	0.03	+
FII C2 biomass								
					No. of *R. cuneata*	0.44	0.05	+
					FII omnivores	0.68	0.006	+
Station P31								
	Winter				**Spring**			
FII tot. biomass								
	total phyto biomass	0.86	0.008	+				
	Cyccho biomass	0.74	0.01	+				
FII tot. troph units								
	NH_3 load	0.78	0.008	–				
	22/23 NH_3 ratio	0.61	0.04	–				
	P+N load diff.	0.53	0.04	–				
	Secchi	0.77	0.04	+				
FII herb biomass								
	sChlor *a*	0.57	0.05	+	P+N ratio diff.	0.6	0.02	+
	Total phyto biomass	0.88	0.005	+	Secchi	0.5	0.05	+
	Cyccho biomass	0.8	0.006	+				
FII omn biomass								
FII C1 biomass								
					sNH_3	0.78	0.003	+
FII C2 biomass								
	sPO_4	0.99	0.0001	–	Cyccho biomass	0.42	0.05	–
	sChlor *a*	0.48	0.05	–				
	Secchi	0.93	0.0005	–				
	sanom O^2	0.57	0.05	–				

(continued)

Table 7.2 (continued) Regressions Run with Annual Averages of FII Total Biomass, FII Total Trophic Units, FII Herbivore, Omnivore, C1 Carnivore, and C2 Carnivore Biomass as Dependent Variables taken in Perdido Bay (Stations 23, 31, 40)

Dependent Variable	Independent Variable	r^2	Significance	Sign	Independent Variable	r^2	Significance	Sign
	Summer				**Fall**			
FII tot. biomass								
	sChlor *a*	0.55	0.05	+	sanom O^2	0.61	0.02	–
	Cyccho biomass	0.7	0.01	+	Promin biomass	0.72	0.02	–
FII tot. troph units								
	NH_3 load	0.78	0.008	–	PO_4 load	0.65	0.01	–
	P+N load diff.	0.6	0.04	–	P+N load diff.	0.41	0.05	–
	22/23 NH_3 ratio	0.61	0.03	–	Cyccho biomass	0.68	0.02	–
	Secchi	0.59	0.04	+				
FII herb biomass								
	sChlor *a*	0.48	0.05	+				
	Cyccho biomass	0.8	0.006	+				
FII omn biomass								
FII C1 biomass								
					sanom O^2	0.68	0.01	–
FII C2 biomass								
	sChlor *a*	0.48	0.05	–	sanom O^2	0.8	0.002	–
	sanom O^2	0.57	0.05	–				
	sPO_4	0.99	0.0001	–				
	Secchi	0.93	0.0005	–				
Station P40								
	Winter				**Spring**			
Dependent Variable								
FII tot. biomass								
	No. of blooms	0.65	0.02	–				
	Raphid biomass	0.58	0.05	–				
	Hetaka biomass	0.48	0.05	–				
FII tot. troph units								
	No. of blooms	0.48	0.05	–	22/23 NH_3 ratio	0.51	0.05	–
	Raphid biomass	0.68	0.02	–	P+N ratio diff.	0.61	0.02	–
	Hetaka biomass	0.66	0.02	–	No. of blooms	0.48	0.05	–
					Total phyto biomass	0.71	0.02	–

					Cyccho biomass	0.85	0.003	–
					Gymspp biomass	0.47	0.05	–
FII herb biomass								
FII omn biomass								
	No. of blooms	0.54	0.05	–				
FII C1 biomass								
	No. of blooms	0.67	0.02	–				
	Raphid biomass	0.69	0.02	–				
FII C2 biomass								
	No. of blooms	0.46	0.05	–	No. of blooms	0.54	0.03	–
	Hetaka biomass	0.57	0.05	–	Total phyto biomass	0.47	0.05	–
					Total diatom biomass	0.47	0.05	–
		Summer				**Fall**		
FII tot. biomass								
	P+N ratio diff.	0.4	0.05	–	P+N load diff.	0.56	0.05	–
	sanom O^2	0.42	0.05	–	No. of blooms	0.69	0.04	–
FII tot. troph units								
	No. of blooms	0.5	0.05	–				
FII herb biomass								
	Total phyto biomass	0.52	0.04	+	Total diatom biomass	0.6	0.05	+
	sChlor *a*	0.44	0.05	–				
FII omn biomass								
					P+N load diff.	0.63	0.05	–
FII C1 biomass								
	P+N ratio diff	0.48	0.05	–	NH_3 load	0.75	0.02	–
	Raphid biomass	0.58	0.02	–	Cyccho biomass	0.61	0.04	–
	Hetaka biomass	0.58	0.02	–	Hetaka biomass	0.79	0.01	–

Note: The data were grouped by station by season, with 3-month averages of the independent variables run against quarterly dependent variables. Independent variables included salinity [s], dissolved oxygen [s,b], total organic carbon [sTOC], ammonia [s], orthophosphate [s], P22/P23 orthophosphate ratio, P22/P23 ammonia ratio, Elevenmile Creek orthophosphate loading, Elevenmile Creek ammonia loading, Perdido River orthophosphate loading, Perdido River ammonia loading rates, differenced ammonia + orthophosphate loading rates, differenced ammonia + orthophosphate ratios, silica [s], pH [s], Secchi depth, cumulative number of phytoplankton taxa, cumulative phytoplankton Shannon diversity indices, bloom numbers, numerical abundance of nanoplankton groups, and numerical abundance of *Cyclotella choctawhatcheeana* (Cyccho), *Prorocentrum minimum* (Promin), *Merismopedia tenuissima* (Merten), *Miraltia throndsenii* (Mirthr), *Heterosigma akashiwo* (Hetaka), *Synedropsis* sp. (Synsp), *Chattonella c.f. subsalsa* (Chasub), and *Gymnodinium* spp. (Gymspp). Only significant ($p \leq 0.05$) regressions are given.

Table 7.3 Regressions Run with Annual Averages of Feeding Modes[a] as Dependent Variables Taken in Perdido Bay (Stations 23, 31, 40)

Dependent Variable	Indep. Variable	r^2	Significance	Sign
Station P23				
Deposit feeders	No. of *C. subsalsa*	0.48	0.05	+
	Chlorophyll *a*	0.37	0.03	+
Filter feeders	No. of *C. subsalsa*	0.48	0.05	+
	Chlorophyll *a*	0.37	0.03	+
Engulfers	Chlorophyll *a*	0.48	0.009	+
	scolor	0.43	0.02	+
	Secchi	0.51	0.05	–
Scrapers	sDO	0.35	0.03	+
	P+N%diff. load	0.36	0.03	–
	P+N%diff. ratio	0.34	0.04	–
	No. of blooms	0.65	0.008	–
Browsers				
Piercers				
Station P31				
Deposit feeders	No. of *C. choctawhatcheeana*	0.58	0.03	+
	No. of *C. subsalsa*	0.61	0.03	+
Filter feeders	*C. choctawhatcheeana*	0.59	0.03	+
	No. of *C. subsalsa*	0.6	0.02	+
Engulfers				
Scrapers	nanoc	0.95	0.0001	+
	No. of *C. subsalsa*	0.7	0.02	+
	No. of *U. eriensis*	0.98	0.0001	+
Browsers				
Piercers				
Station P40				
Deposit feeders	pH	0.73	0.0002	–
	11 mc PO_4 load	0.36	0.03	–
	P+N%diff. load	0.48	0.008	–
	No. of *H. akashiwo*	0.40	0.05	–
	Bloom numbers	0.73	0.003	–
Filter feeders	pH	0.52	0.005	–
	11 mc PO_4 load	0.40	0.02	–
	22/23/PO_4 ratio	0.34	0.04	–
	P+N%diff. ratio	0.39	0.02	–
	No. of blooms	0.48	0.04	–
	No. of phyt. taxa	0.73	0.01	+
	Phyt biomass	0.47	0.04	–
Piercers	P+N%diff. ratio	0.31	0.05	–
	bDO	0.30	0.05	+
Engulfers				
Scrapers				
Browsers				

[a] Deposit feeders, filter feeders, engulfers, scrapers, browsers, piercers.

Note: The data were grouped by station by season, with 3-month averages of the independent variables run against quarterly dependent variables. Independent variables included salinity [s], dissolved oxygen [s,b], total organic carbon [sTOC], ammonia [s], orthophosphate [s], P22/P23 orthophosphate ratio, P22/P23 ammonia ratio, Elevenmile Creek orthophosphate loading, Elevenmile Creek ammonia loading, Perdido River orthophosphate loading, Perdido River ammonia loading rates, differenced ammonia + orthophosphate loading rates, differenced ammonia + orthophosphate ratios, silica [s], pH [s], Secchi depth, cumulative number of phytoplankton taxa, cumulative phytoplankton Shannon diversity indices, numerical abundance of nanoplankton groups, and numerical abundance of *Cyclotella choctawhatcheeana*, *Prorocentrum minimum*, *Merismopedia tenuissima*, *Miraltia throndsenii*, *Heterosigma akashiwo*, *Synedropsis* sp., *Chattonella c.f. subsalsa,* and *Gymnodinium* spp. Only significant ($p \leq 0.05$) regressions are given.

The primary bloom species responded rapidly to the reductions of nutrient loading. The key toxic bloom species (*P. minimum, H. akashiwo, M. tenuissima*) are the main effectors of the loss of secondary production, although such effects are superimposed over other important habitat factors such as dissolved oxygen. The primary water quality indicators of bloom activity are DO, pH, and nutrient concentrations although such indices, by themselves, do not explain the range of biological changes that take place during bloom sequences. The food web changes cannot be ascertained by after-the-fact water quality assessments. These impacts are subtle and cannot be determined without adequate scientific information. Population distribution represents the hands on the clock, and is not necessarily a reliable indicator of bloom impact. Bloom effects can thus go on without outward symptoms until the collapse of a useful fishery. By this time, the habitat changes mediated through impacts on sediments may lead to long-term effects that cannot be determined through the usual modeling efforts that are currently so popular with state and federal regulatory agencies. The food web indices proved to be sensitive indicators of the effects of blooms on secondary producers.

The recovery of upper Perdido Bay from the effects of nutrient loading may take time, and the exact mode of recovery is uncertain. Because the bay has been in a severe drought for the past 4 years, the delay of food web recovery is consistent with the results of the long-term analyses of Apalachicola Bay where secondary productivity was low during later phases of the 1981–1982 drought. In addition, the assimilative capacity of Perdido Bay was reduced during the drought, when the continuous loading of nutrients from the pulp mill, even at relatively low levels, created conditions (i.e., high orthophosphate and ammonia gradients) that encouraged phytoplankton blooms. The pulp mill plans to move the effluent out of Elevenmile Creek and into a marsh treatment system. This should be a positive move for the creek as it does not have the assimilative capacity to process the current loading levels of effluents.

V. FUTURE ANALYTICAL DIRECTIONS

Plankton bloom species respond to qualitative and quantitative changes of nutrient loading due to human activities. The form and timing of nutrient loading with respect to individual nutrient species has an influence on the response of phytoplankton assemblages in receiving systems. Harmful algal blooms (HAB) response is modified by both seasonal and interannual drought–flood cycles. Bloom incidents affect specific state habitat variables such as DO, pH, nutrient concentrations, and sediment quality, but there are no clear relationships between the incidence and magnitude of plankton blooms and chlorophyll *a* concentrations. Generation of plankton blooms is clearly related to state changes of phytoplankton community indices such as biomass, numerical abundance, species richness, species diversity, and population evenness. Plankton blooms also affect invertebrate/fish population dynamics, overall secondary production, and the trophic organization of coastal communities.

The use of linear statistical models is clearly not well suited for analysis of what are essentially nonlinear biological functions such as the relationships of nutrient loading and phytoplankton blooms and the interactions of such blooms with complex food web processes (X. Niu, personal communication). The stochastic approach views a time series event at a specific spatial location as a random sequence and attempts to separate patterns in the series from noise. Univariate linear models (see, e.g., Box and Jenkins, 1976; Pankratz, 1991) or nonlinear models (Tong, 1990) can be built for the time series. Forecasting procedures based on the models have been well established. Space–time linear models suitable for describing the evolving random fields in environmental systems have also been developed. Another approach views the ecosystem as a deterministic dynamic system. For a specific spatial location, the observations over time form a "chaotic" time series. Available procedures for identifying and analyzing chaotic time series center on estimating Lyapounov exponents and dimensions, and building a univariate, nonlinear deterministic model from earlier data (X. Niu, personal communication). Then, model predictions can be made with later data (see, e.g., Farmer and Sidorowich, 1987; Wales 1991; Elsner, 1992).

In the future, we intend to use both linear and nonlinear models to analyze the databases on which the current analyses are founded. The individual (species-specific) bloom sequences will constitute the dependent variables to be run against independent variables representing different configurations of loading patterns and habitat factors. In the same way, food web variables will be run against the plankton bloom data as independent variables. We will use dynamic regression models, as discussed by Pankratz (1991), to investigate the relationship between plankton blooms and such physicochemical variables as river flow characteristics and nutrient loading trends (X. Niu, personal communication). Dynamic regression models, also called transfer function models, will be used to characterize how the biological variables such as individual blooms respond to changes in the environmental characteristics and nutrient loading. Univariate time series and dynamic regression models can be developed with regard to eigenvectors from the Principal Components Analysis (PCA) using the long-term data sets. Time series models (see, e.g., Box and Jenkins, 1976; Pankratz, 1991) have been widely applied to biological data analyses. For instance, Livingston et al. (1997) used dynamic regression models to study the effects of reduction of freshwater flow on biological productivity in the Apalachicola Bay System. Niu et al. (1998) developed time series models for several environmental variables, including salinity, river flow, rainfall, water level, wind speed and direction, in the Apalachicola Bay area. We will build dynamic regression models to explore relationships between individual plankton blooms and water quality predictors.

The database taken in the Perdido system includes three multiyear time periods where there were (1) a period (1988 to 1991) of relatively low orthophosphate and ammonia loading in which mill-derived nutrients did not upset the phytoplankton community structure through the development of blooms; (2) a period from 1992 to summer 2000 during which mill nutrient loading increased leading to seasonal and interannual successions of plankton blooms; and (3) a period from fall 2000 to the present, when nutrient loading by the mill has returned to levels noted during the first few years. The phytoplankton database, consisting of quantified, species-specific changes in population and community variables, can be analyzed for responses to the long-term, detailed nutrient-loading regime. By determination of long-term trends of river flow, nutrient loading, water and sediment quality, and the response of the phytoplankton assemblages (including bloom outbreaks) and associated food webs along with secondary production, the interactions of the blooms, altered phytoplankton community variables, and corresponding changes in associated trophic processes can be mathematically determined. In this research, we will develop space–time nonlinear and nonparametric models for different components in the Perdido system and in other coastal systems of the NE Gulf of Mexico.

Nonlinear and nonparametric techniques for environmental data analysis have been studied extensively in many different contexts (Chen and Tsay, 1993; Niu, 1995). In the future, we will use statistical properties of the new models for estimating both nonlinear and nonparametric functions (X. Niu, personal communication). Developing new space–time nonlinear models and computational techniques for understanding mechanisms of coastal systems poses many challenges to statisticians, especially the need for taking into account high-dimensional inputs/outputs, nonlinear covariate effects, and trend estimation. Using the extensive databases taken over the past 31 years in the NE Gulf of Mexico, we will test the possible nonlinear relationships among multiple components in an ecosystem for modeling nutrient loading/bloom interactions, the long-term effects of HABs on coastal food webs, and determination of realistic water quality indicators of bloom activity.

CHAPTER 8

Comparative Aspects of Trophic Organization

I. OVERFISHING AND POLLUTION IN THE NE GULF

An important goal of coastal ecosystem research is to use comparative data in different systems to test generalities and assumptions concerning key processes such as primary and secondary productivity, nutrient effects on phytoplankton production, phytoplankton bloom formation, and food web dynamics. The coastal systems of the NE Gulf of Mexico provide an opportunity to compare baseline (i.e., natural) conditions in relatively unpolluted areas such as Apalachee Bay (sea grass beds) and Apalachicola Bay (phytoplankton dominated) with data from severely damaged drainage systems (Fenholloway, Perdido, Pensacola, and Choctawhatchee Bay systems) (Livingston, 1984, 2000). Eyre and Balls (1999) noted that, in near-pristine coastal systems, there were general patterns of nutrient changes along salinity gradients. Primary productivity and phytoplankton biomass usually increased within temperate estuarine basins during runoff events due to water column stratification. In natural temperate and tropical coastal systems, increased dissolved inorganic phosphorus (DIP) concentrations seaward were considered a characteristic of near-pristine areas not affected by anthropogenous runoff. Despite the relatively low human population in north Florida, there have been serious problems in terms of nutrient loading and toxic wastes in estuaries that border population centers such as Pensacola and the Destin–Niceville arc of urban development (see Figure 2.5; Livingston, 2000). To the east, low numbers of people, combined with little agricultural and industrial activity, have contributed to the relatively pristine conditions from Apalachicola Bay to Apalachee Bay. In addition, there has been relatively little in the way of overfishing in the eastern end of the Gulf coastal environment.

Recently, Jackson et al. (2001) outlined the hypothesis that ecological extinction due to overfishing takes precedence over other forms of human impacts including pollution, water quality degradation, and anthropogenous climate changes. According to this idea, species that are not fished assume the ecological role of overfished species of the same general trophic orientation. These generalizations were based on a review of paleoecological, archaeological, and historical records and recent ecological data. According to Jackson et al. (2001), overfishing in the past (sometimes the distant past) led to "simplified coastal food webs" in kelp forests, the Gulf of Maine, coral reefs, and various estuaries. The argument was made for "top-down" control due to the loss of benthic suspension feeders that pre-dated "bottom-up" effects of increased anthropogenous nutrient loading associated with eutrophication problems. The Chesapeake Bay and Pamlico Sound areas were given as examples of reduced sea grasses and benthic diatoms due to farming during the 19th century with subsequent increased phytoplankton populations during the 1930s due to the loss of filter-feeding bivalue mollusks. Corresponding increases in the flux of organic matter led to widespread anoxia and hypoxia. Reduced oysters due to overfishing during the 20th century thus led to hypereutrophication. Decline of water quality was secondary to overfishing as a cause of the loss of the bivalve populations and subsequent losses of other species in addition to destruction of sea grass beds and other sensitive habitats. If true, restoration activities such as aquaculture of bivalve mollusks would constitute an important part of pollution abatement. Jackson et al. (2001) thus concluded that early overfishing represented the precondition to the present-day "collapse we are witnessing," and basic changes in our approach to restoration will be required based on the new paradigm if we are to reverse the effects of eutrophication.

Overfishing has caused adverse effects not only on the subject species, but also on associated food webs due to both top-down and bottom-up food web impacts. However, it is possible that recent reviews have overgeneralized the importance of overfishing relative to the current increase of nutrient loading and eutrophication in coastal systems. If overfishing is postulated as the primary factor in the collapse of such systems, such generalizations concerning top-down effects could provide a political excuse to do nothing about anthropogenous nutrient loading. If these hypotheses are wrong, the end result could be further damage to as-yet-unaffected coastal areas. For example, the top-down enthusiasts have provided an alternative hypothesis for the proliferation of macroalgal blooms on coral reefs in Jamaica and southeast Florida. Hughes et al. (1999) attributed these impacts to top-down predation on key populations that graze plants. Loss of such populations, according to this line of reasoning, leads to expansion of plant population and eventual reef destruction. However, Lapointe (1997, 1999) pointed out that these authors ignored much of the field evidence that implicates bottom-up (nutrient enrichment) causative factors in the deterioration of the reefs. If Hughes et al. (1999) are wrong, this could lead to further losses of a unique coastal habitat.

Nutrient loading is having increasingly damaging effects on a series of bays along the NE Gulf coast. In Perdido Bay, massive kills of bivalve mollusks (*Rangia cuneata*) followed the occurrence of plankton blooms that resulted from increased nutrient loading from a paper mill (Livingston, 2000). There was no bivalve fishery in the Perdido system. A similar occurrence was noted in Escambia Bay. Upper Escambia Bay, with reduced circulation and increased nutrient loading from newly urbanized areas, was subject to massive fish kills during the late 1960s and early 1970s (Livingston, 1997a, 1999, 2000). An active oyster fishery was destroyed during this period and most of the sea grasses of the Pensacola Bay system disappeared due to the effects of hypereutrophication. In this case, top predators and filter feeders such as oysters were destroyed by habitat deterioration associated with nutrient imbalances and phytoplankton blooms. Local sports fisheries were no longer viable. The oysters were not subject to overfishing. The fish kills led to restrictions on nutrient loading and eventual (partial) recovery of water quality. However, the oyster industry and sea grass beds have not fully recovered to this day. There were no observable "top-down" effects in the Pensacola system. Non-point-source (urban) pollution continues to play a role in the current problems in Escambia Bay, and sewage discharges and industrial discharges continue to cause serious water quality problems in parts of the Pensacola Bay system (Livingston, 1999). Storm water outfalls are particularly numerous in western sections of the Escambia system, and septic tanks continue to cause problems in lower Escambia Bay. Replacement of the oysters as a form of restoration (Jackson et al., 2001) remains a practical impossibility for this system.

The Choctawhatchee Bay system is represented by a series of complex habitats. Habitat quality is strongly influenced by freshwater input (river, bayou drainages) and salinity stratification (Livingston, 1986a,b, 1987b). Over the past 30 to 35 years, there has been a major increase of urbanization in western parts of the bay (Livingston, 1986a,b). Eastern sections remain relatively undeveloped. According to Livingston (1987c), discharges of sewage and storm water associated with unrestricted land development in western sections of Choctawhatchee Bay were responsible for water quality deterioration in a series of bayou areas in the northern sections of the bay and in Destin Harbor (Old Pass Lagoon) to the south (see Figure 2.2C). Macroinvertebrates and fishes, the backbone of the limited sports and commercial fisheries in the Choctawhatchee system, are concentrated at the river mouth, in the unpolluted (eastern) bayous, and in the shallow sea grass habitats of western sections of the bay. Oyster populations have been relatively restricted prior to and after the recent damaging blooms in the bay. Livingston (1987) noted that discharges of sewage and storm water associated with unrestricted land development in western sections of Choctawhatchee

Bay were responsible for water quality deterioration in a series of bayou areas in northern sections of the bay and in Destin Harbor (Old Pass Lagoon) to the south. Nutrient loading, together with a lack of flushing, contributed to periodically severe water quality problems (hypoxia, anoxia) and associated reductions of natural productivity in the lagoon. This situation was indicated by high nutrients (N, P) in the water column, silt-laden sediments (especially in eastern sections of the lagoon), high phytoplankton productivity, low numbers of herbivores (zooplankton, meroplankton), a depauperate infaunal invertebrate community through central/deep portions of the lagoon, and relatively few epibenthic invertebrates and fishes (Livingston, 1986a,b). Recent uncontrolled urban development accompanied by periodic nutrient loading to the bay have been associated with novel extensive red tide (*Gymnodinium breve*) blooms both in the bay and in offshore areas. Blooms were accompanied by the deaths of sea turtles and over 100 "bottlenose dolphins" in the Choctawhatchee system and along the coast between Choctawhatchee Bay and St. Andrews Bay from August 1999 to mid-January 2000 (K. Spencer, *Destin Log*, 1999–2000). The deaths of fish-eating terrestrial mammals (raccoons and foxes) in the region were also observed in various parts of western Choctawhatchee Bay. Although there are no definitive answers concerning causation of the mammal deaths, researchers found red tide toxins in the dolphins' systems (K. Spencer, *Destin Log*, 1999–2000). Deaths of predators occurred after bloom formation, thus making a "top-down" explanation for recent events highly unlikely.

The Apalachicola system has been a major source of oysters (*Crassostrea virginica*) for centuries (Livingston, 1984). Research on the extensive Apalachicola oyster reefs dates back to the work of Swift (1896) and Danglade (1917). The Apalachicola estuary accounts for about 90% of Florida's commercial fishery (Whitfield and Beaumariage, 1977). This has not changed in recent times. Analyses of oyster distribution (Livingston et al., 1999) indicated no differences when compared to detailed reef descriptions by Swift (1896). During fall 1985, two hurricanes struck the Apalachicola Bay system, a center for oyster production in the northeast Gulf of Mexico (Livingston et al., 1999). The first storm, Hurricane Elena, physically destroyed the major oyster-producing reefs in the Apalachicola estuary in early September. Subsequent oyster growth was substantial, with full recovery of the oyster stock noted within a 12-month period. The timing and nature of the disturbances relative to the natural history of the oyster were crucial to the overall recovery pattern of the population. Effects of the storm increased habitat availability and reduced direct competition and predation such that the oyster population benefited from successful recruitment. The observed response of the Apalachicola oyster population to successive disturbances has significant meaning in terms of the long-term ecological stability of estuarine populations and the evolutionary aspects of such biological response to temporally unstable habitats. In this case, oyster populations are resilient under even the most extreme conditions of natural physical instability. Commercial oystering can have a significant impact on the density and population structure of oyster reefs. However, strict controls (i.e., no dredging, restricted oystering periods) have allowed the relatively stable and productive Apalachicola oyster industry to continue to the present time. Oystering has had only a small effect on the distribution of oysters in the Apalachicola system (Livingston et al., 1999).

A variety of organisms prey on oysters in the Apalachicola system (Menzel and Nichy, 1958; Menzel et al., 1966), including boring sponges, polychaete worms, gastropod mollusks (*Thais haemastoma* and *Melongena corona*), crustaceans (*Callinectes sapidus* and *Menippe mercenaria*), and fishes (*Pogonias cromis*). During periods of high salinity, oyster predation is enhanced, and can be considerable. Menzel et al. (1966) found that the near total demise of the St. Vincent bar resulted from increased predation due to anthropogenous physical alterations in the salinity regime of the bay following the opening of Sikes Cut in the mid-1950s. Experimentally determined mortality data (Livingston et al., 2000) indicated that the overall mortality of oysters in the Apalachicola system was related to salinity and the geographic position of the reefs relative to the natural (East Pass, West Pass, and Indian Pass) and human-made (Sikes Cut) openings to the Gulf. The Apalachicola Bay system has remained free of toxic agents and high anthropogenous nutrient loading (Livingston, 1983). There are no indications of phytoplankton blooms (Iverson et al., 1997). Combinations of "top-down" and "bottom-up" effects are implicated in the history of the Apalachicola estuary. Clearly, overfishing as a primary "top-down" factor (Jackson et al., 2001) does not apply here. The real threats to continued productivity of the Apalachicola system include reduced freshwater input from the river due to urban and agricultural activities upstream (Livingston et al., 1997; Chanton and Lewis, 1999, 2002) and increasing urban development of the bay area (see Chapter 10).

Apalachee Bay remains in almost pristine condition due to the almost total lack of industrial, urban, and agricultural development in mainland drainage basins (Livingston, 2000). Bay wetlands have been purchased for preservation by the state of Florida. There are no signs of toxic agents in water or sediments of this bay (Livingston et al., 1998b), and the only source of pollution to Apalachee Bay is a pulp mill on the Fenholloway River (Livingston et al., 1998b). The dimensions of the impact of the mill have been well established (Livingston, 1975a, 1981b, 1982a, 1984d, 1985a,b, 1997a; Livingston et al., 1998b). Inshore areas are not commercially fished due to rocky outcrops. These areas are open to recreational fishing, but low human populations and the inaccessibility of the Apalachee Bay area have restricted fishing pressure. Long-term studies of the area (1971 to the present; see below) indicate that the sea grass beds in areas unaffected by mill effluents remain in a natural state. The only plankton blooms noted have been associated with pulp mill nutrient loading (see Chapter 9), which represents the primary current threat to Apalachee Bay. The "top-down" hypothesis (Jackson et al., 2002) does not apply here in the largely unaffected sea grass beds of Apalachee Bay.

Dissolved oxygen (DO) can be taken as an important indicator of water quality in coastal areas. Portnoy (1991) found that channelization contributed to oxygen stress by substituting deep organic sinks for what were originally shallow habitats. Reyes and Merino (1991) found that sewage releases and dredging adversely affected the DO in a lagoon in the Mexican Caribbean. Breitburg (1990) indicated that wind-driven salinity alterations were associated with severe hypoxia at depth. Stanley and Nixon (1992), working in the Pamlico River Estuary, found that hypoxia developed only under conditions of vertical water-column stratification and warm temperatures. Stratification and low DO events were associated with freshwater discharge and wind stress. Over the 15-year period of observation, there was no trend toward lower bottom water DO, and there were no demonstrated cause-and-effect relationships among nutrients, algal abundance, and bottom water DO levels. Proposed reductions of nitrogen were not expected to change the natural hypoxia and anoxia in the Pamlico system. Parker and O'Reilly (1991) found that there was a spatial extension of the subpycnoclinal hypoxia eastward in Long Island Sound in terms of both severity and frequency of hypoxia. Improvements in parts of the system were associated with upgrades in several sewage treatment plants in the drainage area. Welsh and Eller (1991) found that the ultimate control of the hypoxic conditions in Long Island Sound was salinity stratification with even a weak pycnocline effective in maintaining low oxygen at depth.

The distribution of DO in the various coastal systems of the NE Gulf have been well studied (Livingston, 2000). Depth and salinity stratification are important qualifying agents in the incidence of hypoxia in the NE Gulf estuaries with the presence of pulp-mill discharges (Perdido and Fenholloway estuaries) and urban runoff (Escambia and Choctawhatchee systems) as important determinants of the spatiotemporal extent of hypoxic conditions at depth. DO remains high above the halocline in all areas from Apalachee Bay to Perdido Bay. Perdido Bay and the Choctawhatchee Bay system are subject to winter hypoxia (DO < 4.0 mg/l) in areas that are strongly salinity stratified. Widespread hypoxia has been noted in the Escambia Bay area. In the Apalachicola estuary, there has been no obvious relationship between winter stratification and DO. Hypoxia has occurred in upland estuarine areas of the Fenholloway estuary relative to the Econfina estuary during winter months, an obvious effect of high DOC loading by the mill. In the spring, there is a recurring, direct correspondence of hypoxic conditions with the level of salinity stratification in the Perdido, Escambia, and Choctawhatchee Bay systems. Although there has been a shift to lower DO in the Apalachicola estuary during spring periods (a temperature effect), there is still no sign of widespread hypoxia. Hypoxic conditions in the Fenholloway estuary intensify during spring months, but are largely restricted to the lower estuary. There is little spring hypoxia in the Econfina estuary. During summer and fall months, hypoxic conditions intensify in the Perdido, Escambia, and Choctawhatchee systems, and are largely associated with salinity stratification. In the Apalachicola estuary, there is increased hypoxia at depth, although there is no clear relationship between the salinity stratification and hypoxia. There is enhanced hypoxia in the Fenholloway estuary, but not the Econfina estuary during summer–fall months. Offshore areas of the Econfina and Fenholloway systems remain well oxygenated.

There is no evidence that overfishing has caused the various effects that have been associated with problems due to nutrient loading as noted above. Selective ignorance of the scientific literature combined with ideological assumptions based on little scientific evidence should not be substituted for facts. Claims in favor of the "overfishing" paradigm should be tested with field experiments before acceptance as a contributing factor to food web deterioration in coastal systems.

II. SPATIAL DISTRIBUTION OF FOOD WEB COMPONENTS

A. Spatial Patterns of Primary Productivity

Spatial distributions of submerged aquatic vegetation (SAV) in the Econfina and Fenholloway systems at intervals over the 1972 through 2002 period are shown in Figure 8.1. During 1972–1973, the Fenholloway system had low levels of SAV compared to reference sites. Econfina SAV was concentrated in western offshore sites during summer–fall months. The lowest concentrations were usually around the Econfina River mouth and in far offshore areas as a result of light and nutrient limitation (Livingston et al., 1998b). During 1992–1993, overall SAV biomass was relatively low in both systems, with most SAV located in offshore areas. The heaviest concentrations in the Econfina were again at western sites. In the Fenholloway area, SAV varied with the highest concentrations between eastern and western offshore areas. During 1999–2001, there was a return to high SAV biomass in the Econfina system, with the highest SAV development in near offshore areas from May through November. Peak SAV development occurred during June. In the Fenholloway system in 1999 through 2001, SAV was concentrated in offshore, western parts of the estuary. Overall, SAV development in the Fenholloway was generally higher from 1999 to 2001, although inshore effects of the river were more extensive than those noted at the reference site. SAV development in Apalachee Bay depends on factors such as depth, light penetration, nutrient distribution, and water/sediment quality (Livingston et al., 1988b).

B. Herbivores

The trophic organization of the Perdido system was mapped during the period before the appearance of phytoplankton blooms. Herbivore biomass was largely restricted to the upper bay with biomass peaks noted off Elevenmile Creek during May, following the spring phytoplankton peaks (Figure 8.2). During the rest of the year, herbivore biomass was concentrated from Elevenmile Creek to the *Vallisneria* beds in western parts of the upper bay. This distribution indicated a close association of the herbivores with phytoplankton and sea grass primary productivity in the upper bay. It also indicated that nutrients from Elevenmile Creek (i.e., the pulp mill) were being incorporated into the Perdido Bay food webs from 1988 to 1991. Herbivores in the Pensacola Bay system were concentrated in upper Escambia Bay during all four seasons, with concentrations in Pensacola Bay and eastern sections of East Bay (Blackwater system) during March, and in various parts of the Blackwater system during June, September, and December. Herbivores in the Choctawhatchee Estuary (Figure 8.2) were concentrated in peripheral areas of western parts of the bay (i.e., areas associated with urban runoff). There was relatively little seasonal variation in this pattern. In the Apalachicola system, herbivores tended to concentrate at the river mouth, in East Bay, and in the primary freshwater drainage (Nick's Hole) off St. George Island. Overall, herbivore development was most pronounced in areas associated with freshwater runoff.

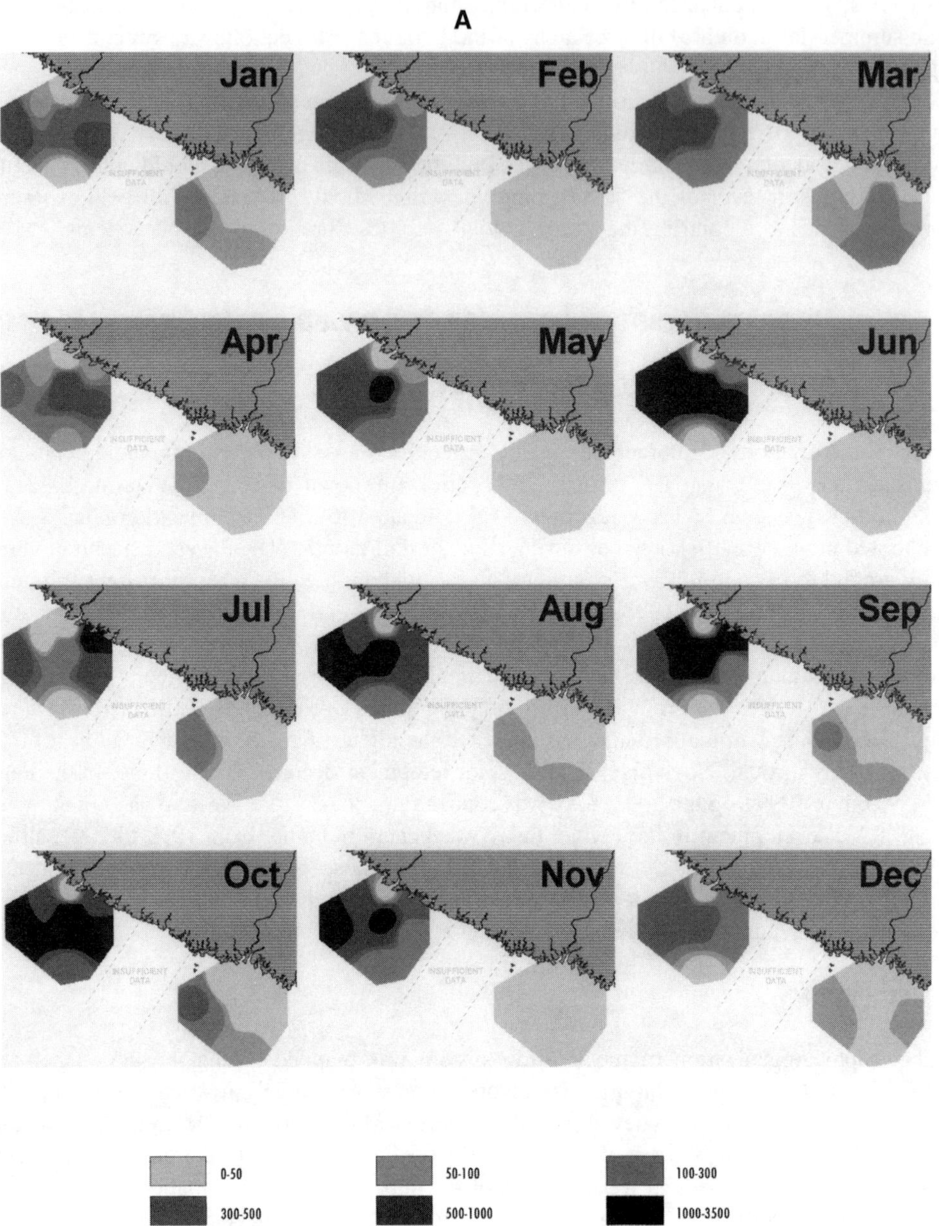

Figure 8.1 (A) Seasonal distribution of submerged aquatic vegetation (g/m^2) in the Econfina and Fenholloway drainages averaged from monthly collections from April 1972 to March 1975.

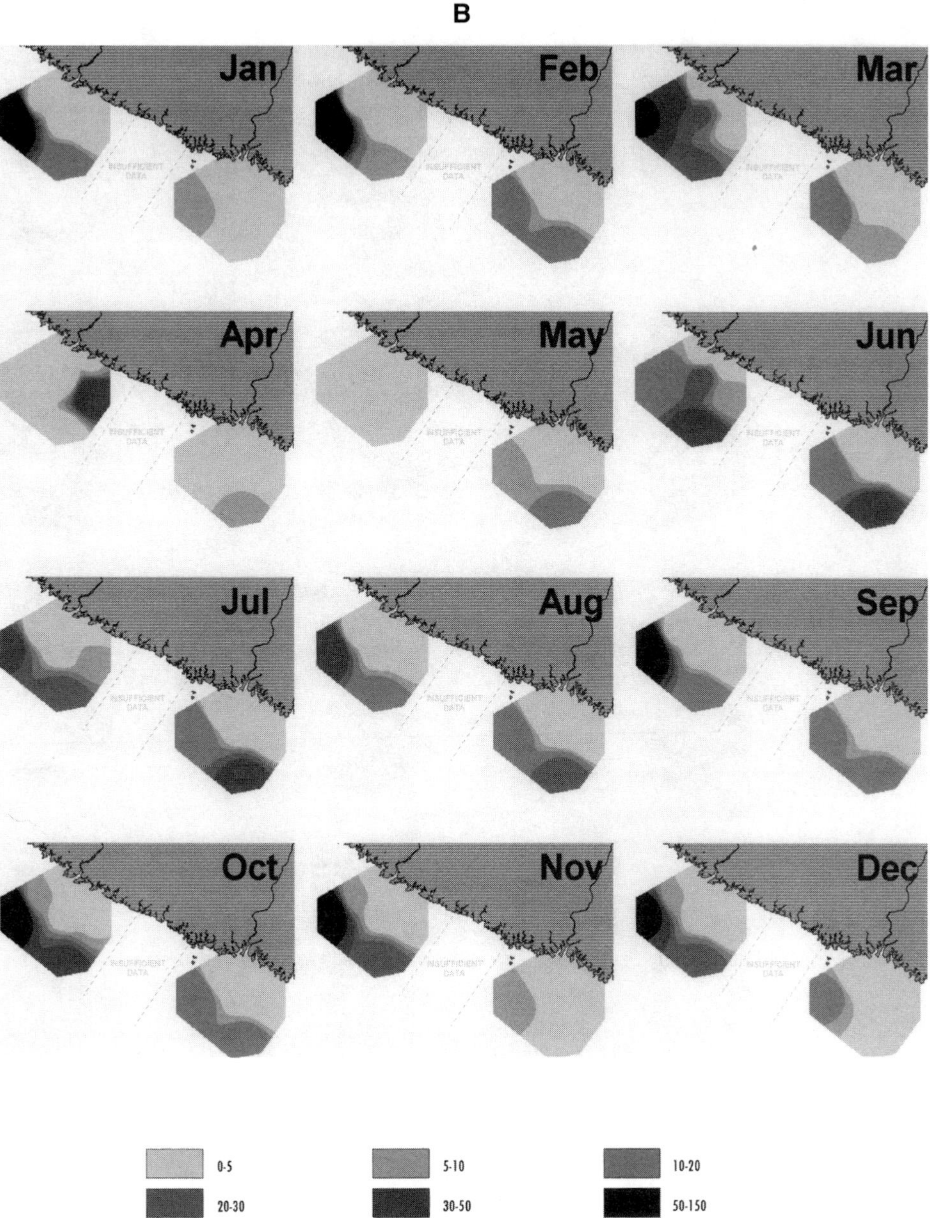

Figure 8.1 (continued) (B) Seasonal distribution of submerged aquatic vegetation (g/m^2) in the Econfina and Fenholloway drainages averaged from monthly collections from April 1992 to March 1993.

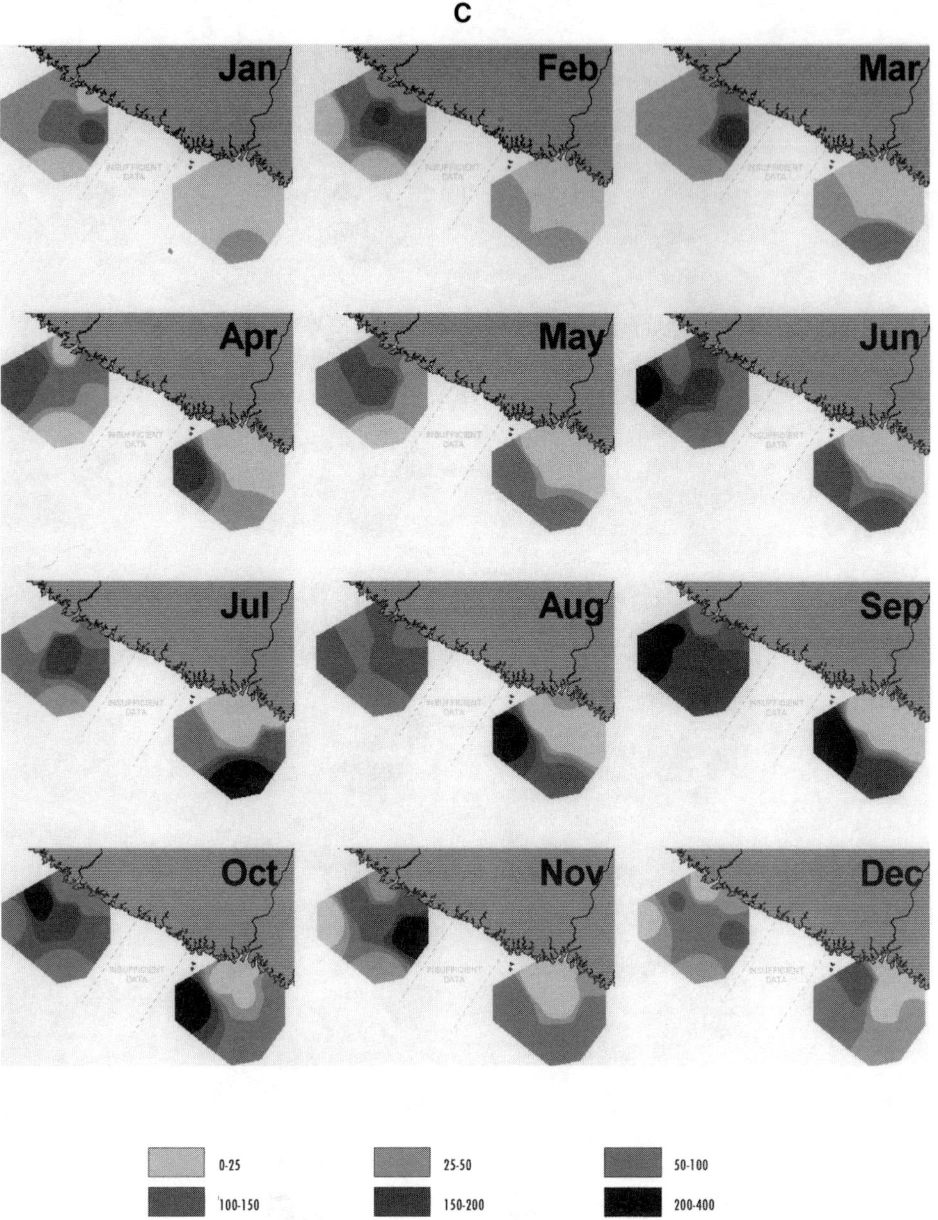

Figure 8.1 (continued) (C) Seasonal distribution of submerged aquatic vegetation (g/m^2) in the Econfina and Fenholloway drainages averaged from monthly collections from April 1999 to March 2001.

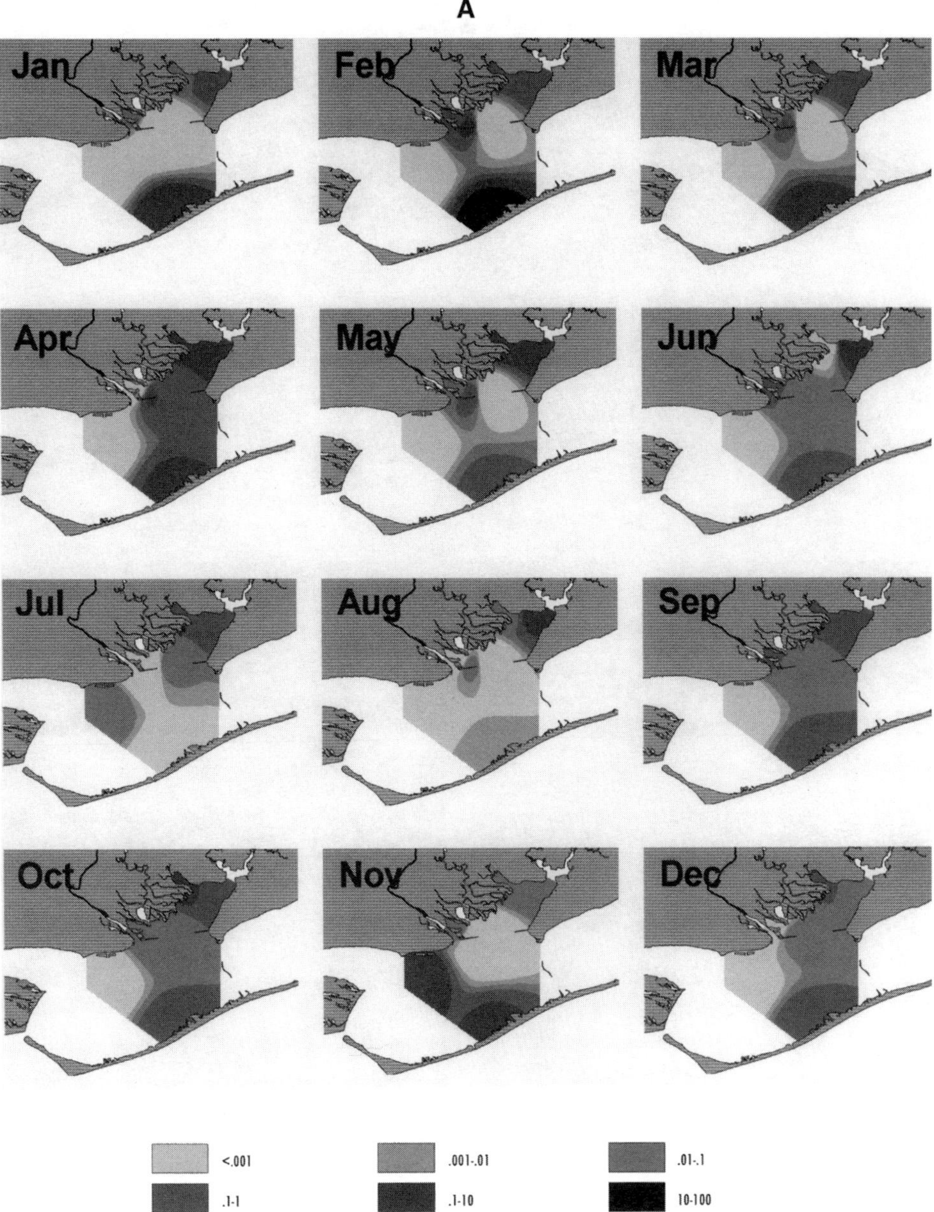

Figure 8.2 (A) Spatial patterns of herbivore biomass (g/m^2) in the Apalachicola system. Data were averaged by month from March 1975 through February 1976.

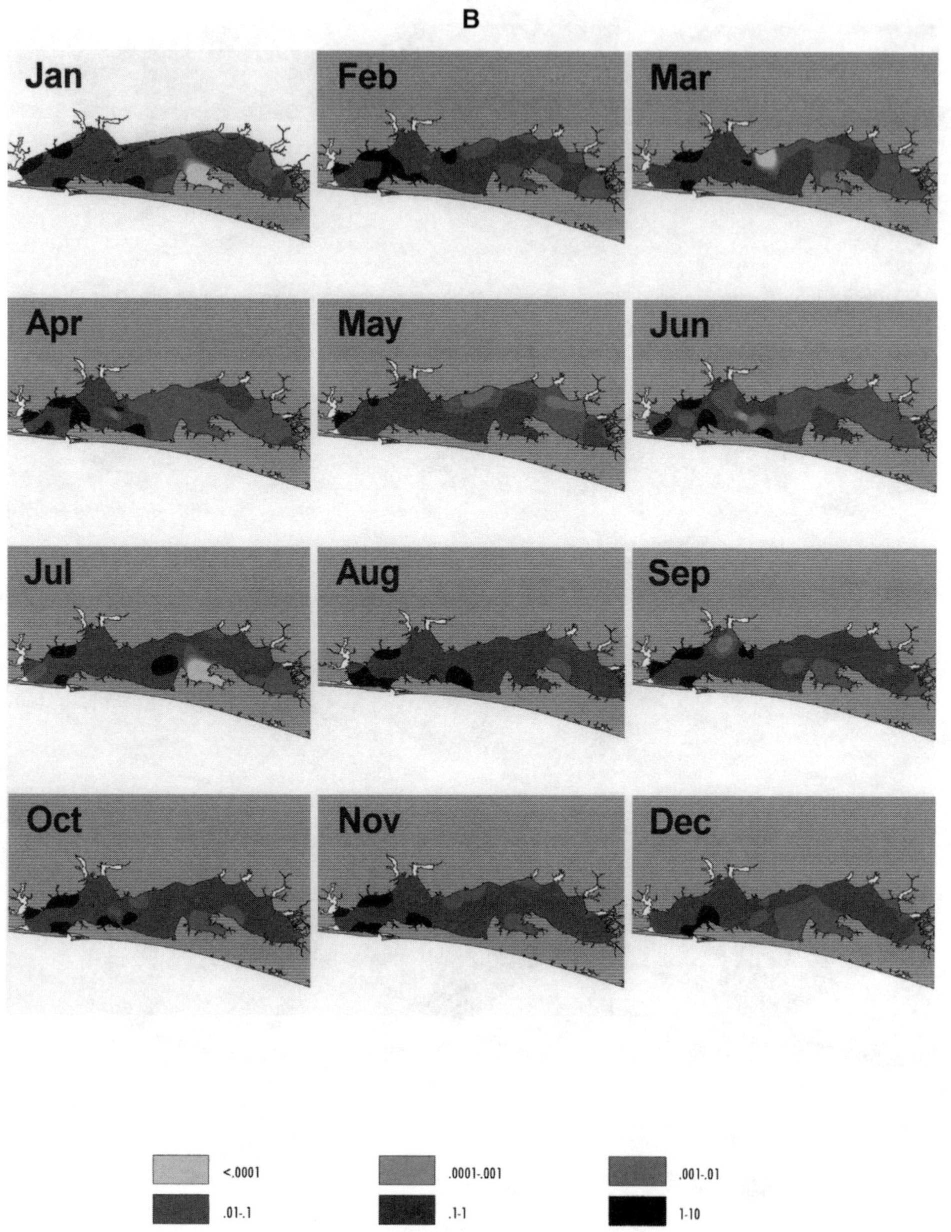

Figure 8.2 (continued) (B) Spatial patterns of herbivore biomass (g/m²) in the Choctawhatchee system. Data were given by month from September 1985 through August 1986.

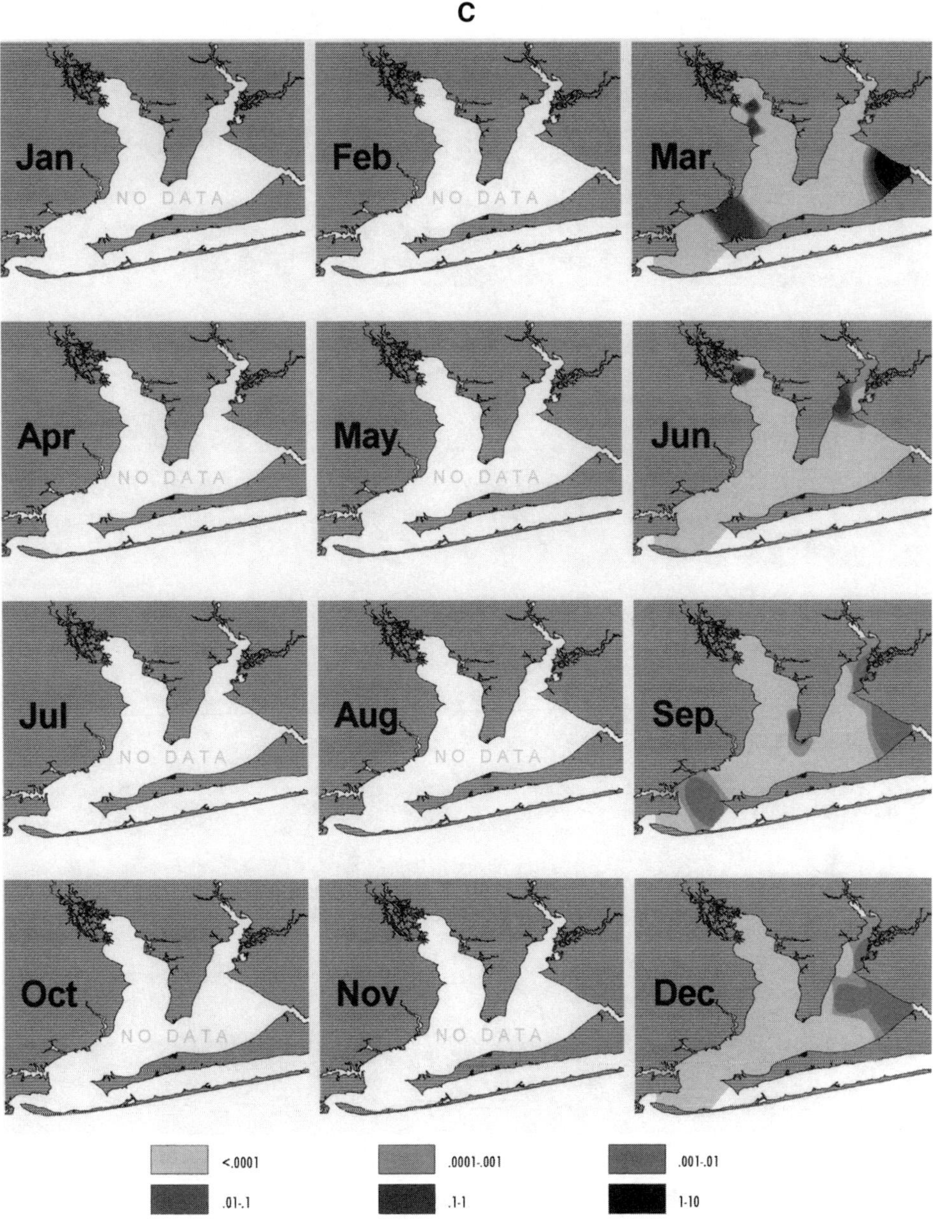

Figure 8.2 (continued) (C) Spatial patterns of herbivore biomass (g/m^2) in the Pensacola system. Data were given by quarter from June 1997 through March 1998.

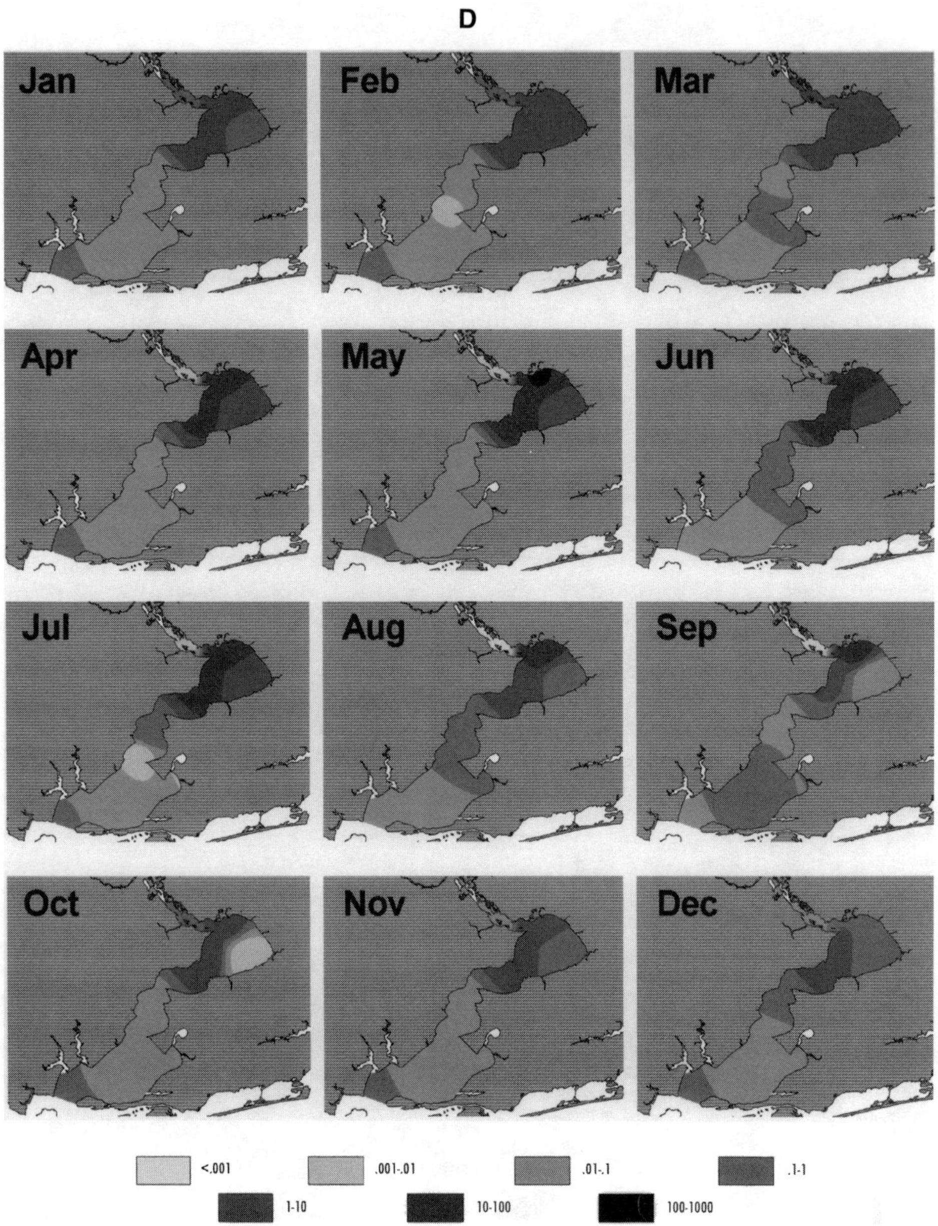

Figure 8.2 (continued) (D) Spatial patterns of herbivore biomass (g/m^2) in the Perdido system. Data were averaged by month from October 1988 through September 1991.

C. Omnivores

Omnivore biomass in the Perdido system (Figure 8.3) was generally low during winter months, and was highest in upper and midbay areas with a July peak in eastern parts of the upper bay. Although the omnivores were more broadly distributed than the herbivores, associations with upper and midbay areas indicated close connections with river flows and the primary producers of the upper bay. During later phytoplankton bloom years, omnivore biomass in the upper bay was positively associated with *Prorocentrum minimum* during the winter, with surface ammonia during the spring, and with bloom numbers during summer (see Table 7.7). In the lower bay, omnivores were positively associated with numbers of winter plankton blooms and with summer N+P loading differences. This indicated that omnivores were prevalent during bloom periods as they took over areas once dominated by herbivores. Omnivore distribution was spread throughout the Pensacola Bay system during various seasons (Figure 8.3). Omnivores in the Choctawhatchee Bay system were concentrated in eastern parts of the bay during spring–summer periods, and in bayou areas in western sections during winter months. In the Apalachicola estuary, omnivores were associated with the freshwater drainages, particularly in the lower bay. Omnivores in the Econfina and Fenholloway drainages were located in offshore sea grass areas with generally low concentrations at the river mouths. Omnivore distribution in the NE Gulf coastal areas was generally associated with freshwater runoff and levels of eutrophication.

D. C1 Carnivores

Primary carnivore distribution in the Perdido estuary varied seasonally (Figure 8.4). During spring, primary carnivores were concentrated in upper (western) and midbay areas. High concentrations of C1 carnivores during August coincided with late summer movements of young adults to the Gulf. The overall distribution of the C1 carnivores was generally connected to herbivore and omnivore associations. However, during bloom years, upper bay herbivores were negatively associated with *P. minimum* biomass during winter periods (see Table 7.7). Because herbivores were located largely in the upper bay, the negative associations with a winter bloom species indicated direct evidence of the adverse effects of blooms on the trophic order of the bay.

The C1 carnivores in the Apalachicola system were more closely associated with biological factors (i.e., predator–prey relationships) since none of the carnivore groups was related directly to any environmental principal component after taking into account the effects of herbivores and omnivores (Livingston et al., 1997) (Figure 8.4). However, statistical analyses of the Perdido data (see Table 7.7) indicated that the C1 carnivores were negatively affected by seasonally changing bloom species and by nutrient loading/concentration factors that caused the blooms. These associations were consistent with the reduction of primary carnivores during periods of peak raphidophyte and dinoflagellate blooms. This divergence represents more evidence of the associations of blooms with disruption of the Perdido Bay food webs when compared to the unpolluted, river-dominated Apalachicola system.

The primary carnivores in the Pensacola system were concentrated in the upper parts of the Escambia and Blackwater Bay areas, with peaks during fall and winter periods (Figure 8.4). The C1 carnivores in Choctawhatchee Bay were concentrated around bayou areas in eastern parts of the bay during most months. In the Apalachicola system, primary carnivores were concentrated around the freshwater runoff areas with some fall (September–October) to winter (December–January) concentrations in the lower bay, coinciding with fall emigration and winter immigration of offshore populations as they entered and left the bay on their annual migrations. In the Econfina and Fenholloway systems, primary carnivores were usually concentrated in offshore sea grass areas. The highest concentrations of this group were usually during summer periods in 1992–1993 and in February during the 1999 to 2001 period. Overall, C1 carnivores were located in areas of high primary productivity and herbivore development, except during periods of plankton blooms.

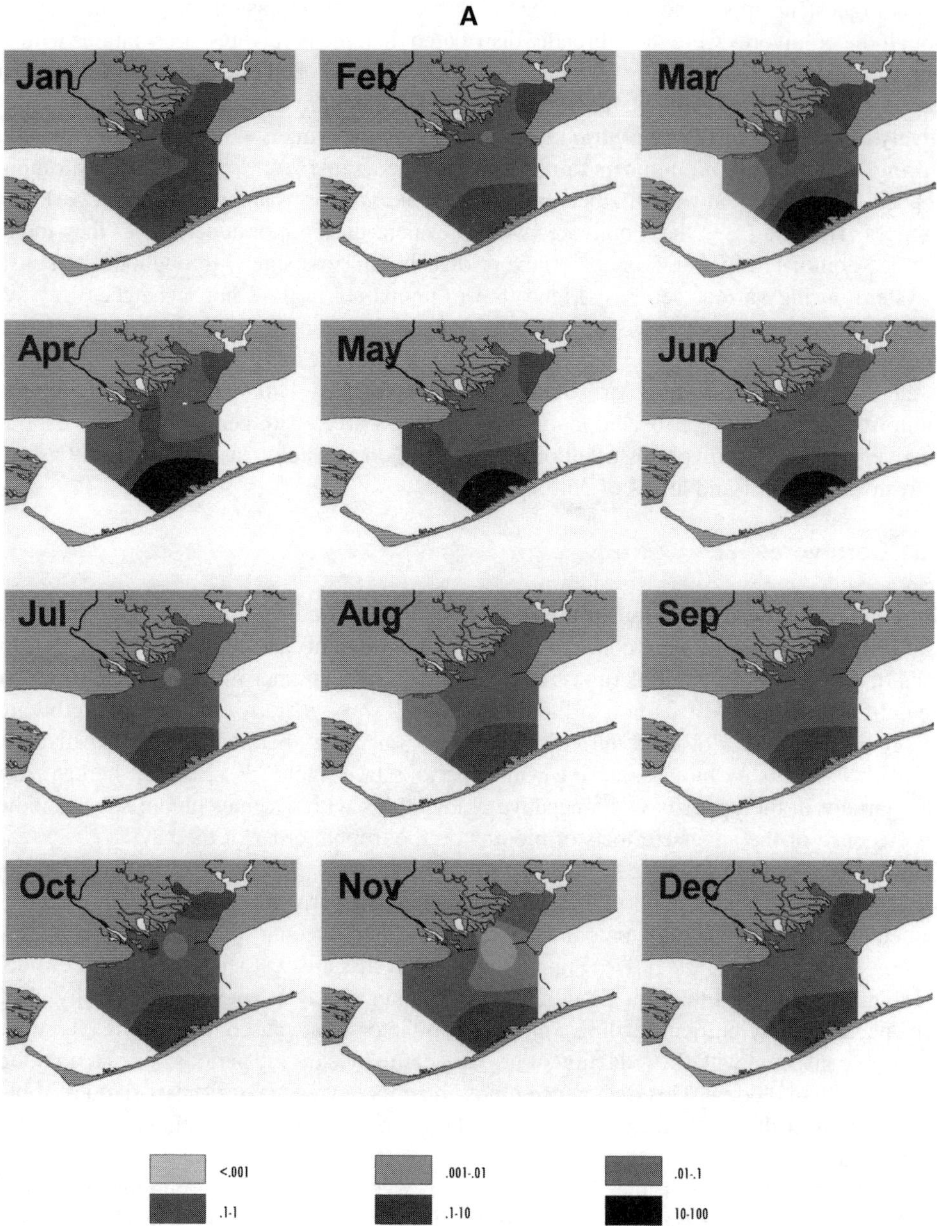

Figure 8.3 (A) Spatial patterns of omnivores in the Apalachicola system. Data were averaged by month from March 1975 through February 1976.

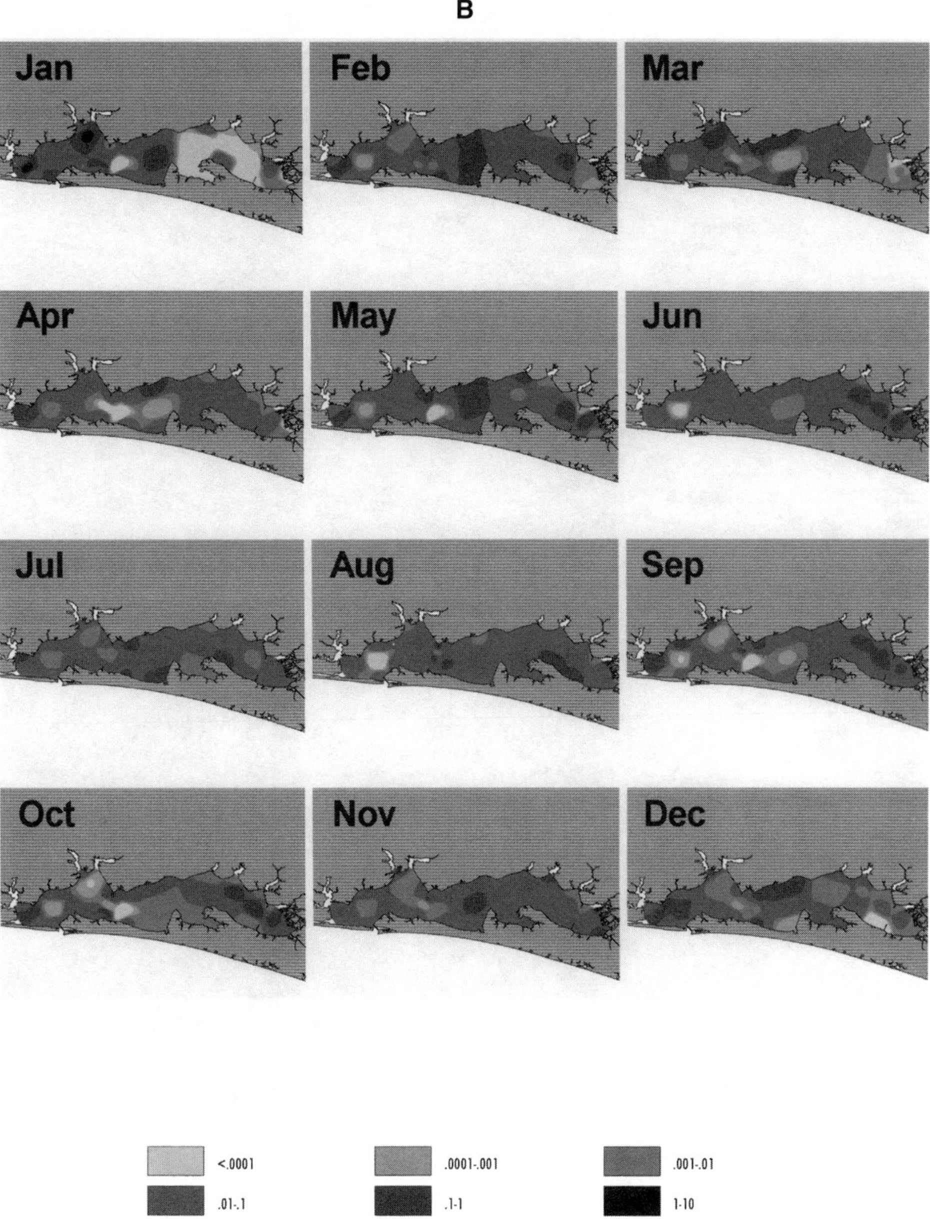

Figure 8.3 (continued) (B) Spatial patterns of omnivores in the Choctawhatchee system. Data were given by month from September 1985 through August 1986.

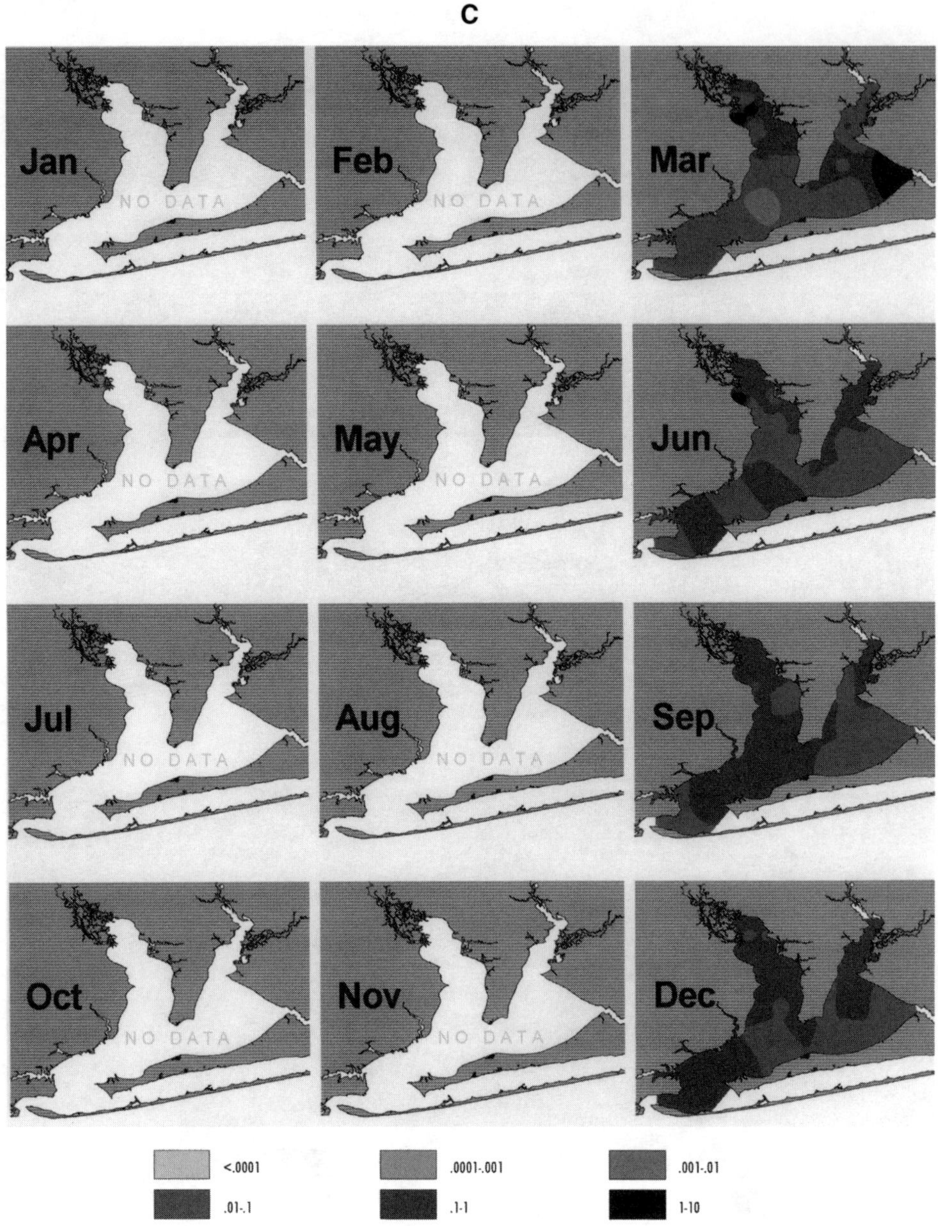

Figure 8.3 (continued) (C) Spatial patterns of omnivores in the Pensacola system. Data were given by quarter from June 1997 through March 1998.

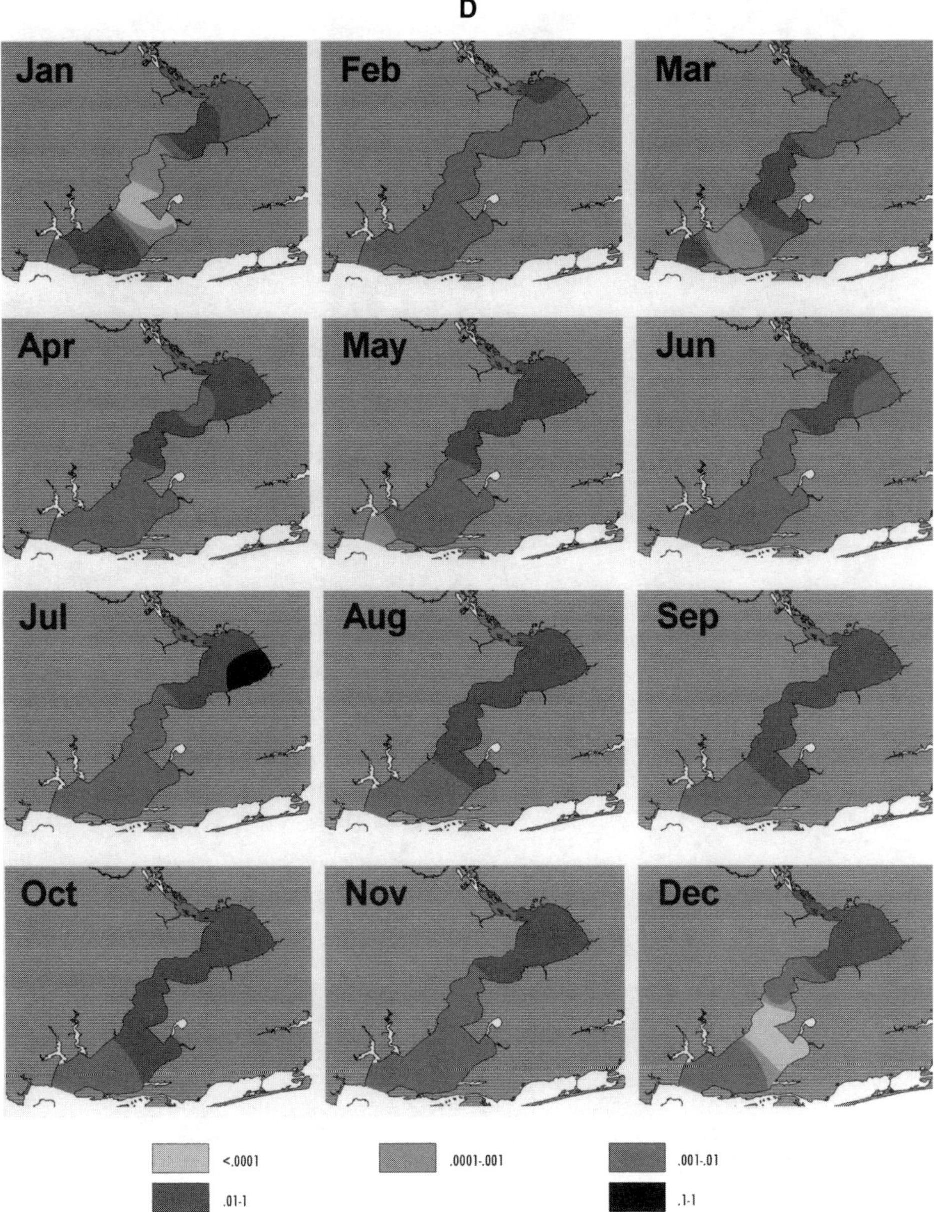

Figure 8.3 (continued) (D) Spatial patterns of omnivores in the Perdido system. Data were averaged by month from October 1988 through September 1991.

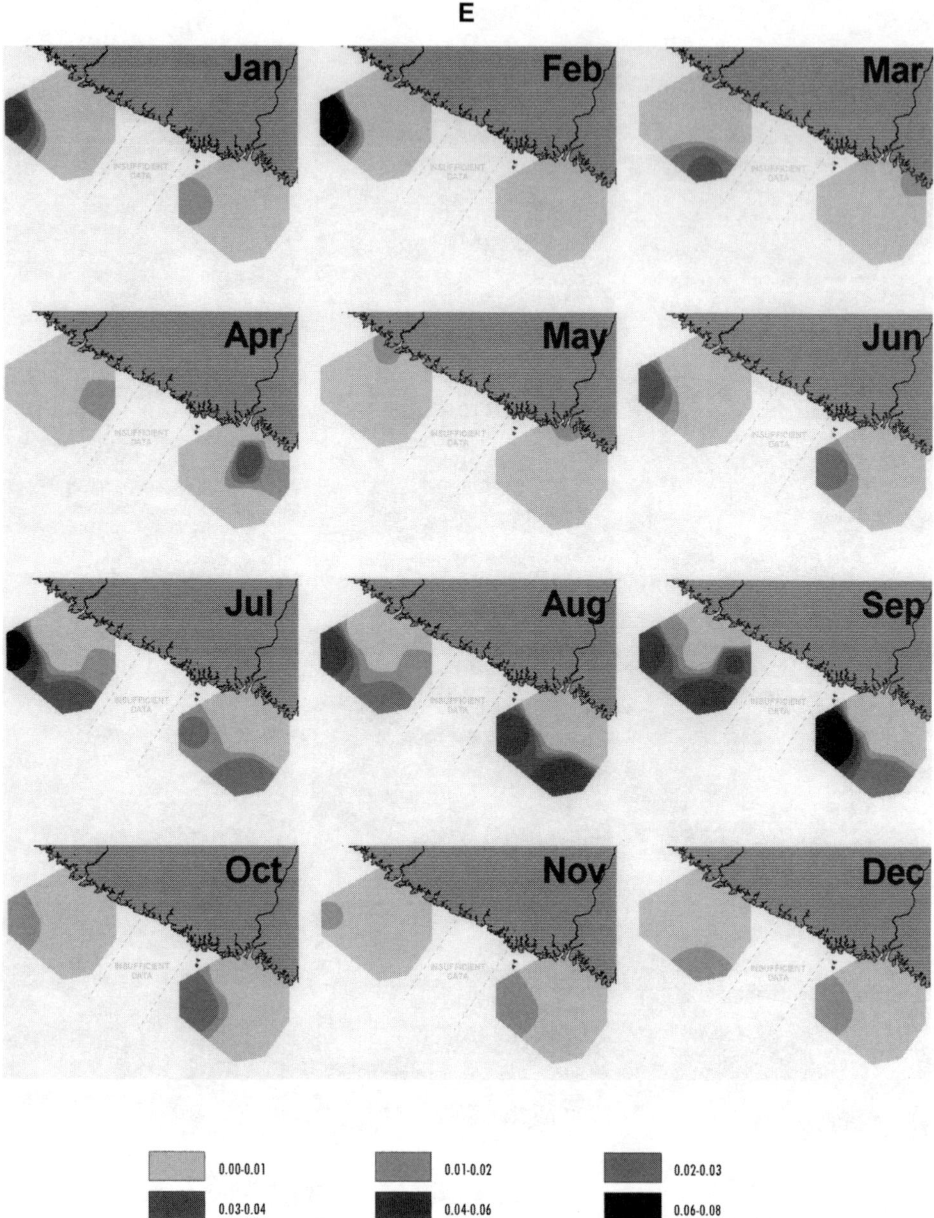

Figure 8.3 (continued) (E) Spatial patterns of omnivores in the Econfina and Fenholloway systems, 1992–1993. Data were given by month from April 1992 through March 1993.

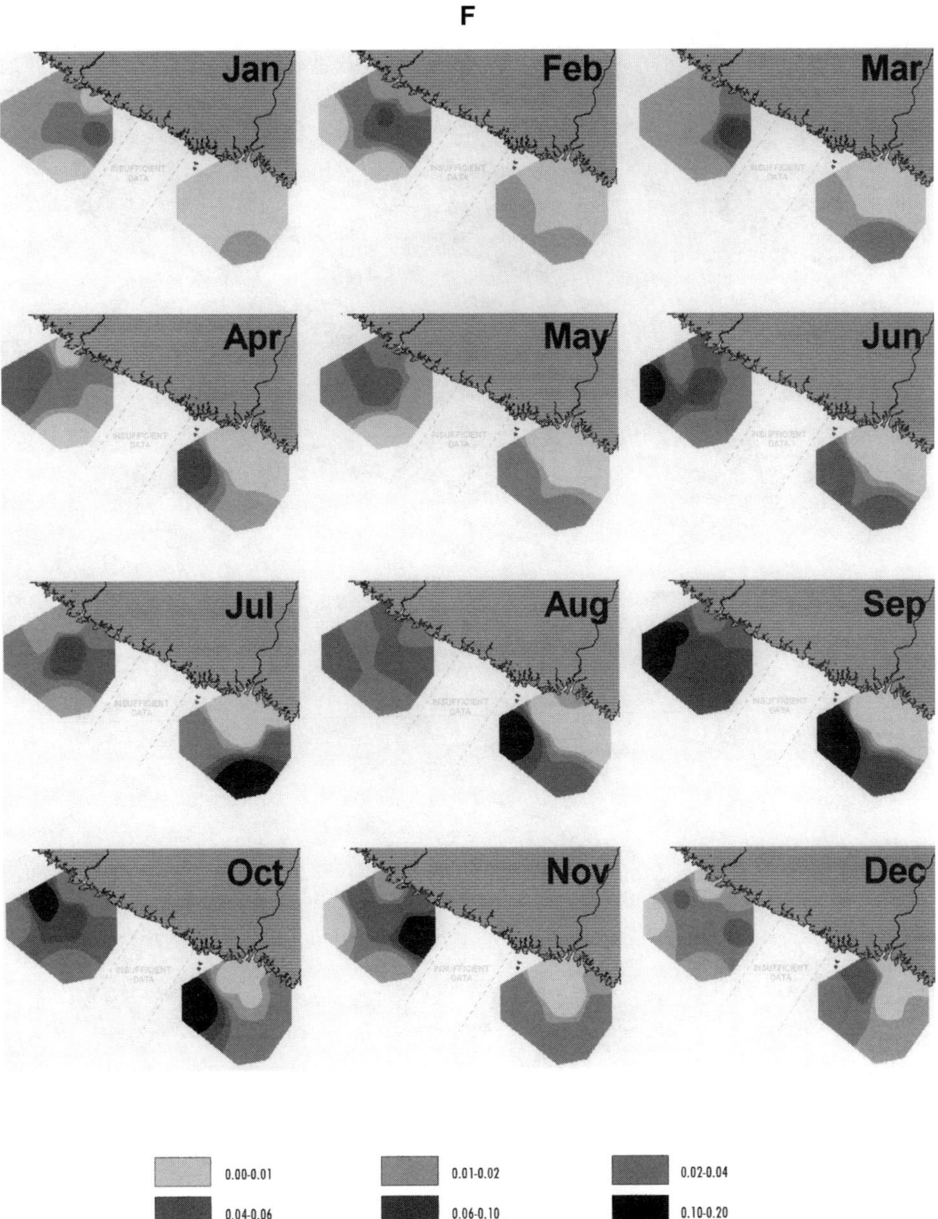

Figure 8.3 (continued) (F) Spatial patterns of omnivores in the Econfina and Fenholloway systems, 1999–2001. Data were given as monthly averages from April 1999 through March 2001.

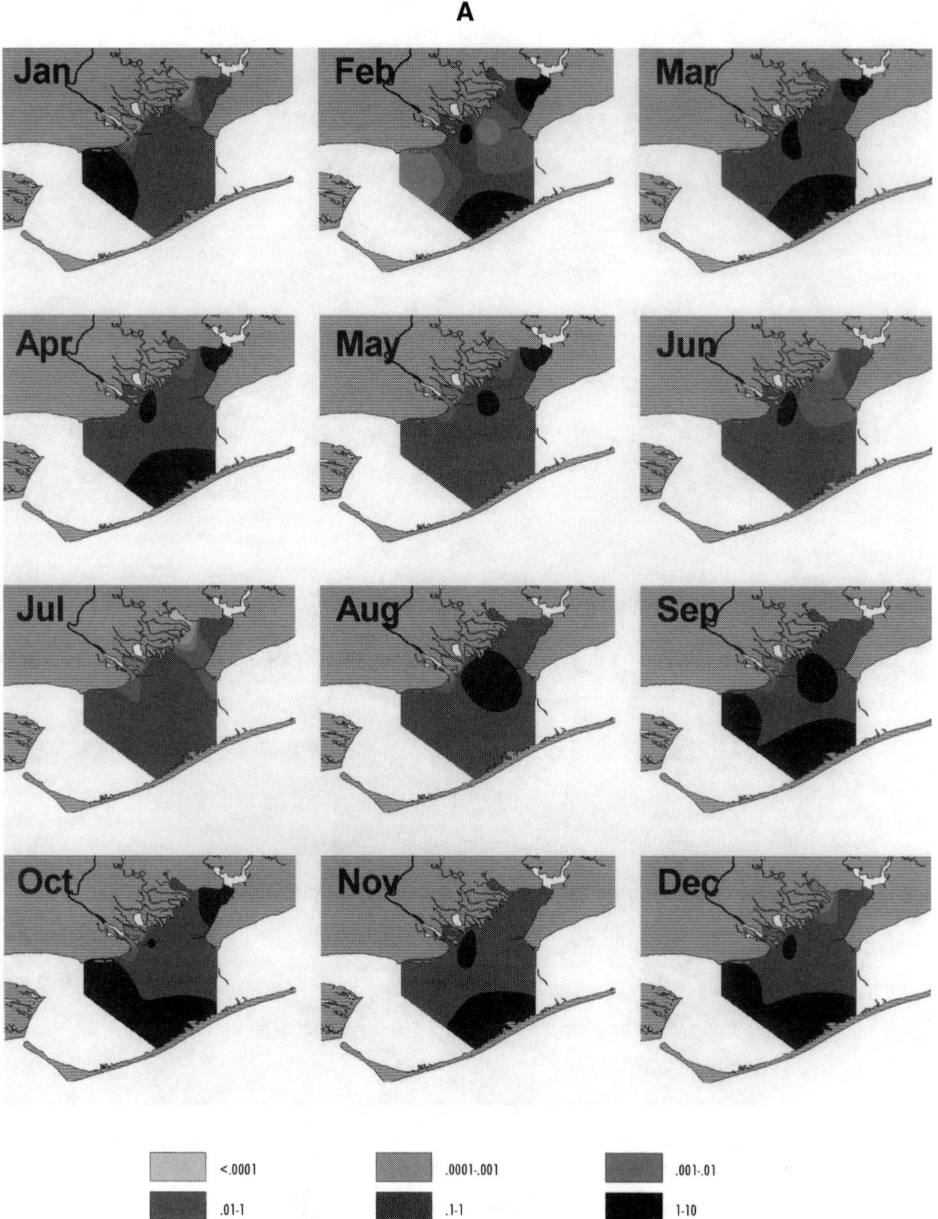

Figure 8.4 (A) Spatial patterns of primary carnivores in the Apalachicola system. Data were averaged by month from March 1975 through February 1976.

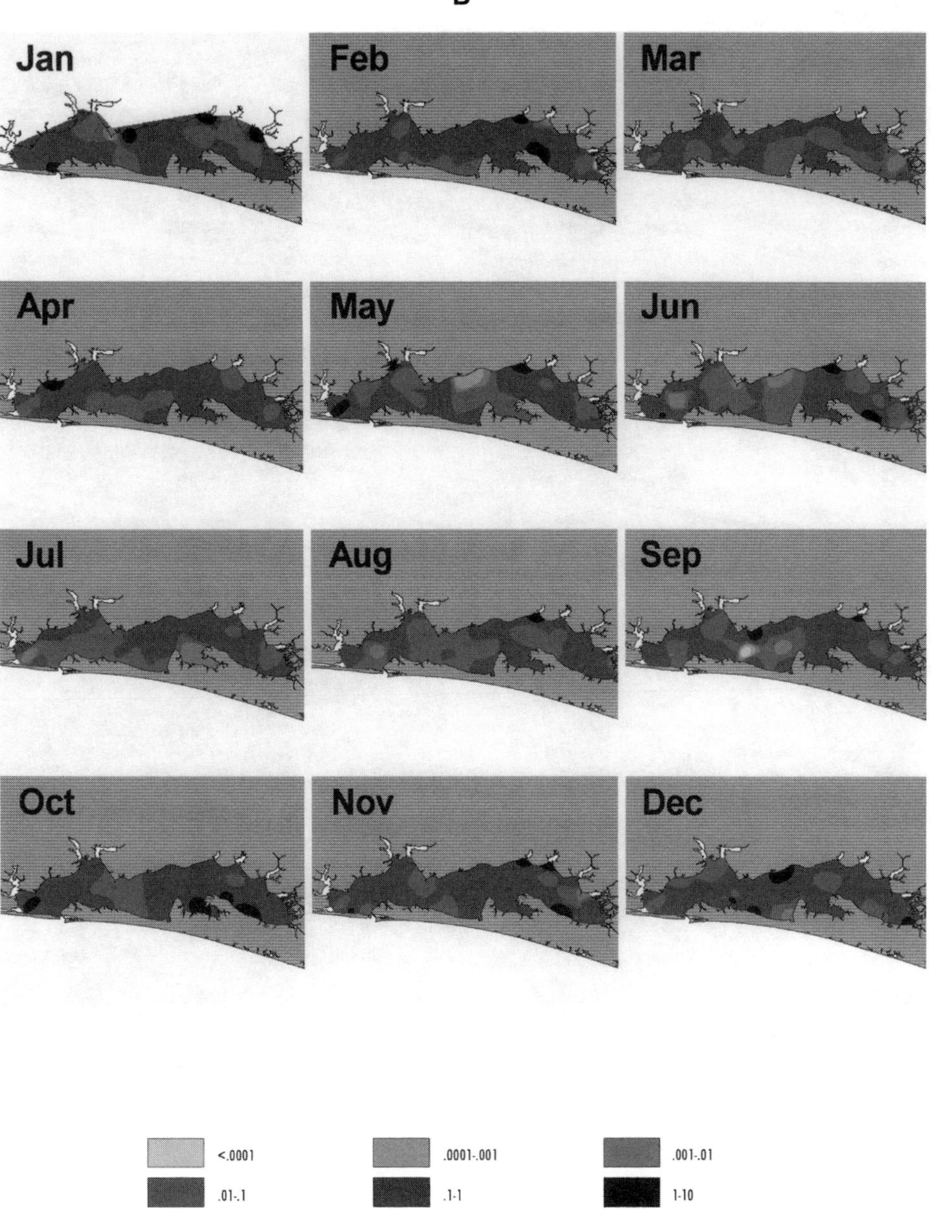

Figure 8.4 (continued) (B) Spatial patterns of primary carnivores in the Choctawhatchee system. Data were given by month from September 1985 through August 1986.

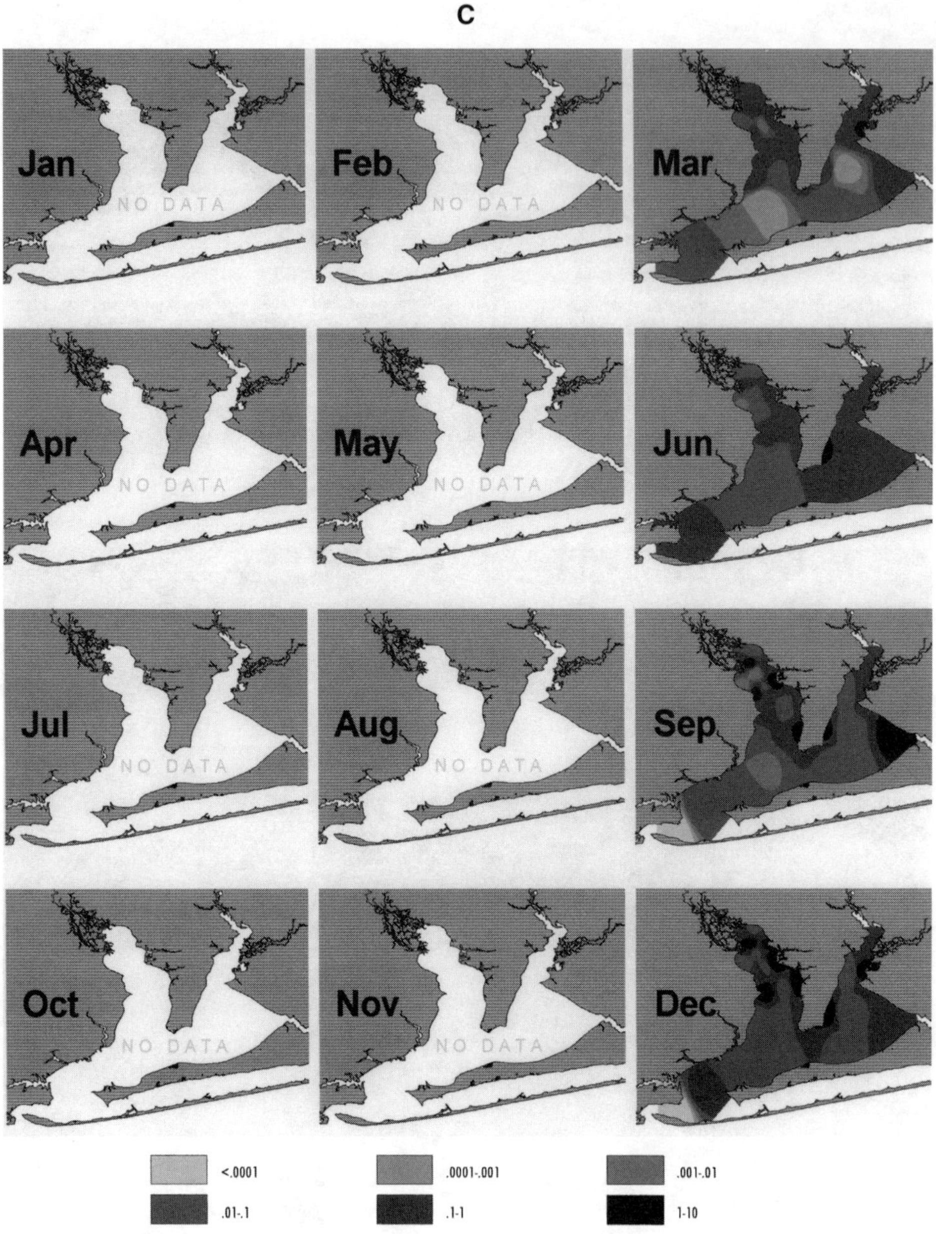

Figure 8.4 (continued) (C) Spatial patterns of primary carnivores in the Pensacola system. Data were given by quarter from June 1997 through March 1998.

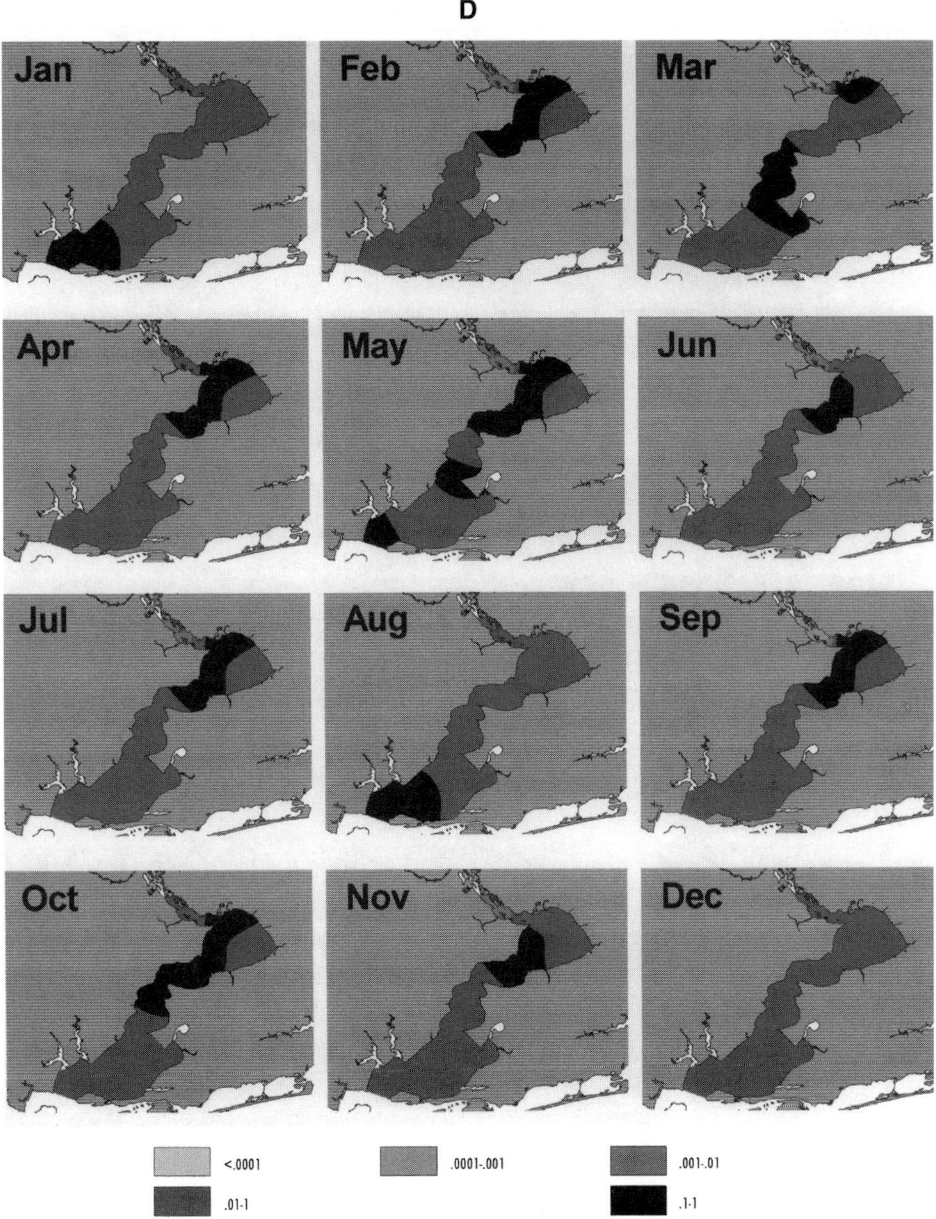

Figure 8.4 (continued) (D) Spatial patterns of primary carnivores in the Perdido system. Data were averaged by month from October 1988 through September 1991.

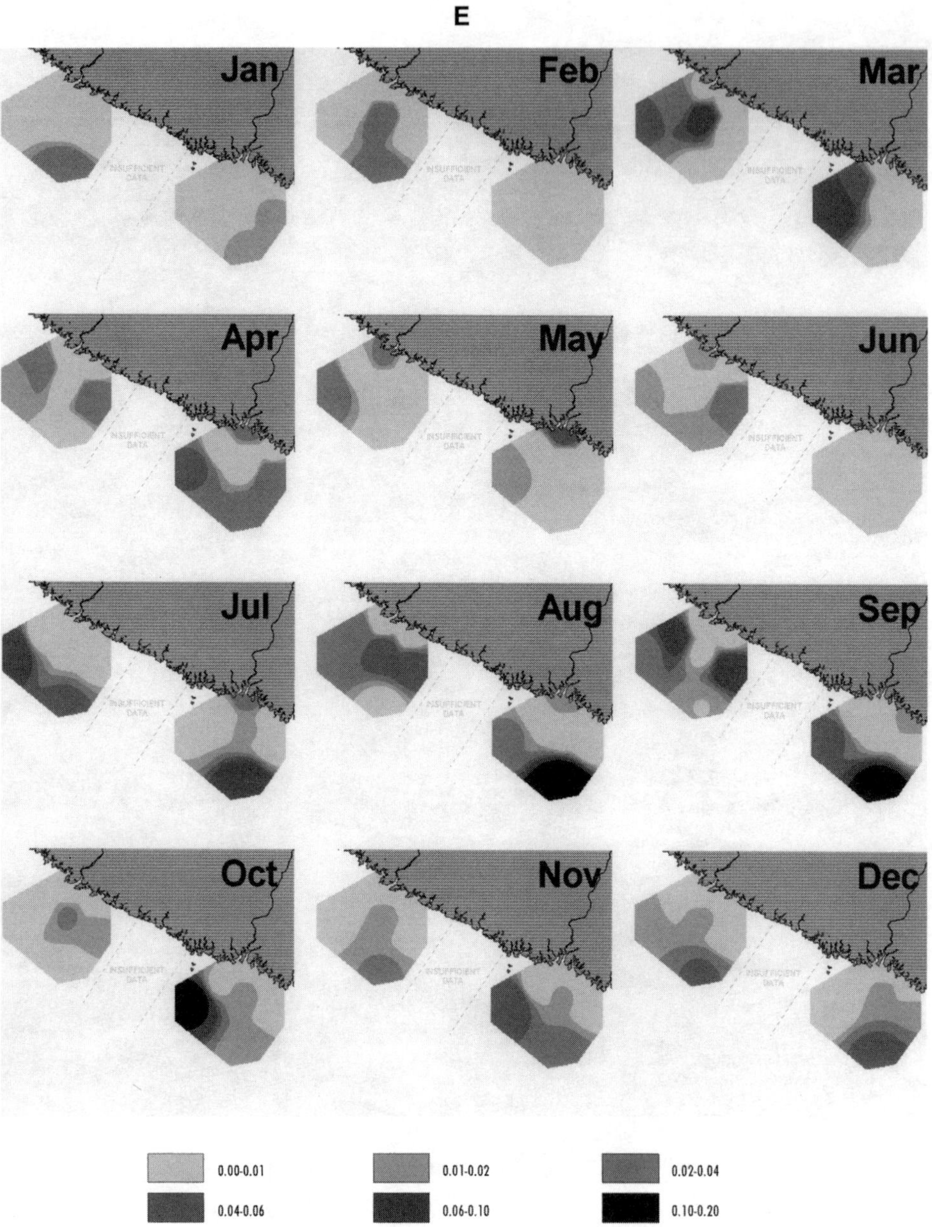

Figure 8.4 (continued) (E) Spatial patterns of primary carnivores in the Econfina and Fenholloway systems, 1992–1993. Data were given by month from April 1992 through March 1993.

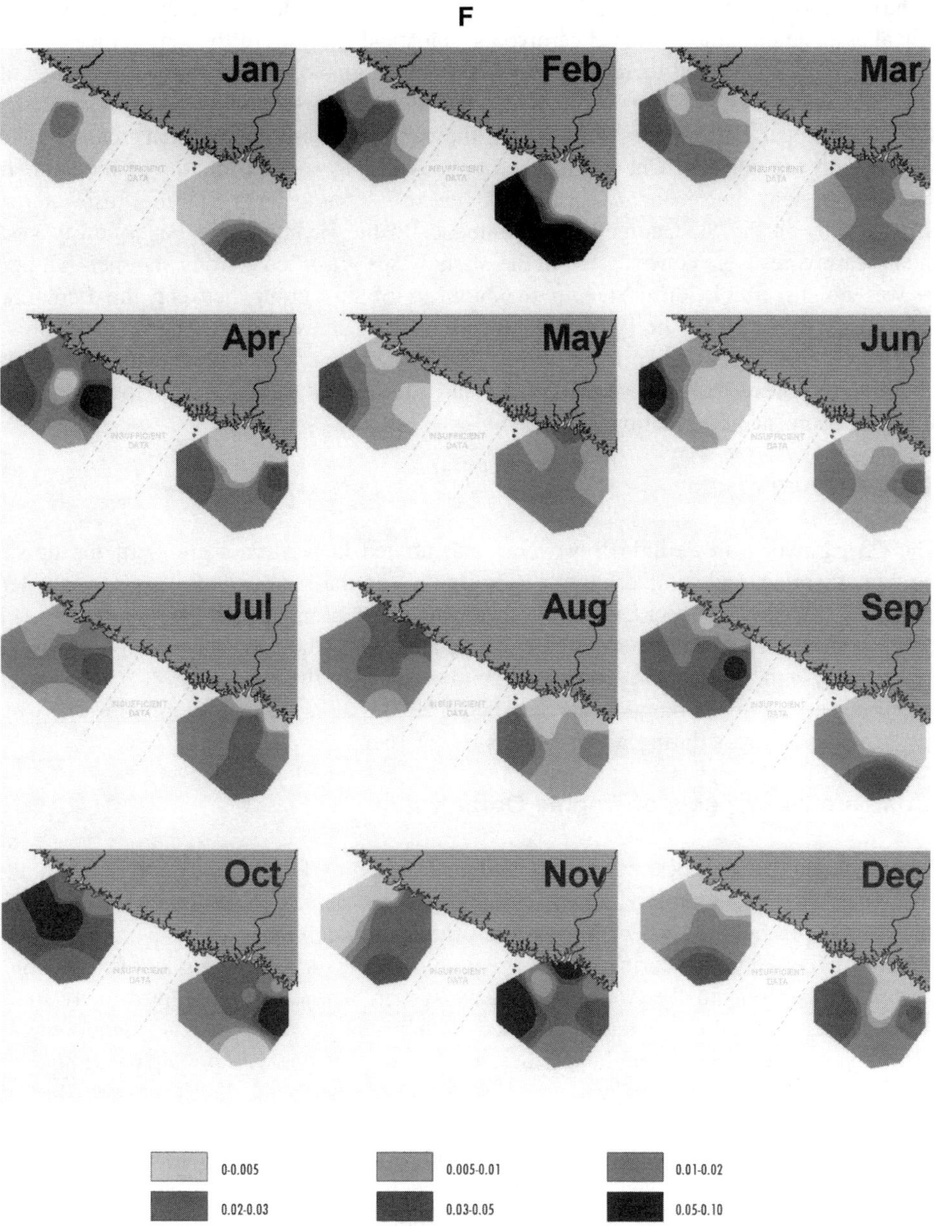

Figure 8.4 (continued) (F) Spatial patterns of primary carnivores in the Econfina and Fenholloway systems, 1999–2001. Data were given as monthly averages from April 1999 through March 2001.

E. C2 Carnivores

Secondary carnivores in the Perdido system during were most prevalent in upper and midbay areas during most months (Figure 8.5). Whereas the Apalachicola data analyses showed mainly biological associations among the C2 carnivores, statistical analyses of the long-term Perdido data (see Table 7.7) indicated that there were negative associations of the C2 carnivores with nutrient loading and phytoplankton blooms. In the Pensacola Bay system, secondary carnivores were mainly located in the upper Escambia system during the different seasons. Secondary carnivores were associated with the northern Choctawhatchee bayous during various months of the year. In the Apalachicola system, the secondary carnivores were usually associated with the freshwater runoff areas, especially in the St. George Island drainage. In the Econfina and Fenholloway systems, secondary carnivores were concentrated in the offshore sea grass beds during summer–fall periods. There was no discernible difference in the biomass of this group between the Econfina and Fenholloway systems during the 1992–1993 and 1999–2001 study periods. The C2 carnivores were thus located in areas of high primary productivity in the various bay systems with seasonal distributions that were related to local habitat and life history patterns. Plankton blooms had a negative effect on such distribution.

F. C3 Carnivores

The C3 carnivores in Perdido Bay were concentrated in western sections of the upper bay during winter, spring, and summer months (Figure 8.6). In the Pensacola Bay system, the tertiary carnivores were located in upper Escambia Bay during different months. Tertiary carnivores (Figure 8.6) were periodically concentrated in areas associated with Choctawhatchee River flows. In the Apalachicola system, this trophic group was mainly located in the lower bay off St. George Island during spring and summer months. In the Apalachee Bay area, tertiary carnivores were mainly located in sea grass areas during warm months of the year.

G. Comparative Aspects of Trophic Organization

Trophic organization is best developed in areas affected by freshwater runoff areas in alluvial systems where phytoplankton productivity is high. Likewise, in sea grass–dominated areas, trophic organization is concentrated where SAV is the primary source of food. Herbivore and omnivore distribution in Perdido Bay during the early (prebloom) years followed relationships noted by Livingston et al. (1997) in the Apalachicola Bay system. These groups were directly linked to physicochemical controlling factors associated with primary productivity. Dynamic regression models in the Apalachicola system linked these two feeding groups with principal component axes representing specific environmental variables associated with Apalachicola River flows. The Perdido River and Elevenmile Creek, which mediate primary production in the upper bay, were directly linked with these trophic components. The preponderance of positive signs associated with the biological coefficients in the dynamic regression models in the Apalachicola system suggested a combination of top-down and bottom-up control of trophic organization in this system (Livingston et al., 1997). It is likely that the same elements prevailed in the Perdido estuary until the phytoplankton blooms appeared in the system.

The Pensacola and Choctawhatchee Bay systems represent areas stressed by dredging, urban development, and hypereutrophication due to nutrient loading from anthropogenous sources (Livingston, 2000). The lack of wetlands combined with hypoxia in deeper parts of the Choctawhatchee system have led to major losses of habitat (Livingston, 1986a,b). The association of herbivores with nutrient-rich runoff from urban areas in the mid-1980s was later translated into plankton blooms and the loss of food web components during the 1990s (Livingston, 2000). Increased herbivore activity during the prebloom 1985–1986 period could have been an indicator of future

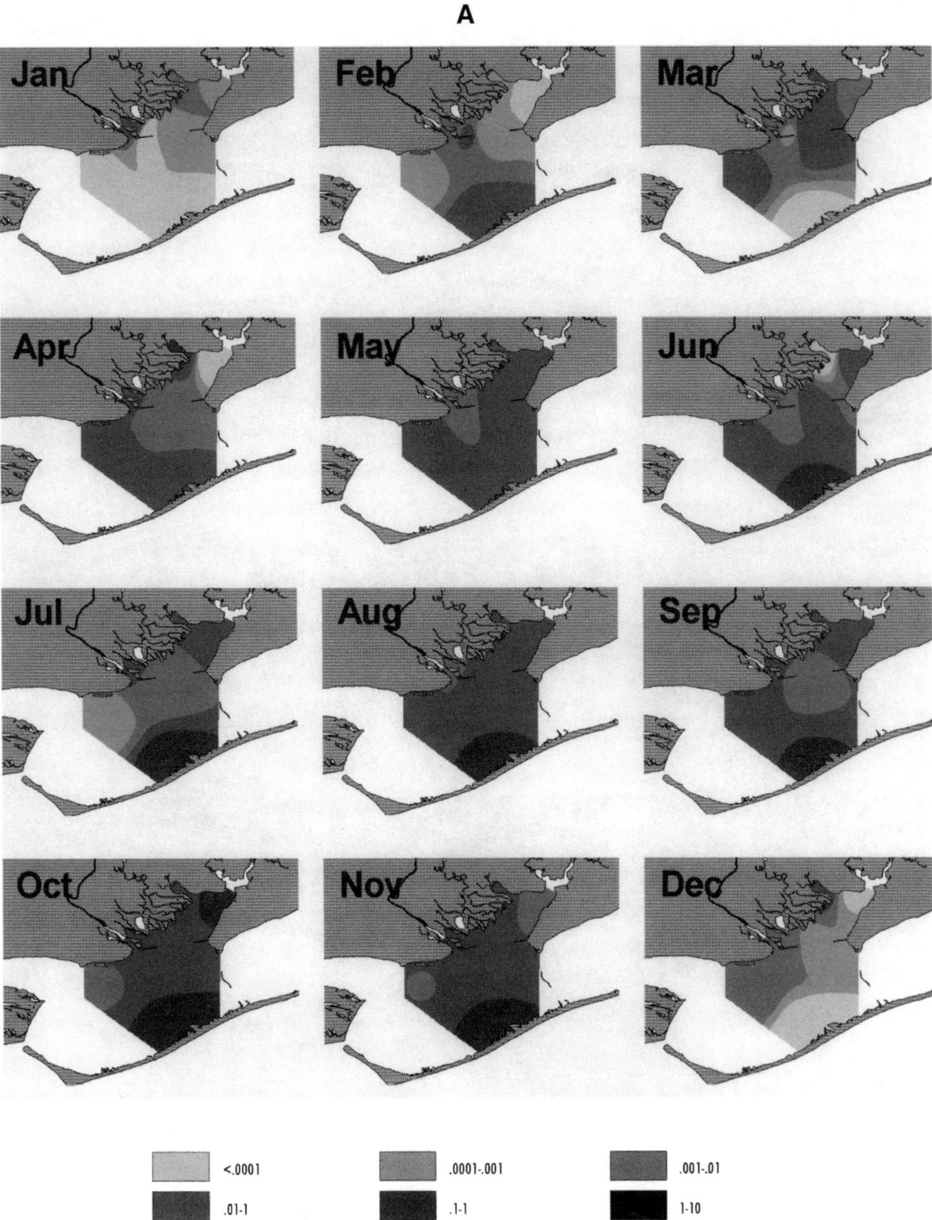

Figure 8.5 (A) Spatial patterns of secondary carnivores in the Apalachicola system. Data were averaged by month from March 1975 through February 1976.

Figure 8.5 (continued) (B) Spatial patterns of secondary carnivores in the Choctawhatchee system. Data were given by month from September 1985 through August 1986.

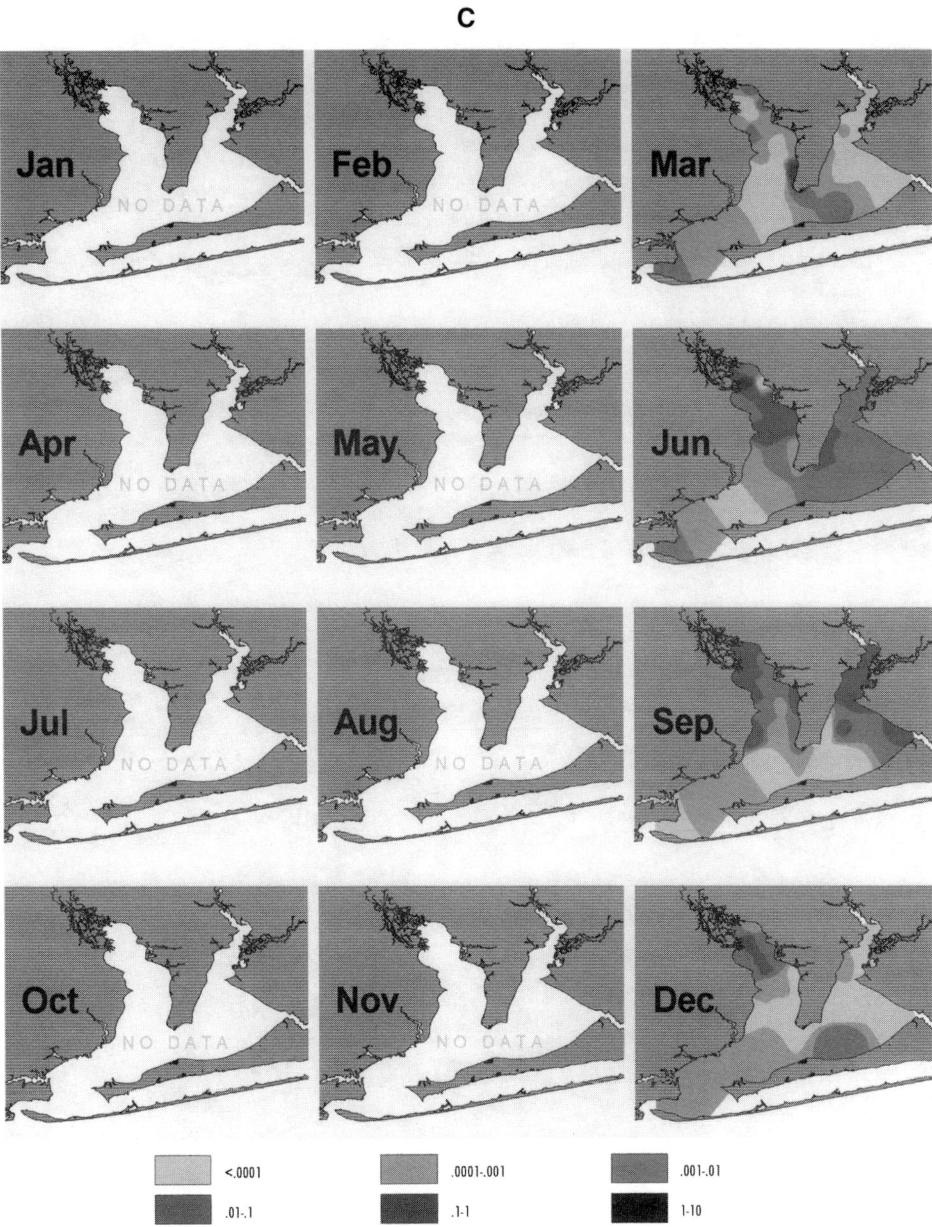

Figure 8.5 (continued) (C) Spatial patterns of secondary carnivores in the Pensacola system. Data were given by quarter from June 1997 through March 1998.

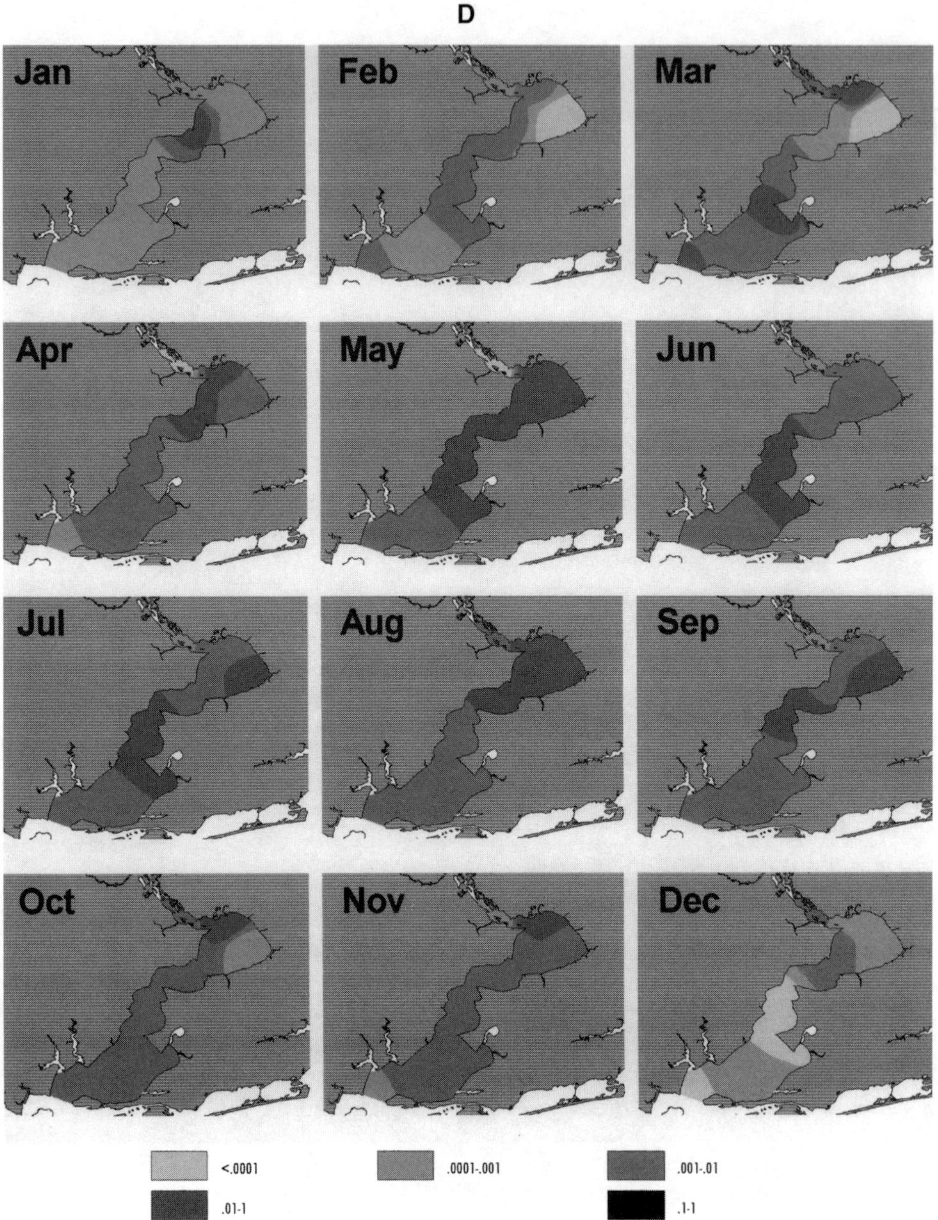

Figure 8.5 (continued) (D) Spatial patterns of secondary carnivores in the Perdido system. Data were averaged by month from October 1988 through September 1991.

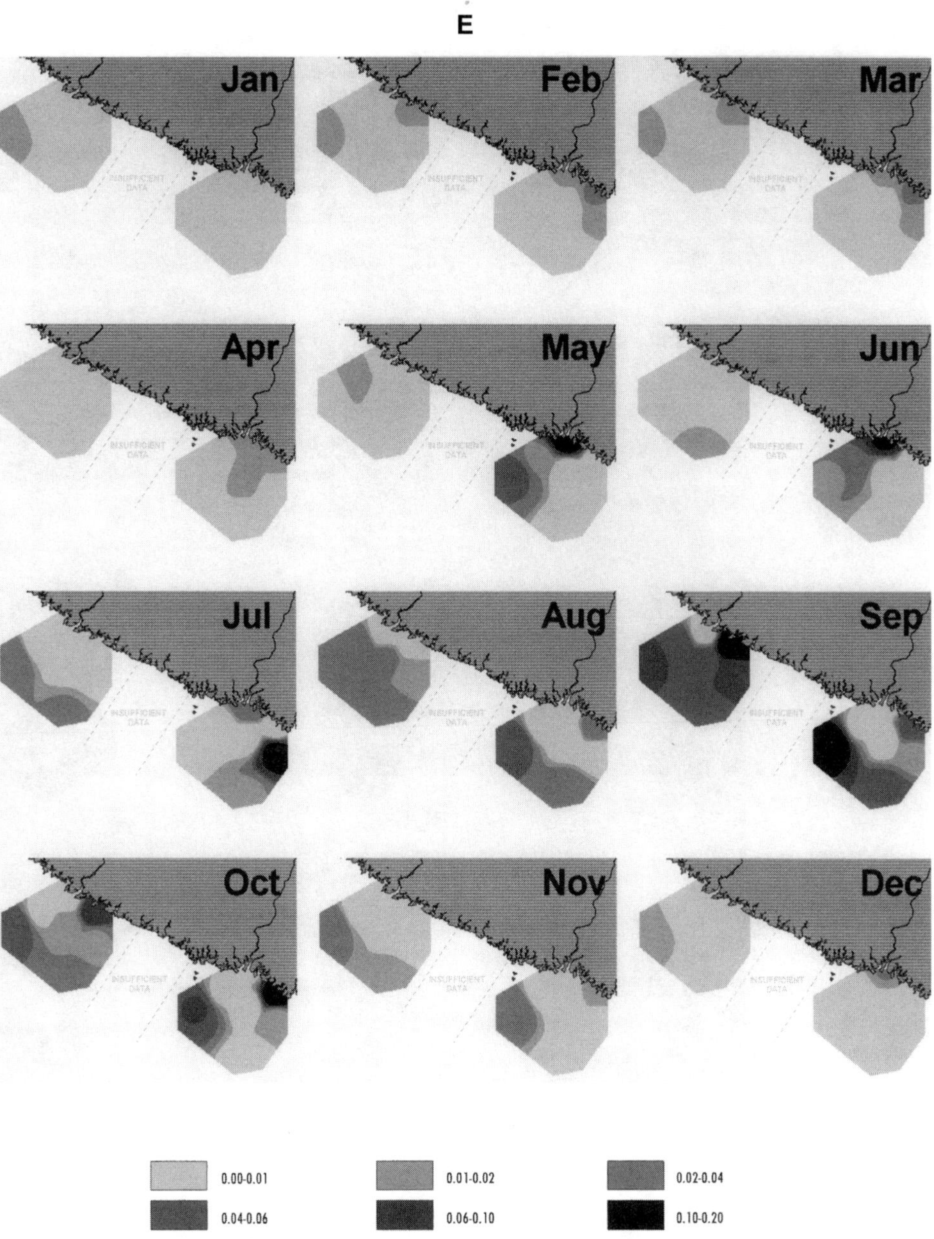

Figure 8.5 (continued) (E) Spatial patterns of secondary carnivores in the Econfina and Fenholloway systems, 1992–1993. Data were given by month from April 1992 through March 1993.

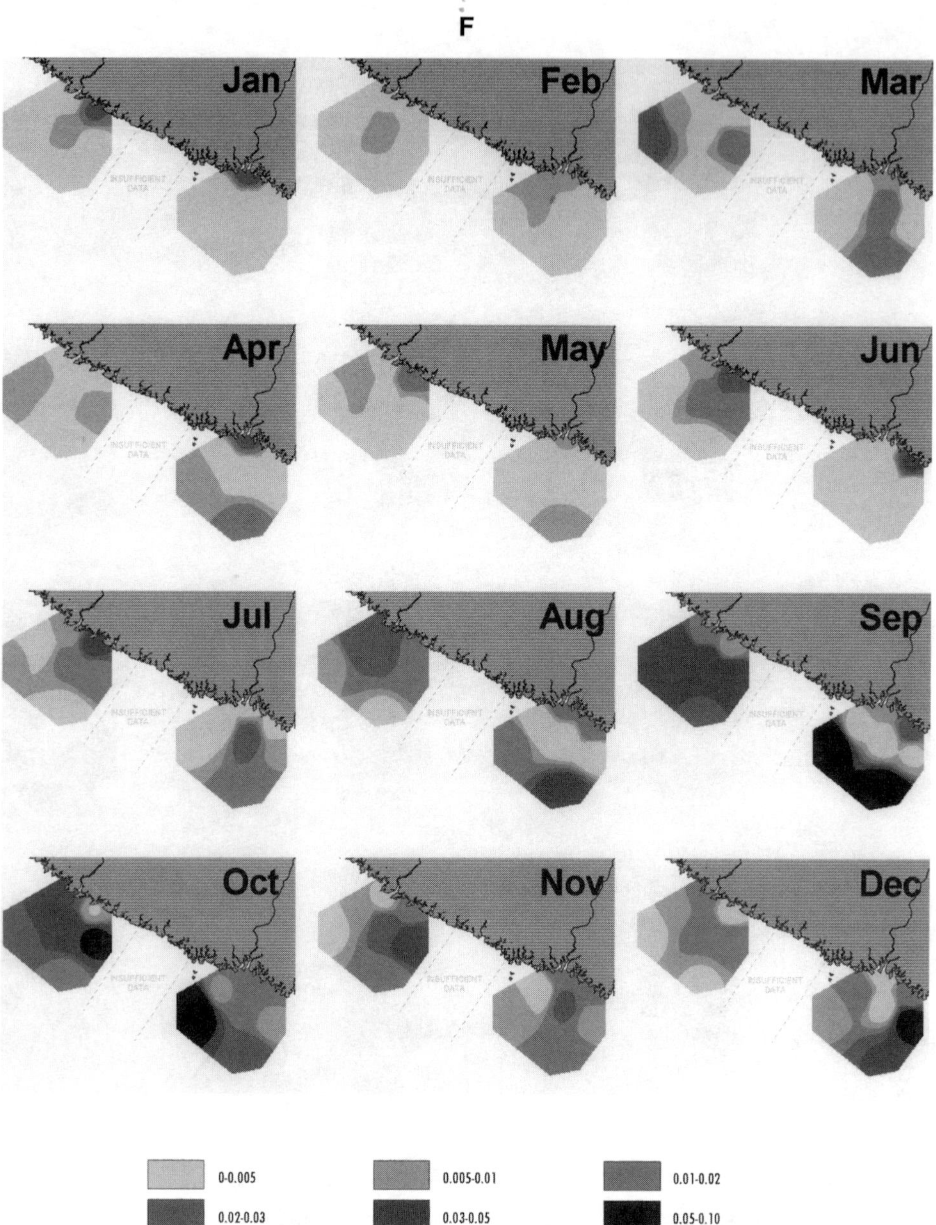

Figure 8.5 (continued) (F) Spatial patterns of secondary carnivores in the Econfina and Fenholloway systems, 1999–2001. Data were given as monthly averages from April 1999 through March 2001.

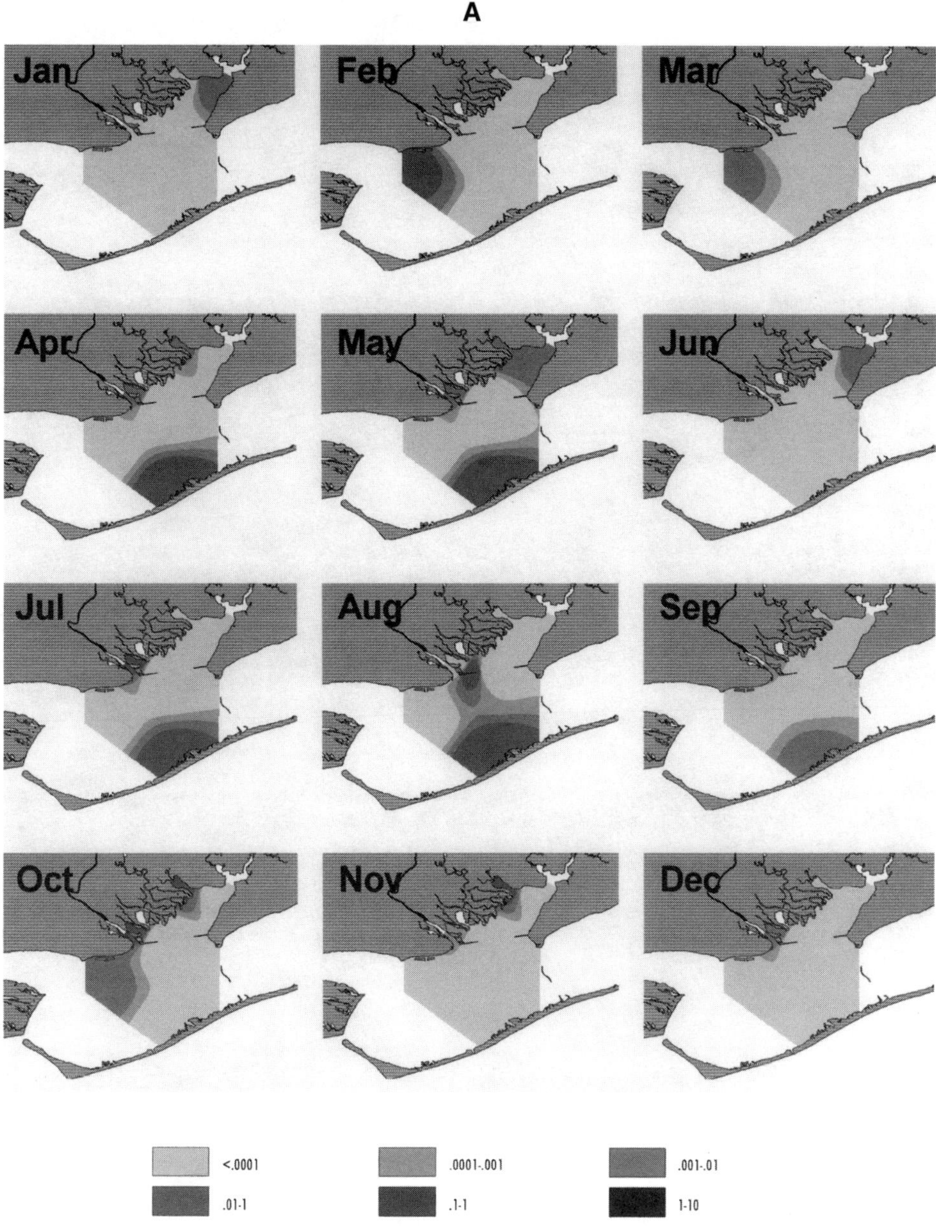

Figure 8.6 (A) Spatial patterns of tertiary carnivores in the Apalachicola system. Data were averaged by month from March 1975 through February 1976.

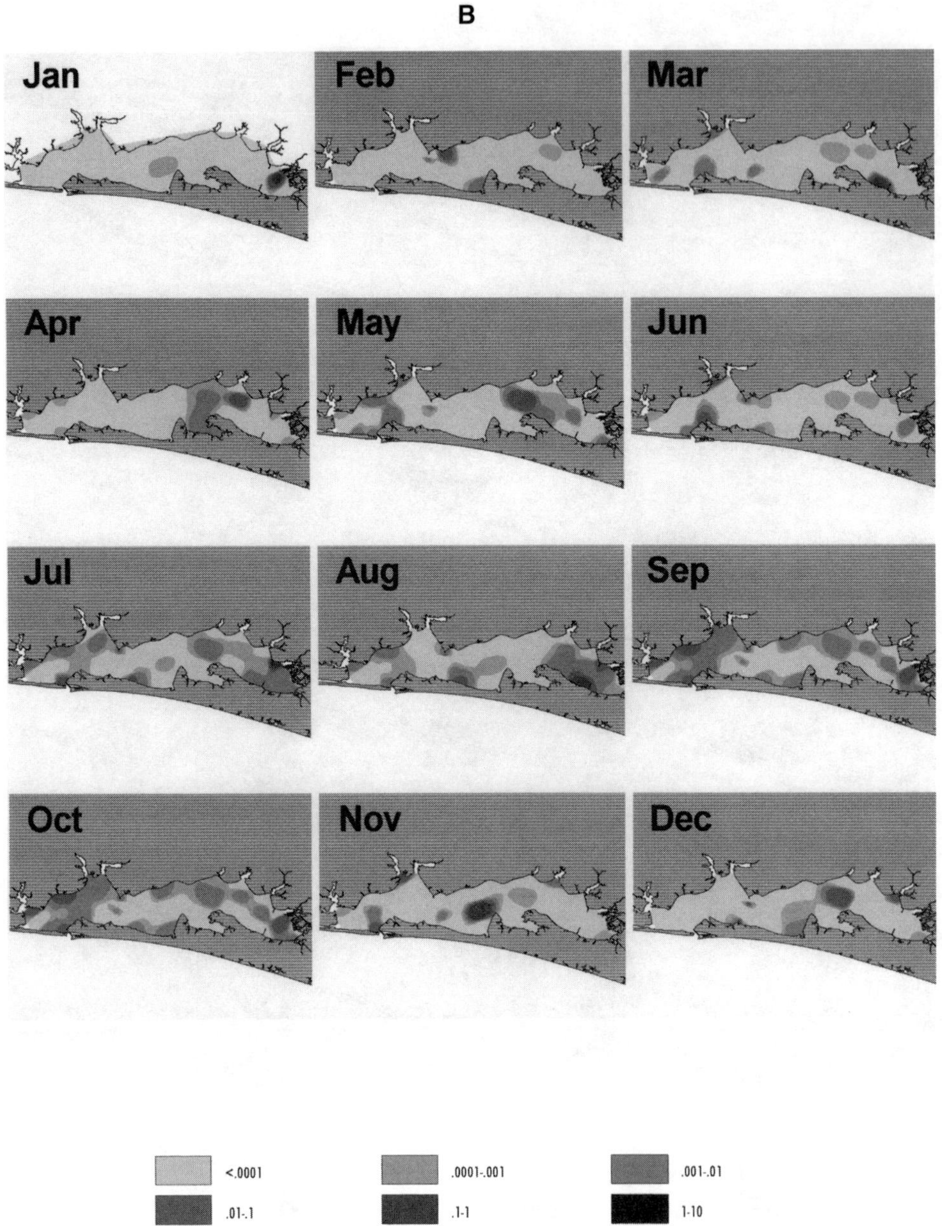

Figure 8.6 (continued) (B) Spatial patterns of tertiary carnivores in the Choctawhatchee system. Data were given by month from September 1985 through August 1986.

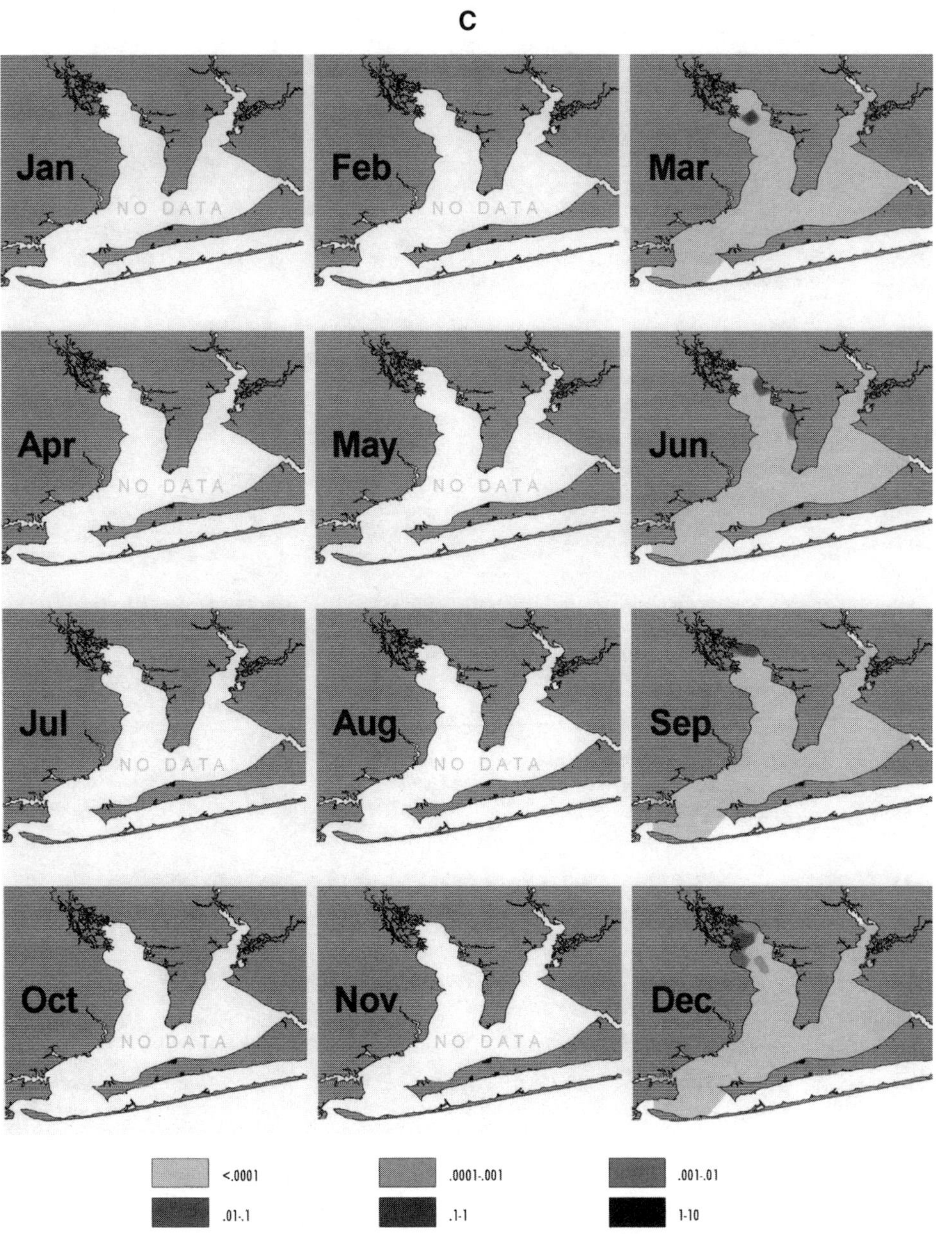

Figure 8.6 (continued) (C) Spatial patterns of tertiary carnivores in the Pensacola system. Data were given by quarter from June 1997 through March 1998.

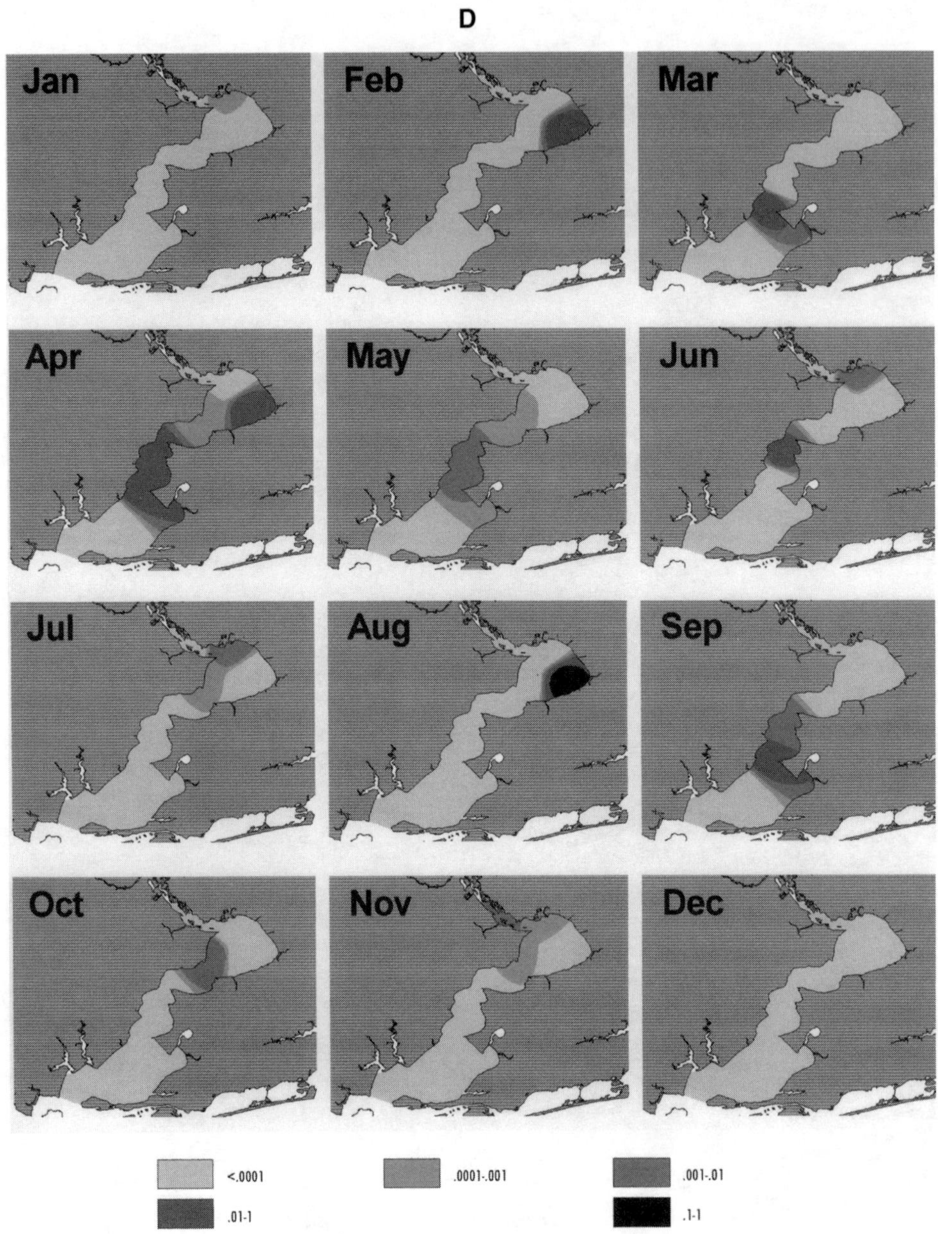

Figure 8.6 (continued) (D) Spatial patterns of tertiary carnivores in the Perdido system. Data were averaged by month from October 1988 through September 1991.

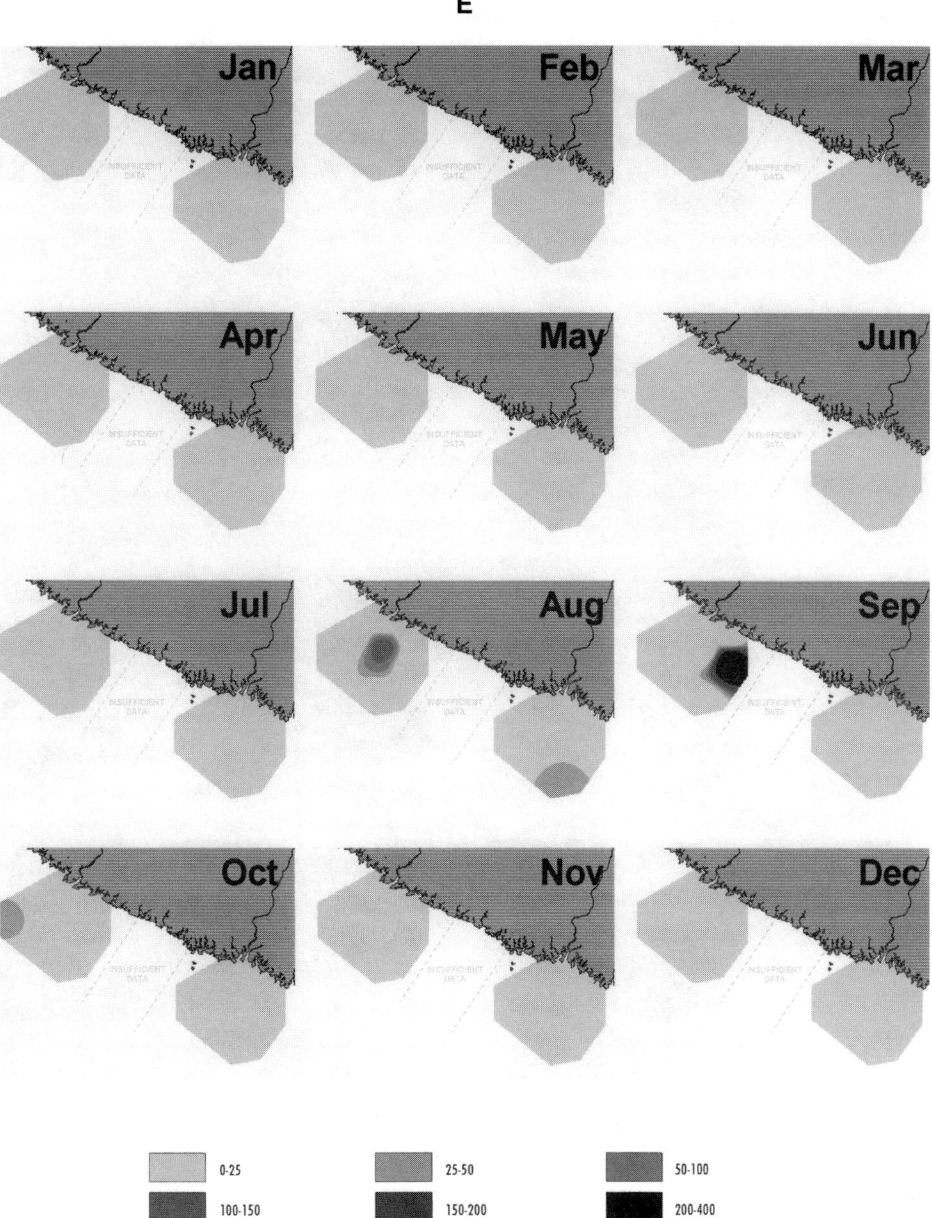

Figure 8.6 (continued) (E) Spatial patterns of tertiary carnivores in the Econfina and Fenholloway systems, 1992–1993. Data were given by month from April 1992 through March 1993.

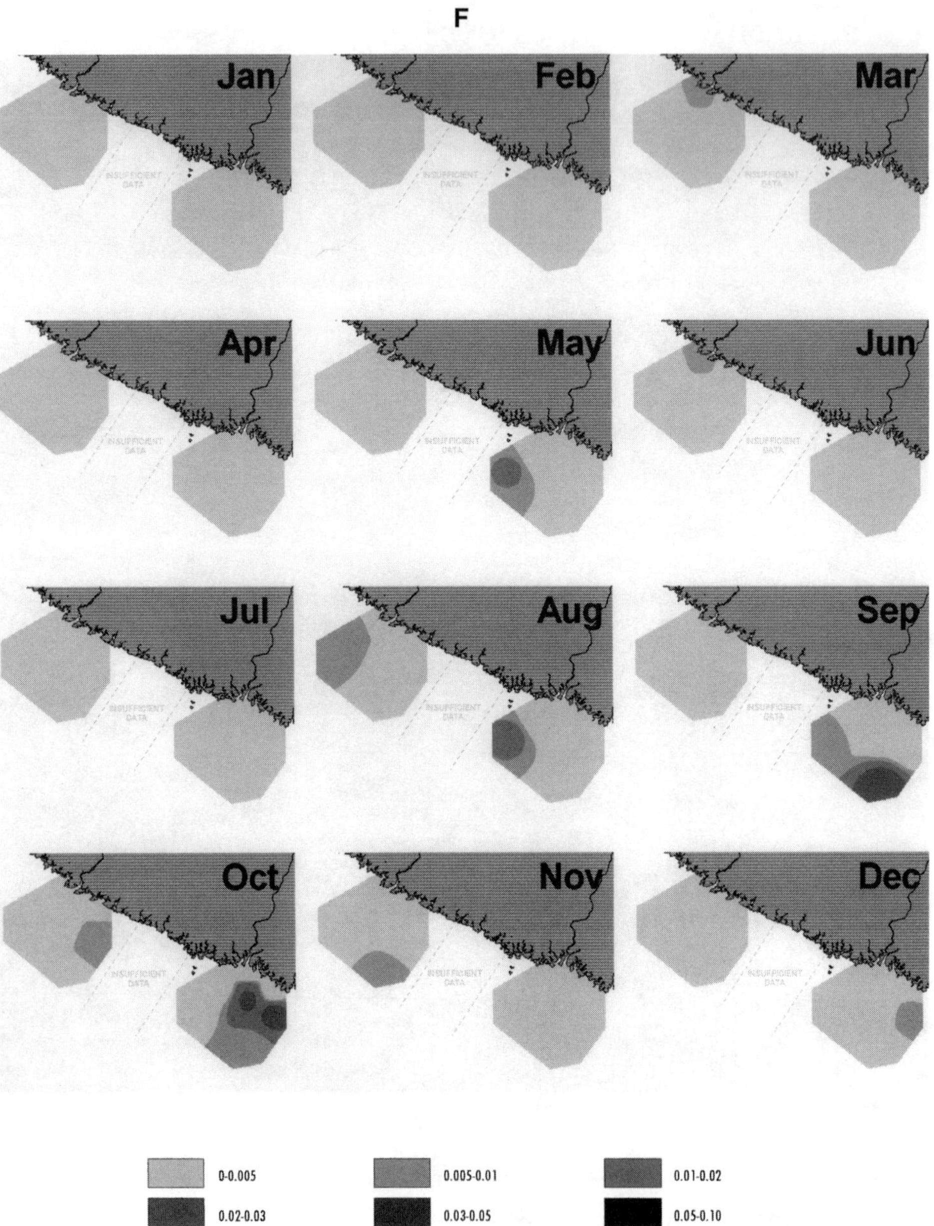

Figure 8.6 (continued) (F) Spatial patterns of tertiary carnivores in the Econfina and Fenholloway systems, 1999–2001. Data were given as monthly averages from April 1999 through March 2001.

hypereutrophication conditions. In any case, both of these systems are characterized by disjunct food webs with losses of secondary production due to a variety of human activities.

In an analysis of data taken during a 7-year period of monthly sampling of fishes and invertebrates of the pristine Apalachicola Bay system (1972 through 1979), Livingston (1984a) noted the spatiotemporal distribution of the dominant organisms in the bay. There was a pattern of association with freshwater runoff areas that included the Apalachicola River (main channel and East Bay) and the Nick's Hole drainage of St. George Island (see Figure 2.2D). Dominance of each species had a specific season of the year that was associated with individual life histories. Winter dominants included anchovies (*Anchoa mitchilli*) and Atlantic croaker (*Micropogonias undulatus*). Spring dominants were spot (*Leiostomus xanthurus*). Spring–early summer dominants included blue crabs (*Callinectes sapidus*) and sand sea trout (*Cynoscion arenarius*). Summer dominants were white shrimp (*Penaeus setiferus*). However, there was continuous interannual variation in the order of the annual dominance hierarchies, with each species experiencing definitive changes in abundance over time.

Trophic organization is best developed in areas affected by freshwater runoff areas in alluvial systems where phytoplankton productivity is dominant, and in sea grass areas where SAV is the primary source of food. Freshwater entry points to alluvial bays in the NE Gulf represent areas of primary productivity and rich development of food web components that follow seasonal life history patterns of component trophic units. This means that freshwater runoff and the delivery of nutrients to receiving bays is crucial for maintenance of the estuarine food webs and the high levels of secondary productivity. Apalachicola Bay represents one of the last alluvial systems along the north Gulf coast that maintains a natural food web with uniformly high secondary production (Livingston, 2000). Trophic organization in the Apalachicola system is concentrated around freshwater runoff areas of the bay. Various trophic units of the different species follow different spatiotemporal patterns, but all are connected by distinctive trophodynamic processes. The history of the Perdido system with the systematic collapse of food web structure after the initiation of phytoplankton blooms is probably the model for the Pensacola and Choctawhatchee systems. The sea grass–based systems in Apalachee Bay follow an entirely different course, with food web structure closely tied to SAV development. In these areas, food web deterioration followed reduced light transmission due to pulp mill effluents with associated changes in food web structure and reduced secondary productivity. Thus, food web patterns were indicative of the importance of natural cycles of unpolluted freshwater runoff to the productivity of coastal systems in the NE Gulf of Mexico, from the sea grass systems of Apalachee Bay to the river-dominated bay areas to the west.

III. TEMPORAL DISTRIBUTION OF TROPHIC UNITS

A. Baywide Trends of Invertebrates and Fishes

The species is an important unit of measure in ecological studies. However, determination of the trophic stages of a given species allows a more precise analysis of the ecological processes associated with the spatial and temporal distribution of organisms in coastal areas. Reorganization of the trophic units into various levels of food web organization is far more realistic than simplistic food web models based on whole-species trophic designations. Long-term distributions of the C1 and C2 fish and invertebrate trophic units are shown in Figure 8.7. There is a continuous shifting of trophic units with time, with community development reflecting continuous changes of the species composition. This reflects natural changes in species populations in the Perdido system over the period of observation. Summation by trophic level avoids such variation, and simplifies interpretations of natural changes vs. anthropogenous impacts. The continuous shifts of individual populations within any given level of food web development thus are not reflected in the more stable food web structure. Trophic organization represents a more stable unit of measure that remains

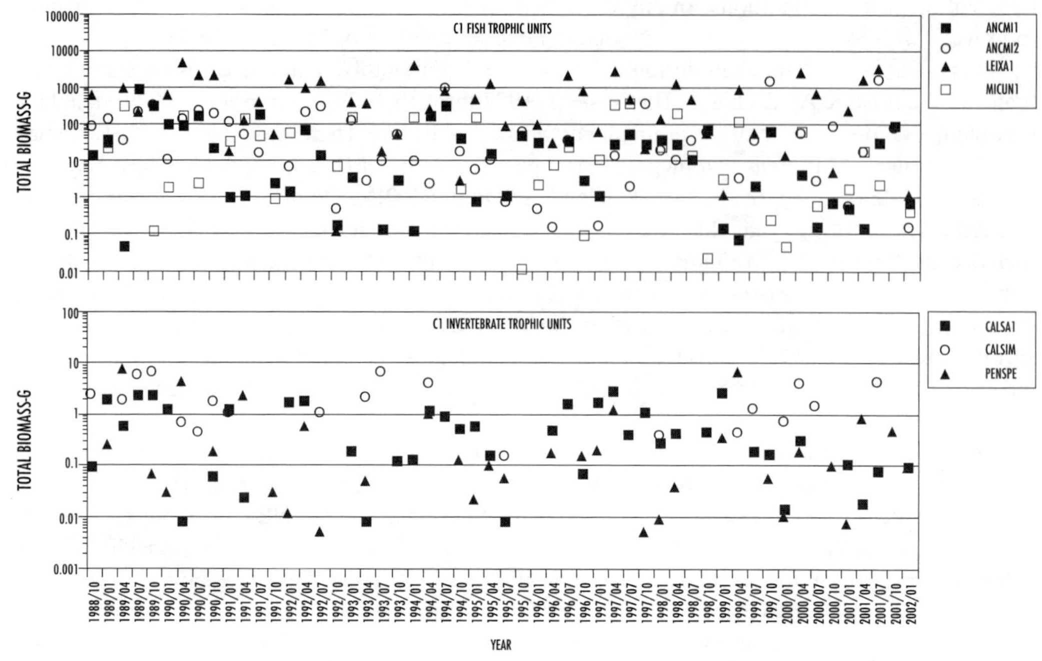

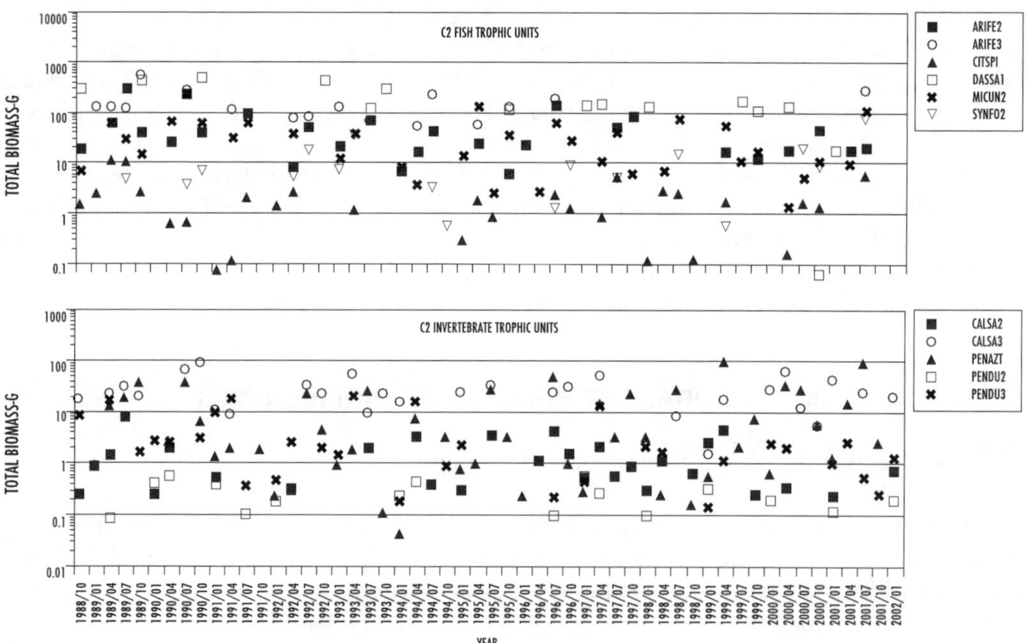

Figure 8.7 Distribution of fish and invertebrate trophic units in the Perdido system (all stations combined). Data were calculated from otter trawl collections taken quarterly from October 1988 through January 2002. Data were organized by C1 and C2 carnivores. Dominant species-specific trophic units include *Anchoa mitchilli* (ANCMI1, ANCMI2), *Leiostomus xanthurus* (LEIXA1), *Micropogonias undulatus* (MICUN1), *Callinectes sapidus* (CALSA1, CALSA2, CALSA3), *C. similis* (CALSIM), *Penaeus* spp. (PENSPE), *P. azatecus* (PENAZT), *P. duorarum* (PENDU2, PENDU3), *Arius felis* (ARIFE2, ARIFE3), *Citharichthys spilopterus* (CITSPI), *Dasyatis sabina* (DASSA1), and *Synodus foetens* (SYNFO2) (see Appendix II).

free of the constant shifts of species distributions in space and time. In a way, species populations represent the hands of the clock, whereas the trophic organization represents the internal clock mechanism whereby the ongoing processes that determine biological response to habitat and productivity changes can be determined. Evaluation of food web changes also facilitates comparisons of biological organization among disparate coastal systems that have differences in species representation. It also permits regional comparisons of the effects of hypereutrophication and toxic substances where, again, the differences in species populations can complicate the interpretation of field data.

B. Trophic Indices

There are various relationships among the different trophic levels that can be quantified by trophic indices. As shown in Figure 8.8, the herbivore/omnivore ratios were generally higher in the upper bay (where there were more herbivores due to spatial patterns of primary productivity) during early (bloom-free) years of sampling. There was a general downward trend of such indices as the

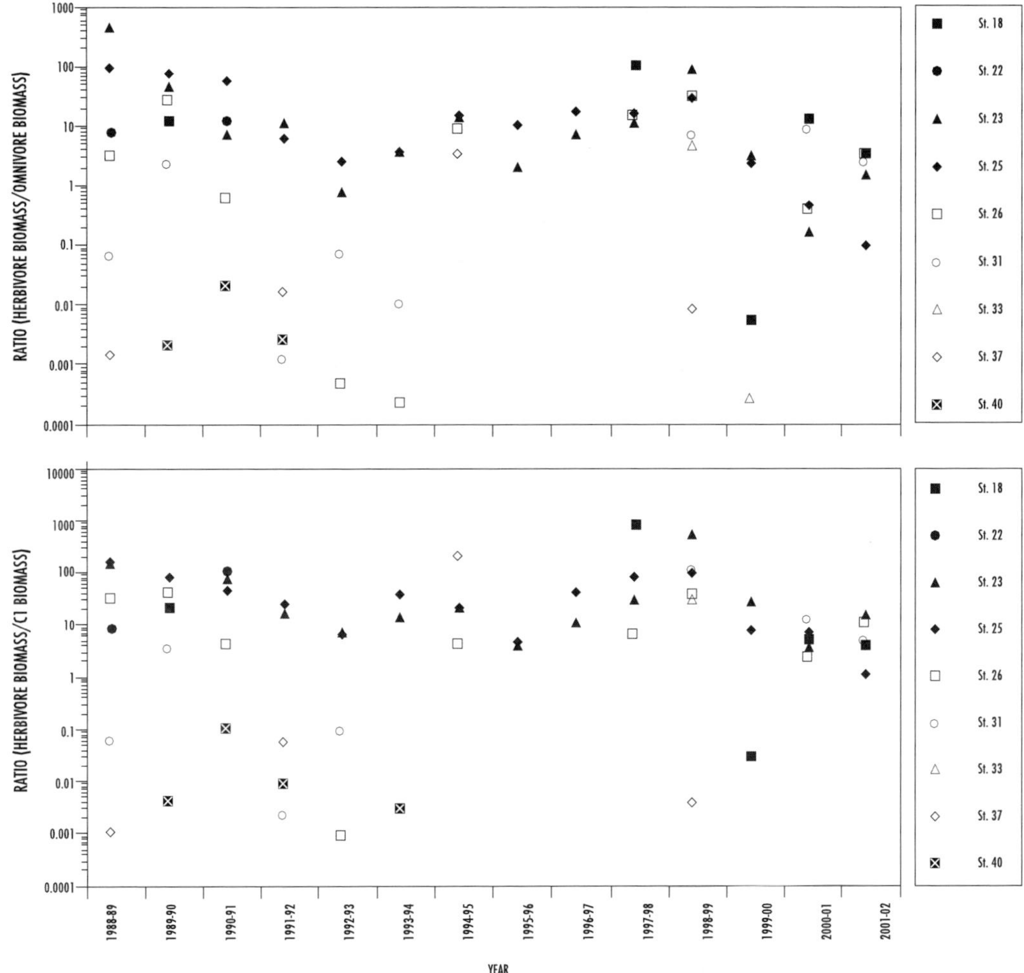

Figure 8.8 Distribution of ratios of FII herbivore biomass: FII omnivore biomass, and FII herbivore biomass: FII C1 biomass in the Perdido system. Data were calculated from collections taken with otter trawls quarterly (October 1988 through January 2002).

omnivores replaced the herbivores with the appearance of plankton blooms. The disappearance of the lower bay indices during the bloom period was another indication of bloom effects on the trophic organization of Perdido Bay, which was consistent with data showing the influence of bloom successions in the upper bay on lower bay areas. With the brief cessation of orthophosphate loadings in late 1997–1998, the reduced blooms were associated with increased herbivore/omnivore ratios. The extended drought and periodically high ammonia and orthophosphate loadings from the mill from 1999 to 2001 were accompanied by reduced herbivore/omnivore ratios. The herbivore/primary carnivore ratios (Figure 8.8) followed this pattern. In both cases, these ratios gave a simple, yet accurate account of the long-term changes of the Perdido Bay trophic organization and the response of disparate biological associations in different habitats to long-term nutrient loading patterns and associated changes in the phytoplankton communities.

CHAPTER 9

Regulation and Food Web Factors

I. INTRODUCTION

Human population growth in coastal areas of the conterminous United States will continue to increase in the future, placing more and more stress on estuarine and coastal habitats. Baird (1996) reviewed the problems with the current state and federal regulatory infrastructure, and he outlined responses appropriate to the scale, urgency, and complexity of the problems. Baird (1996) viewed resource management and regulation in coastal areas as an activity that was essentially mediated by politics. Discharges of nutrients and toxic substances have increased in recent decades. Unfortunately, scientific inquiry has not kept pace with the effects of various human activities on coastal systems. The lack of the institutional infrastructure for research and planning/management activities continues to hinder progressive management of coastal resources.

Knowledge of coastal processes such as trophic organization is an important component of successful management efforts. However, such knowledge will be useless without the political will to enforce regulations associated with adverse impacts on water and sediment quality. Baird (1996) suggested a holistic approach to regulation and management that is not currently popular in terms of funding opportunities or scientific career paths. A relatively high level of human tolerance and political acceptance of environmental degradation has not helped in what is currently a downward spiral of ecosystem health, particularly in coastal areas where population pressure is so great. The regulatory role in all this is complicated, and is prone to forces, economic and political, that remain outside the usual scope of research and management initiatives. Nevertheless, appropriate enforcement policies should be based on objective scientific information, and, unless such information is generated, there is virtually no chance of maintaining what remains of our coastal resources.

II. PULP MILL EFFLUENTS AND FOOD WEB DYNAMICS (APALACHEE BAY)

Maestrini et al. (1984a,b) outlined various problems involved in attempting to make associations of water quality and nutrient factors with descriptive microalgal data. Differences in essential nutrients, the physiological state of the indigenous algal populations, and various limiting factors all contribute to the complexity of causal relationships between chemical factors and phytoplankton associations in estuarine and coastal systems. Natural interactions of physical and biological factors in the determination of phytoplankton population distributions in estuarine and marine systems along the northern Gulf coast of Florida remain relatively poorly documented.

The discharge of industrial wastes may lead to the replacement of indigenous phytoplankton with populations better suited to cope with the stress imposed by pollution. Through such selection, the biotic interrelationships within the community can be modified, and a substantial alteration of the biotic assemblages often results (Rainville et al., 1975). Stockner and Costella (1976) showed that algae can adapt to relatively high concentrations of pulp mill effluent and may be the trophic level least affected by effluent discharge in coastal marine waters. Population genetics change in *Skeletonema costatum* (Gallagher, 1980) indicated that changing environmental conditions are correlated with change in the genetic composition of phytoplankton populations. Axenic culture studies with *S. costata* demonstrated the ability of this species to adapt and exhibit normal growth in relatively high concentrations of pulp mill effluents. Phytoplankton are also affected both by quantity and quality of light. Stockner and Cliff (1976) compared the absorption spectra for phytoplankton with absorbance curves for various concentrations of kraft mill effluent in British Columbia. They found that the absorbance of blue light (400 to 500 nm) coincides with one of the two absorbance peaks for chlorophyll *a* and most peaks of accessory pigments. Thus, a significant positive relationship existed between subsurface light and phytoplankton production across an effluent gradient. Stockner and Cliff (1976) also found strong attenuating properties of concentrated effluent notably in the 400 to 500 nm wavelength. They found *S. costatum* and *Thalassiosira* spp. as dominant phytoplankton species in the effluent system. Their laboratory experiments suggested that, given sufficient time, it is possible for phytoplankton to adapt to relatively high effluent concentration, if the pH remains normal.

A. Phytoplankton Organization in Apalachee Bay, Florida

1. Introduction

Apalachee Bay (see Figure 2.2) is one of the least polluted coastal systems in the conterminous United States (Livingston, 1975, 1982a, 1984b, 1997a, 2000). This shallow bay is dominated by sea grass beds that are part of a massive series of underwater vegetation that ranges from north Florida to Florida Bay. Apalachee Bay receives flows from a series of relatively unpolluted streams, one of which, the Econfina River–estuary, has served as a reference area for studies of the impact of a pulp mill on the Fenholloway River–estuarine system since 1971 (Figure 9.1).

Bitaker (1975) found that the Fenholloway offshore system was characterized by higher nutrients (phosphate, nitrite, nitrate), turbidity, and color than the corresponding Econfina system. There were also shallower Secchi disk readings and higher vertical extinction coefficients. Increased nutrient loading had little effect on phytoplankton production in inshore waters. The effects of mill effluent on sea grass (submerged aquatic vegetation, SAV) beds have been determined (Zimmerman and Livingston, 1976a,b, 1979; Livingston et al., 1998b); these effects are associated with increased loading of dissolved organic carbon (DOC), water color, and nutrients (ammonia and orthophosphate). Combined field and laboratory experiments and descriptive field data indicate that reduced photic depths, reduction of specific wavelength distributions, and altered water/sediment quality are associated with the loss or debilitation of over 28 km^2 of sea grass beds (Livingston et al., 1998b). Qualitative and quantitative light transmission characteristics represent the primary determinants of SAV distribution in offshore areas of the NE Gulf coast of Florida (Livingston et al., 1998b). Livingston (1985a) showed that the offshore Fenholloway system was characterized by altered food web components that in some ways resembled alluvial river estuaries (enhanced phytoplankton production, reduced SAV, increased plankton feeding, domination by soft sediment organisms and food webs, reduced trophic diversity).

Basic questions to be answered concerning recent changes in the Fenholloway system include the following:

- What is the ultimate fate and effects of color and nutrient loading in offshore Fenholloway areas relative to the reference Econfina site?
- How is phytoplankton productivity related to light transmission conditions and nutrient loading in the Econfina and Fenholloway systems?
- How has the Fenholloway offshore system responded to recent reductions of water color and associated increases of light penetration?

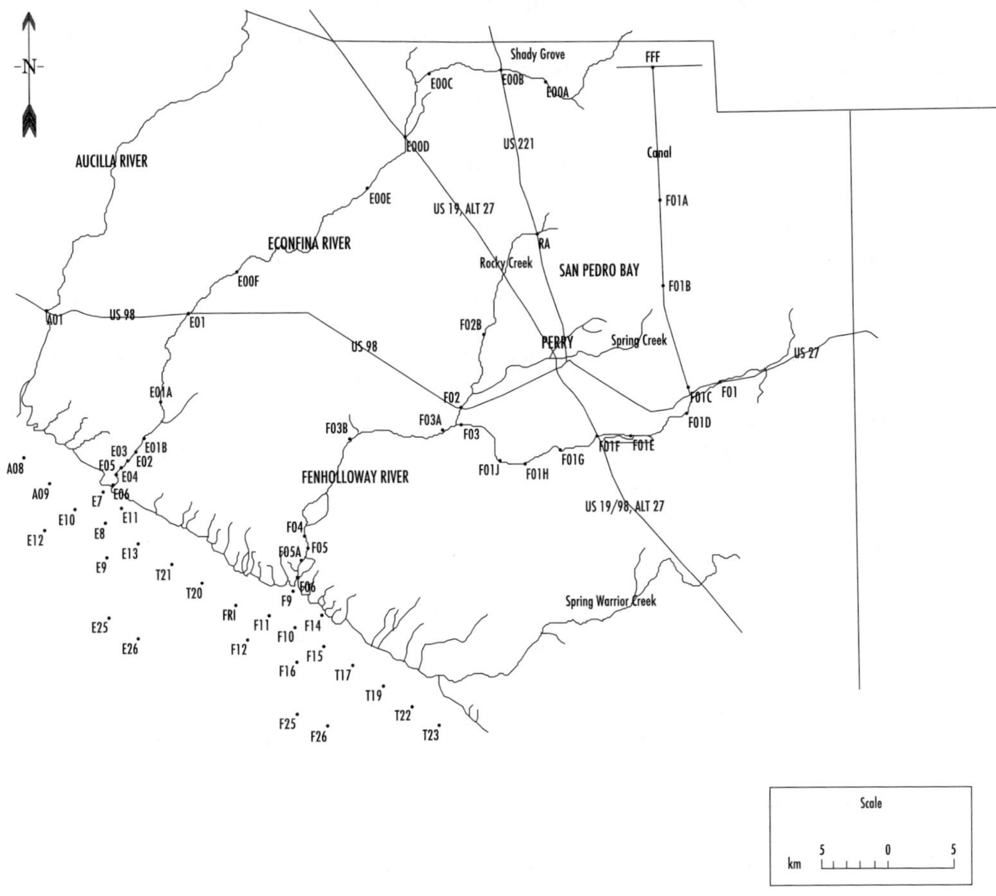

Figure 9.1 Apalachee Bay showing permanent, long-term sampling stations (1971 through 2002) in the Econfina and Fenholloway offshore systems.

A series of phytoplankton studies was carried out in the Econfina and Fenholloway systems in 1992–1993, 1999, and 2000 to answer the above questions.

2. *Physicochemical Factors and Light Distribution*

Previous studies (Livingston et al., 1998b) indicated that water color and DOC were high in the Fenholloway estuary and near-shore area (stations F06, F09, F10, F11, and F14) relative to the Econfina system (Figure 9.1). Both factors tended to become equivalent between stations F10/E08 and F16/E09. Comparative color levels were found somewhere around 3 km offshore. Light transmission trends tended to follow these distributions. Extinction coefficients were higher in the Fenholloway estuary and inshore portions of the Fenholloway system than the Econfina cognates. There was recovery in the Fenholloway system (in terms of light extinction coefficients) at the offshore sites (stations F16 and F25). Compared to the Econfina cognate stations, estuarine and inshore areas of the Fenholloway system had significantly shallower euphotic depths with recovery at stations F16 and F25 along the north/south transect. According to the spectroradiometric data, light penetration was most affected by the mill effluent in the Fenholloway estuary and along inshore portions of the Fenholloway Gulf coast within 3 km of the shore and along shore for distances ±4 to 5 km NW or SE from the mouth of the Fenholloway River.

Livingston et al. (1998b) found that chlorophyll *a* concentrations were relatively low in both systems, but were consistently higher in the Econfina estuary than the Fenholloway estuary. This situation was reversed just offshore where the Fenholloway chlorophyll readings were often higher than those at the cognate Econfina stations. The highest mean concentrations of chlorophyll in both systems were sometimes found displaced to the east in inshore areas. The mean isopleth of 4 μg l^{-1} chlorophyll *a* in the Fenholloway offshore system reached an area 3 to 5 km offshore, whereas the 4 μg l^{-1} isopleth in the Econfina system occurred up to 1 km offshore (Livingston et al., 1998b).

In 1998, the pulp mill installed a color reduction program that, together with a record-breaking 3-year drought from 1998–2002, was associated with a return of SAV in the previously unvegetated Fenholloway system (Livingston, unpublished data). However, relatively steep nutrient gradients (ammonia and orthophosphate) still occurred in inshore areas of the Fenholloway drainage (Figure 9.2), the result of continuously high nutrient loading by the pulp mill. There were also relatively steep water color gradients in the Fenholloway estuary that overlapped the nutrient gradients. Water color in offshore areas was reduced during the 1998 to 2000 period relative to the earlier sampling period (Livingston et al., 1998b). This was attributed to the drought and the mill's color reduction program. A comparison of light transmission changes between the 1992–1993 and 1999–2000 sampling periods (Figure 9.3) indicated that there were increases in PAR photic depths and reductions of light extinction coefficients in both systems during the most recent analysis. Reduced extinction coefficients were pronounced in the Fenholloway estuary, and occurred in areas characterized by the steep nutrient gradients. Secchi depths remained relatively stable in time because many Secchi readings went to the bottom and were therefore relatively useless in this analysis. Chlorophyll *a* concentrations were generally low in both systems (Figure 9.2), although some increases were noted, especially in the Fenholloway system. The highest such concentrations were noted in the Fenholloway estuary and along inshore areas of the Fenholloway coast.

3. Phytoplankton Distribution

During 1992–1993, a total of 264 taxa of phytoplankton were identified representing the following groups: Bacillariophyce (diatoms) (203), Dinophyceae (dinoflagellates) (26), Cyanophyceae (blue-green algae or cyanobacteria) (10), Cryptophyceae (cryptophytes) (2), Chlorophyceae (green algae) (2), Chrysophyceae (golden-brown algae) (2), Prasinophyceae (3), Prymnesiophyceae (haptophytes) (3), and Euglenophyceae (euglenoids) (4). The bulk of the identified nanoflagellates included small cryptophytes with close affinity to Komma and Falcomonad species (Table 9.1). An unidentified pico-nannoplankton community that was difficult to identify was placed under two categories on the basis of presence or absence of flagella, "unidentified nanococcoid" and "unidentified nanoflagellates." The dominant component in the "unidentified nanococcoids" category was a coccoid unicellular organism (without flagella in Lugol's preserved samples) approximately 5 μm in diameter. These ultraplankton, even though not verified with fluorescence microscopy and transmission electron micoscopy, appeared to be related to coccoid blue-green algae described by Waterbury et al. (1979) and Johnson and Sieburth (1979).

During 1992–1993 (before the pulp mill's color reduction program), phytoplankton numerical abundance and species richness were generally higher in the Econfina estuary than cognate areas of the Fenholloway (Table 9.1). The data indicated phytoplankton inhibition in the Fenholloway estuary, probably the result of reduced light availability due to high color (Livingston et al., 1998b). Recovery, in terms of species richness, occurred at station F16 in the Fenholloway; this recovery coincided with spatial patterns of light penetration (Livingston et al., 1998b). Cell numbers peaked in near-shore areas of both systems,

Most of the dominant phytoplankton populations in the subject estuaries and near-shore waters are euryhaline. Salinity was generally not statistically associated with phytoplankton numbers during the 1992–1993 study period. In both systems, numbers of phytoplankton cells were positively associated with temperature (Econfina r^2 = 0.2, p = 0.006: Fenholloway r^2 = 0.23, p = 0.0007).

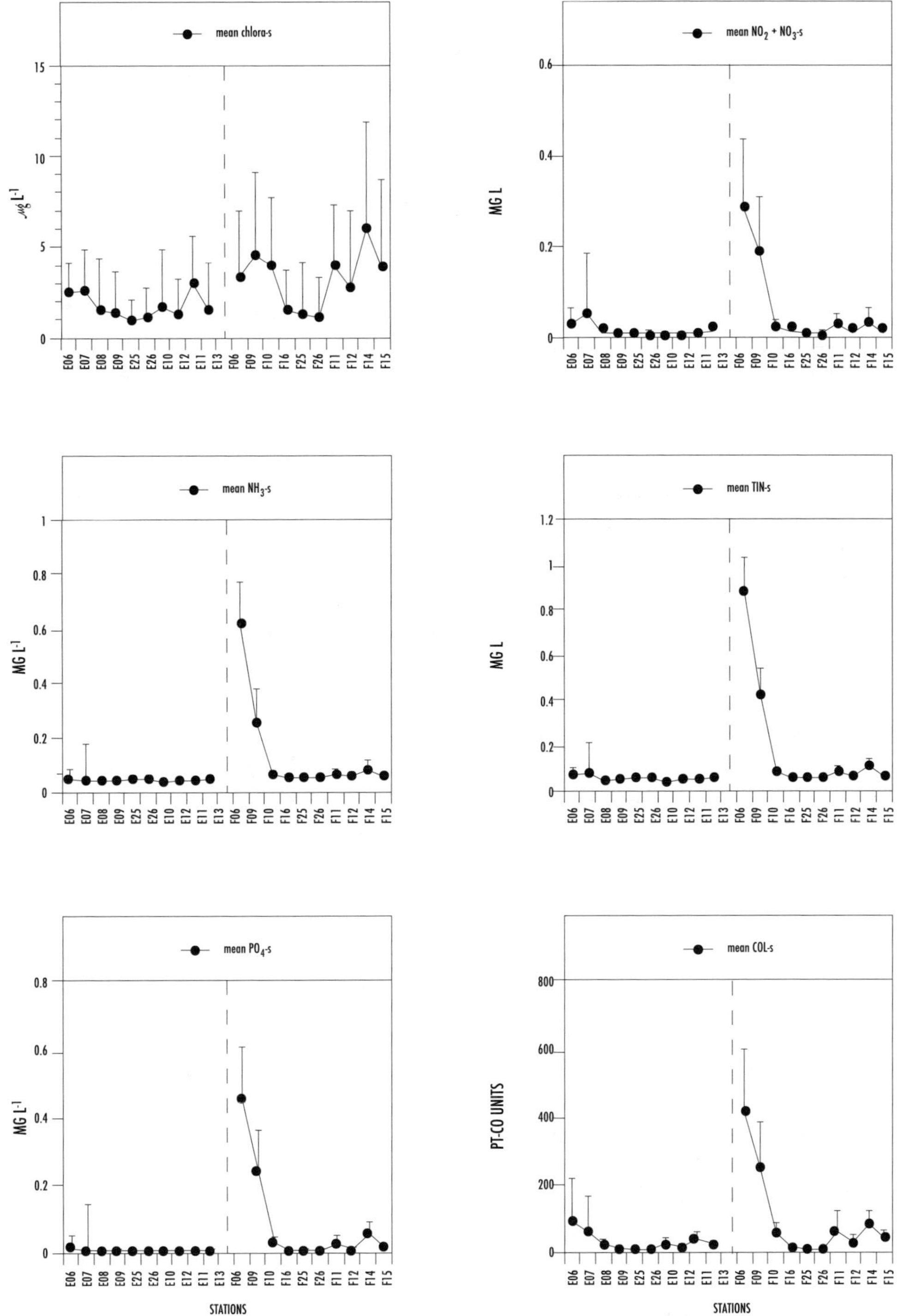

Figure 9.2 Comparisons of water quality factors between the Econfina and Fenholloway offshore systems. Monthly data from April 1999 through October 2000 were averaged and are shown as means (+1 standard deviation). Data include chlorophyll *a*, nitrite and nitrate, ammonia, total inorganic nitrogen (TIN), orthophosphate, and water color.

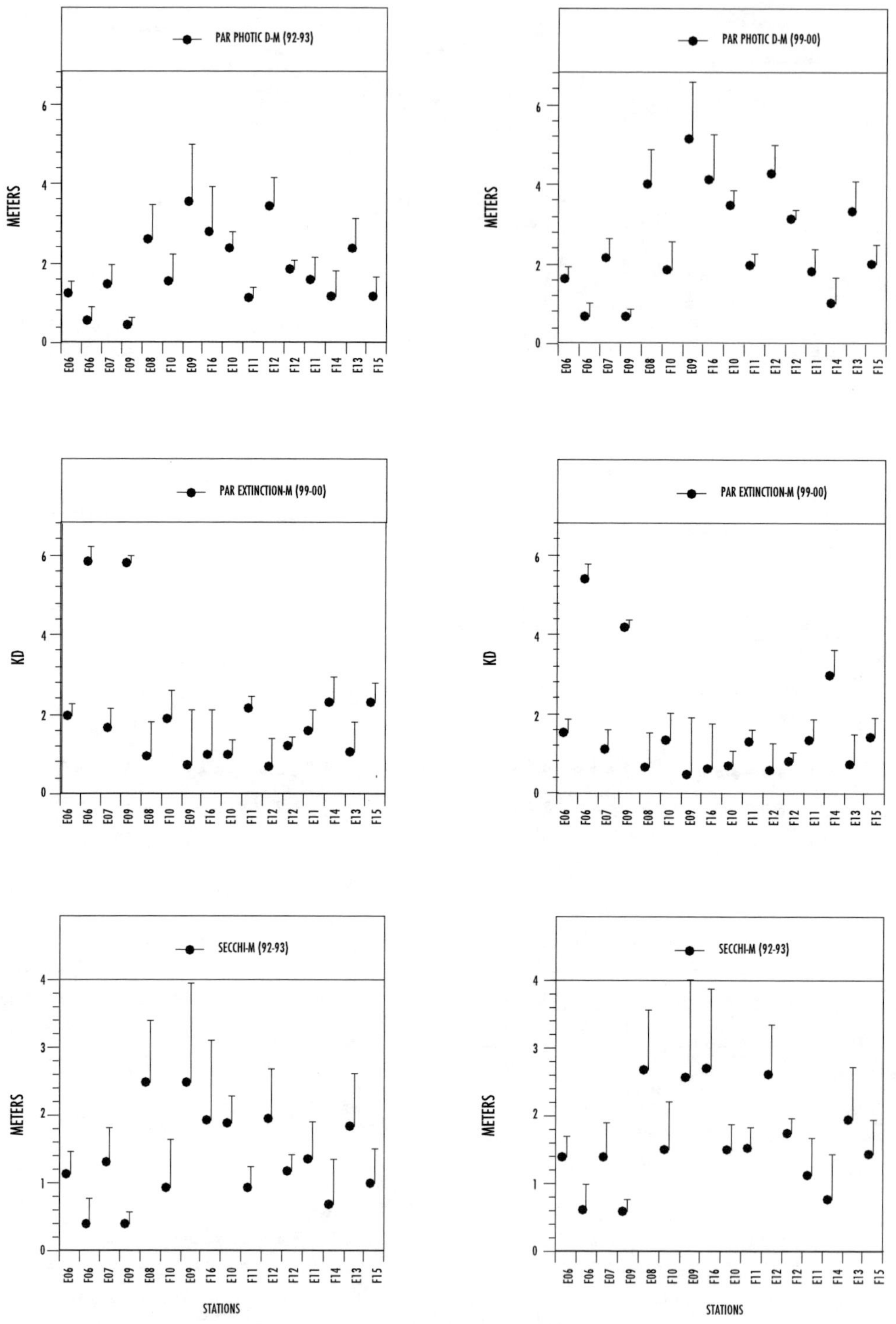

Figure 9.3 Comparison of light transmission in 1999–2000 sampling periods. Monthly data were given as means (+1 standard deviation) over the respective sampling periods (April 1992 through May 1995; April 1999 through October 2000). Data include PAR photic depths, PAR extinction coefficients, and Secchi depths.

Table 9.1 Numerically Dominant Phytoplankton Species, Species Richness, Shannon-Wiener Diversity and Evenness of Phytoplankton Taken with 25-μm Nets and Whole-Water Samples in the Econfina and Fenholloway Systems from April 1992 through March 1993, from April 1999 through December 1999, and from June through October 2000

Whole-Water Phytoplankton: Cells/l

Econfina (April 1992–March 1993)

Taxon	E06	E07	E08	E09	E10	E11	Total	%
Unidentified nanoflagellates	53169	52856	75818	52448	74278	53643	362212	19.57
Unidentified cryptophytes	21340	54642	63058	23810	122674	58941	344464	18.61
Unidentified nanococcoids	20147	17059	34011	13753	18963	26670	130602	7.06
Cyclotella sp. 17 (5–10 μm)	0	0	0	0	91390	0	91390	4.94
Thalassionema nitzschioides	3191	4950	32104	2914	13357	5933	62450	3.37
Pyramimonas spp.	8464	10172	16559	2298	6845	13169	57506	3.11
Navicula spp.	20147	8567	4919	5245	6475	7144	52497	2.84
Unidentified pennate diatoms	13459	14531	4904	2881	8140	6721	50635	2.74
Chaetoceros spp.	694	4829	15606	2864	12284	6872	43148	2.33
Cylindrotheca closterium	12266	10823	4950	2015	3719	8840	42611	2.3

Whole-Water Phytoplankton: Cells/l

Fenholloway (April 1992–March 1993)

Taxon	F06	F09	F10	F16	F11	F14	Total	%
Unidentified nanoflagellates	35893	47377	57017	53327	56407	46737	296757	19.52
Skeletonema costatum	555	1816	116883	4657	52836	95804	272552	17.93
Unidentified cryptophytes	21337	23249	82436	32402	29452	34382	223259	14.69
Unidentified nanococcoids	14934	18012	24883	12697	21997	34016	126539	8.32
Navicula spp.	18383	16711	12007	8492	8159	11156	74906	4.93
Pyramimonas spp.	1748	4541	11526	5387	10656	23710	57568	3.79
Nitzschia section pseudonitzschia spp.	0	394	3811	1807	42735	133	48879	3.22
Cylindrotheca closterium	2778	16166	8584	2567	8862	9024	47981	3.16
Minutocellus scriptus	3968	5086	1517	466	1628	8858	21523	1.42
Nitzschia spp.	4259	3845	2294	1693	2442	5511	20043	1.32

(continued)

Table 9.1 (continued) Numerically Dominant Phytoplankton Species, Species Richness, Shannon-Wiener Diversity and Evenness of Phytoplankton Taken with 25-µm Nets and Whole-Water Samples in the Econfina and Fenholloway Systems from April 1992 through March 1993, from April 1999 through December 1999, and from June through October 2000

25-µm Net Phytoplankton: Cells/l

Econfina (April 1992–March 1993)

Taxon	E06	E07	E09	Total	%
Unidentified nanoflagellates	300	166	183	649	12.74
Cylindrotheca closterium	245	94	66	405	7.95
Navicula spp.	142	127	76	345	6.77
Skeletonema costatum (narrow)	101	155	0	256	5.02
Unidentified nanococcoids	145	32	59	235	4.62
Asterionella japonica	4	163	68	234	4.6
Amphora spp.	49	73	91	212	4.16
Rhizosolenia stolterfothii	4	81	88	173	3.39
Chaetoceros diversus	25	31	75	131	2.57
Thalassionema nitzschioides	33	23	75	131	2.57

25-µm Net Phytoplankton: Cells/l

Fenholloway (April 1999–March 1993)

Taxon	F06	F09	F16	Total	%
Skeletonema costatum	453	10388	14	10854	71.35
Unidentified nanoflagellates	265	236	183	684	4.5
Cylindrotheca closterium	150	211	96	456	3
Navicula spp.	176	115	93	384	2.53
Nitzschia section pseudonitzschia spp.	0	0	257	257	1.69
Chaetoceros diversus	3	15	206	224	1.48
Thalassionema nitzschioides	10	17	148	174	1.15
Chaetoceros compressus	21	0	134	155	1.02
Unidentified nanococcoids	85	36	31	152	1
Chaetoceros simplex	0	0	103	104	0.68

Whole-Water Phytoplankton: Cells/l

Econfina (April 1999–December 1999)

Taxon	E06	E07	E08	E09	E10	E11	Total	%
Undetermined cryptophyte	320614	184092	63585	81119	96228	299027	1044665	30.54
Cyclotella choctawhatcheeana	2072	42718	17428	23707	141212	386243	613380	17.93
Undetermined nanoflagellates	86181	79417	61567	58018	40659	75433	401275	11.73
Johannesbaptistia pellucida	2035	39590	58018	102704	2701	1406	206454	6.04
Asterionellopsis glacialis	37	296	2590	138760	74	93	141850	4.15
Dactyliosolen fragilissimus	815	12412	5076	8391	315	77793	104802	3.06
Chaetoceros laciniosus	2186	70665	20343	815	0	334	94343	2.76
Undetermined nanococcoids	2629	5924	22501	7812	46650	3367	88883	2.6
Gymnodinium spp.	4514	10676	9087	16711	9090	28196	78274	2.29
Cyclotella cf. atomus	5665	6106	10476	8936	15695	23757	70635	2.06

Whole-Water Phytoplankton: Cells/l

Fenholloway (April 1999–December 1999)

Taxon	F06	F09	F10	F16	F11	F14	Total	%
Undetermined cryptophyte	358064	270651	138850	70457	229703	195946	1263671	26.33
Leptocylindrus danicus	18755	206469	493799	10028	15985	1554	746590	15.56
Undetermined nanoflagellates	92645	98513	79576	48093	121982	100025	540834	11.27
Cyclotella cf. atomus	59715	54184	41313	352	126789	80402	362755	7.56
Johannesbaptistia pellucida	2206	6697	39886	3293	279721	20805	352608	7.35
Cerataulina pelagica	625	1258	10510	11785	7451	181736	213365	4.45
Undetermined nanococcoids	499	1518	2868	4497	162792	481	172655	3.6
Cerataulina pelagica (veg.)	0	0	0	0	119963	0	119963	2.5
Dactyliosolen fragilissimus	1915	3145	23348	1647	74009	10361	114425	2.38
Rhizosolenia setigera	17402	7552	66382	13988	4289	4367	113980	2.37

(continued)

Table 9.1 (continued) Numerically Dominant Phytoplankton Species, Species Richness, Shannon-Wiener Diversity and Evenness of Phytoplankton Taken with 25-μm Nets and Whole-Water Samples in the Econfina and Fenholloway Systems from April 1992 through March 1993, from April 1999 through December 1999, and from June through October 2000

25-μm Net Phytoplankton: Cells/l

Econfina (April 1999–October 1999)

Taxon	E06	E07	E08	E09	Total	%
Asterionellopsis glacialis	34	72	1392	79216	80714	33.84
Johannesbaptistia pellucida	1607	7512	21770	40634	71523	29.98
Chaetoceros laciniosus	413	24921	17888	556	43778	18.35
Chaetoceros lauderi	3	9374	1041	58	10476	4.39
Bacteriastrum hyalinum	33	2799	2699	108	5639	2.36
Chaetoceros compressus	71	2219	2130	449	4868	2.04
Chaetoceros c.f. subtilis	411	2224	0	0	2635	1.1
Rhizosolenia setigera	42	928	1480	158	2607	1.09
Pseudonitzschia spp.	37	938	940	342	2257	0.95
Peridinium quinquecorne	1066	28	0	3	1097	0.46

25-μm Net Phytoplankton: Cells/l

Fenholloway (April 1999–October 1999)

Taxon	F06	F09	F10	F16	Total	%
Leptocylindrus danicus	2608	16490	38363	1877	59338	27.93
Rhizosolenia setigera	2658	2629	39170	2683	47140	22.19
Falcula hyalina	16081	15966	9	4	32060	15.09
Chaetoceros laciniosus	43	33	538	20690	21305	10.03
Chaetoceros lauderi	10	133	2346	8347	10837	5.1
Chaetoceros compressus	54	11	8745	573	9384	4.42
Johannesbaptistia pellucida	0	28	1604	5925	7557	3.56
Cerataulina pelagica	142	307	987	4521	5956	2.8
Thalassionema nitzschioides	90	173	2073	72	2409	1.13
Ceratium hircus	70	42	1101	177	1390	0.65

Whole-Water Phytoplankton: Cells/l

Econfina (June 2000–October 2000)

Taxon	E06	E07	E08	E09	E10
Undetermined cryptophyte	123,543	1,038,628	254,579	178,489	275,058
Johannesbaptistia pellucida	0	38,961	85,415	99,734	91,576
Cyclotella c.f. atomus	38,961	104,229	34,299	46,787	24,809
Undetermined nanoflagellates	24,975	119,215	44,789	43,124	40,960
Undetermined nanococcoids	0	24,309	67,100	22,645	112,555
Gymnodinium spp.	666	48,785	50,784	31,137	36,797
Cocconeis spp.	4,995	55,445	32,303	9,159	38,296
Pseudonitzschia spp.	666	12,987	73,260	12,988	24,144
Proboscia alata	0	999	7,993	62,438	1,499
Chaetoceros spp.	0	19,481	28,306	11,823	21,813

Taxon	E11	E12	E13	Total	%
Undetermined cryptophyte	682,984	262,071	297,037	3,112,388	38.6
Johannesbaptistia pellucida	3,497	114,219	233,600	667,001	9.6
Cyclotella c.f. atomus	247,420	24,310	33,634	554,448	7.3
Undetermined nanoflagellates	89,078	42,458	57,777	462,375	6.4
Undetermined nanococcoids	33,633	139,861	55,279	455,381	6.4
Gymnodinium spp.	43,291	35,300	82,253	329,013	4.3
Cocconeis spp.	45,455	15,320	8,993	209,965	2.9
Pseudonitzschia spp.	2,166	56,112	11,656	193,978	2.7
Proboscia alata	1,166	500	79,088	153,681	2.4
Chaetoceros spp.	6,494	27,307	13,488	128,712	1.8

(continued)

Table 9.1 (continued) Numerically Dominant Phytoplankton Species, Species Richness, Shannon-Wiener Diversity and Evenness of Phytoplankton Taken with 25-µm Nets and Whole-Water Samples in the Econfina and Fenholloway Systems from April 1992 through March 1993, from April 1999 through December 1999, and from June through October 2000

Whole-Water Phytoplankton: cells/l

Fenholloway (June 2000–October 2000)

Taxon	F06	F09	F10	F16	F11
Leptocylindrus minimus	0	2,137,860	2,601,396	360,306	102,731
Undetermined cryptophyte	78,921	594,072	175,492	384,283	170,830
Proboscia alata	102,231	364,302	549,950	282,384	170,164
Johannesbaptistia pellucida	0	215,118	165,501	190,976	201,798
Thalassiosira minima proschikinae complex	0	321,345	0	0	0
Chaetoceros spp.	333	79,254	101,566	71,263	141,859
Rhizosolenia setigera	1,332	42,291	68,765	50,117	52,615
Undetermined nanoflagellates	9,324	79,587	45,788	56,777	43,125
Dactyliosolen fragilissimus	666	666	99,567	666	1,333
Gymnodinium spp.	5,661	5,661	22,645	51,949	49,618

Taxon	F14	F12	F15	Total	%
Leptocylindrus minimus	234,932	1,288,544	0	6,725,769	37.4
Undetermined cryptophyte	590,909	175,159	138,030	2,307,696	12.8
Proboscia alata	232,767	458,208	41,292	2,201,298	12.3
Johannesbaptistia pellucida	55,112	199,634	164,670	1,192,808	6.6
Thalassiosira minima proschikinae complex	565,934	0	0	887,279	4.9
Chaetoceros spp.	42,291	80,254	7,659	524,478	2.9
Rhizosolenia setigera	29,804	202,964	1,332	449,220	2.5
Undetermined nanoflagellates	114,719	43,791	50,285	443,395	2.5
Dactyliosolen fragilissimus	102,731	103,563	5,495	314,687	1.8
Gymnodinium spp.	21,313	60,940	31,636	249,421	1.4

Secchi disk readings and DOC were significantly (negatively) associated with numbers of cells in the Econfina system (Secchi $r^2 = 0.52$, $p = 0.002$; DOC $r^2 = 0.62$, $p = 0.03$). Phytoplankton numbers tended to decrease with increased ammonia in the Fenholloway system, whereas there was a weak positive relationship between these two factors in the Econfina. Overall, with the exception of temperature and certain light features, phytoplankton numerical abundance was not strongly associated with any of the environmental features of the two systems.

Phytoplankton species richness in the Fenholloway system was weakly (negatively) associated with DOC ($r^2 = 0.07$, $p = 0.03$), ratio turbidity ($r^2 = 06$, $p = 0.04$), ammonia ($r^2 = 0.16$, $p = 0.0006$), orthophosphate ($r^2 = 0.16$, $p = 0.0003$), and silicate ($r^2 = 0.19$, $p = 0.0004$). Weakly positive (statistically significant) associations in the Econfina system were found between species richness and temperature ($r^2 = 0.22$, $p = 0.0001$) and DOC ($r^2 = 0.26$, $p = 0.0001$). The weak r^2 values reduced the significance of these associations, although the general trends indicated that phytoplankton species richness decreased with reduced light transmission, high nutrients, and high DOC in the Fenholloway system. Major differences in the phytoplankton assemblages were associated with the differing light transmission characteristics of the two systems. Phytoplankton numbers and species richness varied inversely with distance from shore in both areas, with light and nutrient alterations directly implicated in the differences noted between the respective phytoplankton biotas.

The data indicated that, during 1992–1993, high nutrient loading, combined with increased light penetration, accounted for differences in the phytoplankton distributions in the Econfina and Fenholloway systems (Livingston et al., 1998b). Light transmission characteristics were important, and the dominant wavelength at most of the stations in both systems was between 575 and 580 nm. Both Gulf drainages were affected by river-derived dissolved substances (*gelbstoff*) that were associated with the loss of light at lower wavelengths. This *gelbstoff* effect was stronger in the Fenholloway estuary and near-shore waters (F06, F09, F11, F14) than cognate stations in the Econfina system. Phytoplankton numbers in both systems peaked in near-shore waters characterized by increasing light penetration and high nutrients. High nutrient loading from the Fenholloway River was most apparent in these near-shore waters. Farther offshore, there was general decrease of phytoplankton cell numbers and species richness that was directly related to reduced nutrient concentrations. Light and nutrients thus accounted for the spatial distributions of phytoplankton numbers and species richness. Light was limiting in the Fenholloway estuary; in inshore areas, light became less a factor and increased nutrient loading took over as the primary limiting agent relative to the reference system. Farther offshore, nutrients became limiting in both systems with corresponding decreases of phytoplankton numbers and species.

Skeletonema costatum is common worldwide in coastal waters where it is highly adaptable to changes in temperature and salinity. *Skeletonema costatum* was a constituent of the phytoplankton community throughout the year. This species is often dominant due to its relatively high growth rates under a wide range of temperature and light conditions. There is also evidence that *S. costatum* grows well under high nutrient conditions, and it can use organic phosphorus as a source of growth. This species can also assimilate organic molecules such as urea (Round, 1981). It grows in New York Bight waters where massive amounts of chemicals were dumped (Young and Barber, 1973). High numbers of *S. costatum* represented the primary difference between the near-shore Fenholloway and Econfina stations in 1992–1993. In the Fenholloway drainage, this species was weakly (negatively) associated with DOC ($r^2 = 0.10$, $p = 0.02$) and positively associated with percent luminance ($r^2 = 0.07$, $p = 0.04$). This species was thus successful in the near-shore areas of the Fenholloway system where there was relatively poor light transmission and high nutrient concentrations.

Inshore Econfina areas were dominated by nanoflagellates, cryptophytes, nanococcoids, and diatom species such as *Navicula* spp. during 1992–1993. Overall, the nanoflagellates and cryptophytes were generally higher in the Econfina estuarine and near-shore waters than in similar areas of the Fenholloway, whereas the nanococcoids were evenly distributed in the two study areas. The diatom *Chaetoceros fragilis* Meun was dominant in offshore areas of the Econfina system. Inshore stations in the Fenholloway were well represented by species such as *Minutocellus scriptus*, Hasle, Stosch and Syvertsen, *Cryptomonas* spp., *Cylindrotheca closterium* (Ehrenberg) Reimann and Lewin, and various forms of nanoflagellates. These data followed other indications that the estuarine Fenholloway phytoplankton assemblages were relatively depauperate compared with the Econfina estuary.

4. Color Removal and Bloom Generation

A comparison was made of the phytoplankton communities in 1992, 1999, and 2000. Data were analyzed from June through October for each of these years since summer–fall represented the peak bloom period (Figure 9.4). During 1992–1993, the inshore (estuarine) whole-water phytoplankton in the Fenholloway system (F06, F09, F11) showed significant reductions of phytoplankton cumulative species richness compared to the Econfina areas (Table 9.1; Figure 9.4). Numerical abundance was relatively low in both systems during this period. Species diversity and evenness were also generally lower in the Fenholloway estuarine areas. There were indications of increased abundance and species richness farther offshore in the Fenholloway system.

During 1999, species richness increased in both systems relative to 1992–1993. This index remained low at inshore Fenholloway stations. During this period, phytoplankton numbers went

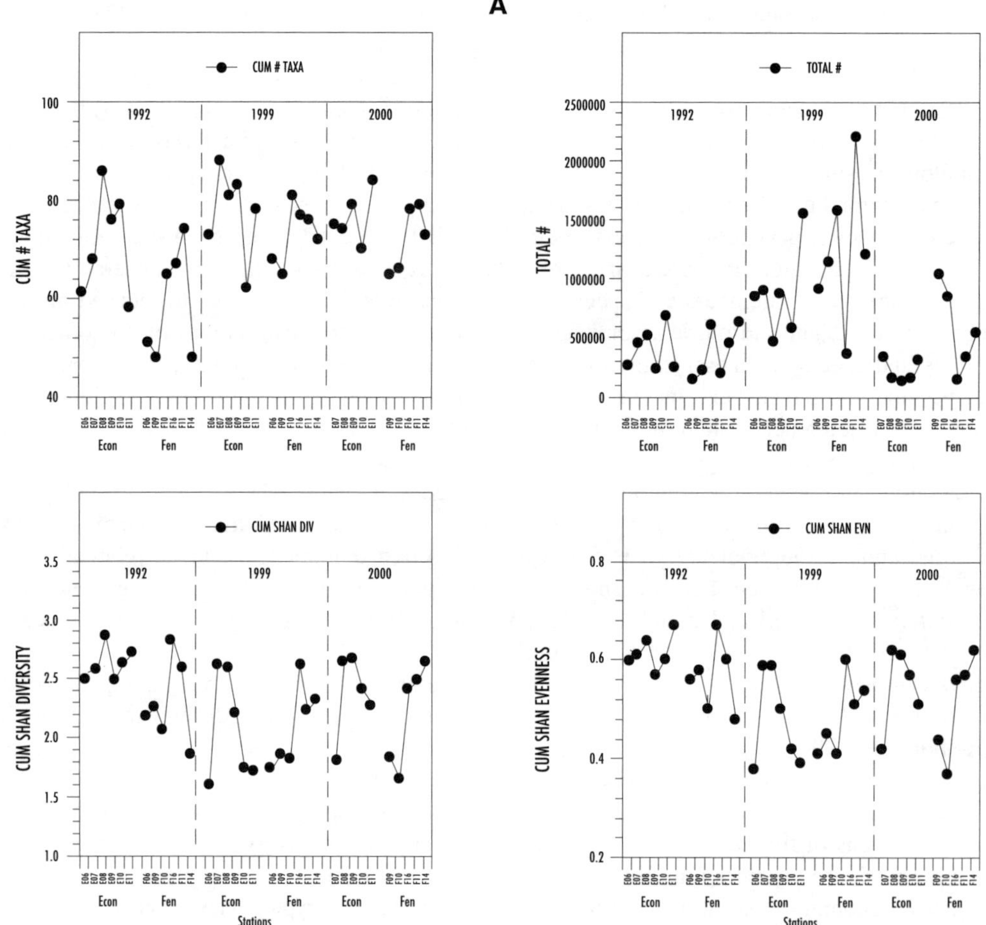

Figure 9.4 Whole-water phytoplankton taken in the Econfina and Fenholloway systems from June through October during the years 1992, 1999, and 2000. Data are presented as averages per station over the respective time periods. (A) Numbers of cells per liter, cumulative species richness, cumulative Shannon-Wiener diversity, and cumulative Shannon evenness. (continued)

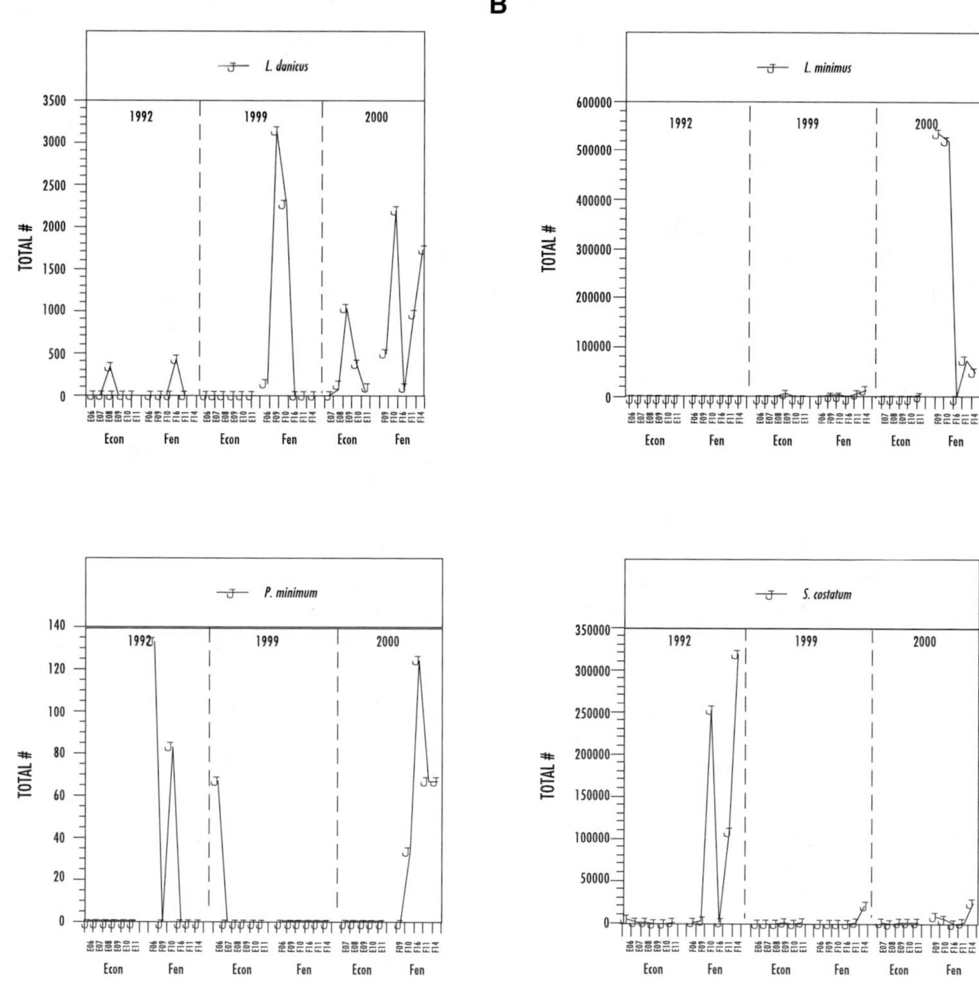

Figure 9.4 (continued) Whole-water phytoplankton taken in the Econfina and Fenholloway systems from June through October during the years 1992, 1999, and 2000. Data are presented as averages per station over the respective time periods. (B) Numerical abundance of bloom species that include *Leptocylindrus danicus, L. minimum, Skeletonema costatum*, and *Prorocentrum minimum.*

up in both systems with particularly high numbers at stations F10 and F11. Species diversity and evenness in both systems were low in inshore areas in 1999. By 2000, Econfina phytoplankton numbers were down due, possibly, to drought effects on nutrient loading. Species richness in the Econfina remained high during 2000. In the Fenholloway system during 2000, inshore numbers remained high. Species richness, diversity, and evenness tended to be low where such high numbers occurred (F09, F10, F14). These community changes with time reflected the incidence of phytoplankton blooms in the Fenholloway system during 1999 and 2000.

The peak phytoplankton numbers in the Fenholloway system (Figure 9.4) were associated with phytoplankton blooms (number cells l^{-1}, $>10^6$) of *Leptocylindrus danicus* (1,831,582 cells l^{-1}, station F09, September 1999; 4,390,902 cells l^{-1}, station F10, September 1999), *Johannesbaptistia pellucida* (2,111,220 cells l^{-1}, station F11, September 1999), and *Cerataulina pelagica* (1,625,299 cells l^{-1}, station F14, October 1999). The diatom *Leptocylindrus danicus* has a worldwide distribution in a variety of tropical coastal environments, and has adverse impacts on various fish species (see above). *Leptocylindrus danicus* was a key pioneer bloom species in the Perdido Bay system during the initial blooms of 1994. *Johannesbaptistia pellucida* is a filamentous blue-green alga with a mucous sheath. It has a worldwide distribution and is not considered a noxious species (A. K. S. K. Prasad, personal communication). *Cerataulina pelagica* is a filamentous diatom that is a known bloom species that closely resembles *L. danicus* in body type and natural history (A. K. S. K. Prasad, personal communnication). These were the first blooms noted in either of the Apalachee study areas, and they occurred during fall months of a drought when water temperature and salinity in the Fenholloway system were relatively high and water color was low. Light attenuation coefficients were also relatively low during this period. Orthophosphate and ammonia loading from the Fenholloway River was relatively high during fall 1999.

During 2000, there were blooms during July in the Fenholloway estuary and in the offshore Fenholloway system (station F15). The primary bloom species was *L. minimum* (Figure 9.4). There was also high dominance of *J. pellucida*, *Rhizosolenia setigera*, and dinoflagellates (*Gymnodinium* spp.). These blooms were not as diverse as the previous year. The trends of both species of *Leptocylindrus* were somewhat different with high numbers of *L. danicus* in estuarine areas during 1999 and more generalized distributions during 2000. *Leptocylindrus minimus* only appeared in inshore Fenholloway waters during 2000. The species *S. costatum,* a primary dominant during 1992, virtually disappeared from the Fenholloway system during the bloom periods. There was a more general distribution of the bloom species *Prorocentrum minimum* in the Fenholloway during 2000. Changes in the Econfina phytoplankton community during 1999–2000 could be ascribed to the naturally altered habitat conditions associated with the drought (i.e., increased light penetration, salinity, water temperature, and reduced nutrient loading). Reduced diversity in the Econfina system during 1999 could be ascribed to increased dominance by various species and general increases in overall numbers of phytoplankton during this period. The trends of the net phytoplankton in the Econfina system tended to follow those of the whole-water phytoplankton.

The data thus showed that during the period of increased light penetration, there were increased phytoplankton numbers in both systems. However, with time during the drought, numerical abundance of phytoplankton in the Econfina system decreased, which could be explained by the Apalachicola trends during the drought of the early 1980s (Livingston et al., 1997). The increased dominance by the bloom species *L. danicus* noted at stations F09 and F10 in 1999 and the related species *L. minimus* during 2000 suggests the possibility that increased light penetration coupled with continued high ammonia and orthophosphate loading in the Fenholloway system led to an overlap of deeper photic depths with the high nutrient gradients. This situation was similar to that of the 1994 initiation of *L. danicus* blooms in Perdido Bay under similar habitat conditions (drought plus high nutrient loading).

During fall 1993, *Falcula hyalina* was dominant in the Perdido system. Dominance by this species in the Perdido system was followed by the *L. danicus* blooms in January–February 1994. Net phytoplankton data supported a basic shift of phytoplankton species composition and abundance in the Fenholloway system during 1999 (Table 9.1). The trends of the Fenholloway phytoplankton data thus indicated that the combination of high nutrient loading (ammonia and orthophosphate), drought conditions, and increased light penetration caused basic changes of the phytoplankton in the Fenholloway estuary with the appearance of potentially damaging blooms as a product of continuously high nutrient loading. If this pattern continues along the lines of that experienced in the Perdido estuary, these preliminary diatom blooms could eventually be replaced by more destructive plankton species such as raphidophytes and dinoflagellates. The more general offshore distribution of *P. minimum* could be an indication of future blooms by this species. There could be adverse effects on offshore sea grass beds with interannual progressions to more toxic bloom species. The effects of the recent blooms on near-shore biota in the Fenholloway system are currently being analyzed.

Chlorophyll *a* concentrations in both the Econfina and Fenholloway systems were generally low, and were not indicative of phytoplankton blooms or the effects of the blooms on associated phytoplankton assemblages. Since both state and federal regulatory agencies use chlorophyll *a* as the primary index to evaluate phytoplankton blooms in coastal systems, there has been no regulatory response to nutrient loading in Apalachee Bay. According to a recent study of the Econfina/Fenholloway system by the U.S. Environmental Protection Agency (EPA, 2000):

> "No phytoplankton blooms were observed during the four sampling periods. Field assessments for pH and dissolved oxygen supported this observation.…Chlorophyll *a* concentrations were low in both the rivers and estuaries.…The maximum chlorophyll *a* concentration measured during the four sampling periods was 10 μg/L in the Fenholloway estuary in June. Concentrations above 15 to 20 μg/L could indicate blooms activity." (U.S. Environmental Protection Agency, 1975; Livingston, 1998)

The EPA sampling occurred during December 1998, June 1999, and August–September 1999. The peak blooms in the Fenholloway system occurred during September–October 1999 (see above). However, there were no phytoplankton analyses carried out in the EPA study, a common failing in most studies of this kind. This constitutes direct evidence that the usual methods used by regulatory agencies to evaluate potentially damaging blooms in coastal systems, with insufficient water quality indicators and no accounting for the most important group of organisms involved in the blooms, are not only flawed but are based on false assumptions of how blooms are initiated and maintained. Unfortunately, this conclusion can be applied to much of the work being carried out by federal programs where long-term, multidisciplinary projects are usually avoided. As noted by Livingston (2000), ignoring the species composition of phytoplankton and using false assumptions concerning so-called indicators of blooms will lead to serious omissions as shown in the EPA studies of the Fenholloway drainage.

This general disregard for adequate regulation of nutrient loading problems in coastal areas is common throughout the United States. The silent and largely overlooked impacts of anthropogenous nutrient loading (urban runoff, agricultural waste discharges, point-source pollution) will continue to contribute to ongoing deterioration of coastal resources in the United States until more attention is paid to the phytoplankton assemblages that represent the first-line response to such nutrient loading. Unless scientifically sound indices are identified whereby the causes and solutions to hypereutrophication can be evaluated, the problem will continue. As shown previously by Livingston (2000), the trophodynamic approach to studying bloom phenomena provides a realistic and objective appraisal of the impacts and solutions to nutrient loading in coastal areas. The lack of adequate funding for ecosystem-level projects, together with outright dishonesty in the review process by state and federal agencies, will eventually be translated into continued and accelerating losses of coastal productivity in the United States.

III. MERCURY AND THE PENOBSCOT RIVER–ESTUARY, MAINE

A. Scientific Literature

Mercury represents a serious threat to aquatic systems in the United States (U.S. Environmental Protection Agency, 1997). In recent years, mercury concentrations in the environment have increased in many aquatic systems of the country. Aerial deposition from coal-fired power plants, incinerators, and industrial activities along with point-source discharges from chloralkali plants are the primary sources of mercury to the environment. Recent studies (U.S. Environmental Protection Agency, 1997) indicate that aerial burdens of mercury have increased fivefold during the industrial era. Aerial deposition, sometimes hundreds or even thousands of miles away from the source, has led to severe contamination of various aquatic systems (U.S. Environmental Protection Agency, 1997).

Mercury does not degrade once it is deposited in aquatic habitats, but it can be altered chemically as it is sequestered in sediments. The inorganic form of mercury can be transformed by microbial activity to the extremely toxic compound methyl mercury. This process is influenced by environmental factors such as available DO, pH, sulfur, and sediment type. Low DO enhances the methylation process. Methyl mercury is able to pass through biological membranes, and its high chemical stability and slow excretion rate from aquatic animals leads to rapid bioconcentration. It is also concentrated up food webs as biomagnification, a process unique to mercury among metals (Regnell and Ewald, 1997). Various studies have illustrated different pathways for bioconcentration and biomagnification of mercury. There is evidence of complex modes of mercury transport in aquatic systems (Locarnini and Presley, 1996; Campbell et al., 1998). The methylated form of mercury is environmentally persistent, lasting for decades to centuries in aquatic systems. This compound is bioconcentrated in individual aquatic organisms and biomagnified up aquatic food webs with top predators such as large fishes, birds, and mammals (including humans) having the highest concentrations. The ecological significance of mercury in the aquatic environment is thus related to food web dynamics.

Because of the chemical characteristics of methyl mercury, acute and chronic toxicological aspects of this compound are important to an understanding of the implications of increasing mercury contamination. Relatively low concentrations of mercury can affect aquatic organisms at various levels of biological organization (Muessig, 1974; Drake, 1975; Zillioux et al., 1993; Burgess et al., 1998; Wolfe et al., 1998) with the most severe effects usually at the top of the food web. The relatively extensive scientific literature concerning toxic effects of mercury on aquatic species (U.S. Environmental Protection Agency, 1997) has defined the limits of the impacts of mercury on river–estuarine systems. Some of these effects include the following:

- Inhibition of algal growth
- Reduced egg deposition and hatching success in fishes
- High embryo-larval mortality and limb deformities in frogs
- High embryo and duckling mortality in black ducks
- Impaired sperm generation in fishes
- Inhibition of fish growth
- Reduced hatching of common terns
- Damaged gill/digestive tissue of oysters and other mollusks
- Impaired kidney function and disruption of fish endocrines
- Impaired immune systems of fishes
- Reduced serotonin concentrations of fishes

Wildlife at the top of aquatic food webs is particularly at risk due to mercury pollution (Meyer et al., 1998). Fish-eating birds and mammals are exposed to both lethal and sublethal effects of mercury as a function of food web processes. These effects include damage to the central nervous system, and impacts at early life history stages that include a range of physiological and behavioral anomalies associated with damage to the central nervous system. Relatively low concentrations of methyl mercury can cause severe effects on embryos and newly born birds and mammals. Mercury has been implicated in the deaths of great white herons in Florida (South Florida Water Management District, 2000). High concentrations observed in bald eagles and Florida panthers in Florida have been associated with problems of survival of these species (South Florida Water Management District, 1994, 2000).

Recent evidence indicates that mercury in riverine drainages can last a very long time. The behavior of mercury and the dynamic habitat changes that are associated with the methylation process have been ascribed to specific conditions found in rivers and estuarine systems. Parks et al. (1989), studying mercury contamination of a Canadian river (the Wabigoon River system) by a chloralkali plant, found that net production of methyl mercury was related to water column equilibrium conditions of methyl mercury and inorganic mercury. The actual generation of methyl mercury is a complex chemical process that involves both the sediment–water interface and overlying water column equlilibria. Continuous methylation in areas distant from the source helped to explain why the methyl mercury concentrations in the river remained high 10 years after cessation of releases of mercury from the plant. These results indicated that the answer to impact evaluations and possible remediation of mercury contamination lies in downstream sediment–water processes and habitat conditions that contribute to production of methyl mercury.

Rada et al. (1986) found that river sediments in Wisconsin were highly contaminated with mercury 20 to 30 years after contamination from human sources had been eliminated. Biological availability was enhanced by rapid methylation of the mercury in surficial sediments even though substantial amounts of the metal were buried. Mean sediment total mercury concentrations from 0.3 to 0.8 $\mu g\ g^{-1}$ (ppm) dry weight with ranges of 0.2 to 1.8 $\mu g\ g^{-1}$ were considered high when compared to reference areas with concentrations of 0.04 to 0.05 $\mu g\ g^{-1}$. Rada et al. (1986) found that fishes had relatively high methyl mercury concentrations that were attributed to uptake by feeding and respiration. Overall, mercury concentrations in depositional areas were considered as mercury sinks. Continual disturbance and resultant exposure of sedimentary mercury to the water column in other areas due to natural river habitat conditions was noted as the source of continuing methyl mercury exposure years or even decades after mercury contamination of the river was eliminated. Increased bioavailability of mercury was thus associated with physical disturbance of the sediments in addition to organic loading and high temperatures. These data indicated that transformation and transport of the mercury in riverine systems is highly dynamic. Once contaminated by loading from an industrial source, the mercury was available to the biota for relatively long periods (years to decades) due to resuspension and subsequent methylation of mercury-contaminated sediments.

Various studies have shown that riverine and coastal wetlands are areas that may be a significant source of methyl mercury due to the combination of habitat conditions that favor the methylation process (Zillioux et al., 1993). A recent proposal by a Crown-owned utility called Hydro-Quebec to construct dams along rivers flowing into the James Bay–Hudson Bay system in Canada led to studies concerning previous dam operations in this area. Habitat changes encountered with flooding of wetlands due to dam construction are well known (Rosenberg et al., 1987). Flooding associated with new impoundments caused reductions of DO that, in turn, provided appropriate habitat conditions for methylization of mercury that had been deposited in the previously unflooded wetlands. The newly flooded wetlands were thus associated with increased methyl mercury in the aquatic food webs. Eventually, mercury was concentrated in top predators such as fishes that for centuries had provided a primary source of food for natives in the area (Brouard et al., 1990). It was estimated that it could take up to 50 years for methyl mercury levels in top predators to return to background levels.

Only recently has the wetland connection been studied as part of the ecosystem response to aerial mercury deposition. The contamination of the Florida Everglades with mercury is a wetland phenomenon where mercury has eliminated consumption of game fishes (South Florida Water Management District, 2000). Wiener and Shields (2000) reviewed the transport, fate, and bioavailability of mercury in the Sudbury River in Massachusetts. The Sudbury system was contaminated by an industrial complex that operated for decades with continuous release of mercury until 1978, High mercury concentrations in sediments and fishes led to studies associated with the Superfund program of the EPA. The authors reported that methyl mercury concentrations in aquatic biota did not parallel the concentrations of total mercury in sediments to which they were exposed. Instead, it was found that contaminated wetlands 25 km downstream from the point source of mercury "produced and exported methyl mercury from inorganic mercury that had originated from the site" (Wiener and Shields, 2000). In depositional areas along the river, mercury was eventually buried, but continuous methylization of inorganic mercury took place in floodplain wetlands of the Sudbury River system. Whereas the relatively high concentrations of mercury in depositional areas such as impoundments were eventually buried with resultant removal of mercury from biological contamination, natural processes associated with the "lesser contaminated" wetland habitat farther downstream remained available for the methylation process. The eventual and continuous transport and entry of methyl mercury into downstream food webs was thus associated with wetland areas far distant from the original source.

These studies explained the processes that have contributed to long-term mercury contamination of various drainage areas. The results showed that mercury transported to wetlands far removed from the origin of the contamination were more problematic than the mercury levels at the industrial site in terms of long-term, adverse effects on the system. Accordingly, remedial efforts at the contaminated site would be relatively meaningless because the real problems of methylation, transport, bioavailability, bioconcentration, and biomagnification of mercury were far removed from the site of mercury origin.

The effects of mercury contamination in aquatic systems of the United States are not restricted to aquatic biota. There is a human dimension to the problem. Because of the recent increases of mercury loading to the atmosphere, the EPA has issued fish consumption advisories in 40 states (U.S. Environmental Protection Agency, 1997). Advisories have been most common for lakes and rivers in states in the Midwest and New England (Maine, Vermont, New Hampshire). The highly conservative nature of mercury (it does not break down) that contributes to its environmental persistence, the ability of methyl mercury to bioconcentrate in aquatic organisms and to biomagnify up food webs, and the high propensity of this form of mercury to cause a broad range of acute and chronic toxic effects, all contribute to the serious nature of the mercury problem (U.S. Environmental Protection Agency, 1997).

Because of the high toxicity at relatively low concentrations, a small amount of mercury can cause significant problems in aquatic systems. For example, the release of less than 23 kg of mercury a year can lead to the contamination of 5000 walleye (2- to 3-lb class) with average tissue concentrations of 0.5 ppm, a level sufficient to trigger fish consumption advisories. Current EPA levels of safe exposure for humans are set at 0.1 $\mu g\ kg^{-1}$ (body weight) d^{-1} with particular concern for pregnant women and children. A review panel of the National Academy of Sciences (NAS) recently endorsed the strict standards of the EPA (0.1 $\mu g\ kg^{-1}$ body weight d^{-1}) based on established low-level effects of methyl mercury on children (Kaiser, 2000). This follows a previous NAS study that estimated that 60,000 newborns each year may suffer developmental damage due to fetal mercury exposure, mainly from mothers' consumption of contaminated fishes. Recent evidence (Houlihan and Wiles, 2001) indicates that about 10% of women of childbearing age have blood methyl mercury levels above the dose considered a risk for adverse neurological effects for fetuses. If the currently inadequate Food and Drug Administration's recommended fish consumption levels are followed, it could expose one fourth of all fetuses to potentially harmful doses of methyl mercury (Houlihan and Wiles, 2001).

B. Food Web Dynamics of Mercury in the Penobscot System

The Penobscot River–estuary (Figure 9.5) is the largest river–estuarine system in Maine with a watershed that exceeds 8500 square miles. Mean discharge rates of the Penobscot River average around 8200 $ft^3\ sec^{-1}$. Tidal ranges approximate 8 to 13 ft during spring tides. The combination of high tidal fluxes and high river flows contribute to relatively dynamic processes of particulate transport. There is no organized body of scientific information concerning the Penobscot system. No comprehensive, long-term ecological studies have been undertaken here.

A chloralkali plant on the Penobscot River near Orrington (Figure 9.5) has been permitted by state and federal agencies to discharge over 7 kg of mercury per year for more than three decades. Wastes from the plant include brine purification muds, mercury cell process water, mercury-contaminated substances, waste paint and solvents, PCBs, acetone, methyl ethyl ketone, and waste oils (Camp Dresser and McKee, Inc., 1998). Limited sediment mercury monitoring around the plant indicated that extremely high mercury residues (460 $mg\ kg^{-1}$) occurred in sediments near the plant. When adjusted for grain size, the distribution of sediment mercury indicated that the chloralkali plant was the most significant source of past and ongoing mercury pollution to the Penobscot system (Maine Department of Environmental Protection, 1998). However, no comprehensive analyses were performed concerning mercury concentrations in sediments and aquatic biota of the lower Penobscot River–estuary. The absence of quantitative mercury loading information together with the almost complete lack of ecological data concerning potential impacts of the high sediment concentrations of mercury in the Penobscot system left open questions concerning environmental effects of past and ongoing mercury discharges.

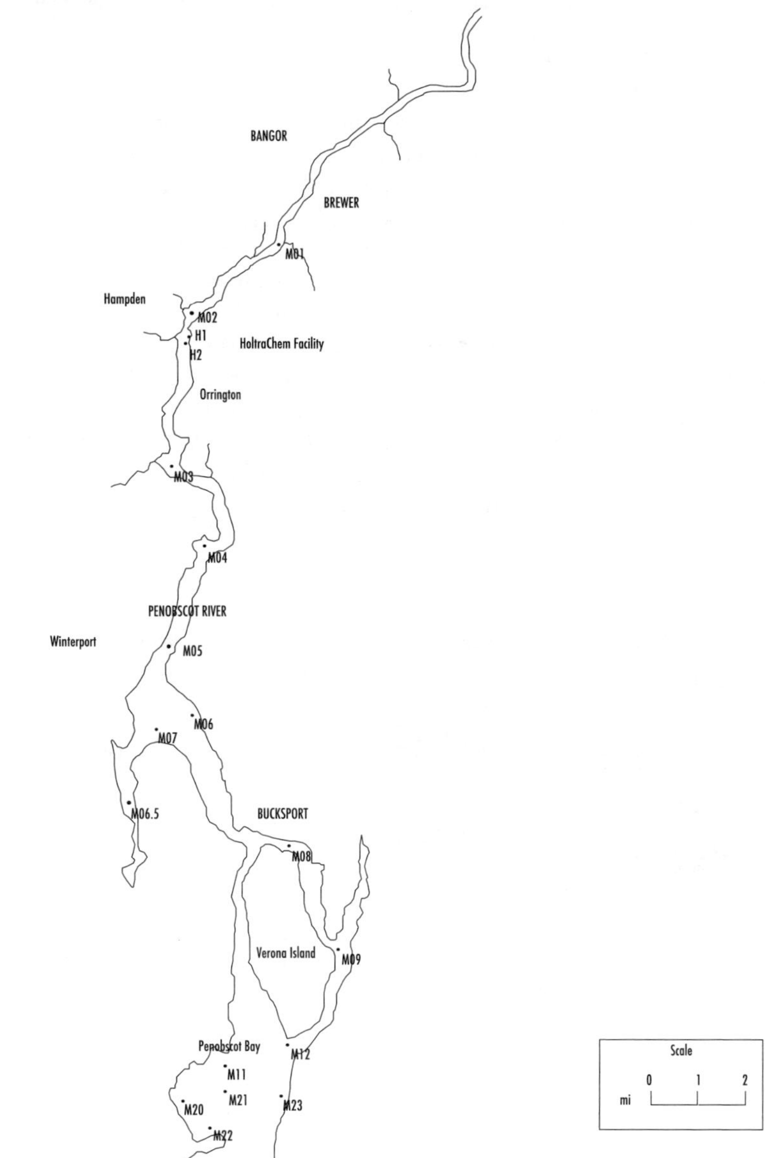

Figure 9.5 Map of the Penobscot River–estuary showing sampling stations for determinations of total mercury and methyl mercury in sediments and mussels.

Based on the lack of mercury data in the Penobscot system, a study was carried out (Livingston, 2001) to answer the following questions:

1. What is the distribution of total mercury and methyl mercury in the Penobscot system?
2. Is there any evidence that discharges from the plant site are affecting the distribution of total mercury and methyl mercury in Penobscot River–estuarine sediments?
3. Is there any evidence that total mercury and methyl mercury in the sediments are contributing to biological concentration of mercury by estuarine organisms?
4. If the chloralkali plant is contributing to the mercury distribution in Penobscot river-–estuarine receiving areas and if such mercury is being methylated in the sediments in concentrations high enough to be deemed ecologically significant, can a restoration program be established to (a) reduce

mercury loading to the river and (b) remove mercury-contaminated sediments that exceed the criteria of state and federal regulatory agencies?

This study was thus limited in scope to a determination of mercury concentrations in sediments and organisms in the Penobscot system (river and bay), and to evaluate the role of the chloralkali plant discharges of mercury on the distribution of mercury in the system.

Methods used in the study are given in Appendix I. After habitat stratification, sampling stations were established along the Penobscot River–estuary (Figure 9.5). Stations were located above and below the plant. Sediment analyses for particle size distribution and sediment mercury/methyl mercury were carried out at the fixed stations. Results of these analyses are shown in Figure 9.6. Sediments at the plant site (H01, H02) were <1 cm deep and were difficult to sample. The substrate here consisted of a thin veneer of sediment on top of a rocky base. Stations M01, M03, M06, and M09 had relatively high percentages of sand whereas all of the other stations were dominated by the silt fractions (Figure 9.6). Clay distribution tended to increase downriver. Bay stations had the highest clay concentrations. Sediment at upper bay stations (M11, M12, M20, M21, M22, M23; Figure 9.5) had relatively high silt concentrations. These areas were clearly depositional. Percent organics was highest at station M06; this station was characterized by variable and occasionally high concentrations of sawdust, a remnant of past logging activities that included log drives and wood processing. Stations with the

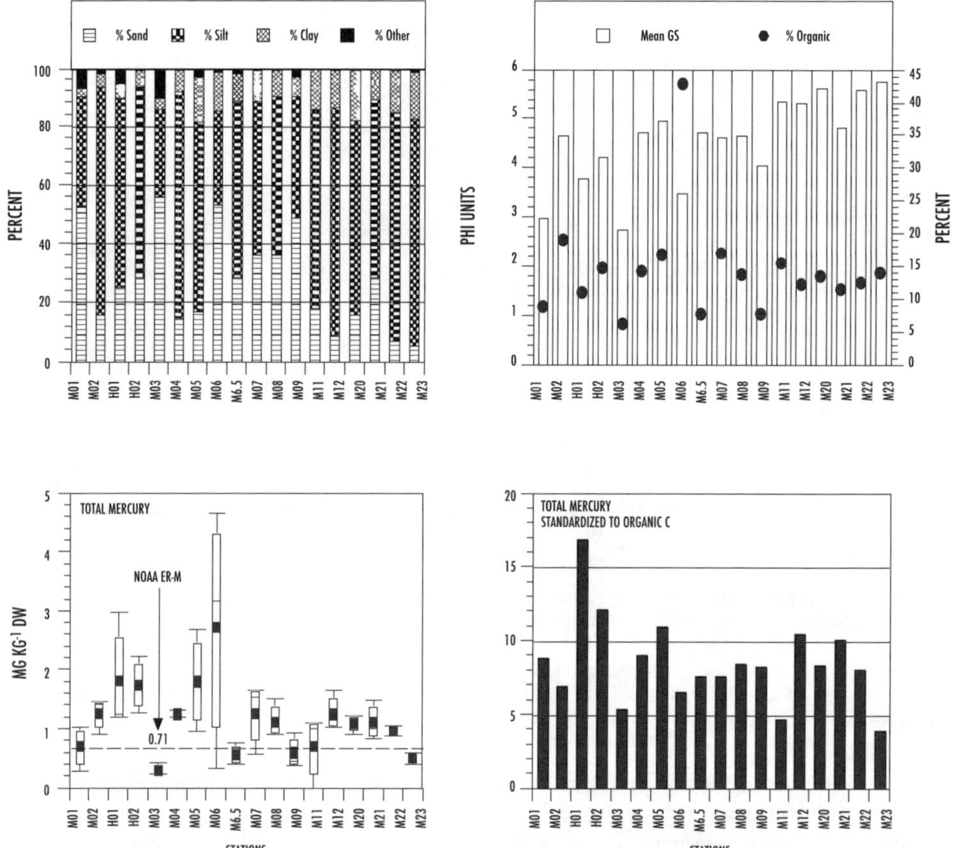

Figure 9.6 Percent sand/silt/clay, mean grain size (phi units), and percent organics in sediments taken in the Penobscot River–estuary during 18–19 October 1999. Also shown are total mercury concentrations in sediment and mean total mercury concentrations standardized to the organic carbon content of the sediments. Total mercury in sediments (mg kg^{-1} dw) is expressed as means with the 10th, 25th, 75th, and 90th percentiles of the replicate samples.

lowest percent organics and the largest particle size distributions included three of the four sandy sites listed above. Bay stations had the most uniformly high phi distributions (i.e., finest sediment particles).

Variance among total mercury and methyl mercury replicates was generally low (Figure 9.6). Methyl mercury sediment concentrations were highest at stations M01, M02, H01, H02, M06, and M07. When taken as means of the data, methyl mercury peaked at station M06 (mean methyl mercury, 0.065 mg kg^{-1} dw; high methyl mercury, 0.12 mg kg^{-1} dw). High concentrations were associated with the high levels of organic matter at this station (42.8%). Mean methyl mercury concentrations were also high between stations M02 (mean methyl mercury, 0.030 mg kg^{-1} dw). and H02 (mean methyl mercury, 0.033 mg kg^{-1} dw). The highest total mercury concentrations were noted at station M06 (4.69 and 3.19 mg kg^{-1} dw). The high variance of total mercury concentrations at this station probably reflects differences in organic deposition. High concentrations of total mercury were found at stations M06, H01, H02, and M05 (Figure 9.6).

Except for the high total sediment mercury concentration at the plant site (460 mg kg^{-1} dw; Maine Department of Environmental Protection, 1998), these data are comparable to previous sediment analyses for mercury in the Penobscot River–estuary. When the data were standardized to sediment organic carbon content (Figure 9.6), the highest standardized values for total mercury were noted at the plant site with generally decreasing values in both directions (upriver and downriver). These results confirmed previous analyses that indicated that the chloralkali plant was the primary source of mercury in the sediments of the Penobscot River–estuary (Sowles, 1997a,b, 1999; Maine Department of Environmental Protection, 1998). In addition, there was a trend of increased standardized mercury values in depositional areas in upper Penobscot Bay (Figure 9.6). The high sediment mercury concentrations at the plant site found in previous sediment mercury surveys of the Penobscot system were not evident in this survey. It is likely that previous sediment mercury was redistributed by turbulent transport due to freshwater flows and tidal action. Total mercury concentrations in mussels (*Mytilus edulis*) taken at station 23 in Penobscot Bay averaged 0.56 mg kg^{-1} dry weight and methyl mercury concentrations in mussels averaged 0.23 mg kg^{-1} dry weight.

Previous average sediment mercury concentrations in the upper Penobscot River–estuary were 17.6 mg kg^{-1} dw (Maine Department of Environmental Protection, 1998). These concentrations represented "the highest levels of mercury in the state and possibly in the country" (Maine Department of Environmental Protection, 1998). Although somewhat lower than those in previous studies, total mercury concentrations in sediments of the Penobscot River–estuary in the current survey (range: 0.25 to 4.69 mg kg^{-1} dw) were generally orders of magnitude higher than concentrations found in sediments from reference areas in Maine (range: nondetectable to 0.14 mg kg^{-1} dw) and unpolluted areas such as Apalachee Bay in the Gulf of Mexico (range: nondetectable to 0.10 mg kg^{-1} dw; Livingston, unpublished data). The concentrations of mercury in the sediments and animals of the Sudbury River system were comparable to those found in the Penobscot system (Livingston, 2001; Zeeman, 2001). These analyses were in agreement that the Penobscot system was seriously contaminated with mercury. Mower et al. (1997) stated that such concentrations "far (100×) exceed the Effects Range-Median (0.71 μg/g dw) suggesting a greater than 75% chance of causing an adverse biological effect." This criterion has been issued by comprehensive studies performed by the National Oceanic and Atmospheric Administration (NOAA).

A literature review was carried out concerning mercury concentrations in aquatic biota from a series of river–estuaries in Maine, including the Penobscot system:

1. Blue mussel (*M. edulis*) tissue mercury concentrations in the Penobscot system were comparable to Mussel Watch data (National Oceanic and Atmospheric Administration) as reported by the Environmental Working Group (Houlihan and Wiles, 2001). The Penobscot blue mussel mercury concentrations were among the highest in the nation.
2. Mercury concentrations in lobster (*Homarus americanus*) hepatopancreas from Penobscot Bay were higher than any mercury concentrations found in lobsters from a series of Maine estuaries in a 1996 analysis (Zeeman, 2001).

3. Average whole-body mercury concentrations of eels (*Anguilla anguilla*) at locations near the chloralkali facility averaged 5 mg kg^{-1} dw, whereas concentrations in fish from an upstream tributary (Kenduskeag River) averaged 0.3 mg kg^{-1} dw (Zeeman, 2001).
4. Zeeman (2001) reported that cormorants from the Penobscot system had higher blood and feather mercury concentrations than mercury in birds taken from eight other Maine estuaries.

The scattered data concerning the concentrations of mercury indicated that the high sediment burdens of mercury in the Penobscot system were getting into the river–estuarine food webs through direct concentration (mussels) and food web biomagnification (cormorants). The data also provided direct proof that the high sediment mercury concentrations in the Penobscot system were being translated into extremely high mercury concentrations in food webs utilized by humans. However, the patch-quilt research effort was inadequate in terms of both the overall scientific complexity of the issue and the potential seriousness of the mercury contamination problem.

A series of additional (though limited) studies was carried out in the Penobscot system. The data were scattered and unrelated to any central focus. A Triad Study of the Penobscot system (Menzie-Cura and Associates, Inc., 2001) indicated the following:

1. Topminnows (*Fundulus heteroclitus*) associated with runoff from Penobscot wetlands had higher mercury tissue concentrations than fish taken in other parts of the system.
2. Sediments taken from areas draining Penobscot wetlands exceeded NOAA effects range-medium (ER-M) criteria that defined a high likelihood of adverse biological effects.
3. Sediments taken from Penobscot areas subject to runoff from wetlands caused significant reductions of growth in test organisms in 100% of chronic bioassays.
4. Qualitative field reviews indicated substantial reductions of wildlife in relatively extensive parts of the Penobscot wetlands.
5. One of two surveys of osprey fledglings in the Penobscot system indicated success rates (numbers of fledglings per active nest) that were below those considered necessary for maintenance of the population.

State and federal agencies determined the criteria for a restoration of the Penobscot systems. The actual data on which restoration projections were based were inadequate at best. For example, despite the above indications of problems with mercury contamination,

1. There has never been a comprehensive analysis of mercury in sediments and animal tissues taken from the Penobscot system.
2. No mercury bioavailability studies have been performed in the Penobscot system.
3. No comprehensive field studies or food web analyses have been undertaken in the Penobscot system.
4. Relatively few sediment toxicity tests have been run with samples from the Penobscot system.
5. There have been no comprehensive benthic community analyses in the Penobscot River–estuary.
6. There has been no comprehensive wildlife evaluation program undertaken in the Penobscot system.
7. There has been no attempt to determine mercury tissue burdens of people who eat contaminated seafood taken from the Penobscot system.

Despite the almost complete lack of scientific data on the Penobscot River–estuary, state and federal regulatory agencies (Maine Department of Environmental Protection, U.S. Environmental Protection Agency) proposed a restoration program whereby sediments exceeding 10.7 mg kg^{-1} dw would be removed from the Penobscot system. This would eliminate remediation everywhere in the Penobscot system with the possible exception of areas immediately adjacent to the chloralkali plant. Aquatic food web considerations and potential human health problems would be simply ignored if such a program were undertaken.

A reasonable restoration program should be based on sound scientific data. Such a program should also be consistent with realistic research objectives, such as (1) to determine answers for questions concerning the potential adverse ecological impact of the mercury contamination on the Penobscot system; (2) to determine the potential for human health concerns regarding the distribution of mercury in Penobscot food webs; and (3) based on the scientific data, to determine if a restoration program such as removal of contaminated sediments would be feasible and, if so, how such a program would be carried out. A mercury restoration program should include the following:

1. Detailed characterizations of fate/transport processes controlling mercury distribution and impacts on sediments and biota of the Penobscot system
2. Determination of the role of Penobscot wetlands in the fate, transport, and bioavailability of mercury in the Penobscot River–estuary
3. Development of a sampling plan with appropriate protocols that would fully characterize current toxic components in sediments of the Penobscot River–estuary
4. Assessment of the bioavailability of mercury in the Penobscot system
5. Development of a comprehensive Triad approach (sediment mercury analyses, benthic community determinations, bioassay experiments with Penobscot sediments) for impact evaluation
6. Analysis of the accumulation of mercury in organisms related to human food webs
7. Evaluation of the feasibility of a mercury restoration program for the Penobscot system

In July 2002, in a lawsuit filed by the Maine's People's Alliance and the Natural Resources Defense Council, Inc. against the owners of the chloralkali plant in the U.S. District Court, Judge Gene Carter filed the following Memorandum of Decision and Order:

> "After hearing the testimony at trial and carefully reviewing the exhibits admitted in evidence, the Court concludes that the methylmercury downriver of the plant, resulting, in part, from Mallinckrodt's actions at the plant site, may present an imminent and substantial endangerment to public health and the environment … The Court found the testimony of Dr. Robert Livingston, Plaintiff's aquatic biologist, particularly credible and persuasive. The evidence clearly demonstrated that the Penobscot River is contaminated with mercury through the mouth of the river and into the Bay … Of particular importance to the Court's conclusion is the data concerning the Frankfort Flats area, which receives drainage from a marsh system and may be the principal area where methylmercury is entering the Penobscot system. Frankfort Flats data showed the highest sediment mercury concentrations, sediment toxicity to benthos, and the highest concentrations of mercury in killifish….Birds that occupy higher trophic levels in the food web feed in the marsh area near Frankfort Flats and the limited survey results demonstrated impaired reproduction of these birds … Downriver areas where mercury is methylating, such as marshes, will continue to supply methylmercury to the lower river unaffected by remediation at the site … Consequently, the methylmercury continually accumulating and biomagnifying in the food web creates a reasonable medical concern for public health and a reasonable scientific concern for the environment downriver of the plant site. The Court will, therefore, order the Defendant Mallinckrodt be responsible for the cost of undertaking a scientific study of mercury contamination downriver of the plant site in the Penobscot River."

The fact that mercury discharges to the Penobscot system were actually permitted by state and local regulatory agencies for decades without even a cursory scientific review of the consequences is not reasonable. The results of the Sudbury Study, an EPA project in a nearby state, have been largely ignored by federal officials who were involved with the Penobscot system. By avoiding the restoration issue, the original error is compounded. The connection of the Penobscot mercury situation with the Apalachee nutrient loading issues is clear. Adequate scientific information is not being generated to solve important environmental issues. There is a general lack of knowledge of fundamental food web issues that are important in the evaluation of the loading of nutrients and toxic wastes into river–estuarine systems.

CHAPTER 10

Trophic Organization and Resource Management

I. INTRODUCTION

The use of food web ecology studied within the context of the historical concept of the ecosystem has not been widely used in planning and management activities in coastal areas. In fact, during the past 30 to 40 years, the story of Florida's aquatic resources has been dismal, with the debilitation or even loss of most of the the primary waterways and river–estuarine systems being the rule. The Kissimmee–Okeechobee–Florida Everglades–Florida Bay–Florida Keys/Coral reef system, the Miami River–Biscayne Bay area, the Indian River system, the St. Johns drainage system, Naples Bay, the Hillsborough River–Tampa Bay area, Sarasota Bay, Choctawhatchee Bay, the Pensacola Bay system, and Perdido Bay are all damaged systems. The loss of productivity in areas such as the Chesapeake Bay system, the New York Bight and New Jersey coast, Penobscot Bay, the Delaware River system, and various other river–estuarine areas throughout the United States has followed a similar pattern. However, not all is lost and there are success stories of progressive planning and management programs that have protected valuable and productive coastal resources. One such case involves the saving of the Apalachicola River and Bay system in north Florida.

The Big Bend area of north Florida (Figure 2.1) is sparsely populated compared to the East and West Coasts of the United States. The Apalachee Bay system represents one of the least polluted such areas in the country, and, with the exception of a pulp mill, is in relatively pristine condition. Likewise, the Apalachicola drainage basin is largely unpopulated with very little in the way of urbanization, agricultural activities, and industrial development. However, in western parts of the Choctawhatchee Bay area, in the Pensacola Bay drainages, and in the Perdido system there have been recent increases in urbanization with accompanying agricultural activity to the north and some industrialization. Planning and management for preservation of aquatic resources in these systems has been virtually non-existent despite the propaganda put out by state and federal agencies and the local news media. Accordingly, the deterioration of these systems has been rapid with losses of habitat and aquatic resources. With the exception of the Perdido drainage system, there has been virtually no ecosystem-level research activity in these areas. Such research has been discouraged by local and state economic interests that control the limited regulation of water quality in the coastal drainages.

During the 1970s and early to mid-1980s, there was some concern for the impacts associated with various forms of human economic development in the Apalachicola drainage system. Most of this concern was focused on the Apalachicola aquatic resources because there were, as yet, no entrenched development interests, and the fisheries in this region provided an economic incentive for good planning. In addition, the political atmosphere at various levels of local, state, and federal government was amenable to both a prolonged and extensive research effort and a series of cooperative interactions that led to the protection of the Apalachicola basin. These intersecting interests included university researchers, local and state politicians, government personnel, and federal agencies that supported a wide range of progressive actions. The public was well informed through various news media with a high level of environmental reporting. There was interest in environmental events by the general public that supported the management initiatives taken in the Apalachicola system. The story of the Apalachicola history over the past 30 to 40 years is instructive in terms of what can be done to preserve natural coastal areas, and how realistic resource planning and management initiatives can make a difference in preventing the loss of aquatic habitats in highly productive coastal areas.

In this chapter, I will outline the history of such an effort.

II. RESOURCE MANAGEMENT: THE PROMISE AND THE REALITY

The promise of the successful application of research results to resource management has not been achieved in many coastal areas of Florida even though there are a number of so-called restoration efforts that have been widely publicized as successful. Most interests have tacitly accepted pollution, habitat degradation, and overfishing, and the result has been widespread system failure even in areas with relatively low human populations. Discussion concerning the role of science in land-planning initiatives and related resource management is carefully controlled, with relatively little in the way of ecosystem-level efforts to understand environmental problems in coastal areas. There has been even less effort to explain the actual application of scientific methods involved in resource manage programs.

These generalizations are applicable to the state of coastal areas countrywide. However, the role of science in this dilemma has not been addressed. In a recent article on the subject of interdisciplinary research at the ecosystem level, Duarte and Piro (2001) stated the following:

> "Insufficient interdisciplinary cooperation in aquatic sciences is due to 1) a deep crisis in the present scientific model that extends beyond aquatic sciences, and 2) a disparity of views as to what interdisciplinary really means."

The exponential increase of scientific information has not really affected ecosystem research. It has actually been part of a recent trend toward fragmentation of the ecological sciences. Increased compartmentalization of environmental research has led to increased specialization and relatively little exchange of information across disciplines; this is antithetical to the very meaning of the definition of the word *ecosystem*. Again, Duarte and Piro (2001) define the basic issue:

> "Contemporary science has approached this kind of problem by piecing together the portions of information provided by the sub disciplines in the hope that this would lead to the understanding of the whole. Unfortunately, there is mounting evidence of failure of efforts to understand the behavior of complex systems by reduction to the analysis of their constituents. Examples of complex problems defeating reductionism abound in the realm of aquatic science: fisheries science is still unreliable as a basis to manage natural stocks despite the growing sophistication of the available population models."

Graham and Dayton (2002) state this observation in another way:

> "Unfortunately, heightened ecological understanding also builds impediments to future progress. Increased specialization and the parallel evolution of seemingly independent sub disciplines generally compel researchers to become increasingly canalized. Specialization also accelerated the expansion of the ecological literature, making it difficult for researchers to track developments in their own sub discipline, let alone the general field of ecology."

The failure of the adequate development of interdisciplinary aquatic research has various causes. Compartmentalization of study efforts by system type, by region, by discipline, by taxon, and by publication opportunity is increasingly evident. Even the catchword "interdisciplinary" is often applied to patch-quilt efforts to paste together isolated studies into some sort of ecosystem context. The inclusion of sociology, economics, and political science in such efforts remains superfluous even though these disciplines play a critical role in the continuing degradation of coastal systems. Relatively little attention has been given to the quality of data that are used in unverified models and endless speculation regarding what exactly constitutes an ecosystem. Commonly accepted assumptions are used to rationalize the lack of empirical data on which ecosystem models are based. The preoccupation with reductionist control of research support and publication continues to hamper the development of realistic interdisciplinary studies. Journals such as *Limnology and Oceanography*, with notoriously biased review procedures, continue to crank out short-term, low-level articles that are not representative of the systems that they purport to understand. As defined by Duarte and Piro (2001):

> "Frequently … interdisciplinary research is obliged to nest between the cracks of research programs and journal scopes, implying that properly judging its assessments becomes a severe challenge for reviewers and journal editors. The resulting publishing difficulties are often quite damaging for the career development of the committed scientists. Interdisciplinary research is, therefore, penalized by the inadequate scientific mechanisms of funding and evaluation."

Funding of long-term interdisciplinary research is uncertain at best, and politically distributed at worst. The basic commitment of many environmental scientists to such research is lacking. The entire system of research funding and publication is antithetical to the successful completion of even modest efforts of ecosystem analyses. It is within the context of the largely uphill struggle to maintain long-term, interdisciplinary studies that the success of such efforts in actual management of aquatic systems can be assessed. Herein lies a primary reason for the loss of valuable aquatic resources since scientific data represent the basis for any effective management program.

As noted in previous chapters, loading of toxic substances and nutrients into coastal systems is directly related to food web processes. The proliferation of human activities in drainage basins and associated coastal areas continues unabated and largely unregulated around the world. In the United States, the twin activities of urbanization and agriculture remain largely unregulated. Bricker et al. (1999) recently noted that 44 estuaries in the conterminous United States, representing 40% of the total estuarine surface area studied, have shown "high expressions of eutrophic conditions" with another 40 estuaries having some signs of harmful eutrophication. Deteriorated water quality and associated biological effects include decreased light availability, high chlorophyll *a* concentrations, increased decomposition of organic matter, high epiphytic/macroalgal growth rates, and changes in algal dominance (diatoms to flagellates, benthic to pelagic dominance). Commercial fisheries have been impaired in 43 estuaries along the east and Gulf coasts of the United States. Bricker et al. (1999) projected that eutrophic conditions would worsen in 86 estuaries by the year 2020. The authors emphasized that, currently, thresholds of nutrient loading that lead to toxic plankton blooms remain largely undetermined, and the relationships of nitrogen and phosphorus to such blooms need "further clarification."

The impacts of increased nutrient loading on plankton community organization and coastal food webs remains generally unknown, even though there are many examples of such impacts on individual populations and fisheries resources. Often, key signs of hypereutrophication go unnoticed until there is an obvious failure of an important fishery. The assimilative capacity of a given coastal system relative to nutrient loading remains undetermined, thus setting the stage for transformation from natural eutrophication processes to hypereutrophication and food web deterioration. Major programs such as ECOHAB continue to fund isolated and limited research efforts while actually discouraging ecosystem-level efforts.

Nutrient discharges and toxic substance loading from point and non-point sources in coastal areas has been documented in various coastal areas (Botton, 1979; Delfino et al., 1984; Postma, 1985; Fonney and Hua, 1989; Anderson and Garrison, 1997; Kennish, 1997, 2001; Bricker et al., 1999; Livingston, 2000). Loading effects include altered water and sediment quality, acute and chronic toxicity to coastal populations, bioconcentration (concentration of toxins by individual organisms), biomagnification (food web concentration) of toxic agents, altered food web processes, debilitated aquatic populations, and adverse effects on important fisheries. There is considerable scientific evidence that receiving systems are stressed due to multiple effects from spills of raw and inadequately treated sewage (Hutchinson, 1973) and from non-point (urban) storm water (Livingston, 2000). Many of the metals and organic contaminants associated with such input are carried by fine particulates that eventually end up in the sediments. Yet, there are still no standards for sediment quality at the state or federal regulatory levels. Biological reworking of sediments along with wind effects and storms make such compounds available to open-water food webs. Trophic interactions lead to food web biomagnification of toxic agents such as organochlorine pesticides and methyl mercury. The fate of these contaminants depends on the persistence of the parent compounds and their by-products, which can be measured over decades.

Hypereutrophication involves complex trophic alterations that include changes in both bottom-up and top-down feeding complexes. However, food web responses to nutrient loading are system specific, and depend on the assimilative capacity of a given system. Reduced production of natural phytoplankton assemblages is often subtle. Detailed, species-specific phytoplankton work is rarely included in ecosystem studies (Livingston, 2000). However, long-term phytoplankton changes culminating in blooms have led to chronic reduction of useful habitat, adverse impacts on key populations, and the loss of important fisheries (Duxbury, 1975; Turner et al., 1987; Livingston, 2000). Unlike often-localized impacts of toxic wastes, altered plankton assemblages can affect relatively extensive coastal areas. For example, the long-term adverse impacts of sewage disposal in the New York Bight have been well documented (Mahoney et al., 1973; Koditschek and Guyre, 1974; O'Connors and Duedall, 1975; National Oceanic and Atmospheric Administration, 1978; Botton, 1979), and include the dissemination of fish pathogens, increased heavy metal concentrations, plankton blooms, and low dissolved oxygen (DO). Hypoxic events associated with the New York Bight are thought to have periodically affected nearly the entire coast of New Jersey. Even so, the actual effects of both offshore and inshore plankton blooms along the New Jersey coast remain largely unknown (Kennish, 2001).

The long-term history of San Francisco Bay is another illustration of impacts of multiple point and non-point pollution sources on a coastal system (CH2M Hill, 1986). For example, from 1993 to 1999, over 270 million gallons of sewage were released into receiving areas of the bay; the spills were accompanied by relatively high concentrations of ammonia and heavy metals (CH2M Hill, 1986). Fecal coliforms, biochemical oxygen demand (BOD), ammonia, cadmium, copper, and zinc were high due to sewage overflows. Long-term evaluations of these and other cumulative impacts have not been subject to detailed, ecosystem-level studies in San Francisco Bay. These discharges are currently common in fresh and estuarine waters throughout the United States and represent common environmental problems in aquatic systems associated with urbanized areas throughout the world.

III. THE APALACHICOLA EXPERIMENT

By an accident of history, the Apalachicola drainage system (Figure 10.1) has escaped major adverse impacts due to human activities. A sparse human population, together with little industrial and municipal development, has been associated with high water quality of the region (Livingston, 1984a). Water quality in the river and bay in recent times has remained in a relatively natural state (Livingston, 1983a), and habitat in terms of wetlands, benthic macrophytes, and bivalve mollusks has continued to be in good condition (Livingston, 1993, 1994a). The current unpolluted state of the Apalachicola River–Bay system is directly related to a series of planning and management actions that were taken during the 1970s and 1980s in direct response to a long-term program of evaluation of sources of productivity and food web processes in the Apalachicola River–Bay system (Livingston, 1984a). The relationships of the estuarine food webs to natural river fluctuations played a role in planning and management approaches used to protect the Apalachicola resource. The history of this effort has been continuously documented in the scientific literature. Basic management

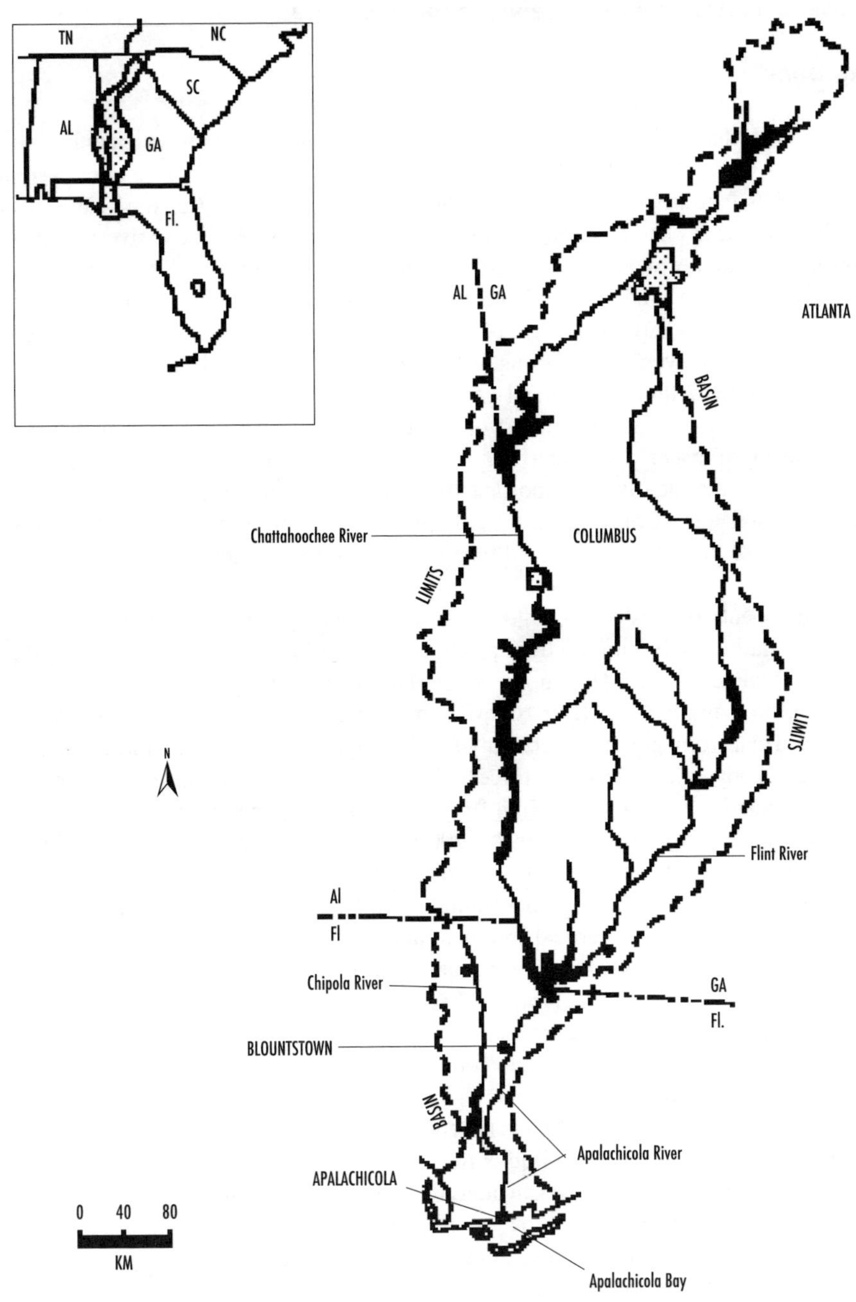

Figure 10.1 The Apalachicola–Chattahoochee–Flint (tri-river) system.

questions have been outlined by Livingston et al. (1974). Subsequent actions, based on a continuous, long-term study of the area from 1972 through 1984, have been documented by Livingston (1975b, 1976b,c, 1977, 1980b, 1982b, 1983a, 1984a, 1985c, 1991b, 2000). Livingston (1991c) reviewed the overall Apalachicola management program.

A. Freshwater Flows and the Apalachicola Resource

1. Background

Florida has more than 1700 streams and rivers (Palmer, 1984), many of which empty into bays, harbors, the Gulf of Mexico, or the Atlantic Ocean (Livingston, 1991a). These river systems have downstream reaches that are affected by ocean tides during at least part of the year. Gulf coastal and peninsular Florida is characterized by Karst topography. Consequently, rivers in these areas receive substantial groundwater inflow, often through seeps and springs. Along the north-central Gulf coast, springs are the predominant source of fresh water in the tidal rivers. McPherson and Hammett (1991) reviewed various aspects of freshwater flows that are specific to Florida. Many of Florida's streams and rivers have downstream reaches that are tidally affected. Flows in tidal reaches result from interactions of channel geometry, ocean tides, wind, density differences, and freshwater inflows. At the downstream end of the tidal reach, seawater mixes with fresh water. Seaward movement of water and waterborne materials is slowed within the tidal reach. This freshwater–saltwater interface is an important focal point for various factors related to primary productivity in coastal areas. Consequently, highly productive tidal rivers act as nurseries for marine organisms and as reservoirs for mixing and transporting the fresh water and nutrients that sustain coastal productivity.

Several components of freshwater inflow to Gulf coastal systems contribute to the estuarine condition: (1) rainfall directly into the tidal reach; (2) basin runoff as river flows; and (3) groundwater seepages (McPherson and Hammett, 1991). The relative magnitudes of these components vary in different drainage basins. River runoff is maximal in coastal areas of the northern Gulf coast of Florida. Tidal rivers are often vertically stratified with the degree of stratification reflecting a balance between buoyant and turbulent forces. Highly stratified conditions exist when freshwater inflow is large in relation to tidal flow. High freshwater inflows generally increase buoyant forces that enhance stratification (Livingston, 1991c, 2000). Turbulent forces of wind and tide increase mixing. At the boundary between the salt and fresh waters is a zone of turbulent mixing that has a net upward circulation (McPherson and Hammett, 1991). Fronts occur in tidal zones due to convergent surface flows, strong vertical motions, and high salinity/density gradients. Fronts are important in governing the transport of fresh water and waterborne materials to the sea and also represent important areas of phytoplankton and bacterial activity and abundance (Livingston et al., 2000). Fronts are often a result of large freshwater inflows, but also may be caused by the interaction of tidal flow with river bathymetry. Channel geometry often affects stratification features of a given estuary, and is important in determination of the location of estuarine fronts. In barrier island estuaries, dredged channels to the Gulf have accentuated the stratification phenomenon, leading to losses of habitat (low DO, generation of liquid mud) and biological productivity (Livingston, 2000).

Human activities, in the form of physical alterations, have significantly altered channel geometry as well as the volume of freshwater inflow entering many tidal reaches in Florida. Stream flow diversions for municipal, agricultural, and industrial uses are also common. The flow in many of Florida's rivers is regulated by control structures. Channel dredging and shoreline construction, by altering the geometric characteristics of a given tidal river, can change the natural productivity of such a system by altering physical and biological processes related to freshwater influxes. Other land-use changes associated with urbanization have increased inflows and shortened periods of runoff. The use of coastal systems as receiving bodies for wastewater adds to biological stress in addition to the increasing tendency for anthropogenous reductions of freshwater flows to the coastal system by direct withdrawals, by construction of canals and control structures, and by groundwater withdrawals. Often, the timing, location, and quality of urban runoff is affected in addition to water quality, which often has much higher loads and concentrations of nutrients and toxic substances. Reduced freshwater inflow can result in increased salinity, associated vegetative changes in wetlands, increased residence time of water and waterborne materials, associated accumulations of nutrients and/or toxic materials in water or sediments, and enhanced conditions for damaging plankton blooms (Livingston, 2000).

There is considerable natural variability of freshwater fluxes in the various sub-basins of the north Florida Gulf region (Livingston, 1984a, 1989). Most of the river systems are composed of a series of tributaries that drain a series of sub-basins into the main stem. The sub-basins often follow hydrological cycles (Livingston et al., 1989, 1990) different from those of the main stem, depending on regional rainfall patterns and the highly variable physiographic conditions in the different regions. Some drainage basins extend into Alabama and Georgia, and the temporal patterns and volumes of river discharge to associated estuaries depend on the cumulative climatological conditions and seasonally varying evapotranspiration rates in associated watersheds. Flooding plays a role during winter and early spring months in the north Florida drainages. Winter–spring flooding in the alluvial streams is a product of reduced evapotranspiration rates during winter rainfall peaks along the heavily vegetated drainage basins (Livingston, 1984a). Reduced summer surface flows are due to increased evapotranspiration in vegetated floodplains. Surface flows are minimal during the fall drought period. Alluvial systems such as the Perdido, Apalachicola, Escambia, and Choctawhatchee have flow patterns that are thus more correlated with seasonal rainfall patterns in Georgia and Alabama (where high percentages of the respective drainage basins are located) than in Florida. The smaller streams along Apalachee Bay are associated with relatively limited estuarine areas, whereas the alluvial river-dominated estuaries to the west are proportionately larger due to higher flow rates. In the case of the Apalachicola system, salinities can be affected by river flow up to 240 km offshore (Livingston, 1984a).

2. *The Apalachicola Drainage System*

In any management program, the resources to be protected should be scientifically defined (Livingston and Joyce, 1977). The Apalachicola River–Bay system is part of a drainage area (the Chattahoochee–Flint–Apalachicola basin; Figure 10.1) of about 48,500 km^2. This system is located in western Georgia, southeastern Alabama, and northern Florida. The Apalachicola River flows 171 km from the confluence of the Chattahoochee and Flint Rivers (the Jim Woodruff Dam) to its terminus in the Apalachicola estuary. The Apalachicola River is 21st in flow magnitude in the conterminous United States, with mean flow rates of ~690 m^3 s^{-1} (1958 through 1980). Annual high flows average 3000 m^3 s^{-1} (Leitman et al., 1982, 1991). The forested floodplain (about 450 km^2) is the largest in Florida (Leitman et al., 1982, 1983). In recent times, much of the floodplain has been owned by timber interests as forestry has been the primary land use (Clewell, 1977). Other activities include minor agricultural and residential use, tupelo honey production, hunting, and sport/commercial fishing (Livingston, 1984a).

The Apalachicola estuary approximates 62,879 ha and is a shallow (mean depth: 2.6 m) lagoon-and-barrier-island complex. The estuary is oriented along an east–west axis, and water movement is controlled by wind currents and tides as a function of the generally shallow depths (Livingston 1984a, Livingston et al., 1999, 2000). Upland marshes grade into soft-sediment areas, fringing sea grass beds, oyster reefs, and a series of passes that are control points for the salinity structure of the system (Livingston, 1984a). The Apalachicola River dominates the bay system as a source of fresh water, nutrients, and organic matter (Livingston, 1984); together with local rainfall, the river is closely associated with the salinity and coastal productivity of the region (Livingston, 1983a, 1984a; Livingston et al., 1997, 1999, 2000).

The problem of reduced flow to river–estuarine systems is of global concern. However, there are very few long-term, ecosystem-level studies that have been designed to answer questions concerning the effects of reduced freshwater flows due to human activities. Riverine development has continued in a largely unregulated environment. This is particularly true in the United States where the U.S. Congress has had a long-standing policy of development of river systems through public works projects executed by the U.S. Army Corps of Engineers. The primary and secondary environmental effects of such development have not been well documented scientifically as part of the devolution of natural river–estuarine systems to human-made ditches.

The Apalachicola system thus represents an unusually important example of a natural alluvial river basin that has remained relatively free of human impacts. Over the past 32 years, the Florida State University Aquatic Research Group (Dr. Robert J. Livingston, Director) has carried out a continuous analysis of the Apalachicola drainage system with most of the primary field work carried out between 1971 and 1984. In recent years, it has become apparent that sources of freshwater flows (i.e., the Chattahoochee and Flint Rivers) to the Apalachicola River–Bay system will be under increasing pressure from municipal and agricultural development, particularly in Georgia.

3. *The Apalachicola Floodplain*

The importance of freshwater flows to the Apalachicola floodplain has been extensively studied. As a consequence of this effort, various aspects of the Apalachicola upland drainage system have been found to be critical to the overall resource management approach for the basin (Livingston and Joyce, 1977):

1. The Apalachicola River is the only river in Florida to go from the Piedmont to the Gulf of Mexico. The Apalachicola drainage basin receives biotic exchanges from the Piedmont, the Atlantic Coastal Plain, the Gulf Coastal Plain, and peninsular Florida. This accounts for the high quality of the terrestrial animal biota of the river floodplain (Means, 1977).
2. The floodplain forests include numerous terrestrial plant species, of which 9 are narrowly endemic, 28 endangered, 17 threatened, and 30 rare (Clewell, 1977).
3. Of all the drainages in north Florida, the Apalachicola River contains the largest number of freshwater bivalve and gastropod mollusks, with high endemism and six rare and endangered species (Heard, 1977).
4. The 86 fish species that have been noted in the Apalachicola River system include three endemics, various important anadromous species, and species that form the basis for important sports and commercial fisheries (Yerger, 1977).
5. The Apalachicola River wetlands remain a center of endemism for various terrestrial species, which includes numerous endangered, threatened, and rare species of amphibians, reptiles, and birds (Means, 1977). Because of the high diversity of wetland and upland habitats, the highest density of amphibians and reptiles in North America (north of Mexico) occurs in the upper Apalachicola basin.

4. *River–Bay Linkages*

a. *Freshwater Wetlands and Receiving Areas*

Linkage between upland freshwater wetlands and the estuarine biota of associated estuaries has been the subject of considerable debate and research in the past (Livingston and Loucks, 1978). The association between freshwater input and estuarine productivity has been indirectly established in a number of river-dominated estuaries (Cross and Williams, 1981). Using data from 64 estuaries in the Gulf of Mexico, Deegan et al. (1986) found that freshwater input was highly correlated ($r = 0.98$) with fishery harvest. Armstrong (1982) found that nutrient budgets in Texas Gulf estuaries were dominated by freshwater inflows. Shellfish and finfish production was a function of nutrient loading rates and average salinity. Funicelli (1984) showed that upland carbon input was in some way associated with estuarine productivity. However, few studies actually evaluated the various facets of the linkage of the freshwater river–wetlands and estuarine productivity (Livingston, 1981a). It is important to understand these connections at the planning stage of any ecosystem operation.

Studies were made concerning the distribution of wetland vegetation in the Apalachicola floodplain (Leitman et al., 1982). Vegetation type was associated with water depth, duration of inundation and saturation, and water-level fluctuation. Stage range is reduced considerably downstream, which indicates a dampening of the river flood stage by the expanding wetlands. Litter fall in the floodplain (800 g m^{-2}) was higher in the Apalachicola wetlands than that noted in many tropical systems and almost all warm temperate systems, which range on the order of 386 to 600 g m^{-2} (Elder and Cairns, 1982). The annual deposition of litter fall in the bottomland hardwood forests of the Apalachicola River floodplain approximates 360,000 metric tons (mt). Seasonal flooding provides the mechanism for mobilization, decomposition, and transfer of the nutrients and detritus from the wetlands to associated aquatic areas (Cairns, 1981; Elder and Cairns, 1982) with a postulated although unknown input from groundwater sources. Of the 214,000 mt of carbon, 21,400 mt of nitrogen, and 1,650 metric tons of phosphorus that are delivered to the estuary over the period of a given year, over half such loading is transferred during the winter–spring flood peaks (Mattraw and Elder, 1982). Reductions of peak Apalachicola River flow rates due to anthropogenous use of fresh water in the Chattahoochee and Flint Rivers would thus eventually threaten and destroy the natural biota (terrestrial and aquatic) that depend on such exchanges.

Various studies (Livingston, 1984a; Chanton and Lewis, 1999, 2002) have noted that delivery of nutrients and dissolved/particulate organic matter are important to the maintenance of the estuarine primary production (autochthonous and allochthonous). There are distinct links between the estuarine food webs and freshwater discharges (Livingston and Loucks, 1978; Livingston, 1981a, 1984a; Livingston et al., 1997). The total particulate organic carbon delivered to the estuary follows seasonal and interannual fluctuations that are closely associated with river flow (Livingston, 1991a; $r^2 = 0.738$). The exact timing and degree of peak river flows relative to seasonal changes in the wetland productivity are important determinants of short-term fluctuations and long-term trends of the input of allochthonous detritus to the estuary (Livingston, 1981a). During summer and fall months, there are no direct correlations of river flow and detrital movement into the bay. By winter, there is a significant relationship between loading microdetritus and river flow peaks. The temporal sequence of such peaks appears to be important because spring river peaks need to be higher than winter peaks to have comparable levels of detrital loading. Alterations of river flooding patterns due to reductions of river flow would thus adversely influence loading of particulate organic carbon to the Apalachicola estuary.

Early field analyses (Livingston et al., 1974, 1976a, 1981a; Livingston and Duncan, 1979) indicated that, in addition to providing the particulate organics that fueled the bay system, river input determined nutrient loading to the estuary, and was a primary factor that controlled phytoplankton production in the bay. As much as 50% of the estuarine phytoplankton productivity, which is the most important single source in overall magnitude of organic C to the bay system, was explained by Apalachicola River flow (Myers, 1977; Myers and Iverson, 1977). Boynton et al. (1982) reported that the Apalachicola system had high phytoplankton productivity relative to other river-dominated estuaries, embayments, lagoons, and fjords around the world. Nixon (1988a,b) stated that the Apalachicola Bay system ranks high in overall primary production compared to other such systems. Wind action in the shallow Apalachicola Bay system has been associated with periodic peaks of phytoplankton production as inorganic nutrients, regenerated in the sediments, were mixed through turbulence into the euphotic zone (Livingston et al., 1974). Nitrogen was limiting during summer periods of moderate to high salinity in the Apalachicola estuary. Light and temperature limitation was highest during winter–spring periods, thus limiting primary production during this time. Nitrogen input to primary production was limited by relatively high flushing rates in the Apalachicola system. Flow rates affected the development of nutrient limitation in the Apalachicola system, with nutrient limitation highest during low-flow summer periods. Recent studies with isotopes (Chanton and Lewis, 2002) noted that both river floodplain detritus and dissolved nutrient-based *in situ* (phytoplankton) productivity were important factors in maintenance of noted high estuarine secondary production. Apalachicola River flow was noted as critical to bay secondary productivity in terms of peak flows (providing allochthonous detritus from river floodplains) and low flows (nutrients for autochthonous phytoplankton productivity).

As a response to the projections of increased anthropogenous freshwater use by the State of Georgia over the next 50 to 100 years (Livingston, 1988), a program was initiated to analyze, quantitatively, the Apalachicola databases generated during the 1970s and 1980s (Livingston et al., 1997, 1999). One of the primary questions of this effort concerns the relationships of reduced Apalachicola River flows to food web relationships and primary/secondary productivity in the Apalachicola River–Bay system.

b. Impacts of Reduced River Flows

Drought/flood periodicity of river input is an important factor in the overall pattern of eutrophication in coastal systems along the NE Gulf coast (Livingston, 2000). The relationships of flow rates, nutrient loading, and the resulting nutrient concentrations appear to be important in the initiation and sustenance of plankton blooms in other (hypereutrophic) river-dominated estuaries (Livingston, 2000). Reduction of Apalachicola River flow, together with nutrient loading from increased land development in the area, could eventually lead to the deterioration of bay productivity due to nutrient-induced plankton blooms (Livingston, 2000). Agricultural and municipal interests along the tri-river system have exerted increasing pressure on the freshwater resources of the tri-river basin during the 1990s (Leitman et al., 1991; Livingson, 2000). Although the reservoirs on this system have not altered the annual flow rates of the Apalachicola River (Meeter et al., 1979), there have been new proposals to reallocate water in the Lake Lanier storage from hydropower to water supply for the rapidly growing Atlanta, Georgia metropolitan area (Figure 10.1). This represents about half of the stored water that is used to augment downstream flows (Leitman et al., 1991). In addition, models indicate that agricultural use of water in the tri-river system will eventually lead to serious depletion of freshwater input to the Apalachicola River from the Georgia area. These changes have been projected to lead to serious problems in the maintenance of Apalachicola Bay productivity (Livingston, 1988).

Apalachicola Bay ecology is closely associated with freshwater input from the Apalachicola River (Livingston, 1984a). A comparison of sciaenid fish collections per unit area along the NE Gulf coast showed that fish standing crop in the Apalachicola estuary was an order of magnitude higher than in other estuaries in the region (Livingston, 1981b). The salinity of the estuary is controlled to a considerable degree by the Apalachicola River (Livingston, 1984a). Seasonal variability of bay salinity is dependent on river flow that, in turn, follows precipitation cycles in Georgia and evapotranspiration rates along the tri-river floodplain (Meeter et al., 1979). Any changes in important habitat variables such as salinity are viewed as serious problems with respect to maintenance of the fisheries of the bay. The distribution of epibenthic organisms in the Apalachicola estuary follows a specific spatial relationship to high river flows. Stations most affected by the river are inhabited by anchovies (*Anchoa mitchilli*), spot (*Leiostomus xanthurus*), Atlantic croaker (*Micropogonias undulatus*), gulf menhaden (*Brevoortia patronus),* white shrimp (*Penaeus setiferus*), and blue crabs (*Callinectes sapidus*). The outer bay stations are often dominated by species such as silver trout (*Bairdiella chrysoura*), pigfish (*Orthopristis chrysoptera*), least squid (*Lolliguncula brevis*), pink shrimp (*P. duorarum*), brown shrimp (*P. aztecus*), and other shrimp species (*Trachypenaeus constrictus*). Sikes Cut, an artificial opening to the Gulf that is maintained by the U.S. Army Corps of Engineers (see Figure 2.2D), is characterized by salinities that resemble the open Gulf. This area is dominated by species such as least squid, anchovies, *Cynoscion arenarius*, *Etropus crossotus*, *Portunus gibbesi*, and *Acetes americanus*.

River flow and rainfall follow a sine curve with peaks of river flow highly correlated with reductions in salinity (Livingston, 1984a). The positive association of important estuarine water quality characteristics and upland rainfall patterns has serious implications for the management of the Apalachicola resources. Cross-correlation analyses indicated that numbers of species of fishes are positively associated with peak river flows. However, there is an inverse relationship between species richness and numerical abundance of infaunal macroinvertebrates (Livingston, 1987b) and litter-associated organisms (Livingston, 1984a). The long-term (14-year) trends of the distribution of invertebrates such as penaeid shrimp indicate that such numbers are associated in various ways with river flow. Fish numbers peak 1 month after river flow peaks, whereas invertebrate numbers are inversely related to peak river conditions with increases during the summer months (Livingston, 1991c). These data are understandable in that top fish dominants such as spot are prevalent in winter–spring months of river flooding, whereas peak numbers of penaeid shrimp usually occur in summer and fall months. Other top dominants such as anchovies reach numerical peaks 3 months before the Apalachicola River floods. Fish biomass has a significant positive correlation with river flow at lags 2 and 3, whereas invertebrate biomass showed a significant positive correlation with river flow peaks at lag 4 (Livingston, 1991c). Cross-correlation analysis also indicates that the various estuarine populations follow a broad spectrum of diverse phase interactions with river flow and associated changes in salinity. River flow, as a habitat variable, is thus a controlling factor for biological organization of the Apalachicola estuary.

c. *Oysters as Sentinels*

Research on the extensive Apalachicola oyster reefs dates to the work of Swift (1896) and Danglade (1917). The Apalachicola estuary accounts for about 90% of Florida's commercial fishery (Whitfield and Beaumariage, 1977). Conditions in the Apalachicola Bay system are highly advantageous for oyster propagation and growth (Menzel and Nichy, 1958; Menzel et al., 1966; Livingston, 1984a; Livingston et al., 1999, 2000) with reefs covering about 7% (4350 ha) of bay bottom (Livingston, 1984a). Mass spawning takes place at temperatures between 26.5 and 28°C, usually from late March through October. Growth rates of oysters in this region are among the most rapid of those recorded with harvestable oysters taken 18 months after spawning (Livingston, 2000). Overall, the oysters in the Apalachicola region combine early sexual development, an extended growing period, and a high growth rate (Hayes and Menzel, 1981) to form one of the most productive oyster industries in the country.

Tropical storms and hurricanes, relatively common along the northern Gulf coast, are often accompanied by storm surges, waves, strong currents, erosion, sedimentation, flooding, altered salinities, and changes in the physiographic structure of inshore waters. On 1–2 September 1985, Hurricane Elena, with winds of approximately 200 km h^{-1} struck the Apalachicola system. A strong storm surge moved in a southwesterly direction along the sound (see Figure 2.2D) and, together with heavy sedimentation, caused physical damage to the most productive Apalachicola oyster reefs in eastern parts of the bay (Livingston, personal observation). On 21 November 1985, Hurricane Kate struck the Florida coast just west of Apalachicola Bay; at landfall, the storm carried 150 km h^{-1} winds with a storm surge exceeding 3.5 m. This hurricane, because of the position of landing and characteristics of the wind distributions as it came ashore, was not associated with observed physical effects on the bay oysters (Livingston, personal observation). Due to the effects of Hurricane Elena, there was no oyster-catching activity in the Apalachicola Bay system from September 1985 through April 1986. In May 1986, several of the best-producing reefs in the eastern section of the bay were again opened to oystering. By September 1986, the bay was fully opened to commercial harvesting. The timing and nature of the disturbances relative to the natural history of the oyster were crucial to the overall recovery pattern of the population.

The storms hit the Apalachicola system during a 2-year study of the oysters, and a detailed evaluation was made of the response of the oysters to the storms (Livingston et al., 1999). Methods for the oyster studies are outlined in Appendix I. Sampling stations were established on both major and minor oyster reefs throughout the Apalachicola estuary (Figure 10.2; Table 10.1). Cat Point, East Hole, and Platform Bars were by far the most productive of the various oyster-producing areas of the bay in terms of oyster numbers and biomass (averaged over all sampling dates from March 1985 through October 1986). High oyster productivity in these areas was due to the extensive areas of the reefs as well as the high mean density and biomass. Western sections of the bay produced relatively fewer oysters than the eastern sections.

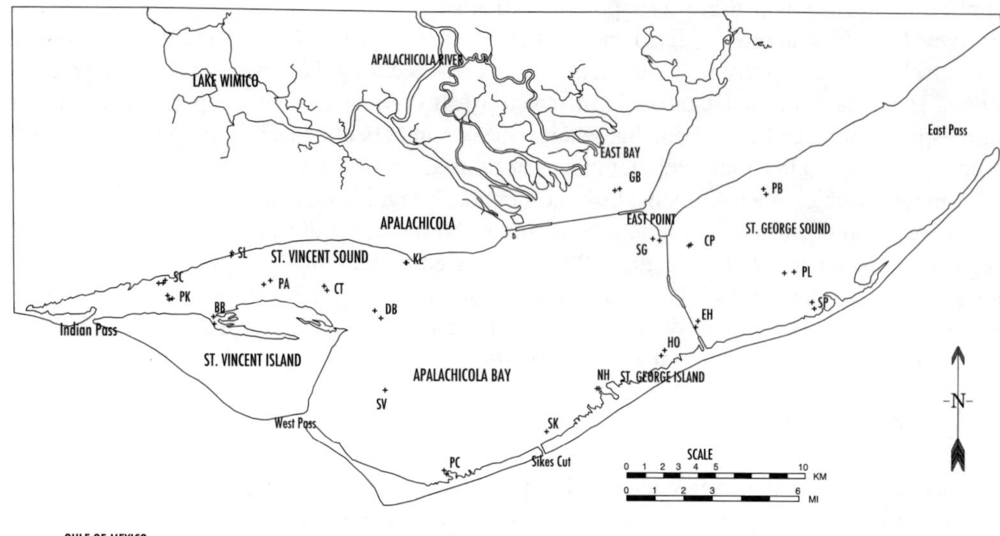

Figure 10.2 Distribution of oyster sampling sites showing general boundaries of each oyster reef. These distributions were based on historic oyster distributions, interviews with commercial oystermen and state agency personnel, and recent field studies (Livingston, unpublished data).

Table 10.1 List of Oyster Reefs Sampled with Estimated Areas of Oyster Distribution at Each Sampling Site

	Reef	Area (ha)	Total N	Total B	N m^{-2}	B m^{-2}
St. Vincent Sound	Scorpion* (SC)	190	46,917	63.7	24.7	33.5
	Schoelles Lease (SL)	85	19,328	13.7	22.7	16.1
	Paradise* (PA)	232	51,030	59.2	22.0	25.5
	Big Bayou (BB)	17	6,584	5.0	38.7	29.2
	Pickalene* (PK)	30	3,143	13.4	15.7	67.2
	Cabbage Top* (CT)	44	14,867	17.4	33.8	39.6
	Kirvin's Lease* (KL)	24	1,812	1.0	7.5	4.1
	Dry Bar (DB)	72	11,699	49.5	16.2	68.7
Apalachicola Bay	St. Vincent (SV)	575	12,423	464.0	2.2	80.7
	Pilot's Cove (PC)	20	3,143	13.4	15.7	67.2
	Sikes Cut* (SK)	1	273	1.4	27.3	43.3
	Nick's Hole* (NH)	14	1,559	3.9	11.1	28.0
	Hotel (HO)	23	1,444	6.2	6.3	26.8
	Sweet Goodson* (SG) 98		25,230	89.1	25.7	91.0
East Bay	Gorrie Bridge* (GB)	67	16,262	34.8	24.3	52.0
St. George Sound	Cat Point Bar* (CP)	514	341,918	1,781.06	6.5	346.5
	Platform (PL)	180	87,455	595.4	48.6	330.8
	East Hole* (EH)	204	97,208	852.5	47.7	417.9
	Porters Bar* (PB)	137	2,883	28.9	2.2	21.1
	Shell Point (SP)	18	125	2.7	0.7	14.8

Note: Reefs are grouped by position within portions of the Apalachicola Bay system. Abbreviations (shown in parentheses) correspond to those sites shown in Figure 10.2. Oyster data are given as total numbers per reef (TotN), total biomass (kg AFDW) per reef (TotB), numbers m^{-2} (N m^{-2}), and biomass (g AFDW) m^{-2} (B m^{-2}) for oysters averaged over the study period. Total numbers per reef and total biomass per reef were divided by 10^3. * = stations used for oyster spat accumulation analysis.

Although various water quality factors were somewhat affected by the hurricanes, the chief effects on the reefs were physical. The sustained winds of Hurricane Elena, moving from northeast to southwest in the Apalachicola region, caused considerable structural damage to the eastern reefs (Porter's Bar, Platform Bar, Cat Point Bar, Sweet Goodson's, East Hole) due to abrasion, sedimentation, and extreme turbulence from wave action and water movement (Livingston, personal observation). There was a decrease in oyster biomass m^{-2}, total numbers, and numbers m^{-2} from August to September 1985 (coincident with Hurricane Elena). There was a major recruitment event just after Hurricane Elena. Consequently, oyster numbers peaked during October–November 1985. From September to October, mean size was reduced to lows for the period of record due to the loss of adult oysters during the hurricane and the subsequent recruitment of small oysters. During fall to early winter, 1985, oyster biomass m^{-2} increased rapidly, and continued to increase to May 1986. Numbers of oysters were relatively stable from December 1985 through April 1986. Overall numbers of oysters in the bay fell in May 1986 during which period oystering was resumed.

As shown by Livingston et al. (1999), the combination of specific attributes of the hurricanes and the timing of these storms relative to the natural history of the oysters defined the nature and extent of the impact and response of this population. Although the storms (especially Hurricane Elena) had an economic impact in virtually destroying the Apalachicola oyster industry, the resilience of the oysters and the fortuitous timing of the storm (before spawning was completed) allowed rapid recovery. Hurricane Kate, by possibly thinning out the newly settled oysters, may have provided improved growing conditions for the survivors by reducing competition. Subsequent increases in biomass were substantial during the winter–spring of 1985–1986 prior to the resumption of oystering in May.

The response to repeated disturbances may take various forms (Westman, 1978) with timing of successive interventions considered crucial to the form of impact. Bohnsack (1983) pointed out that past experimental studies of disturbances have suffered from scaling problems, and relatively few field studies have sufficient data taken prior to the disturbance for an adequate evaluation. Bohnsack (1983) found a high rate of recruitment of juvenile fishes during a cold snap in the Florida Keys; he ascribed this reaction to reduced competition and/or predation, which added to the generally high resilience of reef fishes to regional disturbances. Although the competition/predation explanation is favored by parsimony, other explanations of the observed resilience of the oyster population remain possible. These include the increased success of succeeding spatfall as a result of possible increases of primary productivity, altered current patterns, and increased levels of habitat availability due to as yet unknown mechanisms. Thus, the processes of recovery depend on the timing of the disturbance relative to the life history stage of the subject population. The recovery due to natural disturbance differed from the complete loss of major oyster reefs due to salinity effects after dredging of Sikes Cut in the 1970s (Livingston et al., 2000).

Recent progress in numerical modeling of three-dimensional circulation in estuaries and coastal areas provides a powerful tool that can produce information on hydrodynamic characteristics at requisite scales of space and time. These models provide comprehensive hydrodynamic and thermodynamic information in the computational domain. Coupling of such hydrodynamic information with descriptive and experimental biological data can significantly advance the predictive capability of oyster response models to alterations in important environmental factors such as river flow. For this purpose, Livingston et al. (2000) used a modified version of the Princeton Ocean Model (Blumberg and Mellor, 1980, 1987) in conjunction with descriptive and experimental field data on oyster populations in Apalachicola Bay to examine potential oyster mortality under conditions of reduced Apalachicola River flow. Predictors included both discrete (monthly) field observations and time-averaged simulated data output from the hydrodynamic circulation model as defined in Appendix I. The hydrodynamic model output was used to examine potential oyster mortality under conditions of reduced Apalachicola River flow.

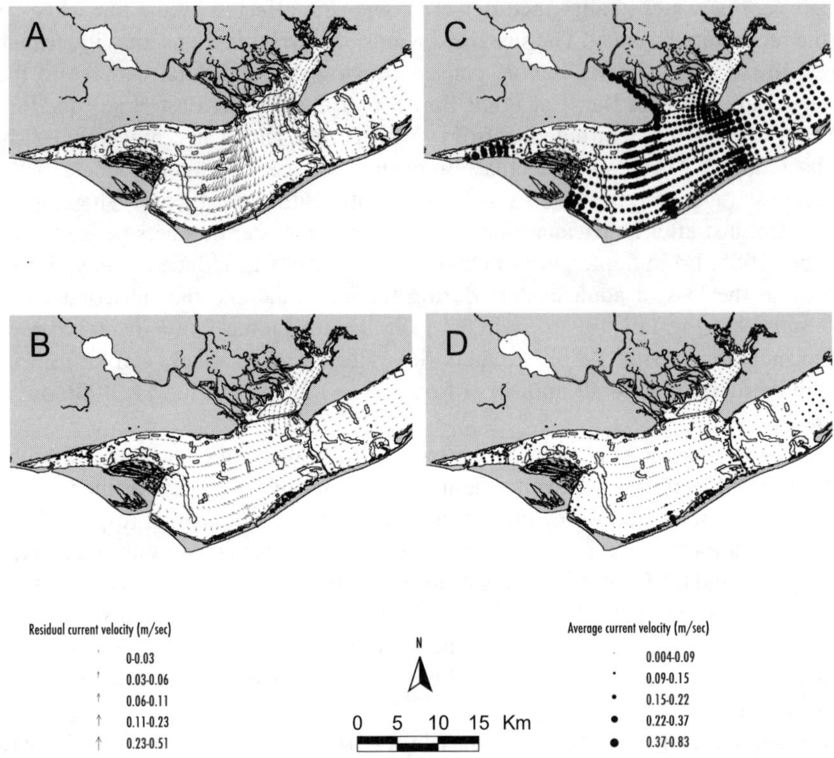

Figure 10.3 Surface and bottom current characteristics in the Apalachicola Bay system during summer (1985) months, including (A) residual surface currents, (B) residual bottom currents, (C) average surface velocity, and (D) average bottom velocity.

Surface and bottom residual current distributions were calculated using the hydrodynamic model for the summer months of 1985 and the winter months of 1986. Current patterns were nearly identical during these two periods. Thus, only the summer pattern is shown (Figure 10.3). Surface currents coming out of East Bay, emanating primarily from Apalachicola River distributaries, meet the westward trending surface currents of St. George Sound in the vicinity of the primary eastern oyster bars (Cat Point and East Hole). Runoff from summer rains in Tate's Hell Swamp contributes periodically to this pattern. The eastern oyster bars are thus aligned in a convergence of surface currents during the different seasons of the year. In Apalachicola Bay, residual surface flows are directed primarily to Sikes Cut and West Pass. St. Vincent Sound exhibits minimal residual surface current action. Residual bottom currents (Figure 10.3) flow in a westward direction from St. George Sound with some incursion of sound water into East Bay. Residual bottom currents show a northward direction from Sikes Cut into interior sections of the bay. Due to the physiographic configuration of Apalachicola Bay, most of the freshwater inflow comes from the northern side of the estuary and the Gulf waters bound it from the south. A two-layered circulation pattern is thus established across the estuary. Average surface current velocities (Figure 10.3) were highest in the river, around the various passes, and in the vicinity of the oyster reefs in eastern portions of the bay. Average bottom current velocities (Figure 10.3) were highest at the passes and in the eastern region of the bay, with the highest average bottom velocities at Sikes Cut. Overall, the surface and bottom velocities in the Apalachicola system were uniquely associated with the passes and the eastern oyster bars that included the highly productive Cat Point, East Hole, and Sweet Goodson oyster reefs.

Oyster mortality due to predation was highest at St. Vincent's Bar, Scorpion, Pickalene, Porter's Bar, and Sikes Cut (Figure 10.4; Livingston et al., 2000). These are the parts of the bay

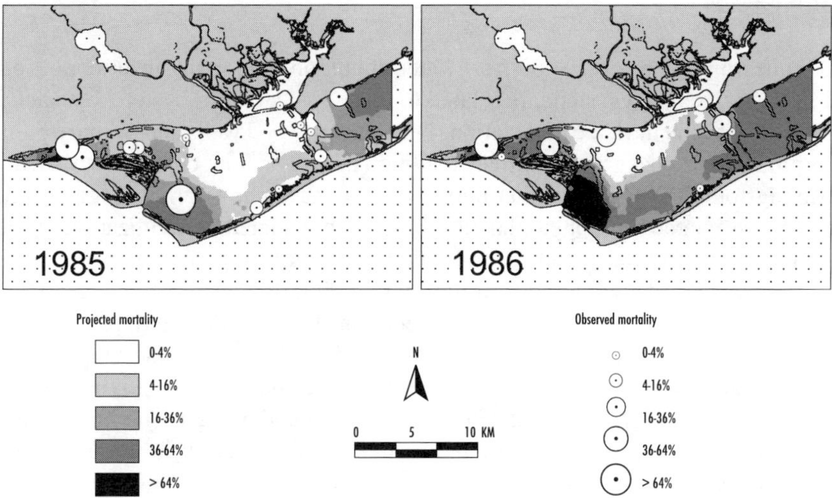

Figure 10.4 Map of projected oyster mortality due to predation in the Apalachicola Bay system based on the statistical model for mortality in 1985 and the hydrodynamic model results for 1985 and 1986. Circles indicate observed mortality values (1985, May through August average; 1986, May). Some areas in East Bay and near the mouth of the Apalachicola River are, due to low maximum salinity values, outside the range of prediction for the model.

distant to river influence (with high salinity). These areas are also in closest proximity to the entry of oyster predators from the Gulf through the respective passes. Oyster mortality was generally low at the highly productive reefs in the eastern part of the bay (Cat Point, East Hole). Such mortality tended to increase up the salinity gradient from Cat Point to Platform Bar. Statistical analyses indicated that overall oyster mortality was positively associated with maximum bottom salinity and surface residual current velocity. Mortality was inversely related to oyster density, bottom residual velocity, and bottom salinity variation. When average monthly percent mortality was regressed against time-averaged predictors derived from the hydrodynamic model, the following regression model was obtained:

$$\sqrt{\%\ \text{mortality}} = -38.5 + 1.59\ (\text{surface salinity maximum}) - 1.84\ (\text{surface salinity standard deviation})$$

Sample size was 13 stations and the r^2 for the model was 0.79 with $p < 0.001$. As the model was derived from data where the salinity maximum ranged from 27 to 35 and the standard deviation ranged from 1.3 to 7.3, it was not applied to data outside these boundaries.

The distribution of mortality during 1985 as a moderately low river flow year was highest in areas directly affected by high salinity. Such mortality was also near the entry points of oyster predators (St. Vincent Bar, Scorpion, Pickalene, Porter's Bar). Mortality on the primary eastern oyster bars was relatively low. Projections of oyster mortality for 1986 (Figure 10.4), a drought year characterized by much lower river flow than 1985, were considerably higher, especially on the highly productive bars in eastern sections of the bay. Experimental oyster mortality data taken during May 1986 confirmed the model projections. During 1986, the projected mortality on highly producing bars such as Cat Point, East Hole, Platform, and Sweet Goodson were extensive. The effect of river flow, as an indirect determinant of oyster mortality through primary control of salinity regimes, was thus a factor in the development of oysters in Apalachicola Bay. Model results indicated that reductions of river flow would be accompanied by substantial reductions in oyster stocks. This model is currently being used to determine appropriate river flow levels necessary to sustain the Apalachicola oyster industry.

d. Long-Term Cycles

Questions concerning the effects of freshwater input on important invertebrate and fish populations in the Apalachicola system are related to those indicated for oysters, but such effects are more complex due to the intricate associations between bay food webs and primary production as mediated by long-term changes of nutrient loading (i.e., river flow changes during droughts and floods). Livingston (2000) showed that nutrient imbalances due to anthropogenous nutrient loading have been associated with phytoplankton blooms, associated destabilization of plankton populations, unbalanced food webs, and general reductions of useful productivity in coastal areas of the NE Gulf of Mexico. The withdrawal of nutrient loading to the bay via reduced freshwater inflows could have direct and indirect adverse effects on the secondary bay production in the Apalachicola estuary. The effects on salinity increases and predation of nurserying invertebrates and fishes could be analogous to those found with the oyster populations (Livingston, 1997a). Livingston (1997b) found that periodic peak floods and prolonged droughts are important natural events that are related to altered patterns of individual fish distribution in terms of numerical abundance and biomass.

Livingston et al. (1997) found that reduced nutrient loading during drought periods was a cause of the loss of secondary productivity of the Apalachicola estuary during and after a given drought. Recovery of such productivity with resumption of increased river flows was likewise a long-term event. Based on the observed trends in the bay, postulated permanent reductions of freshwater flows due to anthropogenous activities in Georgia could lead to reductions of biological productivity (i.e., invertebrate and fish populations) in the Apalachicola Bay system (Livingston et al., 1997). The long-term data indicated that, with reduction of freshwater flow below a level specific for the receiving system, the physically controlled, highly productive river–estuarine system would become a species-rich, biologically controlled bay with substantially reduced productivity. The extremely valuable penaeid shrimp and blue crab industries would cease to exist and finfishing in the bay would be significantly altered with overall reduction of the currently high biomass of fish (Livingston et al., 1997).

Modeling efforts were undertaken to determine the impact of reduced river flows on oyster productivity. Salinity is a very important limiting factor for oyster populations, but it has been hypothesized that such influence is indirect in that low salinity limits predation by excluding important species such as *Thais* (Menzel et al., 1957, 1966). During periods of high salinity, oyster predation is enhanced and can be considerable. Experiments have shown that oysters over 50 mm in length are rare in unprotected areas of high salinity relative to oysters shielded from predation by baskets at similar salinities (Menzel et al., 1966). These experiments indicated that the overall mortality of oysters in the Apalachicola system was related to salinity and the geographic position of the reef relative to the natural (East Pass, West Pass, Indian Pass) and human-made (Sikes Cut) entry points of predators from the Gulf (Livingston et al., 2000). Livingston et al. (1997) hypothesized that, as drought periods were extended, decreases of secondary production followed as the result of nutrient limitation brought on by lengthy periods of low nutrient loading. Livingston et al. (2000) thus confirmed that additional reduction in oyster productivity could be expected as a result of increased or enhanced mortality due to predation and disease should low river flows during drought periods be extended due to upriver activities associated with urbanization and agriculture.

B. Planning and Management Initiatives in the Apalachicola System

1. Research Program Beginnings

We began a field inventory of the Apalachicola Bay system in March 1972. The lower river and estuary has been the primary economic resource for the region for decades. Penaeid shrimp (*Penaeus* spp.), blue crabs (*Callinectes sapidus*), oysters (*Crassostrea virginica)*, and finfishes (primarily sciaenids) have provided the basis for regional sports and commercial fisheries. A joint research venture was initiated by our Florida State University research group with local (Franklin County) interests. This effort was supported by state and federal agencies during the 1970s and early 1980s. The result was a 14-year research program that provided the scientific basis for the management program (Robertson, 1983). The association between local users and the scientific community played a key role in the successful development of the Apalachicola management program. Findings of the long-term research effort were published widely during this early period (Livingston et al., 1974; Livingston and Loucks, 1978; Livingston, 1983, 1984). The basic plan included continuous monitoring of the lower river and bay (physical, chemical, biological), experimental studies of nutrient loading, nutrient limitation, and food web structure, and determination of the role of freshwater input to the bay as a factor in the extraordinary productivity of the Apalachicola system. Research questions were devised to anticipate future (potential) impacts concerning the quality and quantity of freshwater input to the Apalachicola estuary.

Early results (Livingston et al., 1974; Blanchet, 1978; Edmisten, 1979) indicated that the Apalachicola Bay system was extremely productive as a result of geomorphological basin characteristics, salinity distribution, and nutrient loading relationships. Bay habitats were controlled by the Apalachicola River. The naturally high estuarine phytoplankton primary and secondary productivity depended on nutrient loading from the Apalachicola River system (Estabrook, 1973; Livingston et al., 1984). The seafood resources of the region were directly associated with Apalachicola River flow (Livingston et al., 1974). There were some adverse effects due to anthropogenous activities. Physical changes in the river caused by dredging and damming caused deterioration of the river system (Leitman et al., 1991) and fish kills during extreme droughts. Agricultural activities in the river floodplain caused wetland destruction (Livingston et al., 1974) although the primary riverine wetlands were generally intact. Dredging of the bay included the opening and maintenance of an artificial connection to the Gulf of Mexico (Sikes Cut) that had adverse impacts on oyster production (see Figure 10.1). There were some local pollution sources to the bay (Livingston, 1983a). However, the Apalachicola River–Bay system remained largely intact, with natural river flows, an undiminished freshwater and coastal wetland system, and the general absence of adverse effects from nutrient loading and toxic agents.

2. Dam Construction

During the early 1970s, the U.S. Army Corps of Engineers (Mobile District) proposed the construction of a series of four dams on the Apalachicola River (from Jim Woodruff Dam to the bay; see Figure 10.1). These dams were supported by a congressional mandate that a channel be maintained for shipping interests in Alabama and Georgia even though the economic rationale for such a channel was specious. Corps scientists assumed that nitrogen would not be retained behind the dams as efficiently as phosphorus. Therefore, the dams would have no effect on the flow of nutrients to the estuary as it was also assumed that nitrogen was limiting to estuarine phytoplankton productivity. However, most of the studies that found nitrogen to be the chief limiting factor to estuaries were based on studies of East Coast systems (Ryther and Dunstan, 1971). However, these studies did not necessarily apply to areas of salinity transition in southeastern estuaries (Howarth, 1988). Although phosphorus limitation of phytoplankton productivity is well demonstrated in freshwater systems, there had been no rigorous demonstration of such limitation in a range of estuaries (Hecky and Kilham, 1988). However, there was evidence that some estuaries were limited by phosphorus or combinations of phosphorus and nitrogen (Howarth, 1988). Subsequent studies in the Gulf of Mexico gave substance to the importance of phosphorus to estuarine productivity in coastal areas of the NE Gulf of Mexico (Livingston, 2000).

Unlike many estuaries along the East Coast that have been intensively studied, the Apalachicola was found to have relatively high nitrogen:phosphorus ratios (Nixon, 1988). Relatively high N:P ratios (approximating 36:1) for the Apalachicola estuary suggested phosphorus limitation (Howarth,

1988). Preliminary phytoplankton studies (Estabrook, 1973; Livingston et al., 1974) indicated that there were relatively low phosphorus concentrations in the estuary. Additional experimental work (Myers, 1977; Myers and Iverson, 1977, 1981) showed that phytoplankton productivity in the Apalachicola estuary (and other estuaries along the NE Gulf) was phosphorus limited. More recent work (Hecky and Kilham, 1988; Howarth, 1988; Flemer et al., 1997) indicated that the factors that lead to nutrient limitation in transitional parts of southeastern estuaries are highly complex and somewhat erratic, and that phosphorus has been repeatedly implicated as a limiting factor in such systems. The importance of the shallowness of the bay was indicated (Myers, 1977; Myers and Iverson, 1977), as wind mixing of bottom sediments was correlated with increased nutrients and phytoplankton production in the euphotic zone.

One reason for the importance of phosphorus as a limiting or co-limiting factor for phytoplankton productivity in the Apalachicola Bay system could be that the largely undeveloped Apalachicola drainage basin is naturally low in phosphorus (Livingston, 1984a). The fact that phosphorus was projected by the U.S. Army Corps of Engineers to be lost behind the proposed dams together with evidence that such phosphorus was important to bay productivity led to the conclusion that a series of dams along the Apalachicola River would eventually have a direct, adverse impact on primary productivity, nutrient dynamics, and food web structure in the Apalachicola estuary. Apalachicola nutrient limitation data, along with a complete review of the sources of productivity for the estuary, were used by the author to evaluate the impact of dams on the Apalachicola River. After a long and bitter confrontation, the proposal to dam the Apalachicola River was dropped. Recent research (Iverson et al., 1997) indicated that orthophosphate availability limited phytoplankton during both low- and high-salinity winter periods and during the summer at stations with low salinity. Nitrogen, on the other hand, was limiting during summer periods of moderate to high salinity in the Apalachicola estuary. Chanton and Lewis (1999, 2002) provided recent isotopic evidence that seasonal inputs of organic matter and nutrients from the Apalachicola River fueled bay productivity as outlined by Livingston (1984b). Today, the Apalachicola River remains one of the few alluvial rivers in the United States that has not been dammed. This action had implications for the valuable wetlands along the river.

3. *River Wetlands Purchases*

Documentation of facts concerning the Apalachicola basin (Livingston and Joyce, 1977) provided important information for the overall management approach. The basis for a major effort to protect the enormous intrinsic values of the floodplain forests and river–bay fisheries was thus provided by basinwide, scientific documentation of the Apalachicola resource and the underlying processes that were responsible for the extremely high natural productivity of the system. The linkage between the upland freshwater wetlands and the Apalachicola estuary was established by various studies (Livingston et al., 1974, 1976, 1984; Livingston and Loucks, 1978; Cairns, 1981; Elder and Cairns, 1982; Leitman et al., 1982; Mattraw and Elder, 1982; Chanton and Lewis, 2002; see above).

Based on information that related the river–wetlands to estuarine production, the Florida Department of Natural Resources, as part of the Environmentally Endangered Land Program (Chapter 259, Florida statutes), purchased 30,000 acres of hardwood wetlands in the lower Apalachicola for $7,615,000 in December 1976 (Pearce, 1977). This was to be the first of many wetland purchases in the Apalachicola region (Figure 10.5). Further scientific information provided the basis for other purchases of wetlands in the Apalachicola and Choctawhatchee floodplains. Recently, upland and coastal wetlands surrounding East Bay were purchased by state agencies (Figure 10.5). Currently, the Apalachicola River river–wetland system is one of the few alluvial systems in the United States where riverine and coastal wetlands are almost entirely held by public agencies for preservation and management. The protection of these wetlands provided an important step in maintaining natural (quantitatively and qualitatively) freshwater flows to the bay.

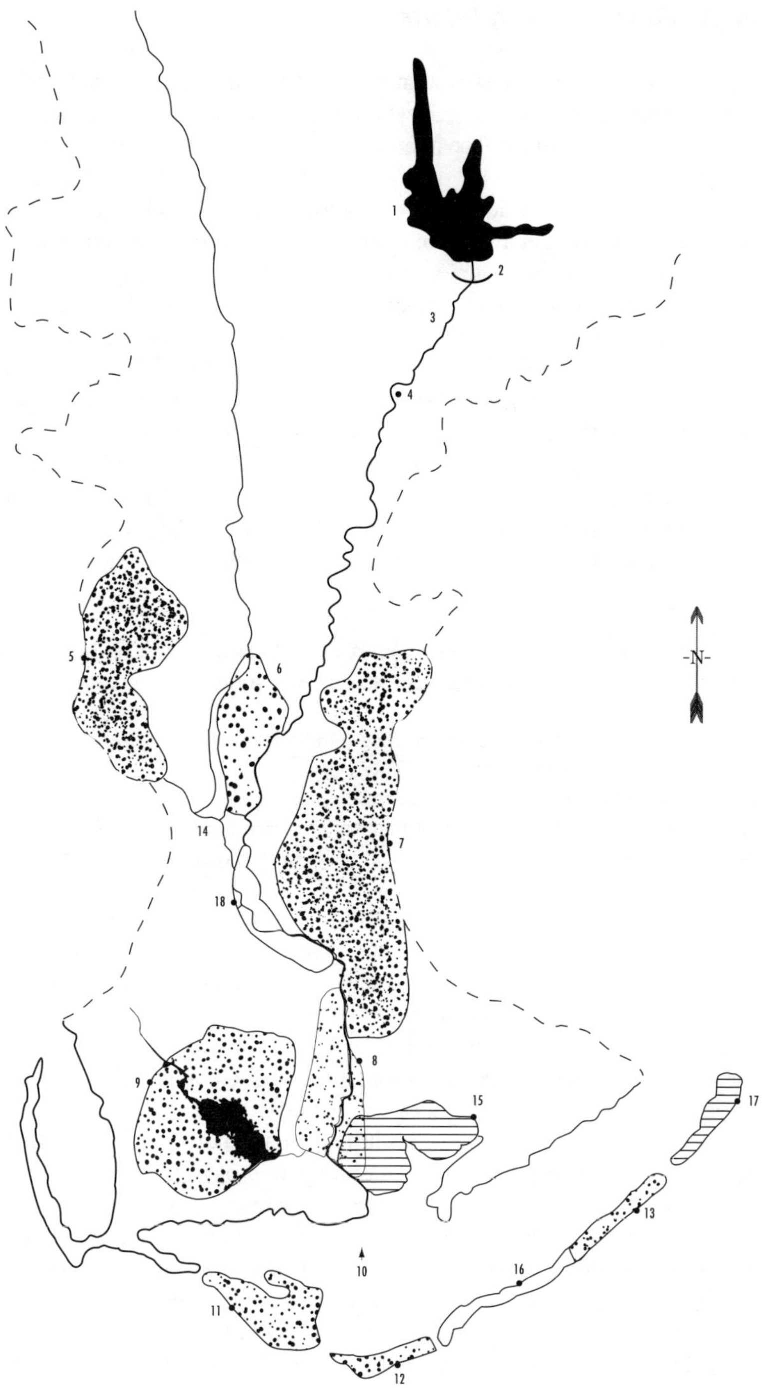

Figure 10.5 The Apalachicola system showing land purchases in the Apalachicola basin as part of the overall planning and management of the aquatic resources of the region. 1, Lake Seminole; Three Rivers State Park. 2, Jim Woodruff Lock and Dam. 3. Jackson County Port Authority. 4, Torreya State Park. 5, Gaskin Wildlife Refuge (private). 6, G.U. Parker Wildlife Management Area (private). 7, Apalachicola National Forest. 8, Environmentally Endangered Land Purchase. 9, Ed Ball Wildlife Management Area (private). 10, Apalachicola Aquatic Preserve. 11, St. Vincent Island National Wildlife Refuge. 12, Little St. George Island EEL Purchase. 13, Bruce State Park. 14, Dead Lake Recreational Area. 15, Estuarine Sanctuary Land Purchase. 16, Unit 4 EEL Purchase. 17, Dog Island. 18, “Save Our Rivers” bottomland hardwood purchase.

4. Local and Regional Planning Efforts

Research results that linked the river wetlands with the estuary were publicized through radio and television shows, newspaper stories, educational tapes, input to secondary school curricula, and various forms of oral presentations. The published results of the long-term bay research program provided the basis for the main elements of other planning initiatives. This work included hydrology (Meeter et al., 1979), the effects of various anthropogenous activities such as agriculture (Livingston et al., 1978), forestry (Livingston et al., 1976; Duncan, 1977; Livingston and Duncan, 1979), and the importance of salinity to the community structure of estuarine organisms (Livingston, 1979). The basic distribution of the estuarine populations was analyzed (Estabrook, 1972; Livingston et al., 1974, 1976, 1977; Livingston, 1976b, 1977, 1981b, 1983; Purcell, 1977; Sheridan, 1979; Sheridan and Livingston, 1979, 1983; Edmisten, 1979; McLane, 1980; Laughlin and Livingston, 1982; Mahoney, 1982; Mahoney and Livingston, 1982). Various studies were also carried out concerning the trophic organization of the estuary (White et al., 1977, 1979a,b; Sheridan, 1978; Laughlin, 1979; Federle et al., 1983; White, 1983). The results of the planning effort were continuously documented (Livingston, 1976b, 1977, 1980b, 1982b, 1983a,b). Efforts were made to provide popularized reviews of the results in the form of an atlas for informed laypeople (Livingston, 1983a).

Based on the scientific data, a regional comprehensive plan was developed that included the following:

1. Designation of the Apalachicola system as an Area of Critical State Concern, a legal (Florida Environmental Land and Water Act of 1972; Chapter 380, Florida statutes) definition of guidelines for development within the region,
2. Creation of cooperative research efforts to determine the potential impact of activities such as ongoing forestry management programs.
3. Provisions for aid to local governments in the development of local comprehensive land-use plans, a function that remains vested primarily at the county commission level in Florida.
4. Continued purchases of environmentally critical lands in the Apalachicola drainage system.

The basis for the overall management effort for the Apalachicola system has been outlined by Livingston (1975). The impact of pesticides on the system was analyzed during this period. The Apalachicola system was not seriously contaminated with organochlorine pesticides (Livingston et al., 1978), although locally high concentrations of DDT were associated with long-term biological effects (Koenig et al., 1976). All spray programs (insects, introduced aquatic plants) were evaluated according to specific criteria that included proximity of runoff to commercially important estuarine populations in space and time. Those programs considered a risk to the estuary were dropped, an unprecedented move in a state that to this day has little regulation of pesticide use. Despite vigorous opposition from state agencies responsible for coastal spray programs, the Franklin County Commission effectively ended spraying in environmentally sensitive areas.

Publication of the Apalachicola database at different levels of technical detail was only one part of the effort. Considerable time was spent in meetings with local, state, and federal officials in an effort to implement an interlocking network of management plans that would eventually be used to protect the range of resources within the Apalachicola basin. There was solid support for the research by public officials such as the head of the Florida Department of Environmental Regulation, Victoria Tschinkel. A series of governors of both parties supported the Apalachicola management effort. However, the complexity of the successful application of reliable scientific data to management questions was illustrated by the history of land development in Franklin County, a process that centered on St. George Island during the 1970s and early 1980s. This narrow sand barrier, about 50 km long, forms the southern border of Apalachicola Bay (Figures 10.1 and 10.5). During the early 1970s, politically powerful real estate developers were pitted against the Franklin County Commission and a handful of technical advisors who were opposed to high-density development in close proximity to the richest oyster bars in Florida. A detailed history of the oyster wars on St. George Island is given by Toner (1975) and Livingston (1976b). As part of continuing pressure on those concerned with the protection of the Apalachicola resource during the 1980s, a real estate developer filed a $60 million law suit against the Franklin County Commission, state and local planners, and the author of this book. In the largest civil rights suit to that time, these officials were charged with illegally interfering with the developer's rights to build condominiums on St. George Island. Four years later, the case was dismissed by a federal judge who took the unusual step of issuing awards of legal fees to the defendants with the following admonition:

> "From the beginning of this case, the litigation was conducted in a manner that indicates the plaintiff's intent was to use the judicial system to harass those who opposed his development plans.... He brought suit against Dr. Robert Livingston ... who advised officials in Franklin County on ways to protect Apalachicola Bay.... The court can only conclude that this litigation is being pursued in bad faith." (Judge William Stafford, Tallahassee, Florida)

The successful outcome of the law suit (from the defendants' point of view) led to the development of one of the most advanced land-use plans in the country for a part of the western section of the island, the so-called Plantation. This plan prevented cutting of trees, enforced low-density development, protected the freshwater drainages of the island, and was part of the overall land preservation effort in the Apalachicola basin (Figure 10.5).

After almost 10 years of acrimonious and costly battles, the Franklin County Commission adopted a strict comprehensive plan that was based on the scientific data accumulated over this period. This plan was designed to permit development in areas that could be serviced by adequate services (sewage treatment plants, storm water treatment). The land–water interfaces were protected by stringent setback rules. The relatively fragile nature of the barrier islands of the Apalachicola system was integrated into the new Franklin County plan.

5. The Apalachicola National Estuarine Sanctuary

As the result of the efforts of a coalition of local and state personnel that instituted land planning in Franklin County in the early 1970s, the scientific database was used to establish the Apalachicola River and Bay Estuarine Sanctuary in 1979. This sanctuary, now called a National Estuarine Reserve, extends up the entire Apalachicola River to the Jim Woodruff Dam, includes about 78,000 ha, and is the largest such reserve in the country. The original designation included $3.8 million for land purchases in the East Bay wetlands that completed the ring of publicly owned property around the major nursery in East Bay (Figure 10.5). In an associated effort (1977), the Florida government purchased Little St. George Island (Figure 10.5). Somewhat later, an area above the East Hole oyster beds was purchased through the efforts of the Land for Public Trust (Caroline Reusch, personal communication; Figure 10.5). St. Vincent Island was already a federal preserve administered by the U.S. Department of the Interior (Figure 10.5). The east end of St. George Island was a state park. Over a period of several years, land purchases on the barrier islands were added to the purchases of the East Bay and Apalachicola River wetlands to complete the ring of publicly owned lands around the most environmentally sensitive areas of the river and bay system (Figure 10.5). These purchases were based almost entirely on scientific reports that prioritized the order of ecological value, in both intrinsic and extrinsic terms, of the far-flung river–estuarine resources. These actions, centered around the new sanctuary (reserve), included protection from storm water runoff and septic/sewage input to major nursery areas such as East Bay, the maintenance of natural drainages that delivered fresh water from protected (fringing) wetlands, and preservation of important habitats (sea grasses) and populations (oysters) that are vulnerable to adverse impacts due to urbanization and agricultural activities.

6. Recent Developments

During the late 1980s and 1990s, there was a reaction to the events of the earlier years. The importance and effectiveness of research in the development and implementation of planning processes in the Apalachicola region had been noted by development interests in the region. There was a gradual shift to development-oriented political and bureaucratic forces as a series of more reactionary governments took over in Florida. A key regional manager was fired at the command of an influential developer. Problems such as the channelization of the Apalachicola River and the maintenance of Sikes Cut were officially ignored. The continued dredging and channelization of the Apalachicola River was officially sanctioned as the slow destruction of one of the last major free-flowing rivers in the country was continued by the U.S. Army Corps of Engineers at the behest of the politicians in Washington, D.C. (Leitman et al., 1991). The impacts of Sikes Cut on the bay were ignored by government officials (Livingston, 2000; Livingston et al., 2000). The research arm of the National Apalachicola Estuarine Reserve was largely eliminated by local, state, and federal officials at the request of local politicians. University grants at the local and federal level (Florida Sea Grant) were canceled through political moves by local land development interests. The deliberate elimination of significant environmental monitoring and research in the Apalachicola system set the stage for initiation of municipal development of the region without being hampered by scientific evaluations of the impact. There was continuing concern regarding freshwater use in Georgia, but even this concern was politicized and distorted by regional news media. The successful management effort of the 1970s and 1980s was largely ignored by an increasingly venal press corps that circumvented even the most basic scientific findings of the past.

The early successes of the Apalachicola management initiatives were due to various unrelated factors. The environmental climate during the 1970s was very different from that of the present time. The close associations between local and state officials and university scientists no longer exist. During the 1970s, the seafood industry was strong, and oyster representatives and associated elected officials contributed significantly to the development of a far-reaching management plan. Today, that is no longer true. However, the symbolic importance of the Apalachicola oyster industry, in terms of historical, cultural, and economic values, continues to be a factor in the water wars of the tri-state interests. Oysters continue to be a rallying point for various actions that could lead to protection of the freshwater resources of the Apalachicola River. That does not, however, lessen the fact that the various elements of resource management and research have been systematically eliminated by changes at various political levels to a point where development interests are now in control of the Apalachicola region. There has been an increasing shift from the open generation and public use of objective scientific data to political and bureaucratic control of information for the advantage of narrow economic interests. This change, together with the strictly controlled release of information by local and regional news media, has led to an increasingly uncertain situation with respect to the maintenance of the Apalachicola system.

Despite recent developments, the Apalachicola experiment remains one of the most advanced and comprehensive land planning efforts in the history of the United States. Most of these advances were based on the relationships of nutrient loading and limitation, primary production, and the related food web network in the Apalachicola estuary. Based on these relationships, continued management of the estuarine resources of the region should include the following:

1. Protection of freshwater and coastal wetlands
2. Maintenance of adequate freshwater input to sustain natural productivity and the important nursery function of the system
3. Minimization of physical alterations of the river–estuary that could lead to adverse changes involving nutrient transfer and salinity distribution
4. Management of local municipal development to minimize input of toxic agents, nutrients, and disease vectors to the estuary

Experience in the Apalachicola estuary has shown that research can have a major impact on the management of an important coastal resource before it is damaged or destroyed. However, a combined descriptive and experimental (ecosystem) approach is effective only when such a program anticipates resource questions that have not been asked (Livingston, 1983) and is of sufficient scope to address systemwide problems (Livingston, 1987). Some form of popularization of the research results is also needed so that informed laypeople (including politicians, bureaucrats, and managers) can understand the scientific issues. However, even if the research is correctly carried out and the information is delivered in an appropriate format, there is still no real guarantee of resource protection unless there is the political will to translate the scientific data into an effective management plan. This includes enforcement of existing environmental laws, which, increasingly, has become a major problem. There is also an ethical basis for the implementation of such a plan that is often ignored. Without an honest appraisal of how coastal systems react to human activities, the deterioration or loss of our coastal resources becomes inevitable.

Since the elimination of local, state, and federal research support in the Apalachicola Bay area by influential real estate developers in the mid-1980s, there has been no organized and effective long-term research in the Apalachicola system despite the fact the Estuarine Reserve was originally established for this purpose. This means that any activities, from bridge building to the current massive increase in real estate development, go on with virtually no objective scientific data to evaluate associated impacts. Even in an area as carefully planned as the Apalachicola drainage system, sustained coastal resource management is increasingly weakened in favor of the economic interests that are gradually taking over the region. Local and state environmental agencies continue to make sure that relatively little environmental information is made available to a public that matches its indifference to environmental issues with a sublime ignorance of the implications of unrestricted economic development. The Apalachicola experiment is still being tested, although the possibility for continued success in protecting the coastal resources of the region has been lessened. Only the future will determine whether or not such resources can be maintained.

CHAPTER 11

Ecosystem Studies: Approaches and Methods

I. ECOSYSTEM RESEARCH VS. PATCH-QUILT ECOLOGY

As noted above, the issue of how to conduct research at what is euphemistically called the ecosystem level has been discussed by scientists for decades. Even the very definition of what constitutes an ecosystem has been called into question (O'Neill, 2001). The definition of ecosystem boundaries remains at issue, and the very purpose of ecosystem research remains debatable. However, by whatever definition, the ecosystem paradigm remains the only viable approach to resource management issues in coastal areas. The need for ecosystem-level research has never been more important even though the exact nature of such research remains undefined and largely ignored.

The development of ecosystem research is not consistent with the usual hypothesis tests that drive most reductionist approaches to environmental science. Even as the definition of what constitutes ecosystem-level research is continuously debated by theoretical ecologists, the purpose of such analyses (i.e., development of a factual basis that is useful for system management) remains clear. Because ecosystem research is often used to answer questions that are not necessarily asked at the beginning of the program, such research must remain open-ended. This precludes the usual pattern of hypothesis development that underlies reductionist approaches to funding and publication. It follows that most so-called interdisciplinary projects remain in the domain of patch-quilt ecological efforts whereby individual projects by subdiscipline remain the primary focus. The ecosystem, as such, is thereby defined by *post hoc* accumulation of information that is often taken in a disparate manner with no real application to the system as a whole. That the overall goals of holistic research are thus not well defined is used to rationalize the lack of application of research results. Accordingly, the usual methodology for ecosystem research, both in terms of data collection and analysis, remains illusory when attempting to answer system-level questions. This blind adherence to time-worn and misplaced reductionist approaches has contributed to the widespread deterioration of coastal systems in the United States in recent times.

The reductionist approach has been the dominant force in the funding and publishing of ecological information for decades. This approach is based on assumptions that are obviously unrealistic when faced with questions regarding resource planning and management. Duarte and Piro (2001) relate the "fragmentation of science" to a proliferation of scientific information that continues to grow at a rapid pace but whose compartmentalization (i.e., increased emphasis on the proliferation of unrelated facts) "is poorly suited to address those problems in nature that involve a large variety of interactions among living (humans included) and non-living components." The reductionist approach attempts to piece together information by subdisciplines in the forlorn hope of understanding the whole as some sort of accumulation of its various parts. This has led to outstanding failures of science to cope with even the most mundane of human effects. By ignoring the relatively complex natural background responses of coastal ecosystems to variation through the multiple temporal cycles of driving factors such as meteorological conditions and associated changes in habitat and assimilative capacity, there is little basis for identification of complex cumulative impacts due to multiple human activities. As a product of the largely reductionist approach, the environmental sciences remain compartmentalized both in terms of funding for research and the publication of research results. The relative lack of consistent, multidisciplinary data sets in coastal systems is complemented by increased emphasis on models for predictive purposes. Because of the paucity of ecosystem-level data, these models are often based on false assumptions and are seldom field verified. Many popular models ignore even the most basic long-term habitat changes, natural and anthropogenous. The current system of funding and publication of field research is thus antithetical to the development of integrated, multidimensional, and long-term research efforts. In addition, there is little incentive for young scientists to participate in ecosystem research. As pointed out by Duarte and Piro (2001), "The resulting publishing difficulties are often quite damaging for the career development of the committed scientists. Interdisciplinary research is, therefore, penalized by the inadequate scientific mechanisms of funding and evaluation."

A comprehensive ecosystem approach requires a very different set of initial research questions from that of the reductionist hypothesis-testing process. Unfortunately, this approach precludes the *modus operandi* that most scientists (who are largely trained as reductionists) follow when faced with the prospect of interdisciplinary research. The absence of adequate scientific data provides uncertainty that is welcomed by political and economic interests that are not really concerned with protection of natural resources. In this sense, the scientific community is as responsible for the loss of natural resources as those forces that have both direct and indirect adverse impacts on such resources. Rather than an emphasis on developing an understanding of the ecosystems on which they are working, reductionists remain dedicated to the repetitious development of nonquestions, the so-called *mu* effect (Dayton, 1979). By missing key interactions of ecosystem processes, members of the scientific community continue to contribute to misleading approaches to questions regarding both protection of natural resources and restoration of damaged coastal systems. This waste of resources for research is matched only by the frantic efforts by bureaucrats and politically influential ecologists to cover up their failures. The situation in the Florida Everglades is one major example of how this works.

In ecosystem research, it is necessary to use a combination of disciplines that include a very different approach from that taken by most ecologists. This approach includes the integration of descriptive field monitoring and laboratory/field experimentation based on comprehensive, long-term observation of the system in question. The process should begin with the development of reasonable research questions concerning system interactions. Ecosystem-level questions should be based on interdisciplinary interactions at various levels of physical and biological organization. These questions should be derived from broad-based preliminary studies. There should be a central focus concerning the integration of the scientific effort with important questions that involve resource management and associated planning policies. In addition, ecosystem-level research has to start at an integrative level with a uniform, broad-based, interdisciplinary database of its own. This research should include monitoring at all levels of system organization in addition to compatible experimental work that takes its hypothesis-testing mechanisms from information taken in the field. The generation of background data should be broad-based enough to provide the background for future (hitherto unasked) research questions. The distinguishing characteristics of an effective ecosystem-level program are thus diametrically opposed to the usual pattern of carrying out a series of uncoordinated, unrelated (reductionist) studies and patching the results together to answer limited questions that are usually unrelated to issues at the ecosystem level.

The basic question of an ecosystem-level study therefore depends on delineation of the spatiotemporal dimensions of the system in question. This involves a precise and comprehensive definition of the basic processes that underlie system function over varying temporal periods. Most coastal systems are composed of physical, chemical, and biological factors that are common to specific classes of such systems. However, in any given river–estuarine/coastal system, such factors interact in very distinct ways, with corresponding differences of primary production and food web architecture. Superimposed over the natural functions and variability of a given aquatic system are the multiple impacts and cumulative effects of human activities in the subject drainage basin. The scope of ecosystem boundaries in coastal areas is often limited to the upland drainage basins on one end and connections to the open ocean on the other. This scope also is within the realm of more general climatological features (e.g., global warming) that need to be taken into consideration. Because of the uniqueness of the combination of state variables that drive individual aquatic systems, the final product of any given ecosystem analysis should also include a comparison with other aquatic systems. To achieve an effective system-level research program thus requires sophisticated levels of planning, consistent, long-term funding, and someone who knows what is going on in the continuous redirection of the research process.

II. AN ALTERNATE APPROACH TO ECOSYSTEM STUDIES

The overall approach to the development of ecosystem-level databases has been carried out by our group in the NE Gulf of Mexico over the past 32 years. Although this approach is certainly not the only way to conduct system research, it has been successful in protecting important coastal resources in Florida. This approach involves the following organization:

A. Tier 1: Establishment of Research Goals

1. There should be a preliminary determination of realistic goals for the long-term research program.
2. One person should be in charge of the program. Central leadership of the research effort is the single most important factor in carrying out a long-term, ecosystem-level research program in a given system. There has to be one person who is familiar with all aspects of the program, and who can keep all of the elements consistent with the stated research goals. Research by committee is self-defeating in long-term, multidisciplinary projects. This aspect goes against most of the current procedures of research funding where competing interests get their piece of the research pie through largely political machinations. However, if an ecosystem project is to achieve its purposes, it must be free of the usual political aspects that dominate research funding today.
3. A multidisciplinary review board should be established that includes representation of the various scientific and engineering subdisciplines, sports and commercial fishing interests, and economists who specialize in resource issues.
4. There should be a complete review of available scientific information concerning the system in question. This information should be lodged in a cataloged research library. Existing data should be summarized as part of the development of the overall research program.
5. There should be a comprehensive determination of habitat distribution in the subject system. This effort should be carried out in conjunction with the development of maps of point and non-point sources of pollution to the system. The current level of GIS technology, together with a wealth of information by local, state, and federal agencies, should facilitate the development of such information.
6. A catalog of anecdotal information should be gathered based on interviews with people who have lived in the system for extended periods, or who have in some way had unique personal knowledge of changes in the system (i.e., fishermen). This effort can be supplemented by a comprehensive long-core sediment analysis to determine the past history of the system.

The operations for the first tier usually take 1 to 2 years depending on the level of effort.

B. Tier 2: Development of the Research Programs

The research effort should include the establishment of a consistent and multidisciplinary database through field monitoring. This program should be based on information gathered during Tier 1 activities, and should be concerned with the definition of the trophic organization of the system in question. The following components should be included in the program:

1. Organization of a central database by a team that includes database management people, computer programmers, statisticians, and modelers for development of a capability for analysis of long-term, multidisciplinary data sets. This team should carry out a continuous analysis of the data with input to the principal investigator and the review team.
2. Development of meteorological databases that include long-term rainfall/river flow (i.e., hydrological) information for production of models based on long-term trends of atmospheric conditions in the basin.
3. Determination of permanent stations for long-term monitoring of water and sediment quality. The monitoring program should be based on a review of Tier 1 results and a preliminary sampling program that stratifies habitat distribution in the system.
4. Development of an adjunct database for three-dimensional modeling of the physical organization of the system.
5. Development of nutrient/pollutant loading models, as derived from stages 1 through 4 (above).
6. Development of a biological monitoring program centered on key populations and communities as indicators of water/sediment quality. This should include all elements of biological organization with an emphasis on the main sources (allochthonous, autochthonous) of primary production and associated food webs of the system. Spatiotemporal food web changes should be integrated with physicochemical and loading programs (common stations, synoptic sampling regimes) so that the long-term changes of the system can be followed. Ideally, species-specific changes of the primary producers (phytoplankton, benthic microphytes, submerged aquatic vegetation, emergent vegetation) should be related to detailed food web responses both with respect to the influence of natural state variables and to impacts of human activities (nutrient/toxic loading, sediment contamination, etc.). The development of a reference system or systems should be contemplated at this stage of research development.
7. Eventual integration of all databases in a series of modeling efforts that include hydrological models, three-dimensional water quality models, water/sediment models, population/community changes, and food web models to answer questions of long-term spatiotemporal changes of systems in question.

Continuity of sampling methods and techniques is essential. All sampling methods should be quantified at the outset of the project. It is advisable to include a broad array of variables at the beginning, with reduction of the overall sampling effort based on analyses of covariance and relevance to established and anticipated research questions. This should include an effort to have consistent taxonomic verification during all phases of the research program. The central organization of the long-term, multidisciplinary research rests on the quality of the scientific effort with appropriate QA/QC programs that ensure database consistency. Continuous review of the database will provide the basis for the development of complementary experimental studies that address specific questions derived from the descriptive field data. Experimental research should be designed to answer ecosystem questions rather than satisfy isolated and often irrelevant questions of individual researchers.

C. Tier 3: Development of Resource Management Programs

Specific studies of different parts of the system should be undertaken to address policy and management questions that include determination of sensitive and/or critical habitats and associated fisheries populations.

1. Management questions should be outlined, based on interviews with policy makers, environmental managers, scientists, and knowledgeable laypersons. Key research questions should be developed that are consistent with associated planning and management objectives of the study.
2. Integration of the scientific activities with policy objectives (political/economic entities) should be a major consideration in the determination of the research program.
3. Short-term analyses designed to answer local and/or regional questions should be developed that are consistent with questions that arise during the course of the study. This part of the project should include applied questions relative to the fisheries potential in the study area.

D. Tier 4: Review of Research Activities

The quality of the database and consistency of the sampling effort should be of primary concern, and should be subject to continuous review.

1. QA/QC reviews should be continuous and comprehensive.
2. There should be periodic reviews of the overall program to ensure that it is consistent with established project goals.
3. Results of the program should be reviewed to determine the need for elimination of established guidelines and/or the addition of new lines of research.

E. Tier 5: Comparison with Other Systems

The ecosystem model outlined above, once implemented, can be expanded to comparative analyses with other coastal systems. These comparisons are vital to an understanding of how any given system works.

F. Tier 6: Development of a Resource Management Program

The management of coastal systems should be based on several features that are known to be key components in the overall productivity and biological integrity of a given system. There should be careful control of freshwater sources to the system. Both qualitative and quantitative aspects of nutrient and toxicant loading should be addressed. There should be an effort to minimize physical alterations of the system (dredging, habitat alteration, wetland destruction) that can lead to increased salinity stratification and associated loss of the nursery function of the system. These processes underlie the basis of coastal productivity and usefulness. Fisheries production and the distribution of individual plant and animal populations depend on species-specific responses to combinations of freshwater input, salinity changes, qualitative and quantitative sources of primary production, associated food web processes, and biological feedback mechanisms such as interspecific predation and competition. Management considerations would include the following:

1. Protection of freshwater and coastal wetlands
2. Maintenance of adequate freshwater input to sustain natural productivity and the important nursery function of the system
3. Minimization of physical alterations of the river–estuary that could lead to adverse changes involving nutrient transfer and salinity distribution
4. Management of local municipal development and agricultural runoff to minimize input of toxic agents, nutrients, and disease vectors to receiving river–estuarine systems

G. Tier 7: Application of Research Activities to Education

1. Development of associated educational activities would include involvement of students from secondary school systems, undergraduate and graduate programs of regional colleges and universities, and public education via the usual media outlets.
2. Analysis, modeling, and publication of project data would include development of symposia and other forms of meetings with publication of project findings that are understandable to the general public.

The development of an ecosystem-level research program requires informed planning and someone in charge who tracks project results while keeping all elements consistent with the project goals and hypotheses. Sampling methods should be quantified before starting the regular field analyses. Many factors should be tested at the beginning, with reduction of redundant sampling after appropriate analyses of the data. The QA/QC program should adhere to detailed written protocols. Field sampling should be established after adequate habitat stratification and sampling patterns that are compatible with both spatial and temporal aspects of project goals. Field collections should be carried out synoptically, and should be frequent enough to meet the demands of the various levels of biological organization of the subject system. The database should be maintained and continuously reviewed by analytical teams of statisticians, modelers, and ecological personnel. Any adjustments made to sampling and/or experimental aspects of the program should be consistent with long-term project goals. Field and laboratory experimental parts of the program should be based on empirical data, and should also answer questions that cannot be supplied by the descriptive field information. All physical, chemical, and biological aspects of the program should be coordinated with field collections and the development of the experimental parts of the project.

CHAPTER 12

Conclusions

1. Over the past 32 years, our research group has carried out a field program combined with long–term analyses of the trophic organization of coastal drainage systems of the Gulf of Mexico. Drainages under study include both alluvial (Perdido, Escambia, Choctawhatchee, Apalachicola) and blackwater (Blackwater, Econfina, Fenholloway) systems. An active field and laboratory experimental program has also been carried out to complement the ongoing field sampling efforts. This effort was based on a series of short- and long-term studies that involved both field descriptive and field and laboratory experimental programs that defined ecosystem processes. The development of food web ecology was simultaneously carried out with physical modeling, chemical determinations based on synoptic sampling efforts, and nutrient loading determinations. Detailed biological analyses were made at various levels of biological organization with particular attention to the connection between producer groups (phytoplankton, benthic microalgae, submerged aquatic vegetation) and secondary producers. Continuous statistical analysis and modeling accompanied the long-term field studies in a series of coastal systems. The data, which are still being processed and published, were applied to various forms of management programs that included both preventative applications and restoration efforts. The central theme remained the response of coastal food webs to both natural and anthropogenous effects.
2. A long-term, systematic analysis of the feeding habits of fishes and invertebrates of coastal regions of the NE Gulf of Mexico indicated that many species undergo a well-ordered progression of changes in food preferences. Although the transitions from one feeding stage to the next are often gradual and not necessarily distinct, such ontogenetic trophic stages are identifiable in various (though not all) species. Important populations of plankton feeders, herbivores, infaunal macroinvertebrate feeders, and benthic predators of various types thus progress through species-specific, ontological changes of feeding habits that are associated with progressive changes in morphology (mouth dimensions, dentition, gut dimension), physiology, and behavior. These species inhabit progressions of habitats based on the ontological feeding units. Elimination of a given habitat due to anthropogenous activities can thus break the progression and alter the development of a given population.
3. The trophic organization of both invertebrates and fishes in a series of coastal environments followed generalized feeding patterns that could be determined through extensive gut analyses. Diets were well mixed with little dependence on one form of food. The presence of plant matter took many forms, and was often size-specific, going from smaller epiphytes to the larger forms of plant matter such as sea grasses with growth of the individuals. Strict herbivory was relatively rare compared with the numbers of omnivores and carnivores. There were ontogenetic feeding trends in most, though not all, species. There was little evidence of purely opportunistic (i.e., random, unspecialized) feeding patterns. There was a common interaction of habitat structural complexity with biological processes, and predator–prey interactions defined exact sequences of feeding progressions in the different coastal areas.
4. Feeding patterns of fishes and invertebrates cannot be generalized at the species level, and should be viewed as progressions of disparate feeding stages that are often more closely related to trophic stages of other species than to their own ontogenetic feeding stages.

5. Trophic components directly linked to phytoplankton and benthic algal production (herbivores, omnivores) are more immediately affected by changes in physical habitat variables than higher trophic levels (carnivores) that are more biologically controlled by prey availability and predator–prey interrelationships. This dichotomy exists with respect to both seasonal and interannual trophic relationships. Herbivore/omnivore populations are thus more closely associated with physical habitat phenomena related to primary productivity, whereas predators follow patterns of prey distribution and competition for feeding grounds.
6. Although species-specific feeding patterns are generally similar among different coastal systems, differences of productivity, habitat, and biological interactions can affect the ontogenetic patterns of a given species. Pollution and habitat alteration can cause changes in associated trophodynamic feeding patterns, and can thus affect the distribution of different populations through habitat loss and breaks in the ontogenetic feeding progressions.
7. Major feeding periods observed in coastal fish and invertebrate communities may be closely related to peaks in prey activity patterns. Diurnal and seasonal dietary shifts are related to differential prey abundance and distribution. Interannual habitat changes can lead to comparable alterations of food web structure through changes in prey activity and distribution. Even slight water quality changes due to pollution that are outside the evolutionary experience of coastal populations can cause serious disruptions of basic habitat structure, energy flow, and community composition through altered food web processes.
8. The species is not always appropriate as a unit of measure when used in quantitative ecological studies. In many instances, more substantial ecological differences exist among life stages of a given species than among similar trophic units of different species. The use of a species in quantitative ecological studies can lead to problems of interpretation concerning the relationships of coastal organisms to complex habitats. Use of the species as a convenient unit of measure substitutes a basically taxonomic entity for ecologically relevant life history stages. Niche breadth of a given species can be so extensive that quantitative determinations of significant ecological processes are difficult to make without the determination of the ontogenetic feeding progressions. Unless there is adequate recognition of the complex ecological stages that characterize coastal organisms, the interpretation of ecosystem-level data remains incomplete.
9. Long-term field data from the various coastal systems were transformed as a function of individual trophic units (infauna, epibenthic invertebrates, fishes) into an empirical food web. Life history factors were reorganized into guild associations that transcended taxonomic boundaries. Similar trophic units of different species were grouped together into food web orders at different levels of specificity. The most generalized reorganization included the following: herbivores (feeding on phytoplankton and benthic algae), omnivores (feeding on detritus and various combinations of plant and animal matter), primary carnivores (feeding on herbivores and detritivorous animals), secondary carnivores (feeding on primary carnivores and omnivores) and tertiary carnivores (feeding on primary and secondary carnivores and omnivores). In this way, the long-term database of the various field collections of infauna and epibenthic macroinvertebrates and fishes was reorganized into a quantitative and detailed trophic matrix.
10. Long-term field data together with predator exclusion experiments with infaunal macroinvertebrates in the Apalachicola estuary suggested that river flow and freshwater runoff influence both abundance and species composition of infauna along spatial and temporal gradients. Physical instability, coupled with high productivity in the upper (river-dominated) parts of the estuary, resulted in high densities of relatively few populations that were adapted to high habitat variation. Areas within even indirect riverine influence had relatively low numbers of species, high numbers of individuals of a few populations, high biomass, and high relative dominance. The adaptive organisms in these physically controlled areas showed little response to predation pressure. In areas of the bay that were outside riverine influence, there were relatively high numbers of species, low numbers of individuals, low biomass, and low relative dominance. Even in the high-salinity environment, however, salinity and temperature had some degree of control of infaunal assemblages. Predation affected relative dominance in high-salinity areas with reduced dominance of individual infaunal species and associated high species richness. Increased species richness in high-salinity situations could be also be the result of higher habitat stability. Lower primary productivity in high-salinity areas could explain the reduced numbers of infauna relative to river-influenced areas. Physical

habitat stressors and biological processes such as predation thus interacted with spatial and temporal trends of productivity and the exclusion of stress-susceptible marine species to result in the observed responses of infaunal assemblages along gradients of riverine influence in the Apalachicola estuary. The high productivity of natural, river-dominated estuaries was thus a product of physical and biological processes that were directly and indirectly related to seasonal and interannual trends of freshwater input to the estuary. The freshwater input was a controlling factor in the definition of food web processes.

11. Food web processes were an important part of the response of coastal populations to impacts of toxic substances. Results of field and laboratory experiments concerning the effects of pentachlorophenol (PCP) on infaunal macroinvertebrates in the Apalachicola system showed that impacts of toxic substances on estuarine organisms should be interpreted within the context of physicochemical habitat variation, predator–prey interactions, and recruitment patterns of dominant populations. The limitations of laboratory tests in predicting effects of toxic agents included problems of interpretation due to the inability of such tests to account for the effects of natural habitat changes and the responses of subject populations to trophic interactions. The same problem was noted when individual species were used as subjects for such tests since laboratory artifacts often had significant effects on time-related test results with individual species. However, simultaneous field and laboratory experiments indicated that some community indicators (numerical abundance, species richness) and specific guild associations (deposit feeding detritivores) were reliable predictors of the effects of toxic substances. Trophic organization thus played an important role in delineation of the effects of toxic substances on coastal systems.

12. By going from detailed trophic studies to more generalized trophic models, a series of models was constructed based on universal attributes of different coastal areas. By quantifying significant portions of the estuarine food webs over prolonged periods, interannual trends of community composition were identified. Specific criteria for the detailed delineation of trophic organization proved to be less important than the consistency of the data on which the trophic models were based. In the Apalachicola system, there was a dichotomy of response of the trophic elements to controlling factors. The results of the long-term analyses of trophic organization in the Apalachicola system suggested that river flow and associated primary productivity were mainly associated with changes in the lower trophic levels, and that the carnivores were associated primarily with the distribution of their prey. Herbivores and omnivores were directly linked to physical and chemical controlling factors that were associated implicitly with the primary productivity of the estuary. The river, which mediates such factors, was thus directly linked to the response of these trophic components. Primary, secondary and tertiary carnivores, on the other hand, were associated more closely with biological factors linked to predator–prey interactions. Physical control of the distribution of eurytopic populations was less important than trophic considerations in the determination of spatiotemporal trends of estuarine populations. Long-term ecological trends in the Apalachicola Bay system indicated that natural interannual variability of freshwater flow is considerable, and that drought/flood cycles altered basic estuarine habitat conditions and trophic organization. Increased primary productivity due to enhanced light penetration was associated with increased herbivore and carnivore activity. With time, however, drought conditions eventually took over and associated low nutrient loading was accompanied by reduced secondary productivity throughout the bay. Natural cycles of river flow and nutrient loading were thus responsible for interannual variation of bay productivity.

13. A 14-year study of the effects of paper mill effluents on the Perdido River–Bay system indicated long-term responses of the estuarine food webs to nutrient loading and phytoplankton community composition. During early years of analysis (1988 through 1991), orthophosphate and ammonia loading from the pulp mill enhanced secondary production in the immediate receiving area of the upper estuary. There were no serious phytoplankton blooms during this period, and the phytoplankton community structure remained diverse. However, during a winter–spring drought in 1993–1994, increased pulp mill orthophosphate loading was associated with the initiation of phytoplankton blooms that were dominated by diatoms. There was an orderly seasonal succession of these bloom species in the bay. Continued high orthophosphate loading and increased ammonia loading from the mill over the next 3 years led to increased phytoplankton bloom frequency and intensity throughout the bay. There were successions of bloom species with replacement of the diatoms by raphidophyte and dinoflagellate populations. Between 1995 and 1998, increased ammonia loading to the upper bay was accompanied by spring–summer blooms. From late summer 1997 through spring 1999, the pulp mill reduced its orthophosphate loading; this was associated with reductions of winter blooms and increased phytoplankton species richness. Increased summer ammonia loading was associated with summer bloom species and associated reductions of phytoplankton species richness. However, by spring 1999, the mill again resumed high loading of orthophosphate to the bay, which was again accompanied by baywide spring and summer blooms, increased dominance of bloom species, and associated reductions of phytoplankton species richness. Subsequent reductions of ammonia and orthophosphate loading by the mill resulted in an almost complete cessation of blooms in the bay.

14. The form and timing of nutrient loading with respect to individual nutrient species had a major influence on the response of phytoplankton assemblages in the receiving system. Bloom response was modified by both seasonal and interannual drought/flood cycles. Bloom incidents affected specific state habitat variables such as dissolved oxygen (DO), pH, nutrient concentrations, and sediment quality. Blooms were associated with increased phytoplankton biomass and numerical abundance, and with reductions of phytoplankton species richness, species diversity, and population evenness. Drought conditions enhanced nutrient concentration gradients in Elevenmile Creek leading to a series of blue-green algae blooms in the creek.

15. Statistical analyses of the 14-year database indicated that plankton bloom species responded to qualitative and quantitative changes of nutrient loading and concentration gradients in upper bay areas. Regressions of the distribution of *Prorocentrum minimum*, a winter bloom species, indicated significant positive associations with orthophosphate concentration ratios in the upper bay. Drought conditions and resulting high orthophosphate concentrations in Elevenmile Creek were related to stimulation of *P. minimum* blooms in the creek and bay. *Prorocentrum minimum* numbers significantly reduced whole-water phytoplankton species richness. Numerical abundance of *Cyclotella choctawhatcheeana*, a spring bloom species, was significantly associated with ammonia ratios in the upper bay and with combined (averaged) differences of orthophosphate and ammonia concentration ratios during spring periods. *Miraltia throndsenii*, a spring bloom species, was associated with nutrient gradients (orthophosphate + ammonia). A summer bloom species, *Heterosigma akashiwo,* was positively correlated with ammonia loading rates, with combined differences of orthophosphate and ammonia concentration ratios, and with differenced N+P loading data. *Heterosigma* blooms were negatively associated with summer phytoplankton species richness indices. *Synedropsis* sp. (a summer bloom species) was significantly associated with orthophosphate ratios, and the N+P differenced data for both concentration ratios and loading. There was also a significant negative correlation of *Synedropsis* sp. with phytoplankton species richness. The key to the identification of bloom associations was the species-specific seasonal progressions of bloom species in the bay.

16. Drought-induced habitat conditions were associated with increased blue-green alga *Merismopedia tenuissima* blooms in Elevenmile Creek. Even though reductions of nutrient loading by the mill led to cessation of bloom events in the bay, drought-induced increases of nutrient concentrations in the form of differenced orthophosphate/ammonia concentration ratios during periods of reduced nutrient loading by the paper mill were associated with increased nanoplankton (cryptophytes and nanoflagellates) cell numbers and biomass in the Perdido system. Surface ammonia concentrations were positively associated with the nanococcoids, which, in turn, were negatively associated with bottom DO in the bay. Nanoflagellate numbers were negatively associated with surface DO.

17. Wolf Bay (lower Perdido system) has been subject to unregulated non-point-source nutrient loading from agricultural areas in Alabama. There was an associated increase of plankton blooms in Wolf Bay, with peaks of most bloom species noted from late 2000 through 2001. The dinoflagellate *P. minimum* was dominant during this period along with blooms of *Skeletonemia costatum,* and *C. choctawhatcheeana*. There were also peaks of *Gymnodinium* spp., *H. akashiwo*, and *Chattonella c.f. subsalsa* during this period. There were associated general declines of water quality features such as bottom oxygen anomalies and Secchi depths that coincided with the bloom peaks. These trends were diametrically opposed to changes in the upper bay, where bloom occurrence declined due to reduced nutrient loading from the point source. The Wolf Bay data indicated that nutrient loading from unregulated non-point sources (i.e., agricultural activities) can be as damaging as that from point sources with the disadvantage that both agricultural and urban runoff are not regulated. However, there were fundamental differences in the form of non-point-source blooms when compared to the more continuous point-source nutrient loading.

18. Phytoplankton blooms in Perdido Bay were associated with reduced numbers, biomass, and species richness of infaunal and epibenthic macroinvertebrates and fishes as a result of disruptions of the food web. Because of the extended analysis of bloom incidence (14 years), specific associations could be made among nutrient loading factors, bloom incidence, and the adverse impacts of individual bloom species on the food web organization of the Perdido system. Upper bay herbivores were significantly (negatively) associated with *P. minimum* biomass during winter periods, and were negatively associated with DO anomalies in the fall. In midbay areas, herbivores were positively associated with nutrient conditions and phytoplankton/diatom biomass during winter and summer periods. In the lower bay, herbivore biomass was positively associated with phytoplankton/diatom indices during summer/fall periods. Because herbivores were located largely in the upper bay, the negative associations with a winter bloom species constituted direct evidence of the adverse effects of blooms on the trophic organization of the bay. Different processes were evident in mid and lower parts of the bay that were probably associated with habitat conditions in these areas. Omnivore biomass in the upper bay was positively associated with *P. minimum* during the winter, and with surface ammonia during the spring. Omnivores were positively associated with bloom numbers in the summer. In the lower bay, omnivores were positively associated with numbers of winter plankton blooms and with summer N+P loading differences. These associations constituted statistical evidence that omnivores were favored by N+P loading and associated plankton blooms during certain seasons. Overall, the seasonal and interannual progressions of blooms had differential effects on different parts of the bay food webs.

19. The loss of bay herbivores was accompanied by reduced levels of carnivore populations. In the upper bay, primary carnivores were negatively associated with orthophosphate ratios, phytoplankton biomass, and bloom species biomass. Primary carnivores were negatively associated with *P. minimum* during summer months. In the lower bay, primary carnivore biomass was negatively associated with winter bloom numbers and with raphidophyte biomass. During summer months, primary carnivores in the upper bay were negatively correlated with N+P ratio differences, and raphidophyte/*H. akashiwo* biomass. During fall, C1 carnivores were negatively associated with ammonia loading, and *H. akashiwo/Cyclotella choctawhatcheeana* biomass. Secondary carnivores were negatively associated with winter bloom numbers, and raphidophyte/*H. akashiwo* biomass. During spring, this group was negatively associated with various nutrient concentration indices, total phytoplankton biomass, and bloom biomass. During summer, C2 biomass was negatively associated with nutrient concentration indices and raphidophyte/*H. akashiwo* biomass and, in the fall, was negatively associated with ammonia loading and *C. choctawhatcheeana/H. akashiwo* biomass. The data thus indicated that C1 and C2 carnivores were negatively affected by seasonally changing bloom species, and by the nutrient loading/concentration factors that caused the blooms. The key toxic bloom species (*P. minimum, H. akashiwo, M. tenuissima*) were the main effectors of the loss of secondary production in Elevenmile Creek and the bay. These effects were superimposed over other important habitat changes such as reduced DO. During the 14th year, mill reductions of nutrient loading were accompanied by a near absence of phytoplankton blooms in Perdido Bay, thus validating the nutrient loading restoration targets. However, bloom cessation was not accompanied by immediate return of the bay food webs, which were probably adversely affected by prolonged drought conditions.

20. Periodic drought occurrences exacerbated bloom occurrences and associated effects on food web processes in the Perdido system. Seasonal and interannual trends of nutrient loading superimposed over these climatological features determined not only the levels of secondary production, but also specific changes of associated food webs in different parts of the bay. Bay recovery as a response to restoration activities was not the mirror image of impact history, and changes in nanoplankton populations and the increased activity of some bloom species indicated complex biological interactions during the recovery phase. There were also some differences between point- and non-point-source nutrient loading with respect to the initiation of plankton blooms. Most point sources are continuous whereas many non-point sources load nutrients mainly during precipitation events, and are therefore intermittent nutrient sources. In the case of the pulp mill, there was continuous nutrient loading during drought as well as flood periods, whereas non-point (agricultural) sources did not load as regularly during droughts.

21. Primary water quality indicators of bloom activity are DO, pH, and nutrient concentrations although such indices, by themselves, do not explain the range of biological changes that take place during bloom sequences. Food web changes cannot be ascertained by after-the-fact water quality assessments. Bloom impacts can be subtle, and cannot be determined without adequate scientific information concerning the phytoplankton associations. The absence of detailed phytoplankton data can lead to serious mistakes in regulatory efforts to control nutrient loading into coastal systems. Analysis of the dependence of regulatory agencies on water quality indices in the Apalachee Bay system provided further proof of the importance of phytoplankton analyses in determinations of hypereutrophication events. The EPA sampled the polluted Fenholloway system during December 1998, June 1999, and August–September 1999 and, based on water quality investigations, indicated no bloom activity. However, simultaneous phytoplankton analyses by our group indicated peak plankton bloom activity in the Fenholloway system during this same period, thus providing evidence that methods used by regulatory agencies to evaluate potentially damaging blooms in coastal systems are based on false assumptions of how blooms are initiated and maintained.

22. The assimilative capacity to absorb anthropogenous nutrient loading varies from system to system, and also varies with time in any given system depending on natural drought/flood sequences and interannual patterns of bloom successions. Food web assessments of the impact of hypereutrophication in coastal areas have some advantages over determinations of changes in individual populations. Population changes are variable and cannot be generalized from system to system. Food web changes due to altered phytoplankton assemblages are more sensitive indicators of bloom impacts on secondary productivity, and can be used in comparative studies of different coastal systems.

23. Overfishing has caused adverse effects not only on overfished species, but also on associated food webs due to both "top-down" and "bottom-up" food web impacts. However, it is possible that recent reviews have overgeneralized the importance of overfishing relative to the current increase of anthropogenous nutrient loading and eutrophication in coastal systems. Overfishing as a primary top-down factor does not apply to most coastal areas of the NE Gulf. Habitat deterioration in this region has been largely due to dredging effects (opening barrier island estuaries to salinity influxes) and hypereutrophication due to anthropogenous nutrient loading from both point and non-point sources.

24. Biological activity and trophic diversity is highest in freshwater runoff areas of alluvial systems, where phytoplankton productivity is high and high dominance of selected eurytopic species prevails. This means that freshwater runoff and the delivery of nutrients to receiving bays are crucial for maintenance of highly productive coastal food webs. Primary and secondary productivity in the Apalachicola system is high due to preservation of the quality and quantity of freshwater runoff. Due to anthropogenous nutrient loading from both point and non-point sources, the Perdido, Escambia, and Choctawhatchee Bay systems have been damaged, with associated alterations of food web structure and sharp reductions of secondary productivity. The sea grass–based systems in Apalachee Bay follow a different course, with food web structure closely tied to the development of submerged aquatic vegetation (SAV). In these areas, food web deterioration was evident due to losses of SAV as a result of reduced light transmission caused by pulp mill effluents. Identification of species-specific changes of primary producers (SAV, phytoplankton) was an important part of the response of food webs to loading of nutrients and toxic agents.

25. The species is an important unit of measure in ecological studies. However, determination of the trophic stages of a given species allows a more precise analysis of the ecological processes associated with the spatial and temporal distribution of organisms in coastal areas. The different trophic levels represent relatively stable units that remain free of the constant shifts of species distributions in space and time. In a way, species populations represent the hands of the clock, whereas food web organization represents the internal clock mechanism whereby the ongoing processes that determine biological response to habitat and productivity changes can be evaluated. Analysis of food web changes also allows comparisons of biological organization among disparate coastal systems that have different species representations. It facilitates regional comparisons of the effects of hypereutrophication and toxic substances where, again, the differences in species populations complicate the interpretation of field data. The use of ratios of trophic levels (i.e., herbivore biomass:omnivore biomass) provided a simple yet accurate index of the long-term changes of the bay food web organization and the response of bay communities to long-term nutrient loading patterns and associated changes in phytoplankton assemblages.

26. Regulatory decisions regarding anthropogenous loading of nutrients and toxic substances to coastal systems would benefit from a more comprehensive review of food web responses. The emphasis of state and federal regulatory agencies on chlorophyll *a* concentrations is spurious as chlorophyll *a* concentrations are not indicative of phytoplankton blooms or the effects of the blooms on phytoplankton assemblages and associated coastal food webs. The silent and largely overlooked impacts of anthropogenous nutrient loading (urban runoff, agricultural waste discharges, point-source pollution) will continue to contribute to ongoing deterioration of coastal resources in the United States until more attention is given to the response of phytoplankton assemblages to nutrient loading. Food web considerations are also important in determination of impacts of chemicals such as methyl mercury. An example of such problems is represented by the Penobscot River–Bay system in Maine where mercury discharges have been permitted by state and local regulatory agencies for decades without even a cursory scientific review of the consequences. The Penobscot system is heavily contaminated with mercury, and the scattered mercury residue analyses indicate that associated food webs (including those used by humans) represent some of the highest concentrations of mercury in the country. By basing a restoration effort on a poor scientific database, the original error has been compounded by these agencies. This is further indication of the importance of food web considerations in the determination of impacts of human activities in coastal areas.

27. By basing resource planning and management in the Apalachicola River–Bay system on protection of qualitative and quantitative aspects of freshwater runoff and associated food webs, this system remains one of the last major alluvial drainages that is largely free of toxic contamination and hypereutrophication due to anthropogenous nutrient loading. The relationships of the estuarine food webs to natural river fluctuations played an important role in planning and management approaches used to protect the Apalachicola resource.

28. The development of ecosystem-level research is not consistent with the usual hypothesis tests that drive most reductionist approaches to environmental science. Even as the definition of what constitutes ecosystem-level research is debated by ecologists, the purpose of such analyses (i.e., development of a factual basis that is useful for system management) remains clear. Because such research is often used to answer questions that are not necessarily asked at the beginning of the program, ecosystem research is often open-ended. This precludes the usual pattern of hypothesis development that underlies reductionist methods of funding and publication. Most so-called interdisciplinary projects remain in the domain of patch-quilt ecological efforts whereby individual projects by subdiscipline remain the primary focus. The ecosystem is defined by *post hoc* accumulations of information that are often not representative of system trends and the effects of human activities. Blind adherence to time-worn and misplaced reductionist approaches has contributed to the widespread deterioration of coastal systems in the United States in recent years.

29. An approach to ecosystem evaluations is proposed that includes many of the elements that have proved successful in the Apalachicola experiment. The process should begin with the development of reasonable research questions concerning system interactions. Ecosystem-level questions should be framed within a network of interdisciplinary disciplines at various levels of physical and biological organization. Research questions should be derived from broad-based preliminary studies. There should be a central focus concerning the integration of the scientific effort with resource management and planning policies. In addition, ecosystem-level research has to start at an integrative level with a uniform, broad-based, interdisciplinary database of its own. This research should include monitoring at all levels of system organization in addition to compatible experimental work that takes its hypothesis-testing mechanisms from information taken in the field. The generation of background data should be broad based enough to provide the background for future (hitherto unasked) research questions. The distinguishing characteristics of an effective ecosystem-level program are thus diametrically opposed to the usual pattern of carrying out a series of uncoordinated, unrelated (reductionist) studies and patching the results together to answer limited questions that are usually unrelated to issues at the ecosystem level.

30. The development of an ecosystem-level research program requires informed planning and someone in charge who keeps track of project results while maintaining the continuity of all project elements. Sampling methods should be quantified before starting the regular field analyses. The database should be maintained and continuously reviewed by analytical teams of statisticians, modelers, and ecological personnel. Any adjustments made to sampling and/or experimental aspects of the program should be consistent with long-term project goals. Field and laboratory experimental parts of the program should be based on empirical data, and should also answer questions that cannot be answered by the descriptive field information.

CHAPTER 13

Summary of Results

The long-term studies of the trophic organization of a series of coastal areas in the NE Gulf of Mexico constitute a different approach to the analysis of spatiotemporal processes in drainage systems. A multidisciplinary approach was used over relatively long periods in systemwide field surveys. A long-term series of trophic studies was undertaken to determine the feeding habits of fishes and invertebrates in the areas of study. These data were then used to reorganize the field information into a series of food webs that represented the complex, ontogenetic progressions of feeding habits of component species. Relatively unpolluted areas representing sea grass–dominated areas and alluvial freshwater runoff were compared to coastal systems affected by anthropogenous activities such as dredging and nutrient loading from point and non-point sources. Seasonal and interannual variation was determined in addition to the efficacy of restoration efforts to reduce nutrient loading.

Sources of primary productivity, habitat quality and distribution, and predator–prey interactions influence the ontogenetic feeding patterns of a given species. Pollution and habitat alteration cause changes in associated trophodynamic feeding patterns, thus altering food web features and the distribution of populations through breaks in ontogenetic feeding progressions. Feeding patterns of coastal fish and invertebrate communities are influenced by complex cyclic interactions. Diurnal, seasonal, and interannual dietary shifts are related to natural and anthropogenous habitat changes and related prey abundance. Interannual habitat changes due to both natural (drought–flood sequences) and anthropogenous (nutrient loading) disturbances thus determine the disposition of coastal food webs. The drought sequences follow 6-year cycles, thus precluding the representative value of more popular 1- to 2-year studies. Even slight water quality changes due to pollution that are outside the evolutionary experience of coastal populations can cause serious disruptions of habitat structure, energy flow, and community composition through altered food web processes. Unless adequate food web data are available, such changes often go unnoticed.

Spatiotemporal gradients of freshwater flows are critical to the productivity of coastal systems and associated patterns of community structure of infaunal macroinvertebrates, which form the basis of the coastal food webs. Areas affected by river flow in the Apalachicola system are characterized by relatively low species richness, high numerical abundance of relatively few populations, high biomass, and high relative dominance. In areas outside of direct riverine influence, there are relatively high numbers of species, low numbers of individuals, low biomass, and low relative dominance. Physical habitat stressors and biological processes such as predation interact with spatiotemporal productivity gradients and the exclusion of stress-susceptible marine species to determine the population distribution and community structure in the Apalachicola Bay system. Trophic diversity and secondary production are highest in areas receiving freshwater runoff (i.e., East Bay, the main stem of the Apalachicola River, the main freshwater drainage of St. George Island; Figure 13.1). Seasonal differences of trophic diversity reflect feeding successions of dominant populations. River flow and associated primary productivity delimit the distribution of herbivores and omnivores,

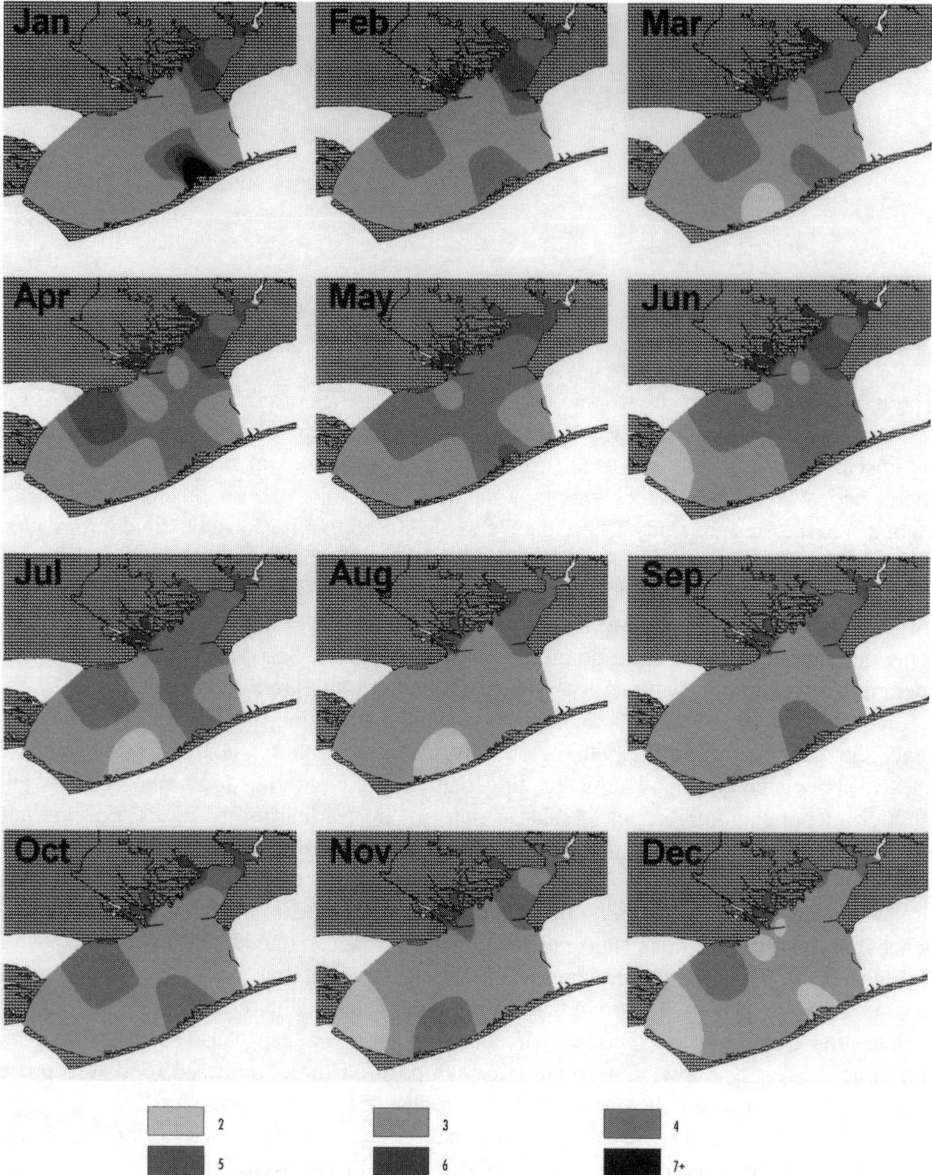

Figure 13.1 Distributions of the number of trophic types in the Apalachicola Bay system taken from data averaged by month from August 1975 through July 1984.

whereas primary, secondary, and tertiary carnivores are more closely associated with biological factors such as prey distribution. Physical control of the distribution of eurytopic carnivorous populations by habitat gradients is less important than trophic considerations in determination of the distribution of coastal assemblages.

Drought/flood cycles alter basic estuarine habitat conditions and primary production in the form of phytoplankton and submerged aquatic vegetation (SAV). Increased primary productivity due to enhanced light penetration during droughts is associated with increased herbivore and carnivore activity as a product of increased microalgal productivity. With time, however, prolonged drought conditions and associated low nutrient loading are accompanied by reduced secondary productivity.

Thus, natural cycles of river flow and nutrient loading are responsible for interannual cycles of bay productivity that approximate the 11- to 12-year drought/flood periodicities.

Alterations of natural patterns of nutrient loading by anthropogenous activities can have major impacts on food web structure through quantitative and qualitative changes of phytoplankton community structure. A 14-year study of the effects of paper mill effluents on the Perdido River–Bay system indicated long-term responses of the estuarine food webs to nutrient loading and associated phytoplankton community composition. Increased orthophosphate loading by the mill was accompanied by a well-ordered succession of phytoplankton blooms that followed species-specific seasonal patterns. There were both seasonal and interannual successions of bloom species during a period of increased orthophosphate and ammonia loading by the mill. Recurrent drought conditions exacerbated bloom occurrence. The form and timing of nutrient loading with respect to individual nutrient species controlled the response of phytoplankton bloom species as a function of relatively stable seasonal occurrences of such populations (Figure 13.2). Individual bloom species responded

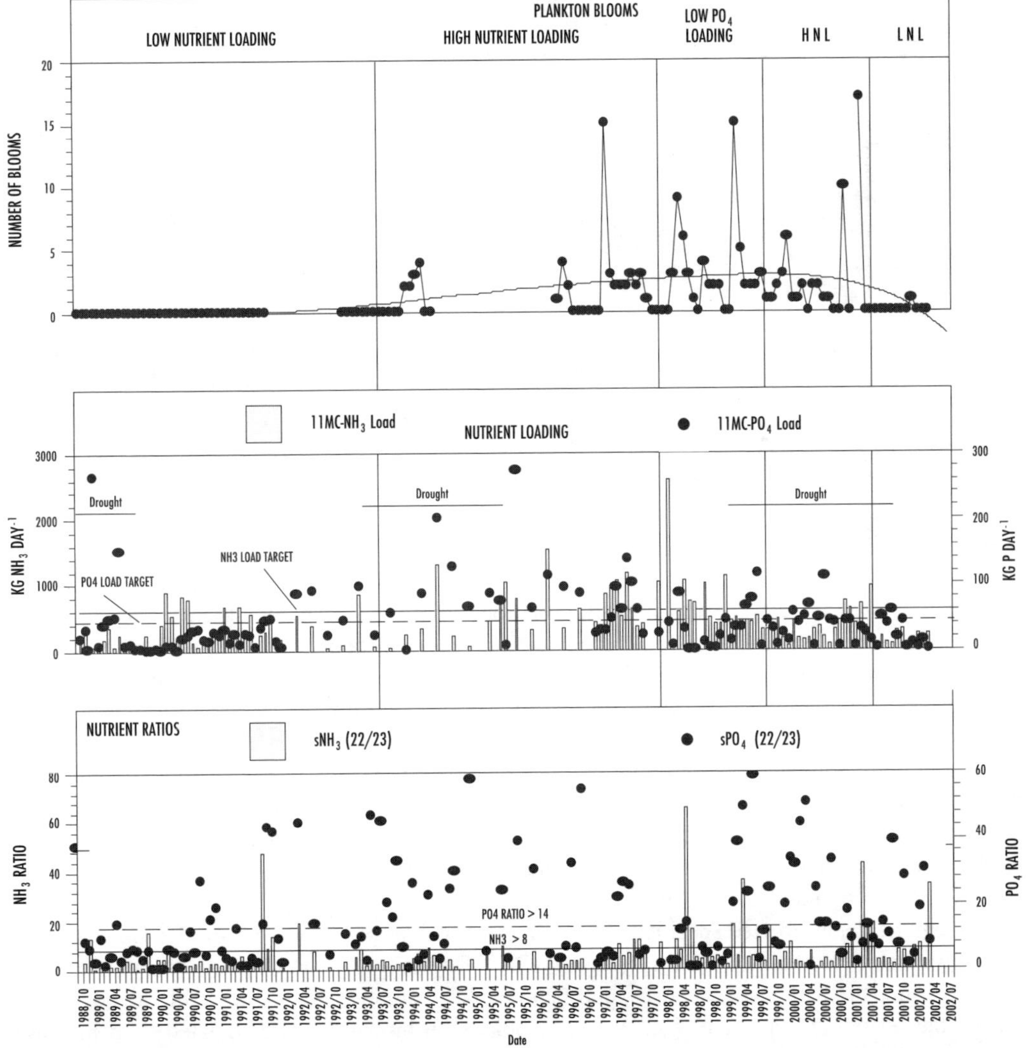

Figure 13.2 Orthophosphate and ammonia loading from Elevenmile Creek and nutrient ratios (P22/P23) taken monthly from October 1988 through May 2002. Also shown are the number of baywide phytoplankton blooms over the sampling period. Nutrient loading targets for the mill restoration program are given as well as nutrient ratios as indicators of possible bloom stimulation.

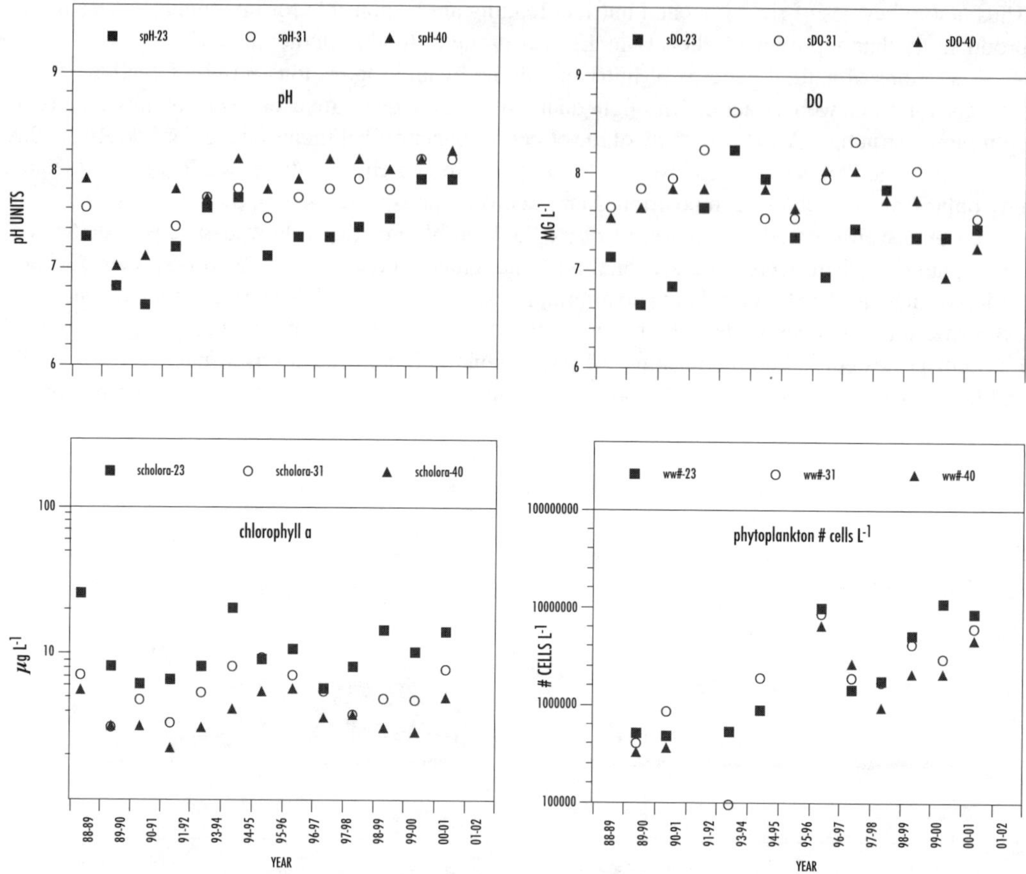

Figure 13.3 Annual averages of pH, DO, chlorophyll *a*, and phytoplankton numbers of cells per liter at three stations in Perdido Bay from 1988–1989 through 2000–2001.

to both seasonal nutrient loading and nutrient concentration patterns as modified by drought cycles. Reduced nutrient loading by the pulp mill in conformance with nutrient loading targets as part of a restoration program was accompanied by cessation of bay phytoplankton blooms, thus supporting the hypothesis that blooms were associated with increased orthophosphate and ammonia loading by the mill. There were significant differences between point- and non-point-source nutrient loading with respect to initiation of the plankton blooms. There were differences between the effects of point-source and non-point-source nutrient loading. Most point sources are continuous, whereas many non-point sources load nutrients during precipitation events. In the case of the pulp mill, there was continuous nutrient loading during drought as well as flood periods whereas non-point (agricultural) sources did not load as regularly during droughts. However, point sources are more easily controlled than non-point sources, which, in many urban and agricultural situations, are usually ignored by regulatory agencies.

Bloom incidents affected specific habitat variables such as dissolved oxygen (DO), pH, nutrient concentrations, and sediment quality (Figure 13.3). Bay recovery after bloom cessation was not the mirror image of impact sequences, and nanoplankton numbers increased as blooms dissipated during the restoration program. Increased phytoplankton biomass was associated with increased pH and reduced DO at depth in the bay. Chlorophyll *a* trends were not indicative of bloom occurrence (Figure 13.4), and followed drought sequences rather than phytoplankton response to nutrient loading. Statistical analyses of the data, based on seasonal occurrences of the bloom species, indicated significant associations with nutrient loading patterns. The distribution of the winter bloom

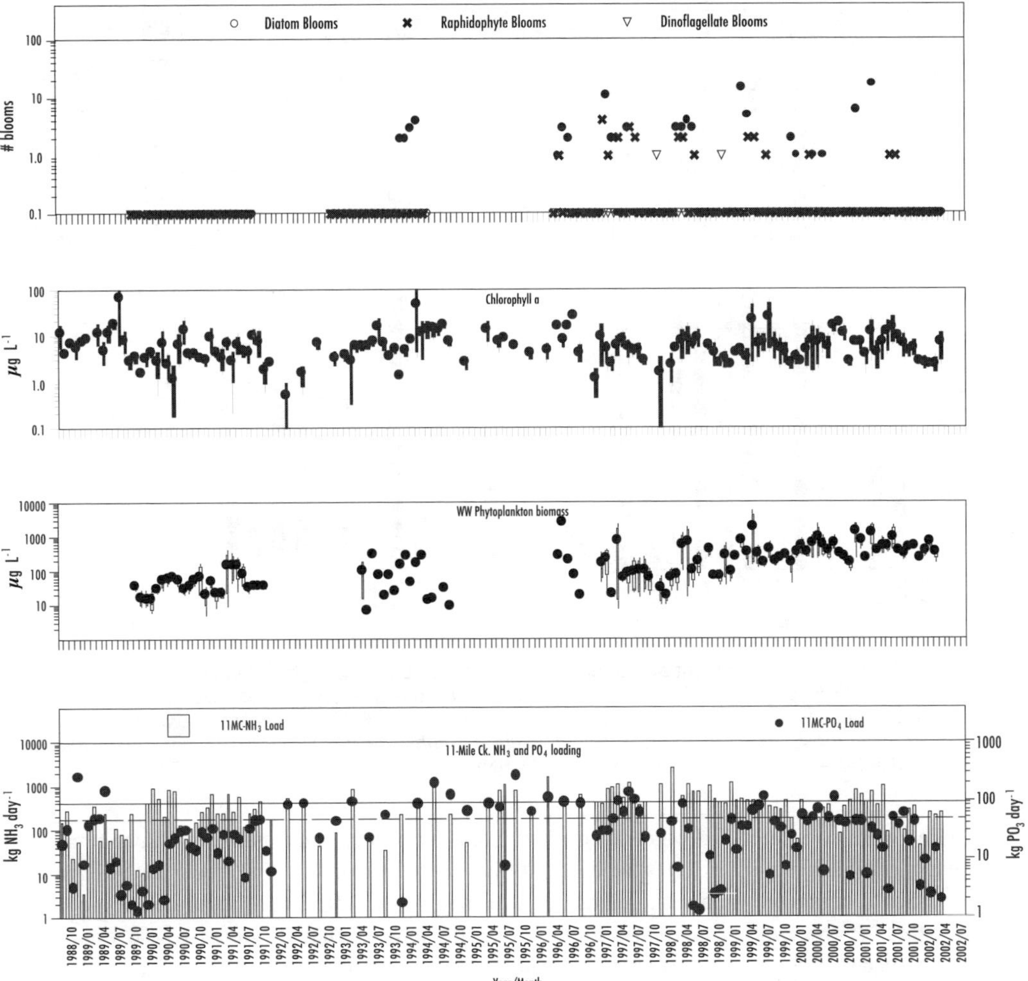

Figure 13.4 Trends of ammonia and orthophosphate loading from Elevenmile Creek, and chlorophyll *a*, whole-water phytoplankton biomass, and blooms of diatoms, raphidophytes, and dinoflagellates in Perdido Bay monthly from October 1988 through May 2002.

species *Prorocentrum minimum* was significantly correlated with orthophosphate concentration ratios. Drought conditions and resulting high orthophosphate concentrations in Elevenmile Creek were related to stimulation of *P. minimum* blooms in the Perdido system. Numerical abundance of *Cyclotella choctawhatcheeana*, a spring bloom species, was significantly associated with ammonia concentration ratios and with combined (averaged) differences of orthophosphate and ammonia concentration ratios taken during spring periods. The spring bloom species *Miraltia throndsenii* was significantly associated with orthophosphate and ammonia gradients. Cell numbers of *Heterosigma akashiwo*, a summer bloom species, were positively correlated with ammonia loading rates, with combined differences of orthophosphate and ammonia concentration ratios, and with differenced N+P loading data. *Synedropsis* sp. (a summer bloom species) was significantly associated with P22/P23 orthophosphate ratios, and the N+P differenced data for both concentration ratios and loading. These patterns of nutrient control of bloom incidence followed the pattern of results from a series of nutrient limitation experiments. Thus, the primary bloom sequences varied both seasonally and interannually, with the cessation of such blooms coinciding with reduced nutrient loading by the mill during 2001–2002 (Figure 13.4).

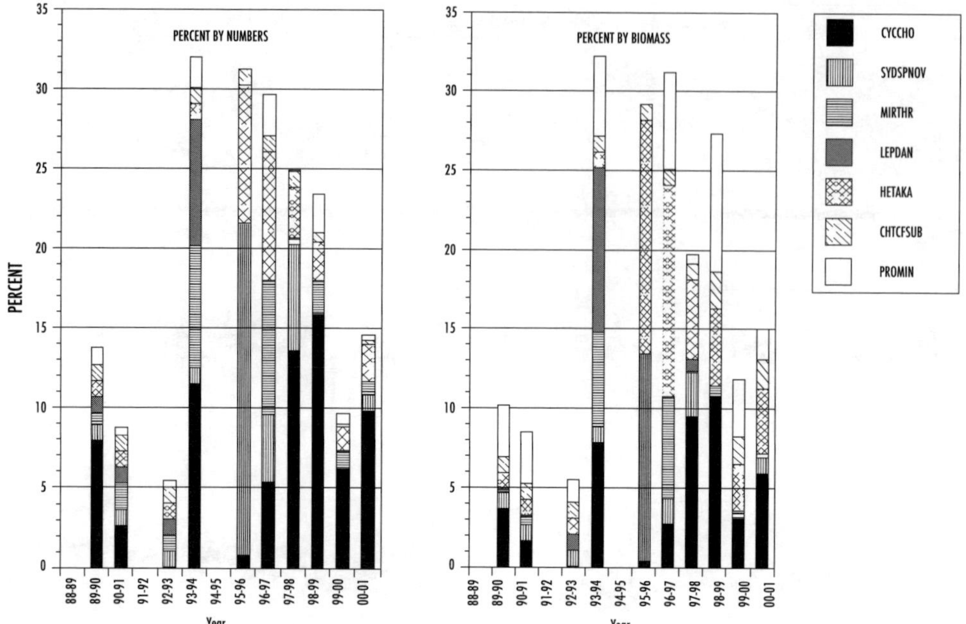

Figure 13.5 Annual trends of percent by numbers and percent by biomass of the main plankton bloom species in Perdido Bay from 1988 through 2001. Species designations are as follows: *Cyclotella choctawhatcheeana* (CYCCHO), *Leptocylindrus danicus* (LEPDAN), *Miraltia throndsenii* (MIRTHR), *Synedropsis* sp. (SYDSPNOV), *Chattonella c.f. subsalsa* (CHTCFSUB), *Heterosigma akashiwo* (HETAKA), *Prorocentrum minimum* (PROMIN).

Blooms were correlated with increased phytoplankton biomass and numerical abundance and with reductions of phytoplankton species richness, species diversity and evenness. Regression analyses of data organized on a seasonal basis confirmed bloom effects on phytoplankton community structure. *Prorocentrum minimum* numbers significantly reduced winter whole-water phytoplankton species richness. *Heterosigma akashiwo* blooms were significantly (negatively) correlated with summer phytoplankton species richness indices. There was also a significant negative correlation of *Synedropsis* sp. with summer phytoplankton species richness. Species that were associated with altered phytoplankton community structure were usually larger than the diatoms that they replaced and were typified by higher overall biomass (Figure 13.5).

Two areas of the bay affected by nutrient loading did not recover over the 14-year period of observation. In Elevenmile Creek, drought-induced increases of nutrient concentrations were associated with increased nanoplankton (cryptophytes and nanoflagellates) cell numbers and biomass, increased *Gymnodinium* spp. cell numbers, and blue-green alga (*Merismopedia tenuissimo*) blooms. Surface ammonia concentrations were positively associated with the nanococcoids, which, in turn, were negatively correlated with bottom DO. Nanoflagellate numbers were negatively associated with surface DO. Wolf Bay (lower Perdido system) was subject to unregulated nutrient loading from agricultural areas in Alabama. There was a general increase of bloom species numbers. Bloom peaks were noted from late 2000 through 2001. The dinoflagellate *P. minimum* was particularly dominant during this period along with blooms of *Skeletonemia costatum* and *C. choctawhatcheeana*. There were associated general declines of water quality features such as bottom oxygen anomalies and Secchi depths during the bloom sequences. These trends were generally opposite to trends in the upper bay where bloom occurrence declined due to reduced nutrient loading from the pulp mill. The Wolf Bay data indicate that nutrient loading from unregulated non-point sources (i.e., agricultural activities) can be as damaging as that from point sources with the disadvantage that such loading is not regulated.

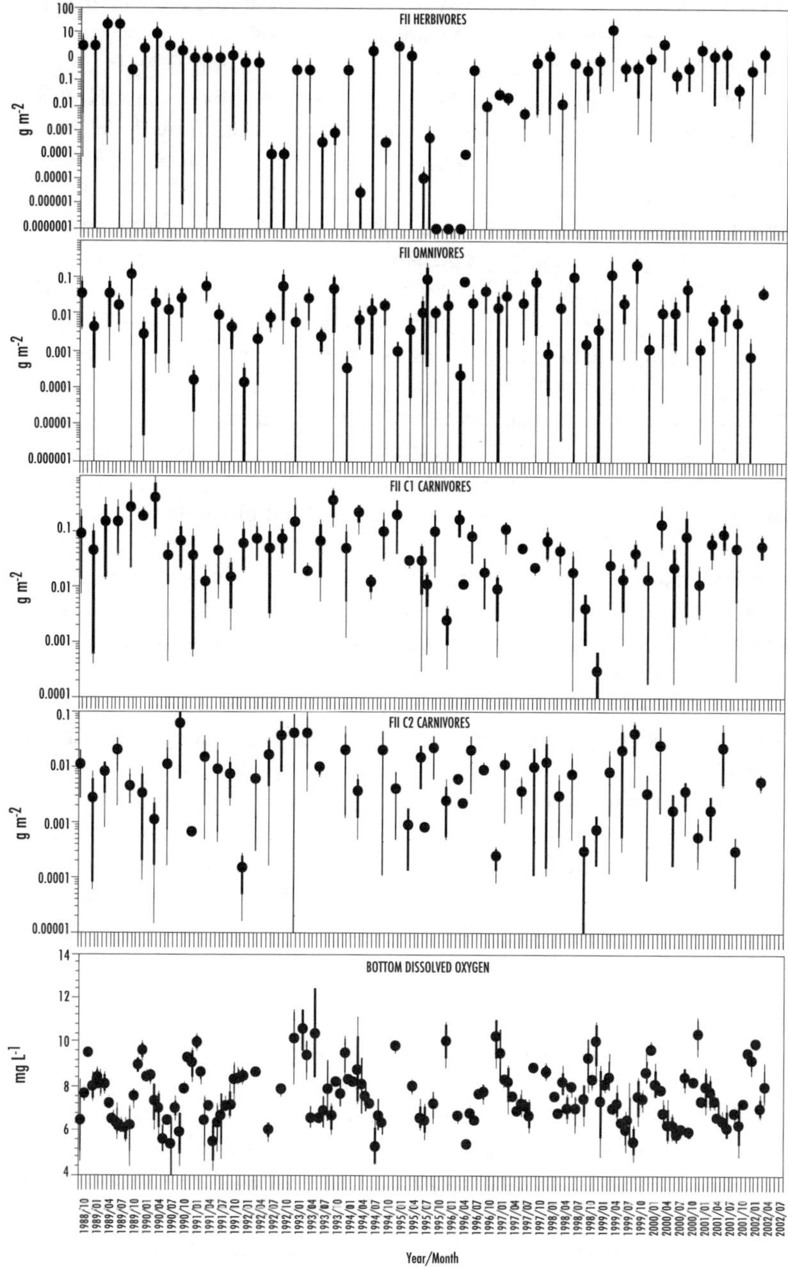

Figure 13.6 Long-term trends of FII herbivore, omnivore, and carnivore biomass in Perdido Bay from October 1988 through April 2002. Bottom DO for the period is also shown. Data are given as means of the long-term sampling stations with the 10th, 25th, 75th, and 90th percentiles also shown.

Phytoplankton blooms were associated with reduced biomass of herbivores, omnivores, and carnivores in Perdido Bay (Figure 13.6). Herbivores were reduced from 1992 to 1997, with pronounced reductions during the bloom period of 1995–1996. After cessation of mill orthophosphate loading from late 1977–1998, there was a brief period of increased herbivore biomass that was followed by another period of decline (1999 to 2001) as nutrient loading from the mill increased. Upper bay herbivores were negatively associated with *P. minimum* biomass during winter periods.

Omnivore biomass was positively associated with *P. minimum* during the winter, surface ammonia during the spring, and bloom numbers in the summer. In the lower bay, omnivores were positively associated with numbers of winter plankton blooms and with summer N+P loading differences. These data indicated that omnivores were favored by nutrient blooms and N+P loading during certain seasons, and possibly replaced herbivores that were displaced by the plankton blooms.

Statistical tests of the food web components with multiple independent factors indicated direct associations of adverse impacts due to bloom occurrences. Primary carnivore biomass decreased during the period of bloom successions. In the upper bay, primary carnivores were negatively associated with orthophosphate ratios, phytoplankton biomass, and bloom species biomass. Primary carnivores were negatively associated with *Prorocentrum minimum* in the upper bay during summer months. In the lower bay, primary carnivore biomass was negatively associated with winter bloom numbers and with raphidophyte biomass. During summer months, primary carnivores in the upper bay were negatively correlated with N+P ratio differences, and raphidophyte (*H. akashiwo*) biomass. During fall, C1 carnivores were negatively associated with ammonia loading and *H. akashiwo*/*C. choctawhatcheeana* biomass. Secondary carnivores were negatively associated with winter bloom numbers, and raphidophyte (*H. akashiwo*) biomass. During spring, C2 carnivores were negatively associated with nutrient concentration indices, total phytoplankton biomass, and bloom biomass. During summer, C2 biomass was negatively correlated with nutrient concentration indices and raphidophyte/*H. akashiwo* biomass. In the fall, this group was negatively associated with ammonia loading, and *C. choctawhatcheeana* and *H. akashiwo* biomass. The key toxic bloom species (*P. minimum, H. akashiwo, M. tenuissima*) were thus the primary effectors of the loss of secondary production, although such effects were superimposed over other important habitat changes such as reduced DO. Bloom cessation was not accompanied by immediate return of the bay food webs. This was compatible with time lags of food web components and the effects of drought on secondary productivity that were noted in the unpolluted Apalachicola system.

The distribution of primary producers (SAV, phytoplankton) determined food web components in Gulf coastal systems. The response of food webs to water quality and nutrient/toxic agent loading was related to long-term productivity trends. Biological activity and trophic diversity were highest in freshwater runoff areas of alluvial systems where phytoplankton productivity was high and high dominance of selected eurytopic species prevailed. This was illustrated in the averaged distributions of C1 carnivores in the various alluvial systems (Figure 13.7). The C1 carnivores represent the prime users of the coastal areas as nurseries. The unpolluted Apalachicola system had the highest concentrations of primary carnivores. As a result of the polluted runoff and bloom activity, the Perdido and Pensacola Bay systems were characterized by sharp reductions of secondary productivity. The relatively high C1 carnivore biomass in the Apalachicola system was located in the major river drainage and freshwater runoff from St. George Island. In Perdido Bay, there was a progressive reduction of upper bay concentrations of C1 carnivores that reflected the temporal progressions of plankton blooms. By 1997 to 1999, carnivores were isolated to the *Vallisneria* beds in western parts of the upper bay. The Pensacola Bay system, bereft of oysters bars and sea grass beds and beset by plankton-related habitat deterioration, had uniformly low levels of C1 carnivore biomass. Thus, the food web components follow patterns of distribution that are associated with freshwater runoff. When either the quantity or the quality of such runoff is adversely affected by human activities, food web structure is disrupted with associated declines of secondary production.

The distribution of dominant coastal species in alluvial systems of the NE Gulf is shown in Figure 13.8. Bay anchovies, as plankton feeders, are located in the upper parts of the respective bays where primary production is highest. However, bloom effects were obvious in the temporal distribution of this species in Perdido Bay. Atlantic croaker were prevalent in river-dominated areas of Apalachicola Bay, but were largely absent from the Pensacola Bay system. This species peaked during the beginning of the blooms in Perdido Bay, but were largely absent during later periods of bloom activity. Spot were absent in the Pensacola system and during later bloom periods in Perdido Bay. White shrimp, again associated with freshwater runoff in Apalachicola Bay where

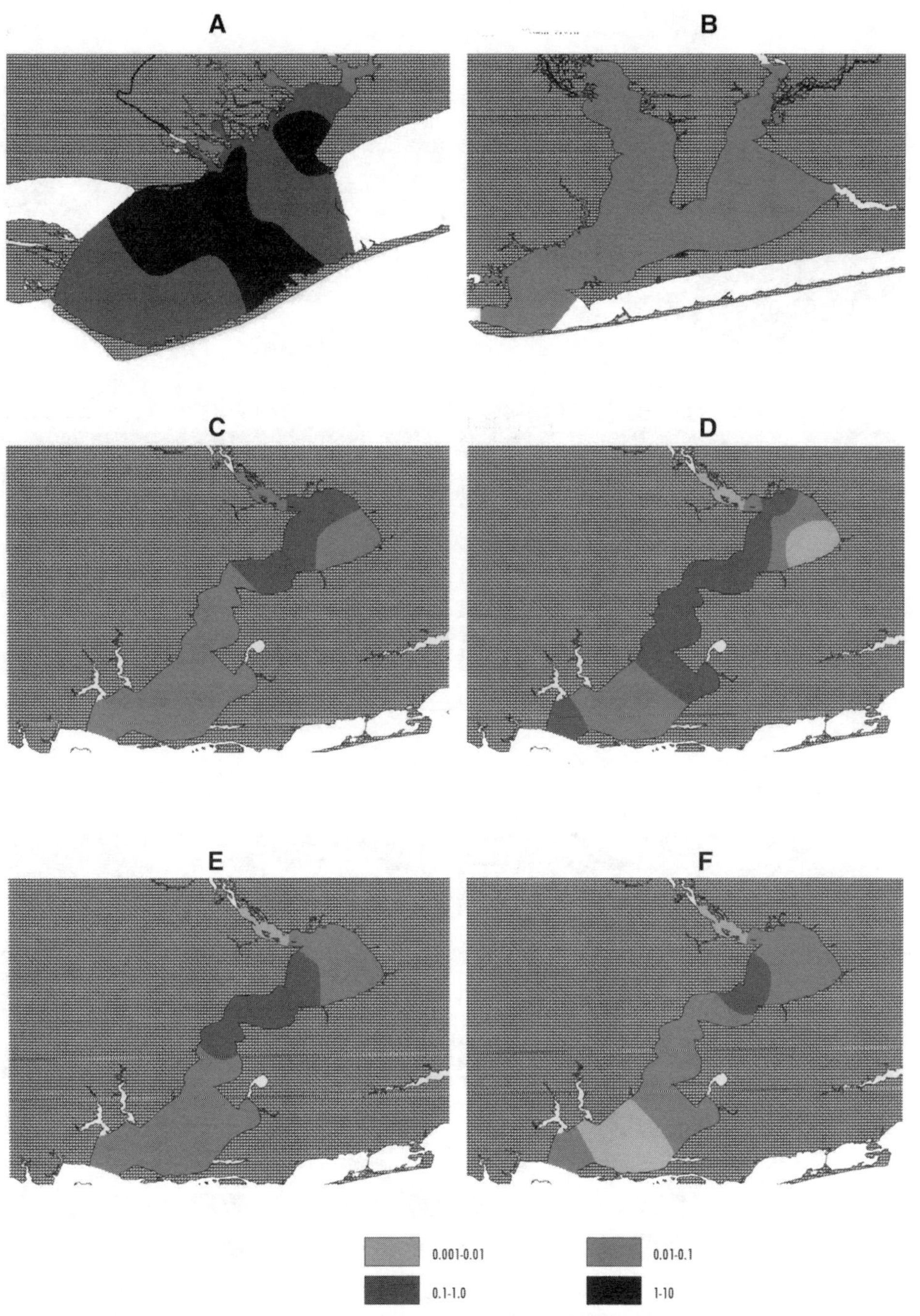

Figure 13.7 Maps of the averaged distributions of primary carnivore biomass for the Apalachicola estuary, 1975–1984 (A); the Escambia–Blackwater systems, 1997–1998 (B); and Perdido Bay, 1988–1990 (C), 1993–1994 (D), 1995–1996 (E), and 1997–1999 (F).

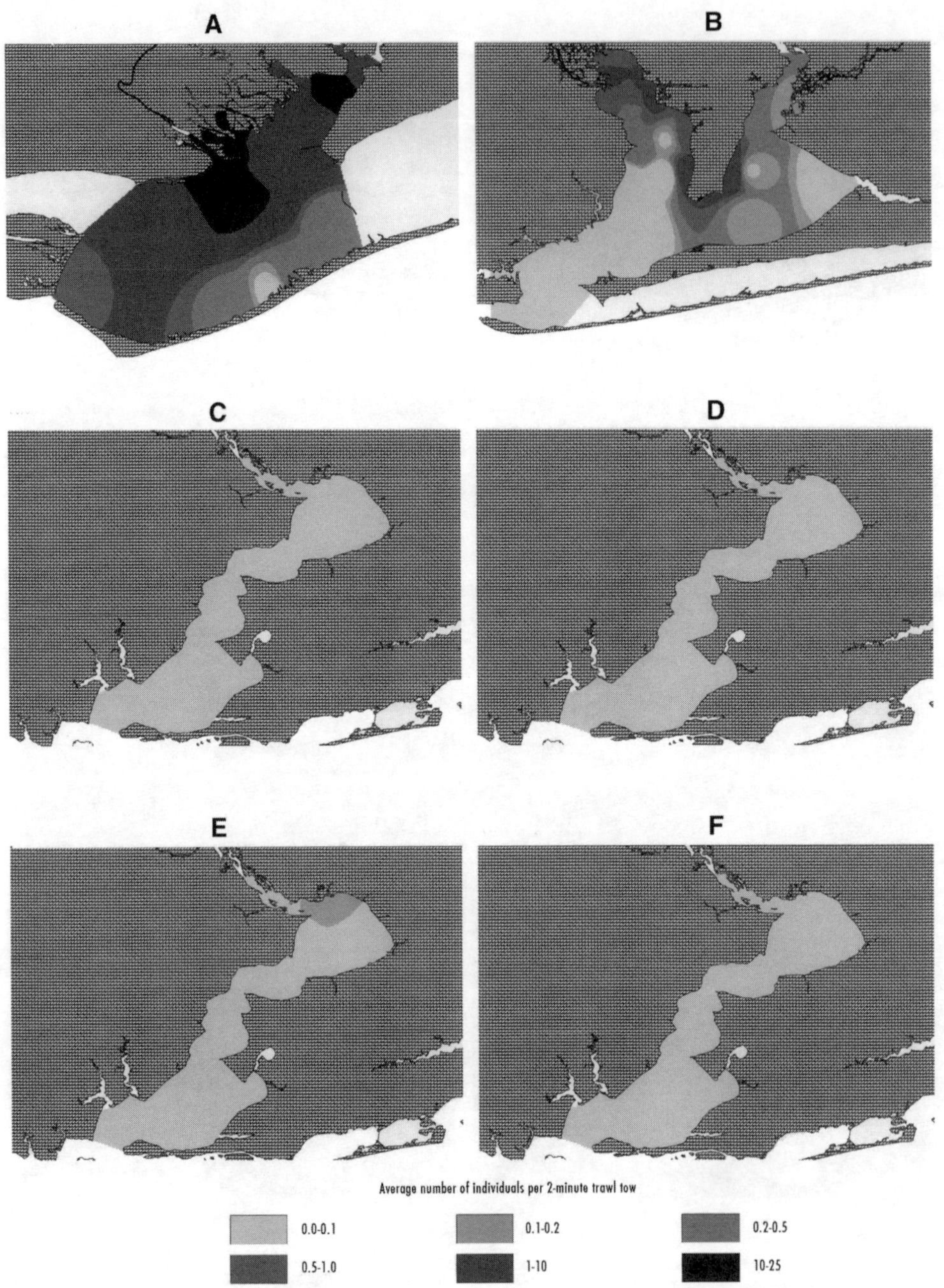

Figure 13.8 Distributions of white shrimp (*Penaeus setiferus*) in the Apalachicola estuary, 1975–1984 (A); the Escambia–Blackwater systems, 1997–1998 (B); and Perdido Bay, 1988–1990 (C), 1993–1994 (D), 1995–1996 (E), and 1997–1999 (F). (continued)

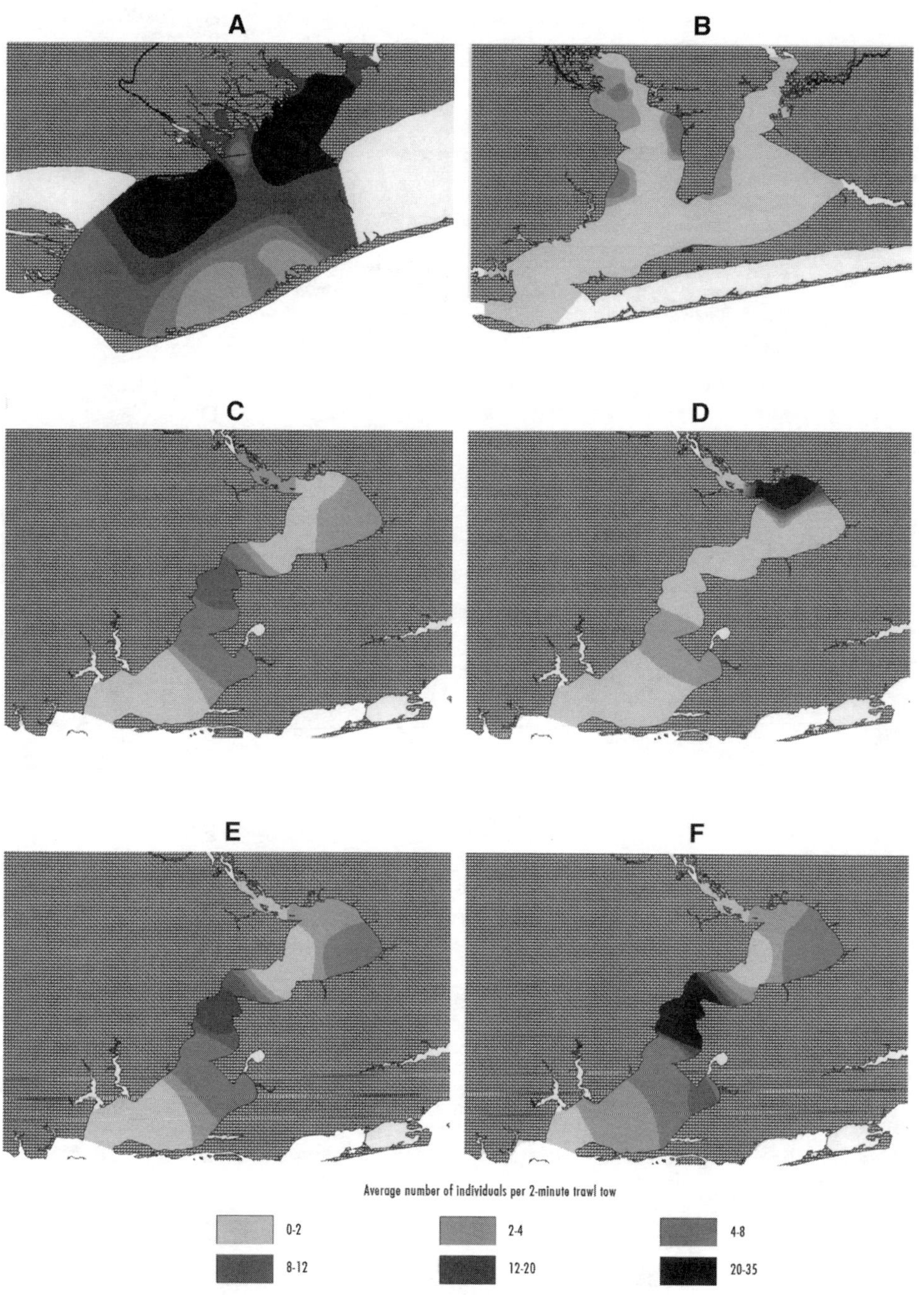

Figure 13.8 (continued) Distributions of Atlantic croaker (*Micropogonias undulatus*) in the Apalachicola estuary, 1975–1984 (A); the Escambia–Blackwater systems, 1997–1998 (B); and Perdido Bay, 1988–1990 (C), 1993–1994 (D), 1995–1996 (E), and 1997–1999 (F). (continued)

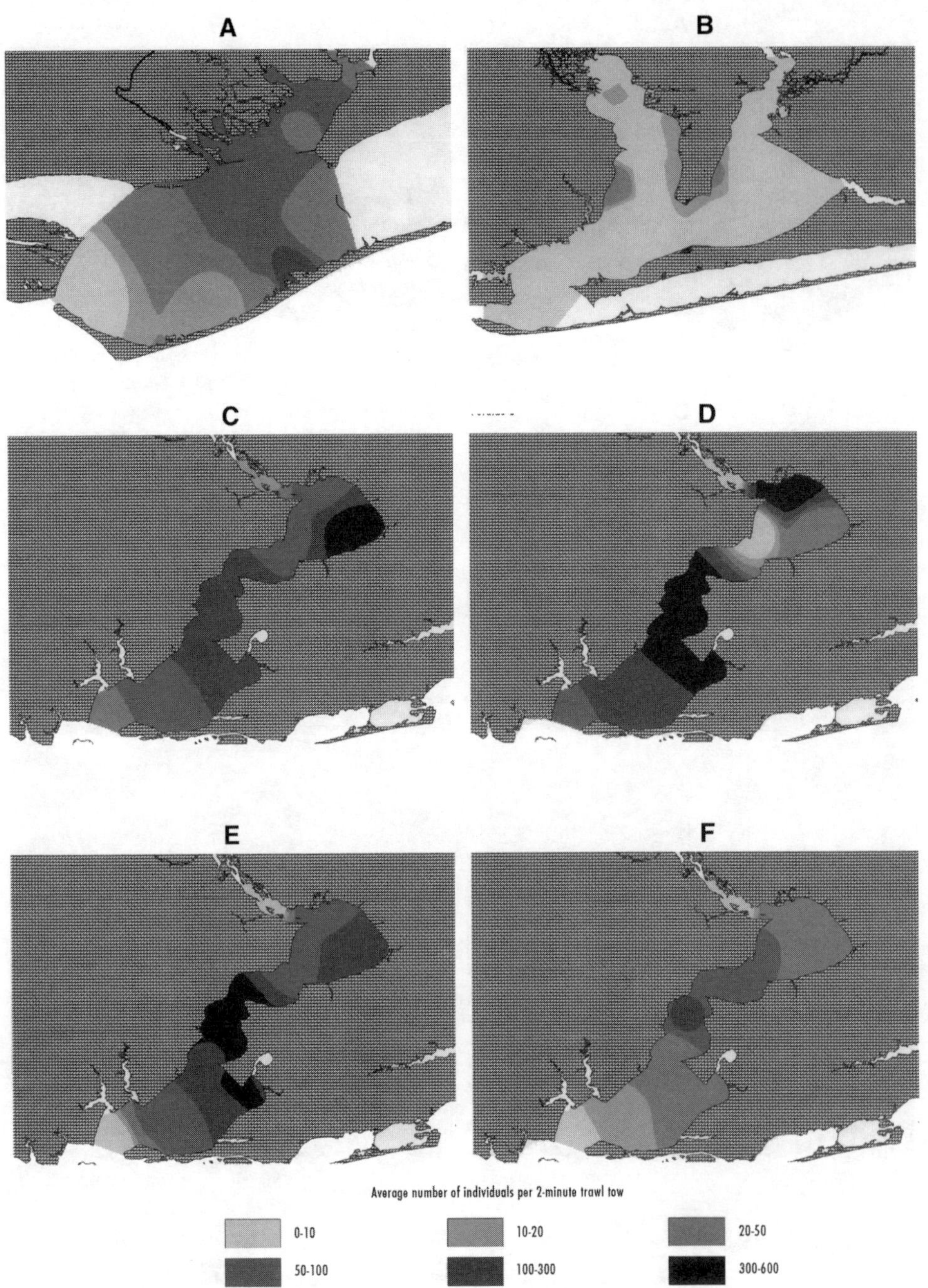

Figure 13.8 (continued) Distributions of spot (*Leiostomus xanthurus*) in the Apalachicola estuary, 1975–1984 (A); the Escambia–Blackwater systems, 1997–1998 (B); and Perdido Bay, 1988–1990 (C), 1993–1994 (D), 1995–1996 (E), and 1997–1999 (F). (continued)

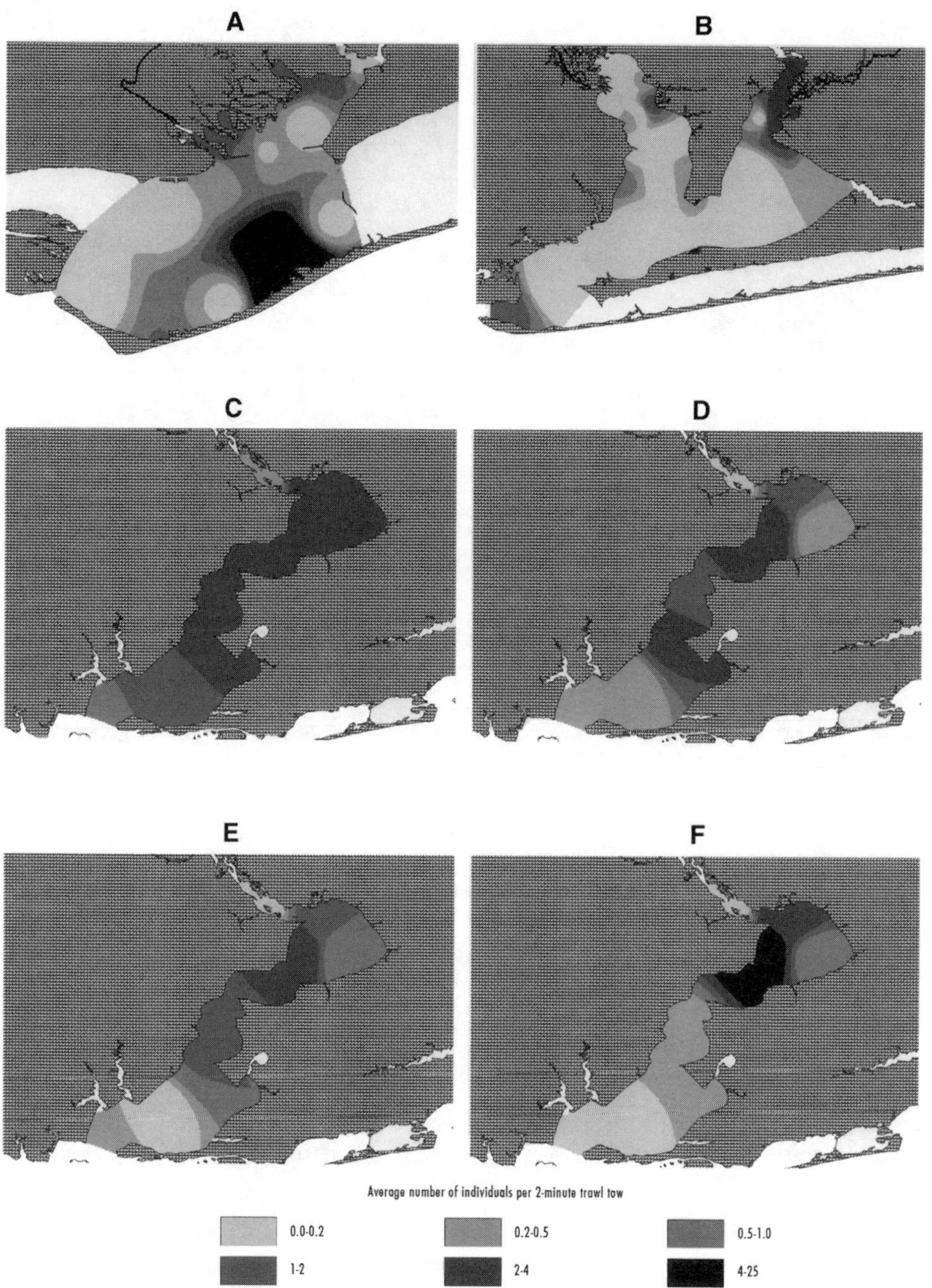

Figure 13.8 (continued) Distributions of pinfish (*Lagodon rhomboides*) in the Apalachicola estuary, 1975–1984 (A); the Escambia–Blackwater systems, 1997–1998 (B); and Perdido Bay, 1988–1990 (C), 1993–1994 (D), 1995–1996 (E), and 1997–1999 (F). (continued)

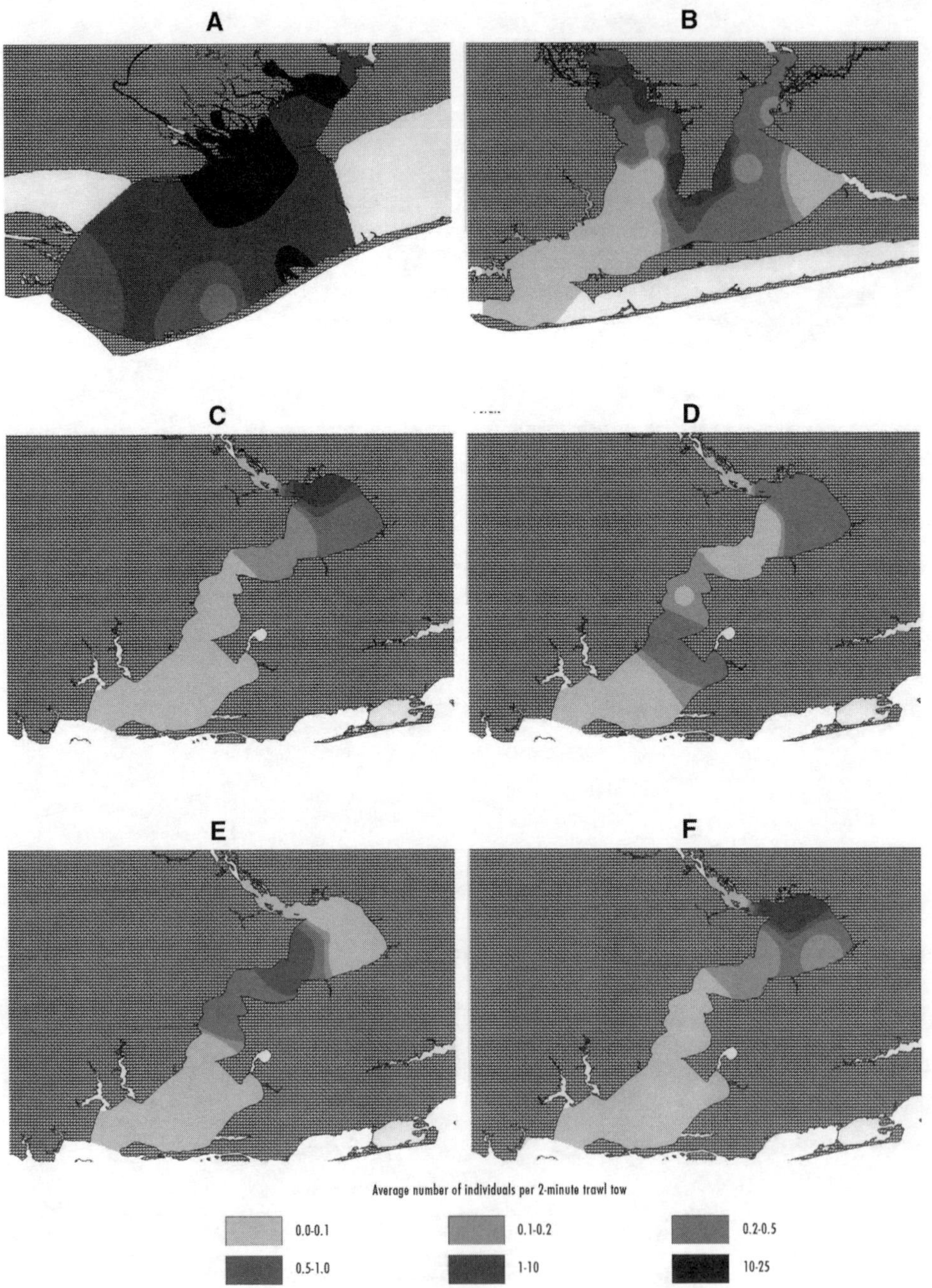

Figure 13.8 (continued) Distributions of blue crabs (*Callinectes sapidus*) in the Apalachicola estuary, 1975–1984 (A); the Escambia–Blackwater systems, 1997–1998 (B); and Perdido Bay, 1988–1990 (C), 1993–1994 (D), 1995–1996 (E), and 1997–1999 (F). (continued)

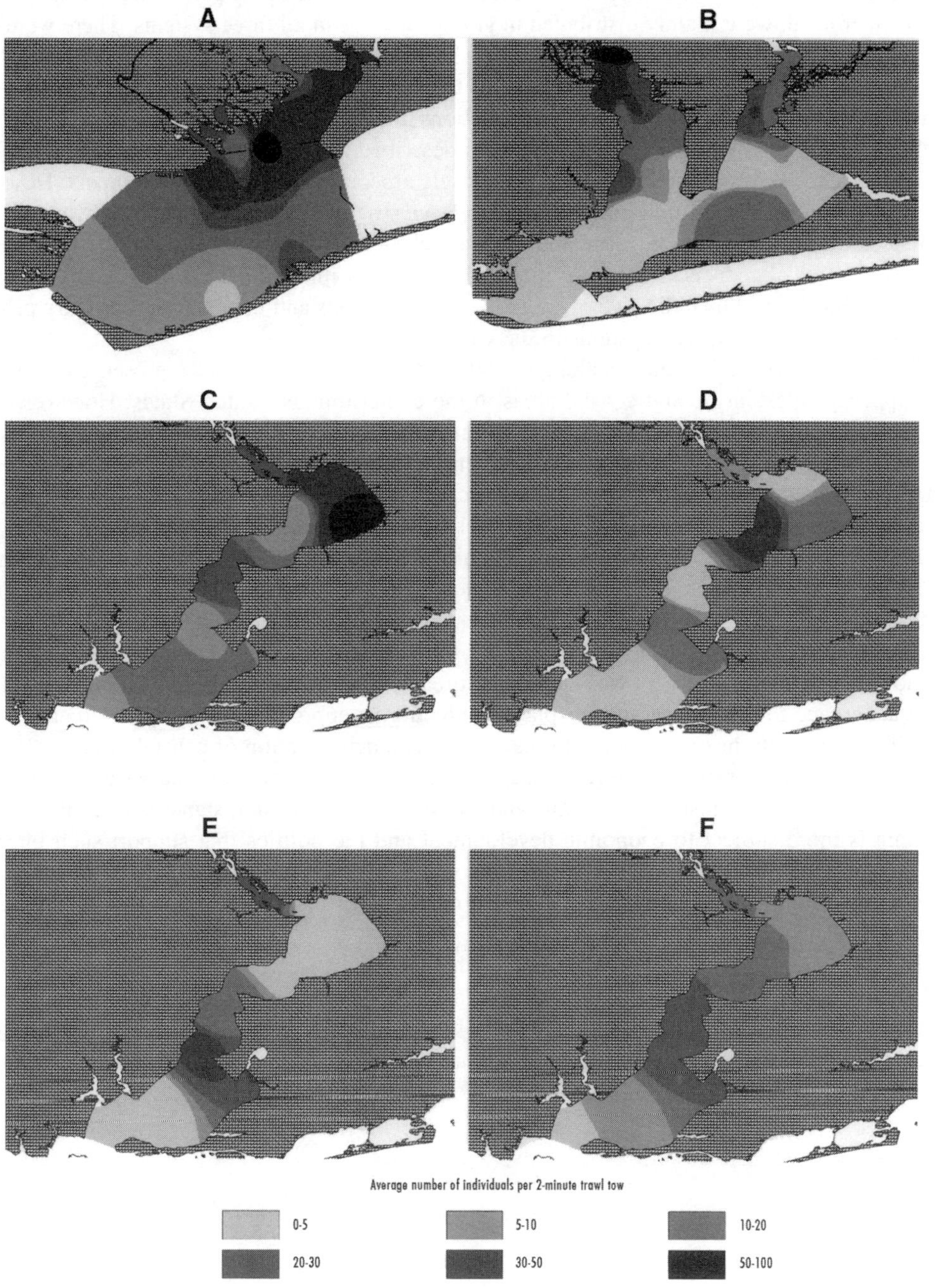

Figure 13.8 (continued) Distributions of bay anchovies (*Anchoa mitchilli*) in the Apalachicola estuary, 1975–1984 (A); the Escambia–Blackwater systems, 1997–1998 (B); and Perdido Bay, 1988–1990 (C), 1993–1994 (D), 1995–1996 (E), and 1997–1999 (F). (continued)

they were numerous, were absent from Perdido Bay and were sparse in the Pensacola system. Blue crab distribution in Apalachicola Bay showed a classic response to freshwater runoff. Blue crabs recovered in Perdido Bay after a period of reduced numbers in the upper part of the system. Pinfish, on the other hand, were mainly distributed in grass bed areas in all three systems. There were thus different trends of distribution of different estuarine species. However, in areas affected by anthropogenous nutrient loading and plankton blooms, these species were adversely affected with severe restrictions of biomass compared to such distributions in the unpolluted Apalachicola system. These results are consistent with the food web patterns described above.

In Apalachee Bay, food web structure was closely associated with SAV development. However, the effects of pollution had similar impacts on food web structure as that noted in the phytoplankton-dominated alluvial systems. In these areas, food web deterioration was evident due to losses of SAV as a result of reduced light transmission due to pulp mill effluents. Food web patterns in Apalachee Bay were thus indicative of water quality conditions and the quality of primary productivity as shown in the alluvial systems to the west.

The losses of coastal resources along the NE Gulf coast of Florida are representative of what has happened to estuarine and coastal areas in the conterminous United States. However, there has been virtually no regulatory response to this situation. Nutrients and toxic substances are still loaded into these systems without the scientific recognition of what is happening because of antiquated ideas concerning ecosystem research. The key to protection of these once productive systems is control of water quality and maintenance of the quality and quantity of freshwater runoff into estuarine systems. Progressive planning and management initiatives taken during the 1970s and early 1980s in the Apalachicola system provides evidence that scientific data can be successfully applied to resource management issues. However, the ecosystem approach taken in the Apalachicola experiment has been largely ignored by the scientific community, and remains unique in that such management was applied before the system was damaged by human activities. Blind adherence to time-worn and misplaced reductionist approaches by major elements of the scientific community has contributed to the widespread deterioration of coastal systems. Reductionist interests that continue to discriminate against long-term, interdisciplinary studies dominate the current system of research funding and publication. In most instances, coastal resource planning is more subject to economic development and the politics that support such interests than progressive efforts to protect natural resources in once-productive coastal areas. These interests now dominate the news media and other forms of communication, which adds to the progressive damage to coastal productivity.

This book represents an alternative approach to coastal research that has been successfully applied to coastal resource management questions. An understanding of coastal food webs is fundamental to this approach, and more emphasis should be placed on how trophic relationships are being affected by the increased level of human activity in ecologically sensitive coastal areas.

APPENDIX I

Field/Laboratory Methods Used for the CARRMA Studies (1971 through 2002)

The research effort in the northern Gulf of Mexico is based on written, peer-reviewed protocols for all field and laboratory operations. Water quality and biological analyses have been taken by personnel from the Center for Aquatic Research and Resource Management (CARRMA, Florida State University, Tallahassee, Florida) and Environmental Planning & Analysis, Inc. (Tallahassee, Florida) with additional work carried out at a series of laboratories in various countries. Water quality methods and analyses and specific biological methods have been continuously certified through the Quality Assurance Section of the Florida Department of Environmental Protection (Comprehensive QAP 940128 and QAP 920101). Various methods that have been used to take field information and run laboratory analyses have been published in the reviewed literature. This includes the following: Livingston et al., 1974, 1976, 1997, 1998a,b; Livingston, 1975, 1976, 1980, 1982, 1984a,b, 1985a, 1987c, 1988, 1992, 1997a; Flemer et al., 1997.

A. SEDIMENT ANALYSES

Sediments were analyzed for particle size distribution and organic composition according to methods described by Mahoney and Livingston (1982) as defined by Galehouse (1971).

1. Particle Size Analysis

Sediment samples were taken with coring devices (Plexiglas: 7.6-cm diameter, 45-cm^2 cross section). Analyses were taken on 10-cm samples for regular percent organics and particle size analysis.

a. Unpreserved sediments were divided into coarse (>62 μm) and silt-clay (<62 μm) fractions.
b. The coarse fraction was analyzed by wet-sieving using 1/2 phi unit intervals.
c. The silt-clay fraction was analyzed using a pipette method in 1/2 phi unit intervals from 4 to 6 phi, and in 1-phi intervals for the finer fractions (6 to 10 phi).
d. Statistical analyses were carried out using phi units and sediment designations as outlined in the Wentworth scale.
e. Statistical treatments of the granulometric results followed the method of moments for calculation of mean grain size (in phi units, with 1 mm = 0.5 phi), skewness (a measure of non-normality of distribution), and kurtosis (a measure of the spread of the distribution). Sorting coefficients represented the measure of grain size as follows: well sorted = 0.50; moderately well sorted = 0.71; moderately sorted = 1.00; poorly sorted = 2.00.

2. Percent Organics

After placing the sediments in a drying oven set at 104°C for a minimum of 12 h, the sediments were ashed in a muffle furnace for 1 h or more at temperatures approximating 550°C. The percent organic fraction of the sediments was calculated as a percent of the dry weight.

Sediment analysis is a two-part process where the first consists of forming a data file and the second is the actual granulometric analysis. The first part may be accomplished either by running the EZseds program to enter and reformat raw data or by exporting previously entered sediment data from one of the EP&A 4th Dimension Data Management System (DMS) sediment data files. EZseds is a FORTRAN program that prepares raw sediment granulometric data for analysis with the MacGranulo program. MacGranulo is a FORTRAN application that performs sediment grain-size analysis on the data in the input file. Results include summary statistics such as mean grain size (as determined by three different methods) and percent organics, particle-size distribution according to the Wentworth classification, and a frequency distribution plot.

3. Mercury Analyses

Sediment samples were collected at a series of sampling sites in the system. Three subsamples were taken at each established station. Sediment samples were placed on ice. The samples were shipped to Enviro-Test Laboratories in Canada. Prior to shipping, all samples were logged into chain of custody forms and a sample logbook. Samples were examined to ensure that accurate labels had been affixed and that all lids were tight. Prior to closing each cooler containing samples, the chain-of-custody form was completed and placed in a reclosable plastic bag that was then placed inside the cooler. The forms were placed in plastic bags. The originator retained a copy of each chain-of-custody form. As a general procedure, samples were taken in order from supposedly clean areas to possible polluted areas. No sampling was carried out in the rain.

Quality control samples (blanks, etc.) were labeled as above. After collection, identification, and preservation, the sample was maintained under chain-of-custody procedures discussed below. A blank consisted of an empty sampling container. For each group of samples (inorganic mercury, methyl mercury), one blank was used for each lot number of jars.

Samples of sediments were taken with corers (the top 2 cm was used for mercury analysis; top 10 cm for sediment particle size and percent organics). A PVC corer was used for the mercury samples. Chemical analyses were run on sediments taken from each of the sampling stations. Three replicate sediment samples were taken at each station for total mercury and methyl mercury analysis. Each of the three replicates was composed of three core subsamples that were cut into squares (all edges "shaved") to eliminate the outer parts of the core sample. The "shaving" of the sample was carried out with an acid (5 to 10% nitric acid) washed plastic knife. Samples were placed in acid-washed 1-pint, wide-neck, Teflon-lined screw-top glass jars. Nitric acid (10%-pesticide grade) was used to wash the jars. Samples were immediately preserved in the field with concentrated nitric acid to pH 2. Care was taken to mix the sediment and check for the proper pH. Each bottle was labeled as indicated above. Acid preservation for each sample was checked with pH paper to make sure that the pH 2 target is maintained. Samples were then placed on ice and bubble wrapped.

Sampling for toxic agents in sediments in other studies followed similar procedures. All sediment samples were taken with corers (the top 2 cm). A stainless steel corer was used for the organic toxicant samples, and a PVC corer was used for the metal samples.

B. PHYSICOCHEMICAL SAMPLING AND ANALYSIS

Designated, fixed stations in a given system were sampled (surface and bottom) for water quality factors. Vertical collections of water temperature, salinity, conductivity, dissolved oxygen (DO), depth, Secchi depth, and pH were taken at the surface (0.1 m) and at 0.5-m intervals below the surface to about 0.5 m above the bottom. Surface and bottom water samples were taken with Niscon bottles for further chemical analyses in the laboratory. All sampling was carried out on a synoptic basis with all samples in a given system taken within one tidal cycle.

A list of variables is given below.

Physicochemical Factors

1. Temperature
2. Salinity/conductivity
3. DO
4. pH
5. Secchi depth
6. Depth
7. Light transmission
8. True color, NCASI colorimetric method
9. True color, spectroscopic method
10. Turbidity, nephelometric method
11. Total dissolved solids
12. Total suspended solids
13. DO, YSI DO meter
14. Particulate organic carbon
15. Dissolved organic carbon
16. Total inorganic carbon
17. Total organic carbon
18. Total carbon
19. Chlorophyll *a*
20. Chlorophyll *b*
21. Chlorophyll *c*

Nutrients

1. Nitrogen nitrate and nitrite
2. Ammonia nitrogen
3. Organic nitrogen and organic phosphate
   a. Total organic nitrogen and total organic phosphate
   b. Dissolved organic nitrogen and dissolved organic phosphate
   c. Particulate organic nitrogen and particulate organic phosphate
4. Total nitrogen
5. Total phosphorus
6. Orthophosphorus
7. Dissolved reactive silicate

Physicochemical data and water samples for chemical analysis were taken according to methods defined by the U.S. Environmental Protection Agency (1983). Temperature and salinity were measured with YSI Model 33 S-C-T meters calibrated in the laboratory with commercial standards. Dissolved oxygen was taken with YSI Model 57 oxygen meters calibrated by the azide modified Winkler technique. The pH was measured with an Orion Model 250A pH meter equipped with a Calomel electrode. Aquatic light readings were taken with a Li-Cor LI-1800UW Underwater Spectroradiometer, which measures the spectral composition of photon flux density at 1 to 2 nm intervals from 300 to 850 nm.

A Beckman DU-64 spectrophotometer was used to analyze true color (APHA, 1989). Both ratio and nephelometric turbidity analyses were carried out. A Ratio Turbidimeter and a Hach model 2100A Turbidimeter were used for turbidity analyses. Quantification of dissolved inorganic nitrogen (nitrite, nitrate) and dissolved inorganic phosphorus (orthophosphate) followed methods outlined by Parsons et al. (1984). Ammonium was measured with an ion electrode (U.S. EPA 1983, method 350.3) after raising the pH to 11. This method had some essential features (e.g., minimal interference from waters highly stained with humic materials and paper mill effluents) that required special attention to the results. Consequently, the level of detection was relatively high (e.g., 2.0 μm ammonium-N) but adequate for this study. Particulate organic nitrogen (U.S. EPA 1983, method 351.4) and particulate organic phosphorus (U.S. EPA 1983, method 364.4) were collected on Gelman glass fiber filters combusted at 500°C and oxidized to inorganic fractions, with organic N measured as ammonia and organic P measured as particulate organic phosphorus. Total dissolved solids (APHA Method 209-B), total suspended solids (APHA Method 209-C), dissolved organic carbon (APHA Method 415.1), chlorophyll *a* (APHA Method 1002-G), and biochemical oxygen demand (APHA Method 405.1) were carried out according to the methods outlined by the American Public Health Association (APHA, 1989). Particulate organic carbon was analyzed according to Parsons et al. (1984) (method 3.1). Nutrient analyses were carried out according to APHA (1989).

C. NUTRIENT LOADING MODELS

The U.S. Geological Survey (Tallahassee, Florida; M. Franklin, personal communication) provided flow data for the Perdido River system, Bayou Marcus Creek, and Elevenmile Creek. Nutrient sampling was carried out in the various rivers leading into the bay. Monthly nutrient concentrations from these rivers were used in the models. Overland runoff from unmonitored areas was estimated using models devised by A. Niedoroda (personal communication). Loading models for the 11-year database were run on monthly, quarterly, and annual intervals.

The nutrient loading models were based on a ratio estimator developed by Dolan et al. (1981). This model corrects for the biases due to sparse temporal sampling. It uses auto- and cross-covariance values of flows and nutrients for correction of the loading calculations. Nutrient data for the models were usually taken monthly, whereas river data were taken daily. Program modifications were made by A. Niedoroda and G. Han (personal communication) whereby (1) due to discontinued river monitoring a Bayou Marcus Creek and the Blackwater River in September 1991, river flows for these areas were estimated from regressions (log/log, Elevenmile Creek and Bayou Marcus Creek; Styx and Blackwater Rivers); (2) stations 9 and 21 were sampled monthly from March 1993 to October 1994 and flows were then estimated by the following, St. 09: (2.52 * St. 05) + (1.01 * St. 04) + (1.10 * St. 07); St. 21: (2.64 * [St. 13 – 32]) + 32. These data are considered questionable at best and should not be used for any serious conclusions.

Dr. Alan Niedoroda and Mr. Gregory Han of Woodward-Clyde designed the loading estimator program. Data were grouped according to methods described in previous reports. Model applications were carried out according to a formula given by Dolan et al. (1981):

$$\bar{\mu}_y = \mu_x \frac{m_y}{m_x} \left(\frac{1 + \frac{1}{n} \frac{S_{xy}}{m_x m_y}}{1 + \frac{1}{n} \frac{S_{x^2}}{m_{x^2}}} \right)$$

where

μ_y = estimated load

μ_x = mean daily flow for the year

m_y = mean daily loading for the days on which concentrations weredetermined

m_x = mean daily flow for the days on which concentrations weredetermined

$$S_{xy} = \frac{1}{(n-1)} \sum_{i=1}^{n} x_i y_i - nm_x m_y$$

$$S_{x^2} = \frac{1}{(n-1)} \sum_{i=1}^{n} x_{i^2} - nm_{x^2}$$

n = number of days on which concentrations were determined
x_i = individual measured flows
y_i = daily loading for each day on which concentrations were determined

This formulation corrects for biases introduced by sparse (temporal) sampling by using auto- and cross-covariance values of the flows and nutrients to correct the loading calculations for variability in the river flow, which influences the nutrient concentrations. This applies where flow data are usually taken regularly (usually daily), whereas nutrient concentrations are usually taken at monthly or quarterly intervals.

D. LIGHT DETERMINATIONS

Light penetration depths were taken using standard Secchi disks. Field light transmission data were taken with a Li-Cor LI-1800UW Underwater Spectroradiometer. The underwater light field was characterized by incident radiant flux per unit surface area as quanta $m^{-2} s^{-1}$. Samples included three to five scans (replicates) taken at 1- to 2-nm intervals that were averaged for each reading. Flux measurements were taken for photosynthetically active radiation (400 to 700 nm; PAR) and individual wavelengths. For any series of collections, field samples were taken during relatively calm conditions between the hours of 1000 and 1400 hours. Multiple air light readings were taken to correct for short-term radiation variability during light measurements.

E. FIELD/LABORATORY METHODS

All sampling efforts were carried out in a similar fashion with respect to sample collection, transfer, and analysis. For all types of organisms, outside corroboration has been carried out by recognized taxonomists continuously. Laboratory specimen libraries have been established and continuous maintained. Chain-of-custody forms are filled out with each transfer with the usual procedure as follows:

1. Samples are delivered (chain of custody signed). Copy of chain of custody form is given to person transferring samples; original is kept with samples. Samples are then stored in a secure area.
2. Samples are processed (if required), date and time recorded (ID No. kept with each sample).
3. Organisms are picked (if required), date and time recorded (ID No. kept with sample).
4. Preparation of specimens are made for identification (ID No. kept with sample).
5. Data are recorded in proper form for reporting to customer.
6. Data, samples, and reference collection are transferred to client (chain of custody signed). Copy of original requested.

1. Phytoplankton and Zooplankton

Net phytoplankton samples were taken with two 25-μm nets (bongo configuration) in duplicate runs for periods of 1 to 2 min. Repetitive (3) 1-l whole water phytoplankton samples were taken at the surface. Phytoplankton samples were immediately fixed in Lugol's solution in its acid version (Lovegrove, 1960). Samples were analyzed by methods described by Prasad et al. (1990) and Prasad and Fryxell (1991).

Zooplankton were taken in various ways. Multiple tows were made with 202-μm nets run over varying time periods that depended on zooplankton density. Sample volumes were recorded so that a quantitative estimate could be made of zooplankton numbers. Another approach, used under certain conditions, involved pumping water (a volume that was determined in the field) through 202-μm nets. Zooplankton preservation was made using 10% formalin. All counts were made to species.

2. Periphyton

A set of slides were anchored at a given stations 2 to 2.5 cm below the surface. After 2 to 3 weeks, we randomly took five slides and placed them in labeled jars filled with distilled water. Jars were transported on ice to the laboratory where they were scraped on both sides and preserved with 5% formalin for a composite sample, which was then stored in screw-cap vials. Each sample was then diluted with distilled water depending on density of algae, thoroughly shaken, and three aliquots of 0.05 or 0.1 ml were placed on each slide, covered, counted, and identified.

3. Benthic Macrophytes

Quantitative sea grass sampling using divers to take eight (randomly distributed) quadrat samples (1/4-m^2) according to methods described by Livingston et al. (1976). Samples were identified to species. Dry weight determinations will be made for macrophytes (rooted and epiphytic) by species for above- and belowground dry weight biomass. Samples were heated in the oven at about 105°C until there was no further weight loss (at least 12 h). Samples were then weighed to the nearest hundredth of a gram.

A map of the offshore sea grasses was developed from just west of the Aucilla River mouth to a point near the Spring Warrior Creek entrance. A set of 166 natural color aerial photographs was taken (AeroMap U.S. Inc., personal communication) along four flight lines between 9:06 and 9:50 A.M. EST on 15 November 1992. Ground-truthing was carried out as a series of diver-transects at 320-m intervals from land to 5 to 7 km offshore: data included visual observations of sea grass species and density, habitat distribution, and bottom type. Sea grass density was estimated as percent cover (bare, no observed macrophytes; very sparse, less than 10% coverage; sparse, 10 to 40% coverage; moderate, 40 to 70% coverage; dense, 70 to 100% coverage). The aerial photography was transferred to a computerized sea grass base map at a scale of 1:24,000. Overlay maps of the transect data were printed at the same scale. The center point of every photograph in the series was then positioned and interpreted using the diver-transect data. The sea grass signature (density, distribution) was identified. Density classification was accomplished by visually comparing the photograph to examples of various density classes from an enlarged crown density scale. The final base map was then photographically reduced in scale and scanned into a computer system. A detailed analysis was carried out concerning the inshore areas of the SAV survey (within 3 km of shore). The survey region was divided into a series of 17 zones along the coast that were 2.5 km wide and 3 km offshore with the river mouths of the Aucilla, Econfina, and Fenholloway Rivers positioned in the center of individual zones. Absolute and relative areas of the different grass densities and other habitat classes, by zone, were then calculated and used for statistical analysis along with water/sediment quality analyses at stations within each zone.

4. Field Collections of Fishes and Invertebrates

All field biological collections were standardized according to statistical analyses carried out early in the program (Hooks et al., 1975; Livingston, 1976, 1987; Greening and Livingston, 1982;

Livingston et al., 1976; Meeter and Livingston, 1978; Stoner et al., 1982). Infaunal macroinvertebrates were taken using coring devices (7.6-cm diameter, 10-cm depth). The number of cores to be taken was determined using large initial samples (40 cores) and a species accumulation analysis based on rarefaction of cumulative biological indices (Livingston et al., 1976). Based on this analysis, multiple (10 to 12) core samples were usually taken randomly at each station with the assured representation of at least 80% of the species taken in the initial sampling. All infaunal samples were preserved in 10% buffered formalin in the field, sieved through 500-μm screens and identified to species wherever possible.

Where roots and the like prevented coring, suction dredges were used as prescribed with multiple sweeps of specific time (number and time to be determined during first field trip). Animals were preserved in 10% buffered formalin, and stored in 100% denatured ethanol.

Epibenthic fishes and invertebrates were collected with 5-m otter trawls (1.9-cm mesh wing and body, 0.6-cm mesh liner) towed at speeds of about 3.5 to 4 km/h for 2 min resulting in a sampling area of about 600 m^2 per tow. Repetitive samples (two or seven) were taken at each site; sampling adequacy of such nets had been determined by Livingston (1976). All collections were made during the same lunar and tidal phase at the first quarter moon. Samples were taken within 2 h of high tide, with day and night trawls separated by one tidal cycle. Larger fishes (>300 mm SL) were taken with 100-m nylon trammel nets and stomachs were removed in the field and injected with a 10% buffered formalin solution.

All organisms were preserved in 10% buffered formalin, sorted and identified to species, counted, and measured (standard length for fishes; total length for penaeid shrimp, carapace width for crabs). Representative samples of fishes and invertebrates were dried and weighed and regressions were run so that data from the biological collections could be converted into dry and ash-free dry mass. The numbers of organisms were then converted to biomass m^{-2}. All biological data (infauna, epibenthic macroinvertebrates, fishes) were expressed as numbers m^{-2} mo^{-1} or biomass m^{-2} mo^{-1}.

5. Trophic Determinations

Get analyses were carried out in the same manner throughout the 30-year field analysis period. Fishes and invertebrates were placed in 5-, 10-, or 20-mm size classes (depending on size), and food items were taken from the stomachs of up to 25 animals in a given size class with pooling by sampling date and station. Determinations similar to those described above were carried out and it was determined that at least 15 animals were needed for a given size class × station × sampling period. Stomach contents were removed and preserved in 70% isopropanol and a dilute solution of rose Bengal stain. In larger animals, we pumped the stomachs out and released the subject. The content analysis was carried out through gravimetric sieve fractionization (Carr and Adams, 1972). Contents were washed through a series of six sieves (2.0- to 0.75-mm mesh) and frequency of occurrence of each food type was recorded for each sieve fraction. As the food items were comparable in size, the relative proportion of each food type was measured directly by counting. Dry weight and ash-free dry weight were determined for each food type by size class. General, mutually exclusive categories were used for the food types that included both plant and animal remains. Stomach contents by food type by size class were then calculated and related back to station/time designations. Based on the long-term stomach content data for each size class, the fishes and invertebrates were then reorganized first into their trophic ontogenetic units using cluster analyses (Czekanowsli [Bray-Curtis] or C-Lamba similarity measure; Flexible Grouping cluster strategy [beta = –0.25]). The basis for this analysis is given by Sheridan (1978) and Livingston (1984).

All biological data (as biomass m^{-2} mo^{-1} of the infauna, epibenthic macroinvertebrates and fishes) were transformed from species-specific data into a new data matrix based on trophic organization as a function of ontogenetic feeding stages of the species found in Perdido Bay over the multiyear sampling program. Ontogenetic feeding units were determined from a series of detailed stomach content analyses carried out with the various epibenthic invertebrates and fishes in the near-shore Gulf region (Sheridan, 1978, 1979; Laughlin, 1979; Sheridan and Livingston, 1979, 1983; Livingston, 1980, 1982, 1984b, unpublished data; Stoner and Livingston, 1980, 1984; Laughlin and Livingston, 1982; Stoner, 1982; Clements and Livingston, 1983, 1984; Leber, 1983, 1985). Based on the long-term stomach content data for each size class, the fishes and invertebrates were reorganized first into their trophic ontogenetic units using cluster analyses. Infaunal macroinvertebrates were also organized by feeding preference based on a review of the scientific literature (Livingston, unpublished data). The field data in Perdido Bay were then reordered into trophic levels so that temporal changes in the overall trophic organization of this system could be determined over the 8-year study period. We assumed that feeding habits (at this level of detail) did not change over the period of observation; this assumption is based on previous analyses of species-specific fish feeding habits that remained stable in a given area over a 6- to 7-year period (Livingston, 1980).

The long-term field data were used to establish a database founded on the total ash-free dry biomass m^{-2} mo^{-1} as a function of the individual feeding units of the infauna, epibenthic macroinvertebrates, and fishes. Data from the various stations in Perdido Bay were used (g m^{-2}) for all statistical analyses. The data were summed across all taxonomic lines and translated into the various trophic levels that included herbivores (feeding on phytoplankton and benthic algae), omnivores (feeding on detritus and various combinations of plant and animal matter), primary carnivores (feeding on herbivores and detritivorous animals), secondary carnivores (feeding on primary carnivores and omnivores), and tertiary carnivores (feeding on primary and secondary carnivores and omnivores). All data are given as ash-free dry mass m^{-2} mo^{-1} or as percent ash-free dry mass m^{-2} mo^{-1}. In this way, the long-term database of the collections of infauna, epibenthic macroinvertebrates, and fishes was reorganized into a quantitative and detailed trophic matrix based not solely on species (Livingston, 1988) but on the complex ontogenetic feeding stages of the various organisms. A detailed review of this process and the resulting feeding categories is given by Livingston et al. (1977).

6. Oyster Studies

Oyster samples (multiple) were taken with full head tongs (16-tooth head; 4.5-m handles) at each station on a monthly basis in Apalachicola Bay from March 1985 through October 1986. The use of tongs was standardized with respect to opening widths and sampling effort. To quantify the sampling effort, a series of 30 standardized, random tong samples was taken at the Big Bayou and Cat Point reefs in February 1985. The cumulative size frequency distribution (in 10-mm increments) was determined and plotted for each sampling site. The number of samples necessary for a specified level of quantification was determined according to a method described by Livingston et al. (1976). The method allowed determination of the number of subsamples necessary to achieve specific levels of size class accumulation when compared to the results of 30 subsamples. Seven subsamples accounted for 80.8% (Big Bayou) and 87.5% (Cat Point) of the sampling variability for the total sample. Accordingly, this number of subsamples (located beyond the asymptote for size class accumulation) was considered to constitute a representative sample taken in each of the test regions. Numbers of oysters per tong were recorded and converted to numbers m^{-2}.

All oysters were measured to the nearest millimeter in the field according to the greatest distance from beak to lip using linear calipers. A total of 140 oysters taken from four stations (n = 35 at each site) was used to determine the relationship of shell length and weight of oyster meat. Four separate length/weight equations were developed to account for known differential growth characteristics in different regions of the bay:

$$\ln(\text{AFDW}) = 2.505 * \ln(\text{LEN}) - 10.980 \text{ (Cat Point, } r^2 = 0.83)$$

$$\ln(\text{AFDW}) = 2.303 * \ln(\text{LEN}) - 10.306 \text{ (East Hole, } r^2 = 0.84)$$

$$\ln(\text{AFDW}) = 2.202 * \ln(\text{LEN}) - 9.125 \text{ (Paradise, } r^2 = 0.90)$$

$$\ln(\text{AFDW}) = 2.465 * \ln(\text{LEN}) - 10.190 \text{ (Scorpion, } r^2 = 0.86)$$

where AFDW is the ash-free dry weight of oyster meat and LEN is oyster shell length in millimeters. F-tests ($\alpha = 0.05$) disclosed that equations from both bars in the eastern bay (Cat Point and East Hole) were significantly different from equations developed for bars in the western bay (Paradise and Scorpion). Although the differences were not significant within the respective regions, east and west, the site-specific equations were used for the transformations of the length data into ash-free dry weights. This allowed the most comprehensive and accurate use of the data for such transformations. All tong data (in terms of numerical abundance and ash-free dry weight) were calculated on a unit m^{-2} basis. These data were then transformed to estimates of total numbers and biomass on each bar based on the estimated size of the bars. Areas were established from computer simulations of oyster-producing areas (Livingston, unpublished data) based on historic records, interviews with oystermen and state environmental agencies, and our past and ongoing field studies (Livingston, 1984).

The tong data (oyster density and length frequency) were standardized by a quantitative comparison with data derived from a series of multiple (0.25 m^{-2}) quadrats taken on the same sampling sites (Cat Point Bar, East Hole, Paradise) over the same period of study by other researchers (Berrigan, 1990). Scattergrams of the respective density and length frequency data bases were log-transformed to approximate the best fit for a normalized distribution. Statistical comparisons were made of the monthly data by station. Independence tests were used to examine each sample for autocorrelation at a number of lags. For each sample, autocorrelation is to be computed for the lesser of 24 or $n/4$ lags, where n is the number of observations. The program then computes the Q statistic of Ljung and Box (1978) for an overall test of the autocorrelations as a group, where the null hypothesis was that correlation at each lag was equal to zero. The parametric F-test for comparison of variances and t-test for comparison of means were used to test the hypothesis that the data came from normally distributed populations. Means testing was carried out with both parametric and nonparametric tests. For independent, random samples from normally distributed populations, the parametric t-test was used to compare the sample means. Where there was serial dependence of data from a given pair of stations, it was removed by differencing the observations and calculating and plotting the autocorrelations of the differences. If the differences were not serially correlated, we applied the Wilcoxon sign-rank test on the differences to compare the two sets of numbers ($p = 0.05$).

Spatfall accumulation was analyzed from a subset of 12 stations. Spat baskets, constructed of plastic-coated wire (25 × 25 × 25 cm; 2.5-cm mesh), were filled with about 20 sun-bleached oyster shells and placed at each site. Bricks were placed at the bottom of each basket so that the oysters remained off the bottom to lessen problems with sedimentation. Samples were retrieved, and new sets of oyster shells were set out at 2- to 3-week intervals. One spat basket was used at each station. Seven shells were randomly chosen from each basket for analysis. Spat counts were made from the inner surface of each valve; this standardization was based on test results that indicated less variance of such adherence on inner surfaces than on outer surfaces.

Oyster data were grouped in two ways prior to analysis: (1) baywide totals and (2) eastern vs. western reefs. In the latter grouping strategy, selected reefs from the eastern bay (Sweet Goodson, Cat Point, East Hole, Porter's Bar) were compared to selected reefs from the western bay (Paradise, Pickalene, Scorpion). These reefs were chosen because of their relative commercial importance to overall oyster production in the bay. Intervention models were used to analyze the effects of the two hurricanes on the monthly total oyster numbers (in thousands) and the monthly average shell length (in mm) of oysters in the bay (Box and Jenkins, 1978; Pankratz, 1991). Three major interventions occurred in the Apalachicola Bay during the study period: Hurricane Elena (combined with the cessation of commercial oystering after September 1985); Hurricane Kate; and the resumption of commercial oystering in May 1986. The models were used to test the possibility that the interventions were significantly associated with level changes in the monthly total oyster numbers $\{N(t)\}$ and the monthly average shell lengths $\{S(t)\}$.

Three indicator variables corresponding to the interventions are defined as

$$X_1(t) = \begin{cases} 0, t < \text{September } 1985 \\ 1, t \geq \text{September } 1985 \end{cases}$$

$$X_2(t) = \begin{cases} 0, t < \text{December } 1985 \\ 1, t \geq \text{December } 1985 \end{cases}$$

$$X_3(t) = \begin{cases} 0, t < \text{May } 1986 \\ 1, t \geq \text{May } 1986 \end{cases}$$

Using B as the backward shift operator such that $BX(t) = X(t - 1)$, the relationship between the monthly total oyster number $\{N(t)\}$ and the three interventions can be described by the following general intervention model:

$$N(t) = \beta_0 + \nu_1(B)X_1(t) + \nu_2(B)X_2(t) + \nu_3(B)X_3(t) + \xi(t) \tag{I.1}$$

where $\nu_1(B)$, $\nu_2(B)$, and $\nu_3(B)$ are polynomials with the typical form:

$$\nu(B) = \omega_0 - \omega_1 B - \ldots - \omega_R B^R$$

where $\{\omega_0, \omega_1, \ldots, \omega_h\}$ are parameters to be estimated.

In Model I.1, β_0 is a constant and $\xi(t)$ is a noise term that is often modeled as a stationary autoregressive moving average (ARMA; p, q) process, where

$$\xi(t) - \phi_1\xi(t-1)\ldots - \phi_p\xi(t-p) = \varepsilon(t) - \theta_1\varepsilon(t-1) - \ldots - \theta_q\varepsilon(t-q),$$

where p is the order of the autoregression (AR) term, and q is the order of the moving average (MA) term. The $\varepsilon(t)$ are assumed to be independent and normally distributed with mean zero and variance σ^2.

Model I.1 extends the traditional linear regression models in two directions. First, the monthly total oyster number series $\{N(t)\}$ may react to an intervention with a time lag. For example, the term

$$\upsilon_1(B)X_1(t) = \omega_0 X_1(t) - \omega_1 X_1(t-1) - \ldots - \omega_h X_1(t-h)$$

in Model I.1 represents that the impact of the first intervention on $N(t)$ is distributed across several time periods. Second, instead of assuming that the errors $\xi(t)$ are independently distributed, ARMA(p,q) models are used for $\xi(t)$, which incorporates possible serial correlations in the response series $\{N(t)\}$. Model I.1 was fitted by using the Linear Transfer Function Identification Method proposed by Pankratz (1991). The relationship between the monthly average shell length series $\{S(t)\}$ and the three interventions was modeled in a similar fashion.

In addition to the overall oyster data, intervention models were fitted to eastern and western sections of the bay using numbers m^{-2} and average shell length (in mm) of oysters from stations in these areas. The four series were denoted by EN(t) and WN(t) for eastern and western densities and by ES(t) and WS(t) for eastern and western shell lengths. Intervention models were fitted to the four series separately.

7. Modeling

Physical processes in Apalachicola Bay were simulated using a time-dependent, three-dimensional model of the Blumberg–Mellor family (Blumberg and Mellor, 1980, 1987). This model solves a system of coupled differential (prognostic) equations for free surface elevation, two components of the horizontal velocity, temperature, salinity, turbulence energy, and turbulence macroscale. The spatial integration is explicit in the horizontal and implicit in the vertical. For its forcing, the model admits a comprehensive database comprised of time-dependent temperature, salinity, surface heat and humidity fluxes, and wind stress, tides, and residual signals, when available. In the horizontal, the model uses a body-fitted curvilinear orthogonal coordinate system that allows one to better adhere to the convoluted coastline. In the vertical, a sigma-coordinate system converts the free surface and seabed into coordinate surfaces, and allows for better resolution of the surface and near-bottom boundary layers. The sophisticated second-order turbulence closure model of Mellor and Yamada (1982), as modified by Galperin et al. (1988), describes the vertical mixing.

The model was calibrated and verified using hydrographic data collected at 0.5-h intervals from 23 instruments located at sites throughout the bay during the 6-month period from June to November 1993 (Wu and Jones, 1992; Jones and Huang, 1996; Huang and Jones, 1997). Measured river inflows, surface wind stress, free surface elevations, temperature and salinity signals at the open boundaries were utilized as time-dependent boundary conditions. The open boundaries comprised five single-point and multipoint grid locations through which Apalachicola Bay connects to the Gulf directly. Freshwater runoff into the bay was drawn from four rivers: Apalachicola River and its tributaries, Whiskey George Creek, Cash Creek, and the Carrabelle River. Flow values measured by the U.S. Geological Survey's Sumatra gauge were used to develop the input from the main stem of the Apalachicola River and its tributaries using a one-dimensional hydrologic model (DYNHYD). Local rainfall data were used as input to hydrologic models (SWMM) for the Whiskey George Creek, Cash Creek, and Carrabelle River as well as the ungauged portion of the Apalachicola River downstream of the Sumatra gauge. Wind forcing was obtained from a continuous record collected at a weather station at the St. George Island causeway. For modeling the 1985–1986 period used in the present study, real-time temperature and salinity signals at the open boundaries were unavailable; composite climatological forcing functions were developed based on Fourier representation of the mean seasonal data in a strategy similar to that of Galperin and Mellor (1990a). Velocity estimates were not calibrated against field observations as currents are more sensitive to precise location. However, the salinity calibration efforts effected an indirect calibration of velocity because salinity is a direct result of velocity-induced advection and diffusivity transport processes.

For this study, model runs were made for entire years 1985 and 1986 in which hourly values were calculated for 715 cells in the Apalachicola Bay system. The model output included elevation and surface and bottom values for salinity, temperature, and components of current velocity (horizontal and vertical). Additional variables were derived from the raw model output and included averages, standard deviations, maximum and minimum values, and residual current velocities and direction. Such variables were computed for specific time periods as described below.

8. Statistical Analyses and GIS Mapping

Dependent variables for all statistical models included larval densities in 1985 and 1986, spatfall from closed sides in 1985 and 1986, average total oysters m^{-2}, new growth (young of the year cohort: October 1985 to April 1986), average growth of whole oyster bars (October 1985 to April 1986), old growth (measured oysters in baskets), and mortality (1985). Two sets of independent variables were developed. The first set of variables was taken from the monthly field database and included surface color, surface turbidity, surface oxygen anomaly, bottom temperature, bottom salinity, bottom dissolved oxygen, Secchi depth, water depth, and salinity stratification. We also used average river flow and total rainfall (East Bay Tower, Florida Division of Forestry) over the 10-day period preceding sampling at each station. The second set of independent variables comprised data derived from the hydrodynamic model and included both surface and bottom values for the average and standard deviation of salinity and temperature, average and residual (resultant) current velocity, and minimum and maximum values for salinity and temperature. Data were calculated for the hydrodynamic model cells that included the geographic location of our stations. Data were time-averaged over the 30-day period preceding the biological collection at each station. Dependent and independent variables were transformed for normality. Associations between the oyster data and the various habitat characteristics were examined using stepwise multiple linear regression. Regression models for the biological variables were developed separately for the monthly field-observed data and the hydrodynamic model output so that a comparison could be made concerning statistical models based on traditional monthly sampling information and statistical models based on continuous data derived from hydrodynamic simulations. A p-value of 0.01 was chosen as the cutoff for including explanatory variables into a final model. Residuals were checked routinely for normality.

We used the best of both groups of statistical models to select a variable (mortality) for testing the application of the hydrodynamic model in the projection of oyster population response under alternative regimes of river flow. Experimental oyster mortality data and the output from the hydrodynamic model were used to develop a seasonal (time-averaged) statistical model. We used a stepwise linear regression method with average monthly percent mortality at 13 stations from May through August 1985 as the dependent variable. As predictors, we used averaged hydrodynamic model output values over the same period. We used the following surface and bottom water predictors: average salinity, standard deviation of the salinities, maximum and minimum salinities, and average water velocity. We also used average surface elevation and salinity stratification. Thus, the resulting statistical model expressed mortality solely in terms of hydrodynamic model output parameters. We then substituted into the regression equation hydrodynamic model values from all of the hydrodynamic model cells to calculate predicted mortality over the whole bay for the May to August period of 1985.

We entered the resulting mortality estimates into an ArcView® Geographic Information System (GIS) database in which latitude and longitude coordinates were input for each of the hydrodynamic model cells. Then we used the GIS software to link each mortality estimate with a geographic location to generate a map surface depicting the spatial distribution of predicted mortality within the bay. An inverse squared-distance weighting method was used for interpolation of the surface. The interpolation was performed for a grid of cells 100 m on a side and incorporated data values from eight nearest neighbors. We applied the same statistical model (developed with 1985 data) to averaged hydrodynamic model output values for the May to August period of 1986 and repeated the GIS mapping process to obtain a prediction surface for potential mortality during 1986.

F. EXPERIMENTAL METHODOLOGY

1. Nutrient Limitation Experiments

We constructed 18-1, all-glass microcosms (width: 20, length: 21.5, and depth: 48 cm) using nontoxic sealant. A wooden rack assembly held microcosms at the surface of large temperature-controlled reservoirs. Bay ambient water temperatures were maintained within ±2°C through use of heating and cooling coils placed in the thermal reservoir. Large volumes of bay water were gently pumped (Ruhl model 1500 rubber diaphragm pump) from approximately 0.5-m depth at each station into polyethylene containers by first passing through a set of nested Nitex plankton nets held in a stand-pipe system to equilibrate water pressure (i.e., a 202-µm net placed inside a 64-m net; 0.5 × 2.0 m) and to remove larger zooplankton.

Sampling was conducted during late afternoon and samples were transported to an experimental "microcosm tank facility" near Tallahassee, Florida, during early evening to reduce heating from daytime peak temperatures. Experiments began the next day. Because a continuous supply of estuarine phytoplankton was logistically infeasible, we used a modified static renewal approach with daily removal of 10% of the volume of each microcosm. A supply of the field sample for each station was maintained as renewal water. This water was filtered through a 25-µ*M* Nitex® net and held at 4°C in the dark. Thus, early experimental artifacts such as production of allelopathic substances would be less likely in such a system compared to a static experimental design. Before use, all collecting vessels, culture tanks, and Tygon air lines were acid washed (10% HCl), rinsed twice with fresh water, rinsed once with deionized water, and air-dried. Microcosms were and sealed with nontoxic silicon sealant. Air was delivered to each microcosm to maintain mixing via Tygon tubing, plastic control valves, and "air stones" from an oil-free aquaculture air pump.

Water collected from a station the previous day was placed in a fiberglass mixing tank during early morning to minimize light shock and was gently mixed with a rubber paddle to ensure homogeneous filling requirements. Water was poured sequentially in small volumes into glass microcosms to ensure homogeneous filling requirements. Reference samples were taken from the mixing tank for chlorophyll *g*, dissolved inorganic phosphate (PO_4), ammonium (NH_3), nitrite (NO_2) and nitrate (NO_2), particulate organic nitrogen (PON), and phosphorus (POP) to characterize initial nutrient conditions and standing stock of phytoplankton. Chlorophyll *a* was collected on Gelman A/E glass fiber filters, extracted with 90% acetone buffered with MgCO, and measured according to Parsons et al. (1984). Quantification of dissolved inorganic N (i.e., (NO_2 and NO_3) and P (PO_4) followed Parsons et al. (1984). Ammonium was measured as NH_3 with an ion electrode (U.S. EPA 1983, method 350.3) after converting NH_4 to NH_3 by raising the pH to 11. This method has some essential features (e.g., minimal interference from waters highly stained with humic materials and paper mill effluents. However, the level of detection typically was relatively high (e.g., 2.0 µ*M* NH_3-N) but adequate for this study. Particulate organic N (U.S. EPA 1983, method 351.4) and POP (U.S. EPA 1983, method 364.4) were collected on Gelman glass fiber filters, combusted at 500°C, and oxidized to inorganic fractions and organic N measured as NH_3 and organic P measured as PO_4.

Single samples were collected daily at approximately 0800 h from each microcosm for chlorophyll analysis to measure changes in phytoplankton biomass. Nutrient enrichment experiments included the following treatments triplicated for each station: three control tanks, three P-enriched tanks at 10 µ*M* PO_4 above ambient, three N-enriched tanks at 50 µ*M* NH_3N above ambient NH_3, and three combined NH_3 + PO_4 (referred to as N+P) above ambient as described for single additions. We used NH_3 enrichments because of interest in potential increased riverine transport into the bay. Although aware of possible suppression of NO_3 uptake by NH_3, this process was judged to be relatively unimportant in bioassays lasting over 5 to 12 days (D'Elia et al., 1986) and at high NO_3 concentrations (Pennock, 1987). Dortch (1990) concluded from a comprehensive literature review that NH_3 suppression of NO_3 uptake under field conditions is quite variable and often undetectable. Nutrient additions were made at time 0 of the experiment. On day 2 and daily thereafter, 18 l of water were collected from each microcosm for chemical and biological analyses. This volume was replaced by an equal volume of renewal water adjusted to ambient temperature before introduction into the microcosms. Hydrographic variables (e.g., surface water temperature (°C) and salinity (8b) (YSI model 33 conductivity meter), and 30-cm all-white Secchi disk readings (m) were collected generally within 1 week of samples for nutrient enrichment experiments.

A randomized block design was used with statistical significance of ($p < 0.05$) within and among stations. The four treatments in each row were randomly selected so that each of the three station-specific experiments had 12 treatments (randomized four treatment by three replicates for three stations). Treatments were oriented in a north/south direction, to minimize shading from changes in the annual sun angle, and to allow detection of a possible blocking effect. A repeated-measures, randomized analysis of variance (ANOVA) was run for each experiment. ANOVA assumptions were checked using residual box plots, the Lilliefors test on the standard residuals and the Bartlett test for equality of variances. A log transformation was necessary to achieve equality of variances. The two-way interaction *p*-values for treatment by day and for block by day were calculated. To determine how treatments differed by day, a randomized block analysis of variance was run for each station, month, and day. As above, a log transformation was necessary to satisfy assumptions. Scheffe's *S*-test was used as a *post hoc* treatment.

2. Diver-Deployed (*in Situ*) Benthic Lander

a. *Introduction*

The various problems and sources of errors with respect to SOD analyses have been reviewed (1987: Delaware Estuary use attainability project, Delaware River Basin Commission, West Trenton, New Jersey). We have analyzed the relatively broad literature on this subject and the method chosen reflects our interpretation of results from the studies of others in addition to our own experience with SOD analyses in our South Carolina work. Replicate samples were used as an indication of experimental error. *In situ* tests have the most potential for error. The laboratory approach, while suffering from the problem of not being carried out in the field, has proved to be the most reliable in terms of variability of results. We followed techniques that have a record of being the most reliable on the basis of intersystem comparisons.

We worked with *in situ* samplers developed and constructed at Texas A&M University (Dr. G. Rowe, personal communication). The diver-deployed Benthic Lander used in the SOD measurements was designed and built by a research team at Texas A&M University under the direction of Dr. Gilbert T. Rowe. A description of the instrument is given in a paper given by Gilbert T. Rowe and Gregory S. Boland ("Benthic Oxygen Demand and Nutrient Regeneration on the Continental Shelf of Louisiana and Texas," abstract for LUMCON meeting, 1991).

b. *Instrument Specifications*

Duplicate samples are taken after the instrument is grounded in sediments at a given station. Dissolved oxygen probes (one for each of the duplicate units) are attached to YSI dissolved oxygen meters that are monitored at 2- to 5-min intervals at the surface. The battery-operated machine has dual stirrers that maintain a regular rate of stirring in each 7-l (v) chamber. Each chamber covers 0.09 m^2 of sediment. Syringe ports are available for removal of water within each chamber for chemical analysis. The Lander is housed in a basic 3/4-in. aluminum frame to which is attached a battery housing with two power Sonic batteries. Ballast weights are used to keep the device in the sediments. Total negative weight is –28.23 kg and the total positive lift (largely from the 13-in. instrument spheres) is 13 kg.

c. *Operation of the Lander*

Laboratory Preparations

1. The YSI meters should be prepared. This includes installation of probe membranes, Redlining the instrument at salinity 0, adjustment of temperature to ambient for calibration, calibrating probes for ambient temperature, setting meter scales to desired dissolved oxygen range, checking calibration with 5 to 10% sodium sulfite solution (0 ppm oxygen). The TEMP should be reset to the expected bottom temperature (based on previous experience). SALINITY should also be set in a similar fashion. Glass and gasket should be cleaned with a solvent (toluene).
2. Make sure that the silicon oil fills the impeller housing.
3. Charge batteries to +13 V and place in housing.
4. Check o-ring on battery and use light treatment of silicon grease on the battery chamber o-rings. Close chamber.
5. Check for chamber volume by measuring distance from the bottom of the rim to the flange (40 mm = 7.0 l).
6. Grease and clean the pins for attachment to the DO batteries (large pin is +). Attach battery wires and start/check stirring assemblies.
7. Secure device in frame and set/secure for travel.

Boat Operations (Deployment)

1. Turn on stirrers by connecting the battery leads. Calibrate the DO meters and place probes in housing.
2. Check syringes for the removal of water.
3. Slide apparatus off boat and let down gently with marker buoy attached to frame.
4. Divers should remove the chamber unit and secure the bungee cords. One diver on each side should lift the apparatus out and away from the frame. Then the divers should flip the chamber unit over to release all air bubbles. Make sure that stirrers are operating.
5. Look for flat area to place chambers and set chambers down gently on sediment, pressing down until the gray flange is sitting on the sediment surface. Again check the stirrers. DO chamber wires should be bundled together.
6. Take water samples (200 ml) with syringes. Filter samples through 0.2-μm filter.
7. Take DO and temperature readings at 5-min intervals, recording data on the sample sheet. Such sheets should be part of the overall data sheet (see BIFS field data sheet summary).
8. After a significant drop in DO has been recorded (1 to 3 h), divers should take a final water sample with the syringes (same as above). The chamber should be picked up and turned upside down to check for the impeller movement (sediment may interfere with such mechanisms). The chamber unit should be placed in the lander housing and attached with bungee cords. Dive lifts should be inflated and divers should escort the apparatus to surface. The apparatus should be carefully placed on the boat and the battery leads detached.

Determination of Oxygen Consumption Rates

$$\text{Rate } R = \frac{(\text{Final O}_2,\ \text{ml / l} - \text{ Initial O}_2,\ \text{ml / l})\ (\text{Volume of chamber } = 7\text{ l})}{(\text{area of chamber},\ 0.091\ \text{m}^2)(\text{Total time, h} - T_{final} - T_{initial})}$$

Conversion of mg/l to ml/l: mg = 0.7 ml

3. Laboratory (*in Vivo*) SOD Determinations

Four replicate samples of undisturbed core samples were taken with the SOD corer apparatus at each of the three stations. Sediment filled the corer to at least half the height of the PVC coring tube. Water filled the rest of the corer. Great care was taken so that there was minimal disturbance to the sample. Detachable cores were capped and placed in racks. The samples were sealed with PVC end caps and placed in an upright position in the rack in a Gott cooler. The seawater bath in the Gott cooler was maintained at the level of water in the cores to lessen pressure on the seals. The samples were taken immediately to the laboratory and equilibrated for 10 h with gentle aeration without disturbing the sediment. Laboratory water temperature was controlled at field levels.

We followed the batch method of analysis. The laboratory SOD measurements were made on undisturbed sediments in the original core tubes. Water circulation was maintained with an internal mixing device. Two replicates from each station set were poisoned with HCN; the remaining two samples were left without any additions or changes. DO oxygen probes were pre- and postcalibrated using the micro-Winkler technique. Temperature and depletion of DO were monitored (without stirring) every 30 min for 4 to 6 h or until 1.5 to 3.0 mg/l DO depletion was obtained. If the initial DO readings in the cores were below 4.0 mg/l, the core water was aerated without sediment disturbance in order to have an initial DO reading of at least 4.0 mg/l.

DO depletion curves were constructed with the cumulative DO depletion data plotted against time. The slope was calculated from the linear portion of the DO depletion curve using the first-degree regression function: the results are to be represented in units of g O_2 m^{-2} h^{-1}.

4. Primary Productivity (*in Situ* Light/Dark Bottle Study)

Primary productivity is the rate at which inorganic carbon is converted to organic carbon via chlorophyll-bearing organisms such as phytoplankton. The plankton respiration term was an important part of evaluating the oxygen-uptake mechanism in the bay. Therefore, we evaluated the phytoplankton net and gross photosynthesis using the classical light and dark bottle oxygen technique (Strickland, 1960). Some unpublished data using the C_{14} uptake method were obtained by personnel of the Environmental Research Laboratory, Gulf Breeze for Perdido Bay. Data taken using this method were comparable to our own.

At present, there is no available method of measuring primary productivity that is free of artifacts. Various authors have recently emphasized the importance of production processes occurring at strong physical gradients (Legendre and Dimers, 1985) and an emerging area of emphasis has concentrated on ecohydrodynamics. The approach in this study examined bay primary productivity in terms of vertical stratification and its relationship to the euphotic zone, the longitudinal salinity gradient in the bays, and temporal patterns over the warm season when hypoxic conditions become prevalent. The scaling of measurements was made mostly at the meso- to macroscale; fine-grain scaling of observations was based on the present design. Integral water column rates of photosynthesis allowed an estimate of system production minus the relatively small contribution from submerged aquatic vegetation and a somewhat more important contribution of phytomicrobenthos.

The approach here is developed from that described in Standard Methods (1989). All procedures involving bottle cleaning, checking for supersaturation, the handling of water samples, and calculations are given in the above citation. The oxygen method will be used whereby clear (light) and darkened (dark) bottles will be filled with water samples and suspended at regular depth intervals for appropriate intervals. The duration of exposure will depend on the rate of net photosynthesis and respiration with a need to avoid bubble formation in light bottles or hypoxia in dark bottles except at depths where hypoxia is present. We will start with sunrise-to-sunset deployment with possible adjustment to 0900 to 1500 hours or some equal distribution of time around noon (CST). Any given bubble development (>0.3-mm diameter) at a single depth will not result in a change in deployment schedule. Bubble development at different depths will be reason to change such deployment based on a initial screening by periodically checking the bottle strings.

a. End Points

Changes in the concentration of DO will be measured with a YSI. Model 58 DO meter equipped with a BOD probe. A magnetic stirrer and stirring bar will be required (a 12-V system will be used). The specifications of this DO unit is ±0.01 mg l^{-1} with an accuracy of ±0.03 mg l^{-1}. These specifications are minimally satisfactory for the warm season in Perdido Bay. More sensitive measurement equipment is not seen to be seaworthy in a small research vessel. If necessary, a more sensitive system will be available at a shore-based system, which will require transport of samples to this laboratory setup. The present method will allow a spatial coverage that is important to the project.

b. Methods

Two numbered light bottles and two dark bottles will be used for each depth at each station. Nonbrass ("marine") wire will be used to attach a strong (plastic or stainless steel) snap-catch to the neck of each 300-ml BOD bottle. Wheaton "400" brand borosilicate (high-purity) glass bottles are to be used. Plastic bottle caps will be used to protect the stoppers. The bottles are to be acid-cleaned (warm 10% HCl) and rinsed with distilled water prior to use. Just before the filling of each bottle, the container will be rinsed with the water to be tested. Phosphorus-containing cleaning agents will not be used. The experimental setup has been constructed so that the supporting lines or racks do not shade the suspended bottles.

c. Calculations

The increase in oxygen concentration in the light bottle during incubation is a measure of the net production (somewhat less than gross production due to respiration). The loss of oxygen in the dark bottle is an estimate of the total plankton respiration.

$$\text{Net photosynthesis} = \text{light bottle DO} - \text{initial DO}$$

$$\text{Respiration} = \text{initial DO} - \text{dark bottle DO}$$

$$\text{Gross photosynthesis} = \text{light bottle DO} - \text{dark bottle DO}$$

Results from duplicates are to be averaged. The calculations of the gross and net production for each incubation depth should be plotted.

$$\text{mg carbon fixed/m}^2 = \text{mg oxygen released/l} \times 12/32 \times 1000 \times K$$

where K = photosynthetic quotient ranging from 1 to 2, depending on the N supply. One mole of oxygen (32 g) is released for each mole of carbon (12 g) fixed. The productivity of a vertical column of water 1 m^2 is determined by plotting productivity for each exposure depth and graphically integrating the area under the curve.

APPENDIX II

Trophic Organization of Infaunal and Epibenthic Macroinvertebrates and Fishes

DESIGNATION

Herbivore
Herbivore (benthic plants)
Omnivore
Herbivore (plankton)
Carnivore
Omnivore (herbivorous)
Non-feeding stage
Omnivore (detritivorous)
Omnivore (carnivorous)
Omnivore (general)
Carnivore (primary)
Carnivore (secondary)
Carnivore (tertiary)
Carnivore (parasitic)

Species Code	Trophic Code	Description	Species Code	Trophic Code	Description
		Fishes			
AAAHDR	H	**Herbivore**	BAGMA3	C2	*Bagre marinus* (>150 mm)
AAAHDR	HB	**Herbivore (benthic plants)**	BAGMAR	C2	*Bagre marinus*
CYPRIN	HB	*Cyprinidae* sp.	BAICH3	C2	*Bairdiella chrysoura* (>60 mm)
CYPVAR	HB	*Cyprinodon variegatus*	BAICHR	C2	*Bairdiella chrysoura*
POECIL	HB	Poeciliidae	BATSOP	C2	*Bathygobius soporator*
POELAT	HB	*Poecilia latipinna*	BOTROB	C2	*Bothus robinsi*
XIPMAC	HB	*Xiphophorus maculatus*	CARBA1	C2	*Caranx bartholomaei* (<70 mm)
AAAHDR	HP	**Herbivore (plankton)**	CARHI1	C2	*Caranx hippos* (<70 mm)
ALOALA	HP	*Alosa alabamae*	CENMAC	C2	*Centrarchus macropterus*
ALOCHR	HP	*Alosa chrysochloris*	CENOCY	C2	*Centropristis ocyurus*
ALOMED	HP	*Alosa mediocris*	CENPHI	C2	*Centropristis philadelphica*
CLUSPE	HP	*Clupeidae* sp.	CENST1	C2	*Centropristis striata* (≤40 mm)
HYPUN2	HP	*Hyporhamphus unifasciatus* (>130 mm)	CENST2	C2	*Centropristis striata* (41–140 mm)
HYPUNI	HP	*Hyporhamphus unifasciatus*	CENSTR	C2	*Centropristis striata*
SARAUR	HP	*Sardinella aurita*	CHAFAB	C2	*Chaetodipterus faber*
AAAHDR	O	**Omnivore**	CHLCH3	C2	*Chloroscombrus chrysurus* (>50 mm)
AAAHDR	OH	**Omnivore (herbivorous)**	CHLCHR	C2	*Chloroscombrus chrysurus*
ADIXEN	OH	*Adinia xenica*	CITSPI	C2	*Citharichthys spilopterus*
DIPHO4	OH	*Diplodus holbrooki* (41–80 mm)	CYNAR3	C2	*Cynoscion arenarius* (>50 mm)
DIPHO5	OH	*Diplodus holbrooki* (>80 mm)	CYNNE1	C2	*Cynoscion nebulosus* (≤40 mm)
DORPET	OH	*Dorosoma petenense*	CYNNE2	C2	*Cynoscion nebulosus* (41–80 mm)
LAGRH5	OH	*Lagodon rhomboides* (>120 mm)	DASAME	C2	*Dasyatis americana*
AAAHDR	OD	**Omnivore (detritivorous)**	DASCEN	C2	*Dasyatis centroura*
MICGU1	OD	*Microgobius gulosus* (<20 mm)	DASSA1	C2	*Dasyatis sabina*
MICGUL	OD	*Microgobius gulosus*	DASSAB	C2	*Dasyatis sabina*
MUGCE1	OD	*Mugil cephalus*	DASSAY	C2	*Dasyatis sayi*
MUGCEP	OD	*Mugil cephalus*	DIPBIV	C2	*Diplectrum bivittatum*
MUGCUR	OD	*Mugil curema*	DIPFOR	C2	*Diplectrum formosum*
MUGSPE	OD	*Mugil* sp.	ELAEVE	C2	*Elassoma evergladei*
SPHMAC	OD	*Sphoeroides maculatus*	ELAOKE	C2	*Elassoma okefenokee*
SPHNEP	OD	*Sphoeroides nephelus*	ELASPE	C2	*Elassoma* sp.
SPHPAR	OD	*Sphoeroides parvus*	ELAZON	C2	*Elassoma zonatum*
SPHSPE	OD	*Sphoeroides spengleri*	ENNGLO	C2	*Enneacanthus gloriosus*
AAAHDR	OC	*Omnivore (carnivorous)*	ENNOBE	C2	*Enneacanthus obesus*
ALUSCH	OC	*Aluterus schoepfi*	ENNSPE	C2	*Enneacanthus* sp.

ANGRO2	OC	*Anguilla rostrata* (>200 mm)	ESOAM1	C2	*Esox americanus* (<301 mm)
ANGROS	OC	*Anguilla rostrata*	ESOAME	C2	*Esox americanus americanus*
ARIFE1	OC	*Arius felis* (≤125 mm)	ESONI1	C2	*Esox niger* (11–160 mm)
BAICH1	OC	*Bairdiella chrysoura* (11–20 mm)	ESONI2	C2	*Esox niger* (161–260 mm)
BREPA1	OC	*Brevoortia patronus* (<50 mm)	ETHBAR	C2	*Etheostoma fusiforme barratti*
BREPA2	OC	*Brevoortia patronus* (≥50 mm)	ETHDA1	C2	*Etheostoma davisoni* (<51 mm)
BREPAT	OC	*Brevoortia patronus*	ETHDAV	C2	*Etheostoma davisoni*
BRESPE	OC	*Brevoortia* sp.	ETHED1	C2	*Etheostoma edwini* (<51 mm)
BRETYR	OC	*Brevoortia tyrannus*	ETHEDW	C2	*Etheostoma edwini*
CHASA3	OC	*Chasmodes saburrae* (>45 mm)	ETHFUS	C2	*Etheostoma fusiforme*
DIPHO3	OC	*Diplodus holbrooki* (31–40 mm)	ETHPRO	C2	*Etheostoma proeliare*
HETFOR	OC	*Heterandria formosa*	ETHSPE	C2	*Etheostoma* sp.
HYBWI1	OC	*Hybopsis winchelli* (<61 mm)	ETHSPN	C2	*Etheostoma* sp. nov.
HYBWIN	OC	*Hybopsis winchelli*	ETHSTI	C2	*Etheostoma stigmaeum*
ICTPUN	OC	*Ictalurus punctatus*	ETHSW1	C2	*Etheostoma swaini* (<51 mm)
LAGRH2	OC	*Lagodon rhomboides* (16–25 mm)	ETHSWA	C2	*Etheostoma swaini*
LEIXA2	OC	*Leiostomus xanthurus* (70–99 mm)	ETRCRO	C2	*Etropus crossotus*
LEIXA4	OC	*Leiostomus xanthurus* (>90 mm)	ETRMIC	C2	*Etropus microstomus*
LEIXAN	OC	*Leiostomus xanthurus*	ETRRIM	C2	*Etropus rimosus*
MICUND	OC	*Micropogonias undulatus*	EUCARG	C2	*Eucinostomus argenteus*
MONCI1	OC	*Monacanthus ciliatus* (16–30 mm)	EUCGUL	C2	*Eucinostomus gula*
MONCI2	OC	*Monacanthus ciliatus* (31–60 mm)	EUCLEF	C2	*Eucinostomus lefroyi*
MONCIL	OC	*Monacanthus ciliatus*	EUCSPE	C2	*Eucinostomus* sp.
NOTTE1	OC	*Notropis texanus* (11–30 mm)	FUNCHR	C2	*Fundulus chrysotus*
NOTTE2	OC	*Notropis texanus* (31–60 mm)	FUNCIN	C2	*Fundulus cingulatus*
NOTTE3	OC	*Notropis texanus* (61–70 mm)	FUNESC	C2	*Fundulus escambiae*
NOTTEX	OC	*Notropis texanus*	FUNGRA	C2	*Fundulus grandis*
OPIOGL	OC	*Opisthonema oglinum*	FUNHET	C2	*Fundulus heteroclitus*
ORTCH2	OC	*Orthopristis chrysoptera* (21–50 mm)	FUNJEN	C2	*Fundulus jenkinsi*
ORTCHR	OC	*Orthopristis chrysoptera*	FUNOL1	C2	*Fundulus olivaceous* (11–40 mm)
STEHI1	OC	*Stephanolepis hispidus* (26–40 mm)	FUNOL2	C2	*Fundulus olivaceous* (41–70 mm)
STEHI2	OC	*Stephanolepis hispidus* (41–80 mm)	FUNOLI	C2	*Fundulus olivaceous*
STEHIS	OC	*Stephanolepis hispidus*	FUNPUL	C2	*Fundulus pulvereus*
SYNFO1	OC	*Synodus foetens* (≤75 mm)	FUNSIM	C2	*Fundulus similis*
SYNSC1	OC	*Syngnathus scovelli* (1–75 mm)	GYMMIC	C2	*Gymnura micrura*
SYNSC2	OC	*Syngnathus scovelli* (76–125 mm)	GYMSAX	C2	*Gymnothorax saxicola*
SYNSC3	OC	*Syngnathus scovelli* (>125 mm)	HAEPL2	C2	*Haemulon plumieri* (>35 mm)
SYNSPE	OC	*Syngnathus* sp.	HIPERE	C2	*Hippocampus erectus*

(continued)

Species Code	Trophic Code	Description	Species Code	Trophic Code	Description
TRIMA1	OC	*Trinectes maculatus* (≤50 mm)	HIPSPE	C2	*Hippocampus* sp.
AAAHDR	OR	**Omnivore (general)**	HIPZOS	C2	*Hippocampus zosterae*
ARCPRO	OR	*Archosargus probatocephalus*	HYPBER	C2	*Hypleurochilus bermudensis*
ARIFEL	OR	*Arius felis*	HYPGEM	C2	*Hypleurochilus geminatus*
CARCY1	OR	*Carpiodes cyprinus* (<176 mm)	HYPHEN	C2	*Hypsoblennius hentzi*
CARCYP	OR	*Carpiodes cyprinus*	LACMAX	C2	*Lachnolaimus maximus*
CARVE1	OR	*Carpiodes velifer* (<601 mm)	LEPAU1	C2	*Lepomis auritus* (11–80 mm)
CARVEL	OR	*Carpiodes velifer*	LEPAU2	C2	*Lepomis auritus* (81–160 mm)
CHASA2	OR	*Chasmodes saburrae* (26–45 mm)	LEPAUR	C2	*Lepomis auritus*
CHASAB	OR	*Chasmodes saburrae*	LEPCYA	C2	*Lepomis cyanellus*
DIPHOL	OR	*Diplodus holbrooki*	LEPGIB	C2	*Lepomis gibbosus*
DORCEP	OR	*Dorosoma cepedianum*	LEPGU1	C2	*Lepomis gulosus* (11–60 mm)
ERIOBL	OR	*Erimyzon oblongus*	LEPGU2	C2	*Lepomis gulosus* (61–120 mm)
ERISPE	OR	*Erimyzon* sp.	LEPGU3	C2	*Lepomis gulosus* (121–160 mm)
ERISUC	OR	*Erimyzon sucetta*	LEPGUL	C2	*Lepomis gulosus*
ERITEN	OR	*Erimyzon tenuis*	LEPMA1	C2	*Lepomis macrochirus* (11–90 mm)
FLOCAR	OR	*Floridichthys carpio*	LEPMA2	C2	*Lepomis macrochirus* (91–180 mm)
ICTSER	OR	*Ictalurus serracanthus*	LEPMAC	C2	*Lepomis macrochirus*
LAGRH3	OR	*Lagodon rhomboides* (26–60 mm)	LEPMAR	C2	*Lepomis marginatus*
LAGRH4	OR	*Lagodon rhomboides* (61–120 mm)	LEPME1	C2	*Lepomis megalotis* (11–70 mm)
LAGRHO	OR	*Lagodon rhomboides*	LEPME2	C2	*Lepomis megalotis* (71–160 mm)
MICGU2	OR	*Microgobius gulosus* (>20 mm)	LEPMEG	C2	*Lepomis megalotis*
MONCI3	OR	*Monacanthus ciliatus* (>60 mm)	LEPMI1	C2	*Lepomis microlophus* (11–80 mm)
MOXPO1	OR	*Moxostoma poecilurum* (<100 mm)	LEPMI2	C2	*Lepomis microlophus* (81–160 mm)
MOXPO2	OR	*Moxostoma poecilurum* (101–300 mm)	LEPMI3	C2	*Lepomis microlophus* (161–220 mm)
MOXPOE	OR	*Moxostoma poecilurum*	LEPMIC	C2	*Lepomis microlophus*
ORTCH1	OR	*Orthopristis chrysoptera* (11–20 mm)	LEPPU1	C2	*Lepomis punctatus* (11–60 mm)
STEHI3	OR	*Stephanolepis hispidus* (>80 mm)	LEPPU2	C2	*Lepomis punctatus* (61–140 mm)
SYNSCO	OR	*Syngnathus scovelli*	LEPPUN	C2	*Lepomis punctatus*
TRIMAC	OR	*Trinectes maculatus*	LEPSPE	C2	*Lepomis* sp.
AAAHDR	C	**Carnivore**	LUTGR1	C2	*Lutjanus griseus* (<50 mm)
CHOCOR	C	*Chologaster cornuta*	LUTSYN	C2	*Lutjanus synagris*
ECHNAU	C	*Echeneis naucrates*	MENAME	C2	*Menticirrhus americanus*
ECHSPE	C	*Echeneis* sp.	MENLIT	C2	*Menticirrhus littoralis*
HISHIS	C	*Histrio histrio*	MENSAX	C2	*Menticirrhus saxatilis*
AAAHDR	C1	**Carnivore (primary)**	MENSPE	C2	*Menticirrhus* sp.
AMECAT	C1	*Ameiurus catus*	MICPUN	C2	*Micropterus punctulatus*

AMENAT	C1	*Ameiurus natalis*	MICSA1	C2	*Micropterus salmoides* (≤70 mm)
AMENEB	C1	*Ameiurus nebulosus*	MICSAL	C2	*Micropterus salmoides*
ANCHEP	C1	*Anchoa hepsetus*	MICSPE	C2	*Micropterus* sp.
ANCLYO	C1	*Anchoa lyolepis*	MICUN2	C2	*Micropogonias undulatus* (≥70 mm)
ANCMI1	C1	*Anchoa mitchilli* (≤40 mm)	MICUN3	C2	*Micropogonias undulatus* (>100 mm)
ANCMI2	C1	*Anchoa mitchilli* (>40 mm)	MORAME	C2	*Morone americana*
ANCMIT	C1	*Anchoa mitchilli*	MORCHR	C2	*Morone chrysops*
ANCNAS	C1	*Anchoa nasuta*	MORHYB	C2	*Morone hybrid (saxatilis × chrysops)*
ANCSPE	C1	*Anchoa* sp.	MORSAX	C2	*Morone saxatilis*
ANGRO1	C1	*Anguilla rostrata* (<200 mm)	MORSPE	C2	*Morone* sp.
APHSA1	C1	*Aphredoderus sayanus* (11–50 mm)	NARBRA	C2	*Narcine brasiliensis*
APHSAY	C1	*Aphredoderus sayanus*	NAUDUC	C2	*Naucrates ductor*
APOTOW	C1	*Apogon townsendi*	OLISA1	C2	*Oligoplites saurus*
ASTSTE	C1	*Astrapogon stellatus*	OLISAU	C2	*Oligoplites saurus*
BAICH2	C1	*Bairdiella chrysoura* (21–60 mm)	OPSBET	C2	*Opsanus beta*
CALARC	C1	*Calamus arctifrons*	OPSPAR	C2	*Opsanus pardus*
CARJUV	C1	*Carangidae* juvenile	OPSTAU	C2	*Opsanus tau*
CHASA1	C1	*Chasmodes saburrae* (<25 mm)	ORTCH4	C2	*Orthopristis chrysoptera* (>100 mm)
CHISCH	C1	*Chilomycterus schoepfi*	PARAL1	C2	*Paralichthys albigutta*
CHLCH1	C1	*Chloroscombrus chrysurus* (≤40 mm)	PARDEN	C2	*Paralichthys dentatus*
CHLCH2	C1	*Chloroscombrus chrysurus* (41–50 mm)	PARFA1	C2	*Paraclinus fasciatus*
CYNAR1	C1	*Cynoscion arenarius* (<30 mm)	PARFAS	C2	*Paraclinus fasciatus*
CYNAR2	C1	*Cynoscion arenarius* (30–49 mm)	PARLE1	C2	*Paralichthys lethostigma* (<60 mm)
CYPCER	C1	*Cyprinella venusta cercostigma*	PARMAR	C2	*Paraclinus marmoratus*
CYPVE1	C1	*Cyprinella venusta* (11–50 mm)	PEPALE	C2	*Peprilus alepidotus*
CYPVE2	C1	*Cyprinella venusta* (51–90 mm)	PEPBUR	C2	*Peprilus burti*
CYPVE3	C1	*Cyprinella venusta* (91–140 mm)	PEPPAR	C2	*Peprilus paru*
CYPVEN	C1	*Cyprinella venusta*	PEPTRI	C2	*Peprilus triacanthus*
DIPHO1	C1	*Diplodus holbrooki* (11–15 mm)	PERFLA	C2	*Perca flavescens*
DIPHO2	C1	*Diplodus holbrooki* (16–30 mm)	PERNI1	C2	*Percina nigrofasciata* (<101 mm)
EUCAR1	C1	*Eucinostomus argenteus* (16—30 mm)	PERNIG	C2	*Percina nigrofasciata*
EUCAR2	C1	*Eucinostomus argenteus* (>30 mm)	PEROUA	C2	*Percina ouachitae*
EUCGU1	C1	*Eucinostomus gula* (21–35 mm)	POGCRO	C2	*Pogonias cromis*
EUCGU2	C1	*Eucinostomus gula* (36–60 mm)	POMANN	C2	*Pomoxis annularis*
EUCGU3	C1	*Eucinostomus gula* (>60 mm)	POMNIG	C2	*Pomoxis nigromaculatus*
FUNCON	C1	*Fundulus confluentus*	PRIRUB	C2	*Prionotus rubio*
GAMAF1	C1	*Gambusia affinis* (<51 mm)	PRISCI	C2	*Prionotus scitulus*
GAMAFF	C1	*Gambusia affinis*	PRISPE	C2	*Prionotus* sp.

(continued)

Species Code	Trophic Code	Description	Species Code	Trophic Code	Description
GAMHO1	C1	*Gambusia holbrooki* (<51 mm)	PRITRI	C2	*Prionotus tribulus*
GAMHOL	C1	*Gambusia holbrooki*	RAJEGL	C2	*Raja eglanteria*
GOBBOL	C1	*Gobionellus boleosoma*	RAJOLS	C2	*Raja olseni*
GOBBOS	C1	*Gobiosoma bosci*	RAJTEX	C2	*Raja texana*
GOBHAS	C1	*Gobionellus hastatus*	RHIBON	C2	*Rhinoptera bonasus*
GOBROB	C1	*Gobiosoma robustum*	RHILEN	C2	*Rhinobatos lentiginosus*
GOBSSP	C1	*Gobiosoma* sp.	SAUBRA	C2	*Saurida brasiliensis*
HAEAUR	C1	*Haemulon aurolineatum*	SCIOC1	C2	*Sciaenops ocellatus* (<100 mm)
HAEFLA	C1	*Haemulon flavolineatum*	SELSET	C2	*Selene setapinnis*
HAEPL1	C1	*Haemulon plumieri* (<35 mm)	SELVOM	C2	*Selene vomer*
HAEPLU	C1	*Haemulon plumieri*	SERSUB	C2	*Serranus subligarius*
HAESPE	C1	*Haemulon* sp.	STELAN	C2	*Stellifer lanceolatus*
HARJA1	C1	*Harengula jaguana*	STRMAR	C2	*Strongylura marina*
HARJAG	C1	*Harengula jaguana*	SYNFL2	C2	*Syngnathus floridae* (76–125 mm)
HYPUN1	C1	*Hyporhamphus unifasciatus* (<130 mm)	SYNFL3	C2	*Syngnathus floridae* (>125 mm)
ICTCAT	C1	*Ictalurus catus*	SYNFLO	C2	*Syngnathus floridae*
ICTMEL	C1	*Ictalurus melas*	SYNFO2	C2	*Synodus foetens* (75–250 mm)
ICTSPE	C1	*Ictalurus* sp.	SYNFO3	C2	*Synodus foetens* (>250 mm)
LABSI1	C1	*Labidesthes sicculus* (<71 mm)	SYNFOE	C2	*Synodus foetens*
LABSIC	C1	*Labidesthes sicculus*	SYNLOU	C2	*Syngnathus louisianae*
LABVAN	C1	*Labidesthes sicculus vanhyningi*	TRACAR	C2	*Trachinotus carolinus*
LACQUA	C1	*Lactophrys quadricornis*	TRAFAL	C2	*Trachinotus falcatus*
LACTRI	C1	*Lactophrys triqueter*	TRILEP	C2	*Trichiurus lepturus*
LAGRH1	C1	*Lagodon rhomboides* (11–15 mm)	TRIMA2	C2	*Trinectes maculatus* (>50 mm)
LEIXA1	C1	*Leiostomus xanthurus* (<70 mm)	UMBPYG	C2	*Umbra pygmaea*
LEIXA3	C1	*Leiostomus xanthurus* (≥100 mm)	UROFLO	C2	*Urophycis floridana*
LUCGOO	C1	*Lucania goodei*	UROREG	C2	*Urophycis regia*
LUCPAR	C1	*Lucania parva*	AAAHDR	C3	**Carnivore (tertiary)**
MENBER	C1	*Menidia beryllina*	ANCQU2	C3	*Ancylopsetta quadrocellata* (>75 mm)
MENMEN	C1	*Menidia menidia*	ANCQUA	C3	*Ancylopsetta quadrocellata*
MENPE1	C1	*Menidia peninsulae*	ATRSPA	C3	*Atractosteus spatula*
MENPE2	C1	*Menidia peninsulae* (>20 mm)	CARACR	C3	*Carcharhinus acronotus*
MENPEN	C1	*Menidia peninsulae*	CARBA2	C3	*Caranx bartholomaei* (>70 mm)
MICCRI	C1	*Micrognathus criniger*	CARBAR	C3	*Caranx bartholomaei*
MICTHA	C1	*Microgobius thalassinus*	CARCRY	C3	*Caranx crysos*
MICUN1	C1	*Micropogonias undulatus* (<70 mm)	CARFUS	C3	*Caranx fusus*
MINME1	C1	*Minytrema melanops* (<261 mm)	CARHI2	C3	*Caranx hippos* (>70 mm)

MINMEL	C1	*Minytrema melanops*	CARHIP	C3	*Caranx hippos*
NOTATH	C1	*Notropis atherinoides*	CARLEU	C3	*Carcharhinus leucas*
NOTCHA	C1	*Notropis chalybaeus*	CARLIM	C3	*Carcharhinus limbatus*
NOTCRY	C1	*Notemigonus crysoleucas*	CARPLU	C3	*Carcharhinus plumbeus*
NOTEMI	C1	*Notropis emiliae*	CARPOR	C3	*Carcharhinus porosus*
NOTFUN	C1	*Noturus funebris*	CENST3	C3	*Centropristis striata* (>140 mm)
NOTGYR	C1	*Noturus gyrinus*	CYCCHI	C3	*Cyclopsetta chittendeni*
NOTHA1	C1	*Notropis harperi* (<61 mm)	CYNARE	C3	*Cynoscion arenarius*
NOTHAR	C1	*Notropis harperi*	CYNNE3	C3	*Cynoscion nebulosus* (>80 mm)
NOTLEP	C1	*Noturus leptacanthus*	CYNNEB	C3	*Cynoscion nebulosus*
NOTPET	C1	*Notropis petersoni*	CYNREG	C3	*Cynoscion regalis*
NOTROS	C1	*Notropis roseipinnis*	CYNSPE	C3	*Cynoscion* sp.
NOTSPE	C1	*Notropis* sp.	ELOSAU	C3	*Elops saurus*
NOTWEL	C1	*Notropis welaka*	LEPOCU	C3	*Lepisosteus oculatus*
NOTZON	C1	*Notropis zonatus*	LEPOSS	C3	*Lepisosteus osseus*
ORTCH3	C1	*Orthopristis chrysoptera* (51–100 mm)	LUTCAM	C3	*Lutjanus campechanus*
POLOCT	C1	*Polydactylus octonemus*	LUTGR2	C3	*Lutjanus griseus* (>50 mm)
PTEHYP	C1	*Pteronotropis hypselopterus*	LUTGRI	C3	*Lutjanus griseus*
SCIJUV	C1	*Sciaenidae* juvenile	MEGATL	C3	*Megalops atlanticus*
SYMDIO	C1	*Symphurus diomedianus*	MICSA2	C3	*Micropterus salmoides* (>70 *mm)*
SYMPLA	C1	*Symphurus plagiusa*	PARALB	C3	*Paralichthys albigutta*
SYNFL1	C1	*Syngnathus floridae* (1–75 mm)	PARLE2	C3	*Paralichthys lethostigma* (>60 mm)
AAAHDR	C2	**Carnivore (secondary)**	PARLET	C3	*Paralichthys lethostigma*
ACAPOM	C2	*Acantharchus pomotis*	PARSPE	C3	*Paralichthys* sp.
ACHLIN	C2	*Achirus lineatus*	POMSAL	C3	*Pomatomus saltatrix*
ACIOXY	C2	*Acipenser oxyrhynchus*	RACCAN	C3	*Rachycentron canadum*
AMICAL	C2	*Amia calva*	RHITER	C3	*Rhizoprionodon terraenovae*
ANCQU1	C2	*Ancylopsetta quadrocellata* (<75 mm)	SCIOC2	C3	*Sciaenops ocellatus* (>100 mm)
APHSA2	C2	*Aphredoderus sayanus* (51–80 mm)	SCIOCE	C3	*Sciaenops ocellatus*
ARIFE2	C2	*Arius felis* (126–250 mm)	SCOMAC	C3	*Scomberomorus maculatus*
ARIFE3	C2	*Arius felis* (>250 mm)	SPHBAR	C3	*Sphyraena barracuda*
BAGMA1	C2	*Bagre marinus* (≤125 mm)	SPHBOR	C3	*Sphyraena borealis*
BAGMA2	C2	*Bagre marinus* (126–150 mm)	SPHGUA	C3	*Sphyraena guachancho*
			SPHTIB	C3	*Sphyrna tiburo*

(continued)

Species Code	Trophic Code	Description	Species Code	Trophic Code	Description
		Invertebrates			
AAAHDR	H	**Herbivore**	ODOSTO	OR	*Odostomia* sp.
AGRSPE	H	*Agrypnia* sp.	OEDICE	OR	Oedicerotidae
ARCHIP	H	*Archips* sp.	OLIGON	OR	Oligoneuriidae
BAGSPA	H	*Bagous* sp. adult	ORCBUL	OR	*Orchesella bulba*
BAGSPE	H	*Bagous* sp.	ORCGRI	OR	*Orchestia grillus*
BANKSP	H	*Banksiola* sp.	ORCUHL	OR	*Orchestia uhleri*
CORIXI	H	Corixidae	ORMSPE	OR	*Ormosia* sp.
ELLMIN	H	*Ellipes minuta*	ORTANN	OR	*Orthocladius annectens*
ESTSPE	H	*Estigmene* sp.	ORTHGD	OR	*Orthocladiinae* genus D
EUKCLA	H	*Eukiefferiella claripennis* group	ORTHOC	OR	*Orthocladius* sp.
EUKISP	H	*Eukiefferiella* sp.	ORTINI	OR	Orthocladiinae
HALIPL	H	*Haliplidae* adult	ORTSPC	OR	*Orthocladiinae* sp. C
HALIPS	H	Haliplidae	OXYETH	OR	*Oxyethira* sp.
HALPUN	H	*Haliplus punctatus*	OXYSMI	OR	*Oxyurostylis smithi*
HELICH	H	*Helichus* sp.	PALPAL	OR	*Palaemonetes paludosus*
HESBRI	H	*Hesperocorixa brimleyi*	PARACH	OR	*Parachironomus* sp.
HESSPE	H	*Hesperophylax* sp.	PARAKS	OR	*Parakiefferiella* sp.
HESVUL	H	*Hesperocorixa vulgaris*	PARALE	OR	*Paraleptophlebia* sp.
HETAME	H	*Heteroplectron americanum*	PARATA	OR	*Paratanytarsus* sp.
HYDRPT	H	*Hydroptila* sp.	PARATE	OR	*Paratendipes* sp.
HYDRSB	H	*Hydroptila* sp. b	PARBAS	OR	*Paratendipes basidens*
HYDSAD	H	*Hydrochus* sp. (adult)	PARCAR	OR	*Parachironomus carinatus*
HYDSEE	H	*Hydrochus* sp. e	PARLIT	OR	*Paranais litoralis*
HYDSPA	H	*Hydrochus* sp. a	PARM/T	OR	*Parachironomus monochromus*
HYDSPB	H	*Hydrochus* sp. b	PARMET	OR	*Parametriocnemus* sp.
HYDSPC	H	*Hydrochus* sp. c	PARNIG	OR	*Paralauterborniella nigrohalterale*
HYDSPD	H	*Hydrochus* sp. d	PARSPA	OR	*Paraphoxus* sp. a
HYDTIL	H	Hydroptilidae	PHAESP	OR	*Phaenopsectra* sp.
LEPIDA	H	*Lepidoptera* genus a	PHOMAC	OR	*Photis macromanus*
LEPIDO	H	Lepidoptera	PHOSP1	OR	*Photis* sp. 1
LEPISP	H	*Lepidostoma* sp.	PHYINT	OR	*Physella integra*
LIPSOT	H	*Lipsothrix* sp.	PHYLOC	OR	*Phylocentropus* sp.
LISLAC	H	*Lissorhoptrus lacustris*	PHYSEL	OR	*Physella* sp.
NECCAN	H	*Nectopsyche candida*	PLADUM	OR	*Platynereis dumerili*
NECEXQ	H	*Nectopsyche exquisita*	PLEGLA	OR	*Pleusymtes glaber*

NECPAV	H	*Nectopsyche pavida*	PODRII	OR	*Podochela riisei*
NECTOP	H	*Nectopsyche* sp.	PODUSP	OR	*Podura* sp.
NOCTUI	H	*Noctuidae* genus a	POLAVI	OR	*Polypedilum aviceps*
NYMSPE	H	*Nymphula* sp.	POLCON	OR	*Polypedilum convictum*
ONYNIG	H	*Onychylis nigrirostris* complex	POLHAL	OR	*Polypedilum halterale* group
ORTHOT	H	*Orthotrichia* sp.	POLILL	OR	*Polypedilum illinoense*
PARAPO	H	*Parapoynx* sp.	POLOPH	OR	*Polypedilum ophiodes*
PARVES	H	*Parenthis vestitus*	POLPED	OR	*Polypedilum pedestre*
PELOPP	H	*Peltodytes oppositus*	POLSCA	OR	*Polypedilum scalaenum*
PELSEX	H	*Peltodytes sexmaculatus*	POLSPB	OR	*Polypedilum* sp. B
PELSQU	H	*Pelenomus squamosus*	POLSPC	OR	*Polypedilum* sp. C
PELTMU	H	*Peltodytes muticus*	POLTRI	OR	*Polypedilum trigonus*
PELTOP	H	Peltoperlidae	POLTRO	OR	Polycentropodidae
PHRLAR	H	*Phryganea* sp.	POLTRT	OR	*Polypedilum tritum*
POLFAL	H	*Polypedilum fallax*	POLYPE	OR	*Polypedilum* sp.
PTISPE	H	*Ptilostomis* sp.	POLYS1	OR	*Polypedilum* sp. 1
PYCNOP	H	*Pycnopsyche* sp.	POLYS2	OR	*Polypedilum* sp. 2
PYRALB	H	Pyralidae genus b	PONSPA	OR	*Pontogeneia* sp. a
PYRALC	H	Pyralidae genus c	POTAMY	OR	*Potamyia* sp.
PYRALD	H	Pyralidae	POTFLA	OR	*Potamyia flava*
PYRALI	H	Pyralidae genus a	POTTLO	OR	*Potthastia longimana* group
SIGARA	H	*Sigara* sp.	PRCSPE	OR	*Prionocyphon* sp.
SPETES	H	*Sperchopsis tessellatus*	PRIACU	OR	*Pristinella acuminata*
STEAES	H	*Stenochironomus c.f. aestivalis*	PRIAEQ	OR	*Pristina aequiseta*
STENOC	H	*Stenochironomus* sp.	PRIBRE	OR	*Pristina breviseta*
TRILOU	H	*Trichocorixa louisianae*	PRILEI	OR	*Pristina leidyi*
TRISPE	H	*Trichocorixa* sp.	PRILOD	OR	*Pristina longidentata*
TRITAR	H	*Triaenodes tardus*	PRILON	OR	*Pristinella longisoma*
XYLPAR	H	*Xylotopus par*	PRILOS	OR	*Pristina longiseta*
AAAHDR	HB	**Herbivore (benthic plants)**	PRIOSB	OR	*Pristinella osborni*
ACLSPE	HB	*Aclis* sp.	PRIPLU	OR	*Pristina plumaseta*
ACMAEI	HB	*Acmaeidae* sp.	PRISIM	OR	*Pristinella sima*
ACMSPE	HB	*Acmaea* sp.	PRISPE	OR	*Pristina* sp.
ACTPUN	HB	*Acteon punctostriatus*	PRKSPA	OR	*Parakiefferiella* sp. A
ALCINS	HB	*Alcirona insulans*	PROATL	OR	*Promysis atlantica*
ALPNOR	HB	*Alpheus normanni*	PROCAM	OR	*Procambarus* sp.
AMNSPE	HB	*Amnicola* sp.	PROCLO	OR	*Procloeon* sp.
ANCYLI	HB	Ancylidae	PROMOR	OR	*Promoresia* sp.

(continued)

Species Code	Trophic Code	Description	Species Code	Trophic Code	Description
APLFLO	HB	*Aplysia floridensis*	PROPEN	OR	*Procambarus paeninsulanus*
APLSPE	HB	*Aplysia* sp.	PROSPI	OR	*Procambarus spiculifer*
APLWIL	HB	*Aplysia willcoxi*	PSECLN	OR	*Baetis* (Pseudocloeon) sp.
ASTLUN	HB	*Astyris lunata*	PSECTR	OR	*Psectrocladius* sp.
BURLEA	HB	*Bursatella leachi plei*	PSEELA	OR	*Psectrocladius elatus*
CAMGEN	HB	*Campeloma geniculum*	PSEUDC	OR	*Pseudochironomus* sp.
CAMLIM	HB	*Campeloma limum*	PSEUSM	OR	*Pseudosmittia* sp.
CAMPSP	HB	*Campeloma* sp.	PSTSPE	OR	*Pagastiella* sp.
CASLUN	HB	*Cassidinidea lunifrons*	PSYCHO	OR	Psychomyiidae
CASOVA	HB	*Cassidinidea ovalis*	PYRSPE	OR	*Pyramidella* sp.
CASSPE	HB	*Cassidinidea* sp.	RHABDO	OR	*Rhabdomastix* sp.
CHISPA	HB	*Polyplacophoran* sp. a	RHEDIS	OR	*Rheotanytarsus distinctissimus* gr.
CHISPB	HB	*Polyplacophoran* sp. b	RHEEXI	OR	*Rheotanytarsus exiguus* group
CHISPC	HB	*Polyplacophoran* sp. c	RHEOCR	OR	*Rheocricotopus* sp.
CHITON	HB	*Chiton* sp.	RHEOTA	OR	*Rheotanytarsus* sp.
CHITUB	HB	*Chiton tuberculatus*	RHESPE	OR	*Rheosmittia* sp.
CYMCOM	HB	*Cymadusa compta*	RIMFRE	OR	*Rimula frenulata*
CYMFIL	HB	*Cymadusa* sp.	RUDNAG	OR	*Rudilemboides naglei*
DENLAQ	HB	*Dentalium laqueatum*	SEMICE	OR	*Semicerura ?* sp.
DENSPE	HB	*Dentalium* sp.	SERDEF	OR	*Seratella deficiens*
DIOCAY	HB	*Diodora cayenensis*	SETODE	OR	*Setodes* sp.
ELIBOY	HB	*Elimia boykiana*	SETSTE	OR	*Setodes stehri*
ELICUR	HB	*Elimia curvicostata*	SIMJON	OR	*Simulium jonesi*
ELISPE	HB	*Elimia* sp.	SIMLA1	OR	*Simulium* sp. 1
EPIDIL	HB	*Epialtus dilatatus*	SIMSNO	OR	*Simulium snowi*
FERHEN	HB	*Ferrissia hendersoni*	SIMUL3	OR	*Simulium* sp. 3
GASSP1	HB	*Gastropod* sp. 1	SIMULI	OR	Simuliidae
GASSP2	HB	*Gastropod* sp. 2	SIMUND	OR	*Simulium underhilli*
GASSP3	HB	*Gastropod* sp. 3	SIMUSP	OR	*Simulium* sp.
GASSP4	HB	*Gastropod* sp. 4	SIMVIT	OR	*Simulium vittatum*
GASTRO	HB	Gastropoda	SINSPE	OR	*Sinella* sp.
GYRSPE	HB	*Gyraulus* sp.	SIPSPA	OR	*Sipunculid* sp. a
HAMSUC	HB	*Haminoea succinea*	SIPSPB	OR	*Sipunculid* sp. b
HEBEXC	HB	*Hebetancylus excentricus*	SIPSPE	OR	*Siphlonurus* sp.
HEBSPE	HB	*Hebetancylus* sp.	SISSPE	OR	*Sisyra* sp.
HELSPE	HB	*Helisoma* sp.	SLAAPP	OR	*Slavina appendiculata*
HYDROB	HB	Hydrobiidae	SMICRI	OR	*Smicridea* sp.

LAEFUS	HB	*Laevapex fuscus*	SMINTH	OR	Sminthuridae
LAESPE	HB	*Laevapex* sp.	SMTDSP	OR	*Sminthurides* sp.
LIBDU1	HB	*Libinia dubia* (<30 mm)	SMTRSP	OR	*Sminthurus* sp.
LIBDU2	HB	*Libinia dubia* (30–60 mm)	SPEJOS	OR	*Specaria josinae*
LIBDUB	HB	*Libinia dubia*	SPHTAY	OR	*Sphaerosyllis taylori*
MICDIL	HB	*Micromenetus dilatatus*	SPICOS	OR	*Spiochaetopterus costorum*
NEOPA1	HB	*Neopanope packardii* (<6 mm)	SPINPL	OR	Spionidae post-larvae
NEOPA2	HB	*Neopanope packardii* (6–20 mm)	STEANT	OR	*Stenelmis antennalis*
NEOPAC	HB	*Neopanope packardii*	STECON	OR	*Stenelmis convexula*
PARCAU	HB	*Paracerceis caudata*	STEDEC	OR	*Stenelmis decorata*
PARINT	HB	*Parviturboides interruptus*	STEEXI	OR	*Stenonema exiguum*
PITAN2	HB	*Pitho anisodon* (11–20 mm)	STEFLO	OR	*Stenacron floridense*
PITANI	HB	*Pitho anisodon*	STEFUS	OR	*Stenelmis fuscata*
PLANOR	HB	Planorbidae	STEINT	OR	*Stenacron interpunctatum*
PLASP1	HB	Planorbidae sp. 1	STELSP	OR	*Stelechomyia* sp.
PLEURO	HB	Pleuroceridae	STEMAR	OR	*Stenoninereis martini*
RETCAN	HB	*Retusa canaliculata*	STEMOD	OR	*Stenonema modestum*
SPHDES	HB	*Sphaeroma destructor*	STEMSP	OR	*Stempellina* sp.
SPHQUA	HB	*Sphaeroma quadridentatum*	STEMXI	OR	*Stenonema mexicanum integrum*
SPHTER	HB	*Sphaeroma terebrans*	STENAC	OR	*Stenacron* sp.
STRSPE	HB	*Strombiformis* sp.	STENAD	OR	*Stenelmis* sp. adult
TURCAS	HB	*Turbo castaneus*	STENEL	OR	*Stenelmis* sp.
TURRSP	HB	*Turritella* sp.	STENMA	OR	*Stenelmis bicarinata*
TURSPE	HB	*Turridae* sp.	STENON	OR	*Stenonema* sp.
AAAHDR	HP	**Herbivore (plankton)**	STENS2	OR	*Stenelmis* sp. 2
AEQMUC	HP	*Aequipecten mucosa*	STEPER	OR	*Stelechomyia perpulchra*
AMYAEQ	HP	*Amygdalum aequalis*	STEPSP	OR	*Stephensoniana* sp.
AMYPAP	HP	*Amygdalum papyria*	STESMI	OR	*Stenonema smithae*
ANABRA	HP	*Anadara brasiliana*	STESPE	OR	*Stempellinella* sp.
ANASPE	HP	*Anadara* sp.	STETAN	OR	*Stephensoniana tandyi*
ANATRA	HP	*Anadara transversa*	STETRI	OR	*Stephensoniana trivandrana*
ANOAUB	HP	*Anomalocardia auberianna*	STICTO	OR	*Stictochironomus* sp.
ANOSIM	HP	*Anomia simplex*	STISPA	OR	*Stictochironomus* sp. a
APLISP	HP	*Aplidium* sp.	STISPB	OR	*Stictochironomus* sp. b
ARCCOR	HP	*Arcinella cornuta*	STNLSA	OR	*Stenelmis* sp. A
ASCCUR	HP	*Ascidea c.f. curvata*	STNLSB	OR	*Stenelmis* sp. B
BALEBU	HP	*Balanus eburneus*	STNLSC	OR	*Stenelmis* sp. C
BALTRI	HP	*Balanus trigonus*	STRATI	OR	Stratiomyidae

(continued)

Species Code	Trophic Code	Description	Species Code	Trophic Code	Description
BARTRU	HP	*Barnea truncata*	STYFOS	OR	*Stylaria fossularis*
BITSPE	HP	*Bittium* sp.	STYLAC	OR	*Stylaria lacustris*
BITVAR	HP	*Bittium varium*	SYNAME	OR	*Synchelidium americanum*
BOTPLA	HP	*Botryllus c.f. planus*	SYNBIF	OR	*Synurella bifurcata*
BRAEXU	HP	*Brachidontes exustus*	SYNCHA	OR	*Synurella chamberlaini*
BRAMOD	HP	*Brachiodontes modiolus*	SYNORT	OR	*Synorthocladius* sp.
BRAREC	HP	*Brachidontes recurvus*	TAESPE	OR	*Taeniopteryx* sp.
BRASPE	HP	*Brachidontes* sp.	TALSPE	OR	*Talitroides* sp.
BUGSPE	HP	*Bugula* sp.	TALTOP	OR	*Talitroides topitotum*
CALANO	HP	Calanoid copepod	TANAID	OR	Tanaidacea
CALCEN	HP	*Calyptraea centralis*	TANINI	OR	Tanytarsini
CARFLO	HP	*Cardita floridana*	TANSP2	OR	*Tanaid* sp. 2
CERISP	HP	*Cerithium* sp.	TANYTA	OR	*Tanytarsus* sp.
CERITH	HP	Cerithiidae sp.	TAPBOW	OR	*Taphromysis bowmani*
CERMUS	HP	*Cerithium muscarum*	TAPLOU	OR	*Taphromysis louisianae*
CERSUB	HP	*Cerithiopsis subulatum*	TAPSPE	OR	*Taphromysis* sp.
CHICAN	HP	*Chione cancellata*	THIEN1	OR	*Thienemanniella* sp. 1
CHIGRU	HP	*Chione grus*	THIEN2	OR	*Thienemanniella* sp. 2
CHILAT	HP	*Chione latilirata*	THIENE	OR	*Thienemanniella* sp.
CHIPAP	HP	*Chione paphia*	THYSPE	OR	*Thyone* sp.
CHOAME	HP	*Chone c.f. americana*	TIPULA	OR	*Tipula* sp.
CHODUN	HP	*Chone duneri*	TIPULI	OR	Tipulidae
CLISPE	HP	*Clione* sp.	TIRTRO	OR	*Tiron tropakis*
CNIHYD	HP	*Unid. hydromedusae*	TOZCAR	OR	*Tozeuma carolinense*
CONLEU	HP	*Congeria leucophaeta*	TRIANO	OR	*Triaenodes* sp.
CORBIC	HP	*Corbicula* sp.	TRIBEL	OR	*Tribelos* sp.
CORFLU	HP	*Corbicula fluminea*	TRICOR	OR	*Tricorythodes* sp.
CREBAR	HP	*Crepidula barrattiana*	TRIFUS	OR	*Tribelos fuscicorne*
CREFOR	HP	*Crepidula fornicata*	TRIJUC	OR	*Tribelos jucundum*
CREMAC	HP	*Crepidula maculosa*	TRIPHA	OR	*Tribelos* sp./*Phaenopsectra* sp.
CREPID	HP	*Crepidula* sp.	TVEDIG	OR	*Tvetenia discoloripes* group
CREPLA	HP	*Crepidula plana*	TVESPE	OR	*Tvetenia* sp.
CYCLOP	HP	Cyclopoid copepod	UNNMUL	OR	*Unniella multivirga*
DINROB	HP	*Dinocardium robustum*	VEJCOM	OR	*Vejdovskyella comata*
DIPSMI	HP	*Diplothyra smithii*	WAPMOB	OR	*Wapsa mobilis*
DOSDIS	HP	*Dosinia discus*	WEBTRI	OR	*Websternereis tridentata*
DOSELE	HP	*Dosinia elegans*	WILNIG	OR	*Willowsia nigromaculata*

ELLIPT	HP	*Elliptio* sp.	XENBRE	OR	*Xenanthura brevitelson*
ENTOPR	HP	Entoprocta	AAAHDR	C	**Carnivore**
GEUDEM	HP	*Geukensia demissa*	ABLABE	C	*Ablabesmyia* sp.
HYDMIC	HP	*Hydroides microtus*	ABLANN	C	*Ablabesmyia annulata*
HYDPRO	HP	*Hydroides protilicola*	ABLAS1	C	*Ablabesmyia* sp. 1
HYDUNC	HP	*Hydroides uncinata*	ABLAS2	C	*Ablabesmyia* sp. 2
LAELAE	HP	*Laevicardium laevigatum*	ABLASP	C	*Ablabesmyia aspera*
LAEMOR	HP	*Laevicardium mortoni*	ABLHAU	C	*Ablabesmyia hauberi*
LIMPEL	HP	*Lima pellucida*	ABLJAN	C	*Ablabesmyia janta*
LUCNAS	HP	*Lucina nassula*	ABLMAL	C	*Ablabesmyia mallochi*
LYOHYA	HP	*Lyonsia hyalina*	ABLMON	C	*Ablabesmyia monilis*
MACFRA	HP	*Mactra fragilis*	ABLPEL	C	*Ablabesmyia peleensis*
MDLMDL	HP	*Modiolus modiolus*	ABLRGP	C	*Ablabesmyia rhamphe* group
MEGPIG	HP	*Megalomma pigmentum*	ACIFRA	C	*Acilius fraternus*
MEGSPA	HP	*Megalomma* sp. A	ACRABN	C	*Acroneuria abnormis*
MOGSPE	HP	*Mogula* sp.	ACRMEL	C	*Acroneuria mela*
MULLAT	HP	*Mulinia lateralis*	ACRONE	C	*Acroneuria* sp.
MUSLAT	HP	*Musculus lateralis*	AESHNI	C	Aeshnidae
MUSPAR	HP	*Musculum partumcium*	AESLAR	C	*Aeshna* sp.
MYSELL	HP	*Mysella* sp.	AGASPE	C	*Agabus* sp.
MYTLEU	HP	*Mytilopsis leucophaeta*	ALLSPE	C	*Alluaudomyia* sp.
ORTMOD	HP	*Orthocyclops modestus*	AMPHIA	C	*Amphiagrion* sp.
PARTRI	HP	*Parastarte triquetra*	ANAJUN	C	*Anax junius*
PINCAR	HP	*Pinna carnea*	ANISOP	C	*Anisoptera*
PINIMB	HP	*Pinctada imbricata*	APHWIL	C	*Aphylla williamsoni*
POLCAR	HP	*Polymesoda caroliniana*	APSESP	C	*Apsectrotanypus* sp.
POMCAE	HP	*Pomatoceros caerulescens*	ARGFUM	C	*Argia fumipennis*
POMSPE	HP	*Pomatoceros* sp.	ARGIA	C	*Argia* sp.
PSEFLO	HP	*Pseudocyrena floridana*	ARGSED	C	*Argia sedula*
RANCUN	HP	*Rangia cuneata*	ARGTIB	C	*Argia c.f. tibialis*
RENREN	HP	*Renilla reniformis*	ARGTRA	C	*Argia translata*
RENSPE	HP	*Renilla* sp.	ATHERI	C	*Atherix* sp.
SABELL	HP	Sabellidae	BASJAN	C	*Basiaeschna janata*
SABFLO	HP	*Sabellaria floridensis*	BECSPE	C	*Beckidia* sp.
SABMEL	HP	*Sabella melanostigma*	BELSPE	C	*Belostoma* sp.
SABSPA	HP	*Sabella* sp. A	BELTES	C	*Belostoma testaceum*
SABSPE	HP	*Sabella* sp.	BERACU	C	*Berosus aculeatus*
SABVUL	HP	*Sabellaria vulgaris vulgaris*	BERINF	C	*Berosus infuscatus*

(continued)

Species Code	Trophic Code	Description	Species Code	Trophic Code	Description
SERPUL	HP	*Serpulidae* sp.	BEROAD	C	*Berosus* sp. adult
SERSP1	HP	*Serpulid* sp. 1	BERSTR	C	*Berosus striatus*
SPISPI	HP	*Spirorbis spirillum*	BIDPUL	C	*Bidessonotus pulicarius*
STYPLI	HP	*Styella plicata*	BOYERI	C	*Boyeria* sp.
TRAEGM	HP	*Trachycardium egmontiarum*	BOYVIN	C	*Boyeria vinosa*
TRAMUR	HP	*Trachycardium muricatum*	BUEMAR	C	*Buenoa marki*
WATSPE	HP	*Watersipora* sp.	BUENOA	C	*Buenoa* sp.
AAAHDR	O	**Omnivore**	CALDIM	C	*Calopteryx dimidiata*
BRACSP	O	*Brachycentrus* sp.	CALLIP	C	*Calliphoridae*
CORYS3	O	*Corynoneura* sp. 3	CALMAC	C	*Calopteryx maculata*
CORYS4	O	*Corynoneura* sp. 4	CALOPT	C	*Calopteryx* sp.
CULERR	O	*Culex erraticus*	CALOSP	C	Calopterygidae
EPOSPE	O	*Epoicocladius* sp.	CELSPE	C	*Celina* sp.
GOELSP	O	*Goeldichironomus* sp.	CERATO	C	*Ceratopogonidae* sp.
LEUCTR	O	Leuctridae	CHADIA	C	*Chaetogaster diaphanus*
MICPED	O	*Microtendipes pedellus*	CHAOBO	C	*Chaoborus* sp.
MOLOSP	O	*Molophilus* sp.	CHAPUN	C	*Chaoborus punctipennis*
MOLSPA	O	*Molophilus* sp. a	CHARAS	C	*Chauliodes rastricornis*
ORTHO1	O	*Orthocladius* sp. 1	CHAULI	C	*Chauliodes* sp.
ORTHO2	O	*Orthocladius* sp. 2	CHELSP	C	*Chelifera* sp.
ORTHT3	O	*Orthocladius* type iii	CHESP2	C	*Chernovskiia* sp. 2
PARASP	O	*Paramerina* sp.	CHESPE	C	*Chernovskiia* sp.
PARCHA	O	*Parachaetocladius* sp.	CHRYSO	C	*Chrysops* sp.
PSEFUL	O	*Pseudochironomus fulviventris* group	CLINOT	C	*Clinotanypus* sp.
RHEROB	O	*Rheocricotopus robacki*	CLINSP	C	*Clinocera* sp.
RHETUB	O	*Rheocricotopus tuberculatus*	COECON	C	*Coelotanypus concinnus*
ZALSPE	O	*Zalutschia* sp.	COELOT	C	*Coelotanypus* sp.
AAAHDR	OH	**Omnivore (herbivorous)**	COENAG	C	Coenagrionidae
GAMMUC	OH	*Gammarus mucronatus*	COENSP	C	*Coenagrion* sp.
MACGLA	OH	*Macronychus glabratus*	COESCA	C	*Coelotanypus scapularis*
MICRPU	OH	*Microcylloepus pusillus*	CONCHA	C	*Conchapelopia* sp.
NEOTE2	OH	*Neopanope texana* (10–20 mm)	COPELA	C	*Copelatus* sp.
NEOTEX	OH	*Neopanope texana*	COPINT	C	*Coptotomus interrogatus*
PAVOLI	OH	*Paraleptophlebia volitans*	COPTOT	C	*Coptotomus* sp.
PITAN1	OH	*Pitho anisodon* (<11 mm)	CORCOR	C	*Corydalus cornutus*
PODRI2	OH	*Podochela riisei* (6–20 mm)	CORDSP	C	*Cordulegaster* sp.
STEHUN	OH	*Stenelmis hungerfordi*	CORDUL	C	*Corduliidae*

STESIN	OH	*Stenelmis sinuata*	CORFAS	C	*Cordulegaster fasciata*
AAAHDR	OD	**Omnivore (detritivorous)**	CORING	C	*Coryphaeschna ingens*
ABRAEQ	OD	*Abra aequalis*	CORMAC	C	*Cordulegaster maculata*
ADESPE	OD	*Adelodrilus* sp.	COROBL	C	*Cordulegaster obliqua*
AELSPE	OD	*Aelosoma* sp.	CRYFUL	C	*Cryptochironomus fulvus* group
ALOSPE	OD	*Alona* sp.	CRYPCH	C	*Cryptochironomus* sp.
AMPGUN	OD	*Amphicteis gunneri*	CULICO	C	*Culicoides* sp.
AMPSPE	OD	*Amphitrite* sp.	CYBFIM	C	*Cybister fimbriolatus*
AMPVAD	OD	*Ampelisca vadorum*	CYBISP	C	*Cybister* sp.
ANCVAD	OD	*Ancyronyx variegatus* adult	DASYSA	C	*Dasyhelea* sp. adult
ANCVAR	OD	*Ancyronyx variegatus*	DASYSP	C	*Dasyhelea* sp.
ANCYRO	OD	*Ancyronyx* sp.	DEROTA	C	*Derotanypus* sp.
ANIPYR	OD	*Anisocentropis pyraloides*	DESGRA	C	*Desmopachria grana* complex
ANISPE	OD	*Anisocentropus* sp.	DESPH?	C	*Desserobdella phalera ?*
ANOBRA	OD	*Anomalocardia brasiliana*	DIDYMO	C	*Didymops* sp.
ANORAD	OD	*Anodontoides radiatus*	DINCAR	C	*Dineutus carolinus*
AONMAY	OD	*Aonides mayaguezensis*	DINCIL	C	*Dineutus ciliatus*
APODAY	OD	*Apoprionospio dayi*	DINDIS	C	*Dineutes discolor*
APOPYG	OD	*Apoprionospio pygmaea*	DINEAD	C	*Dineutus* sp. adult
AQUGIB	OD	*Aequipectin gibbus*	DINESP	C	*Dineutus* sp.
AQUIRR	OD	*Aequipectin irradians*	DINNIG	C	*Dineutus nigrior*
ARICID	OD	*Aricidea* sp.	DINSER	C	*Dineutus serrulatus*
ARIFRA	OD	*Aricidea fragilis*	DOLICH	C	Dolichopodidae
ARIJEF	OD	*Aricidea jeffersii*	DOLTRI	C	*Dolomedes triton*
ARIPHI	OD	*Aricidea philbinae*	DROMOG	C	*Dromogomphus* sp.
ARISPA	OD	*Aricidea* sp. a	DROSPI	C	*Dromogomphus spinosus*
ARISPD	OD	*Aricidea (Acmira)* sp. D	DUGSPE	C	*Dugesia* sp.
ARITAY	OD	*Aricidea taylori*	DYTISA	C	*Dytiscidae* genus a
ARIWAS	OD	*Aricidea wassi*	DYTISB	C	*Dytiscidae* genus b
ARMAGI	OD	*Armandia agilis*	DYTISC	C	Dytiscidae
ARMMAC	OD	*Armandia maculata*	DYTISD	C	*Dytiscidae* genus d
ASASPA	OD	*Asabellides* sp. a	DYTISP	C	*Dytiscus* sp.
ASYSPE	OD	*Asychis* sp.	EMPIDI	C	Empididae
ATRRIG	OD	*Atrina rigida*	ENADIV	C	*Enallagma divagans*
ATRSER	OD	*Atrina serrata*	ENAGEM	C	*Enallagma geminatum*
AULPIG	OD	*Aulodrilus pigueti*	ENALLA	C	*Enallagma* sp.
AULPLU	OD	*Aulodrilus pluriseta*	ENASPB	C	*Enallagma* sp. b
AXIMUC	OD	*Axiothella mucosa*	ENOCIN	C	*Enochrus cinctus*

(continued)

Species Code	Trophic Code	Description	Species Code	Trophic Code	Description
BARPAU	OD	*Barbidvillus paucisetus*	EPIPRI	C	*Epitheca princeps*
BARSPE	OD	*Barbatia* sp.	EPITSP	C	*Epitheca* sp.
BATCAT	OD	*Batea catharinensis*	ERYCON	C	*Erythrodiplax connata minuscula*
BOCLIG	OD	*Boccardia ligerica*	ERYSIM	C	*Erythemis simplicicollis*
BOSSPE	OD	*Bosmina* sp.	FITSPE	C	*Fittkauimyia* sp.
BOTVEJ	OD	*Bothrioneurum vejdovskyanum*	GERSPE	C	*Gerris* sp.
BRAAME	OD	*Branchioasychis americana*	GLOSSI	C	Glossiphoniidae
BRASOW	OD	*Branchiura sowerbyi*	GOMEXI	C	*Gomphus exilis*
BRIFLA	OD	*Brillia flavifrons*	GOMGOM	C	*Gomphus (Phanogomphus)* sp.
CAESPE	OD	*Caecum* sp.	GOMHYB	C	*Gomphus hybridus*
CALPRE	OD	*Callibaetis pretiosus*	GOMHYL	C	*Gomphus (Hylogomphus)* sp.
CANDSP	OD	*Candona* sp.	GOMPHI	C	Gomphidae
CAPACI	OD	*Capitomastus aciculatus*	GOMPHU	C	*Gomphus* sp.
CAPCAP	OD	*Capitella capitata*	GOMROG	C	*Gomphus rogersi*
CAPITE	OD	*Capitellidae* sp.	GYRANA	C	*Gyrinus analis*
CAPJON	OD	*Capitella jonesi*	GYRINU	C	*Gyrinus* sp.
CAPSPE	OD	*Capitella* sp.	GYRPAC	C	*Gyrinus pachysomus*
CARHOB	OD	*Carazziella hobsonae*	HAGBRE	C	*Hagenius brevistylus*
CAUSPE	OD	*Caulleriella* sp.	HARCGA	C	*Harnischia* complex genus A
CERSPE	OD	*Cerapus* sp. *(c.f. tubularis)*	HARCGB	C	*Harnischia* complex genus B
CERTAR	OD	*Ceraclea tarsipunctata*	HARCGC	C	*Harnischia* complex genus C
CHASPE	OD	*Chaetozone* sp.	HARNIS	C	*Harnischia complex*
CHIDEC	OD	*Chironomus decorus* group	HARNSP	C	*Harnischia* sp.
CHISPE	OD	*Chimarra* sp.	HAYSEN	C	*Hayesomyia senata*
CHISTI	OD	*Chironomus stigmaterus*	HAYSPE	C	*Hayesomyia* sp.
CIRFIL	OD	*Cirriformia filigera*	HELCOR	C	*Helopelopia cornuticaudata*
CIRLYR	OD	*Cirrophorus lyra*	HELELO	C	*Helobdella elongata*
CIRRAT	OD	Cirratulidae	HELMAC	C	*Helochares maculicollis*
CIRSPA	OD	*Cirratulus* sp. a	HELOPH	C	*Helophorus* sp.
CIRSPE	OD	*Cirriformia* sp.	HEMSPE	C	*Hemerodromia* sp.
CIRTEN	OD	*Cirriformia tentaculata*	HESPER	C	*Hesperocorixia* sp.
CLADOC	OD	*Cladocera*	HETAER	C	*Hetaerina* sp.
CLASP1	OD	*Cladoceran* sp. 1	HETMER	C	*Hetaerina americana*
CLYSPE	OD	*Clymenella* sp.	HETTIT	C	*Hetaerina titia*
CORDIE	OD	*Corbula dietziana*	HEXSPE	C	*Hexatoma* sp.
CORLOU	OD	*Corophium louisianum*	HOPSPE	C	*Hoperius* sp.
COSDEL	OD	*Cossura delta*	HYDBIM	C	*Hydaticus bimarginatus*

COSSOY	OD	*Cossura soyeri*	HYDCLY	C	*Hydroporus clypealis*
CRAVIR	OD	*Crassostrea virginica*	HYDOBL	C	*Hydroporus oblitus*
CTESER	OD	*Ctenodrilus serratus*	HYDPHO	C	*Hydroperla phormidia*
CUMACE	OD	Cumacea	HYDPOR	C	*Hydroporus* sp.
CUMSP1	OD	*Cumacean* sp. 1	HYDPS2	C	*Hydroporus* sp. 2
CUMSP2	OD	*Cumacean* sp. 2	HYDRAD	C	*Hydroporus* sp. adult
CUMSP3	OD	*Cumacean* sp. 3	HYDROC	C	*Hydrocanthus oblongus*
CYAPOL	OD	*Cyathura polita*	HYDRPH	C	Hydrophilidae
CYCVAR	OD	*Cyclaspis c.f. varians*	HYDVIT	C	*Hydroporus vittapennis*
CYPHON	OD	*Cyphon* sp.	ISCCRE	C	*Ischnura credula*
CYRFRA	OD	*Cyrnellus fraternus*	ISCHAS	C	*Ischnura (Anomalagrion) hastata*
CYRNEL	OD	*Cyrnellus* sp.	ISCHNU	C	*Ischnura* sp.
DASSPE	OD	*Dasybranchus* sp.	ISCPOS	C	*Ischnura posita*
DISUNC	OD	*Dispio uncinata*	ISCVER	C	*Ischnura verticalis*
ECHIUR	OD	*Echiura*	KOESPE	C	*Koenika* sp.
EDOSPE	OD	*Edotea* sp. *(c.f. montosa)*	KRESPE	C	*Krenosmitta* sp.
ELLCOM	OD	*Elliptio complanta*	LABPIL	C	*Labrundinia pilosella*
ENCALB	OD	*Enchytraeus albidus*	LABRUN	C	*Labrundinia* sp.
ENCHYT	OD	Enchytraeidae	LACDAD	C	*Laccodytes* sp. adult
ENCSPE	OD	*Enchytraeus* sp.	LACDSP	C	*Laccodytes* sp.
ENSMIN	OD	*Ensis minor*	LACPAD	C	*Laccophilus* sp. adult
ENTOMO	OD	Entomobryidae	LACPSP	C	*Laccophilus* sp.
EURSPE	OD	*Eurycercus* sp.	LACRUF	C	*Laccophilus fasciatus rufus*
FABSPA	OD	*Fabricia* sp. a	LARSIA	C	*Larsia* sp.
FABSPE	OD	*Fabricia* sp.	LESTES	C	*Lestes* sp.
GLOSSO	OD	Glossoscolecidae	LESVIG	C	*Lestes vigilax*
GRABON	OD	*Grandidierella bonnieroides*	LIBELL	C	Libellulidae
HARRAP	OD	*Hargeria rapax*	LIBELU	C	*Libellula* sp.
HELFAS	OD	*Helichus fastigiatus*	LIBLYD	C	*Libellula lydia*
HELLIT	OD	*Helichus lithophilus*	LIMCAN	C	*Limnoporus canaliculatus*
HETFIL	OD	*Heteromastus filiformis*	LIMNSP	C	*Limnophila* sp.
HOBFLO	OD	*Hobsonia florida*	LIMOSP	C	*Limonia* sp.
ILYTEM	OD	*Ilyodrilus templetoni*	LIOPIL	C	*Lioporius pilatei*
IMMCAP	OD	Immature tubificid w/cap setae	MACILL	C	*Macromia illinoiensis*
IMMTUB	OD	Immature tubificid w/o cap setae	MACROM	C	*Macromia* sp.
ISCREC	OD	*Ischidium recurvatum*	MACTAE	C	*Macromia taeniolata*
ISOFRE	OD	*Isochaetides freyi*	MATLEE	C	*Matus leechi*
ISONYC	OD	*Isonychia* sp.	MATOVA	C	*Matus ovatus blatchleyi*

(continued)

Species Code	Trophic Code	Description	Species Code	Trophic Code	Description
ISOPUL	OD	*Isolda pulchella*	MERFLA	C	*Meropelopia flavifrons*
ISOSAY	OD	*Isonychia sayi*	MERSPE	C	*Meropelopia* sp.
ISOSIC	OD	*Isonychia sicca*	MESMUL	C	*Mesovelia mulsanti*
ISOSPA	OD	*Isonychia* sp. a	MICPAL	C	*Microvelia paludicola*
KINSPE	OD	*Kincaidiana* sp.	MOOMIC	C	*Mooreobdella microstoma*
LAECUL	OD	*Laeonereis culveri*	MOOTET	C	*Mooreobdella tetragon*
LAMSPE	OD	*Lampsilis* sp.	MYXSPE	C	*Myxosargus* sp.
LAMTER	OD	*Lampsilis teres (floridensis)*	NANBEL	C	*Nannothemis bella*
LEIFOL	OD	*Leitoscoloplos foliosus*	NASPEN	C	*Nasiaeschna pentacantha*
LEIFRA	OD	*Leitoscoloplos fragilis*	NATARS	C	*Natarsia* sp.
LEIROB	OD	*Leitoscoloplos robustus*	NEOHYD	C	*Neohydrophilus* sp.
LEISPE	OD	*Leitoscoloplos* sp.	NEOPER	C	*Neoperla* sp.
LEPLEP	OD	*Leptodea leptodon*	NEUALA	C	*Neurocordulia alabamensis*
LEUAME	OD	*Leucon americana*	NEUROC	C	*Neurocordulia* sp.
LEUSPE	OD	*Leuctra* sp.	NIGSER	C	*Nigronia serricornis*
LEVGRA	OD	*Levinsenia gracilis*	NIGSPE	C	*Nigronia* sp.
LIMANG	OD	*Limnodrilus angustipenis*	NILFIM	C	*Nilotanypus fimbriatus*
LIMCER	OD	*Limnodrilus cervix*	NILSPE	C	*Nilotanypus* sp.
LIMHOF	OD	*Limnodrilus hoffmeisteri*	NOTIRR	C	*Notonecta irrorata*
LIMMAU	OD	*Limnodrilus maumeensis*	NOTONE	C	*Notonecta* sp.
LIMMED	OD	*Limnodriloides medioporus*	NOTUHL	C	*Notonecta c.f. uhleri*
LIMRUB	OD	*Limnodrilus rubripenis*	ODONAT	C	Odonata
LIMSPE	OD	*Limnodriloides* sp.	OECETI	C	*Oecetis* sp.
LIMUDE	OD	*Limnodrilus udekemianus*	OPHIOG	C	*Ophiogomphus* sp.
LIMWIN	OD	*Limnodriloides winckelmanni*	PACLON	C	*Pachydiplax longipennis*
LOIMED	OD	*Loimia medusa*	PALPOM	C	*Bezzia/Palpomyia* group
LOIVIR	OD	*Loimia viridis*	PANHYM	C	*Pantala hymenea*
LUMBRC	OD	Lumbriculidae	PANTAL	C	*Pantala* sp.
LUMBSP	OD	*Lumbriculus* sp.	PARABO	C	*Parachironomus abortivus*
LYPDIV	OD	*Lype diversa*	PARACL	C	*Paracladopelma* sp.
MACBAL	OD	*Macoma balthica*	PARAGS	C	*Paragnetina* sp.
MACCAR	OD	*Macrostemum carolina*	PARAUN	C	*Paracladopelma undine*
MACCON	OD	*Macoma constricta*	PARFUM	C	*Paragnetina fumosa*
MACMIT	OD	*Macoma mitchelli*	PARKAN	C	*Paragnetina kansensis*
MACSPE	OD	*Macroma* sp.	PARLOG	C	*Paracladopelma loganae*
MACTEN	OD	*Macoma tenta*	PELCAR	C	*Pelocoris carolinensis*
MAGPET	OD	*Magelona pettiboneae*	PELFEM	C	*Pelocoris femoratus*

MAGPOL	OD	*Magelona polydentata*	PELLAR	C	*Peltodytes* sp.
MAGSPB	OD	*Magelona* sp. B	PELSPE	C	*Pelocoris* sp.
MAGSPE	OD	*Magelona* sp.	PENINC	C	*Pentaneura inconspicua*
MALSAR	OD	*Maldane sarsi*	PENTAN	C	*Pentaneura* sp.
MEDAMB	OD	*Mediomastus ambiseta*	PERDRY	C	*Perlinella drymo*
MEDICA	OD	*Mediomastus californiensis*	PEREPH	C	*Perlinella ephyre*
MEDSPE	OD	*Mediomastus* sp.	PERLES	C	*Perlesta* sp.
MELMAC	OD	*Melinna maculata*	PERLID	C	Perlidae
MERMER	OD	*Mercenaria mercenaria*	PERLSP	C	*Perlinella* sp.
MICHAN	OD	*Microdeutopus hancocki*	PERPLA	C	*Perlesta placida*
MICMYE	OD	*Microdeutopus myersi*	PERSPE	C	*Perithemus* sp.
MICSPE	OD	*Microdeutopus* sp.	PILSPE	C	*Pilaria* sp.
MINPER	OD	*Minuspio perkinsi*	POLYCE	C	*Polycentropus* sp.
MINSPE	OD	*Minuspio* sp.	PROBEL	C	*Procladius bellus*
MONIRR	OD	*Monopylephorus irroratus*	PROBEZ	C	*Probezzia* sp.
MONPAR	OD	*Monopylephorus parvus*	PROCLA	C	*Procladius* sp.
MONSPE	OD	*Monopylephorus* sp.	PROGOM	C	*Progomphus* sp.
MUNREY	OD	*Munna reynoldsi*	PROOBS	C	*Progomphus obscurus*
MYAARE	OD	*Mya arenaria*	PROSUB	C	*Procladius sublettei*
NAISET	OD	*Naineris setosa*	RANAUS	C	*Ranatra australis*
NEMCAR	OD	*Nemocapnia carolina*	RANBUE	C	*Ranatra buenoi*
NEOCOM	OD	*Neoephemera compressa*	RANKIR	C	*Ranatra kirkaldyi*
NEOYOU	OD	*Neoephemera youngi*	RANNIG	C	*Ranatra nigra*
NERREC	OD	*Neritina reclivata*	RHAOBE	C	*Rhagovelia obesa*
NERSUC	OD	*Nereis succinea*	ROBASP	C	*Robackia* sp.
NOTHEM	OD	*Notomastus hemipodus*	ROBCLA	C	*Robackia claviger*
NOTJUV	OD	*Notomastus* sp. juvenile	ROBDEM	C	*Robackia demeijerei*
NOTLAT	OD	*Notomastus latericeus*	ROBSP2	C	*Robackia* sp. 2
NOTSPE	OD	*Notomastus* sp.	SAESPE	C	*Saetheria* sp.
NUCACU	OD	*Nuculana acuta*	SEPSPE	C	*Sepedon* sp.
NUCPRO	OD	*Nuculana proxima*	SIALIS	C	*Sialis* sp.
NUCSPE	OD	*Nuculana* sp.	SIAMOH	C	*Sialis mohri*
OLIGOC	OD	*Oligochaeta*	SOMSPE	C	*Somatochlora* sp.
OLISP1	OD	*Oligochaete* sp. 1	SPHSPE	C	*Sphaeromias* sp.
OLISP2	OD	*Oligochaete* sp. 2	STYLSP	C	*Stylurus* sp.
OPHACU	OD	*Ophelina acuminata*	STYNOT	C	*Stylurus notatus*
OPHANG	OD	*Ophiothrix angulata*	SUPGIA	C	*Suphisellus gibbulus* adult
OPHBRE	OD	*Ophioderma brevispinum*	TABANI	C	Tabanidae

(continued)

Species Code	Trophic Code	Description	Species Code	Trophic Code	Description
OPHELE	OD	*Ophiolepis elegans*	TABSPE	C	*Tabanus* sp.
OPHIUR	OD	*Ophiuroidea*	TANCLA	C	*Tanypus clavatus*
OPHSPE	OD	*Ophelia* sp.	TANNEO	C	*Tanypus neopunctipennis*
ORBRIS	OD	*Orbinia riseri*	TANPUN	C	*Tanypus punctipennis*
OSTRAC	OD	*Ostracoda*	TANYPO	C	Tanypodinae
OWEFUS	OD	*Owenia fusiformis*	TANYPU	C	*Tanypus* sp.
PACTRA	OD	*Pachygrapsus transversus*	TETRSP	C	*Tetragoneuria* sp.
PALINT	OD	*Palaemonetes intermedius*	TETSPE	C	*Tetragnatha* sp.
PALKAD	OD	*Palaemonetes kadiakensis*	THEBAS	C	*Thermonectus basillaris*
PALPUG	OD	*Palaemonetes pugio*	THIEGR	C	*Thienemannimyia group*
PALSPE	OD	*Palaemonetes* sp.	TRAMSP	C	*Tramea* sp.
PALVUL	OD	*Palaemonetes vulgaris*	TREPOB	C	*Trepobates* sp.
PARFUL	OD	*Paraonis fulgens*	TROCOL	C	*Tropisternus collaris striolatus*
PARPIN	OD	*Paraprionospio pinnata*	TRONAT	C	*Tropisternus natator*
PARTEN	OD	*Paracaprella tenuis*	TRONIM	C	*Tropisternus lateralis nimbatus*
PECGOU	OD	*Pectinaria gouldi*	TROPIS	C	*Tropisternus* sp.
PECSPE	OD	*Pectinaria* sp.	UVAGAD	C	*Uvarus granarius* adult
PEOBIP	OD	*Peosidrilus biprostatus*	UVALAA	C	*Uvarus lacustris* adult
PHAMED	OD	*Phallodrilus medioporus*	UVALAC	C	*Uvarus lacustris*
PHAMON	OD	*Phallodrilus monospermathecus*	ZAVREL	C	*Zavrelimyia* sp.
PHASPE	OD	*Phallodrilus* sp.	ZYGOPT	C	Zygoptera
PHESPA	OD	*Pherusa* sp. a	AAAHDR	C1	**Carnivore (primary)**
PHOARC	OD	*Phoronis architecta*	ACEAME	C1	*Acetes americanus*
PHORON	OD	Phoronida	AGLVER	C1	*Aglaophamus verrilli*
PIRROB	OD	*Piromis roberti*	ALBSPE	C1	*Albia* sp.
PISCAS	OD	*Pisidium casertanum*	ALPHET	C1	*Alpheus heterochaelis*
PISCRI	OD	*Pista cristata*	AMBSY1	C1	*Ambidexter symmetricus* (5–10 mm)
PISDUB	OD	*Pisidium dubium*	AMEMAG	C1	*Americonuphis magna*
PISSPA	OD	*Pista* sp. a	AMPROS	C1	*Amphinome rostrata*
PISSPE	OD	*Pisidium* sp.	ANAAVA	C1	*Anachis avara*
POEJOH	OD	*Poecilochaetus johnstoni*	ANASPA	C1	*Anachis sparsa*
POGONO	OD	*Pogonophora*	ANCISP	C1	*Ancistrosyllis* sp.
POLHAR	OD	*Polydora hartmanae*	ANCJON	C1	*Ancistrosyllis jonesi*
POLLIG	OD	*Polydora ligni*	ANCPAP	C1	*Ancistrosyllis papillosa*
POLPIC	OD	*Polyophthalmus pictus*	ANOINS	C1	*Anoplodactylus insignis*
POLSOC	OD	*Polydora socialis*	ANTSPE	C1	*Antinoe* sp.
POLSPE	OD	*Polydora* sp.	ARAIRI	C1	*Arabella iricolor*

POLWEB	OD	*Polydora websteri*	ARASPE	C1	*Arabella* sp.
POLYCA	OD	*Polycirrus carolinensis*	ARRSPE	C1	*Arrenurus* sp.
PORIFE	OD	Porifera	ATRSPE	C1	*Atractides* sp.
PRICIR	OD	*Prionospio cirrifera*	AUTSPE	C1	*Autolytus* sp.
PRICRI	OD	*Prionospio cristata*	BHAGOO	C1	*Bhawania goodei*
PRIHET	OD	*Prionospio heterobranchia*	BRACLA	C1	*Brania clavata*
PRIONO	OD	*Prionospio* sp.	BRAGAL	C1	*Brania gallagheri*
PRIPER	OD	*Prionospio perkinsi*	BRASPA	C1	*Brania* sp. a
PRISYN	OD	*Pristina synclites*	BRAWEL	C1	*Brania wellfleetensis*
PSACON	OD	*Psammoryctides convolutus*	BULOCC	C1	*Bulla occidentalis*
PSYCHD	OD	Psychodidae	BULSTR	C1	*Bulla striata*
PTEDOR	OD	*Pteronarcys dorsata*	BUSCON	C1	*Busycon contrarium*
PTERON	OD	*Pteronarcys* sp.	BUSSPI	C1	*Busycon spiratum*
QUIMUL	OD	*Quistadrilus multisetosus*	CABINC	C1	*Cabira incerta*
RHISPE	OD	*Rhizodrilus* sp.	CALEUG	C1	*Calliostoma euglyptum*
RHYSPE	OD	*Rhynchelmis* sp.	CALSA1	C1	*Callinectes sapidus* (<30 mm)
SAMSPA	OD	*Samythella* sp. a	CALSIM	C1	*Callinectes similis*
SANCRU	OD	*Sanguinolaria cruenta*	CANCAN	C1	*Cantharus cancellaria*
SCOLSP	OD	*Scoloplos* sp.	CANMUL	C1	*Cantharus multangulus*
SCORUB	OD	*Scoloplos rubra*	CANTIN	C1	*Cantharus tinctus*
SCOSPE	OD	*Scolelepis* sp.	CAPPEN	C1	*Caprella penantis*
SCOSQU	OD	*Scolelepis squamata*	CAPREL	C1	Caprellidae
SCOTEX	OD	*Scolelepis texana*	CARTRE	C1	*Carinoma tremaphorus*
SCYPLA	OD	*Scyphoproctus platyproctus*	CERLAC	C1	*Cerebratulus lacteus*
SCYSPA	OD	*Scyphoproctus* sp. a	CHRQUI	C1	*Chrysaora quinquecirrha*
SEMPRO	OD	*Semele prolifica*	CLATSP	C1	*Clathrosperchon* sp.
SEMPUR	OD	*Semele purpurascens*	CLIVIT	C1	*Clibanarius vittatus*
SEMSPE	OD	*Semele* sp.	COLRUS	C1	*Columbella rusticoides*
SMIMAR	OD	*Smithsonidrilus marinus*	CONJAS	C1	*Conus jaspideus stearnsi*
SOLSPE	OD	*Solen* sp.	CONSPE	C1	*Conus* sp.
SOLVIR	OD	*Solen viridis*	DIOCUP	C1	*Diopatra cuprea*
SPARGA	OD	*Sparganophilus* sp.	DOLOME	C1	*Dolomedes* sp.
SPHAER	OD	Sphaeriidae	DORSOC	C1	*Dorvillea sociabilis*
SPIBOM	OD	*Spiophanes bombyx*	DORSPA	C1	*Dorvillea* sp. a
SPINIK	OD	*Spirosperma nikolskyi*	DORSPB	C1	*Dorvilleid* sp. b
SPIPET	OD	*Spio pettiboneae*	DORSPE	C1	*Dorvillea* sp.
SPISET	OD	*Spio setosa*	DRIMAG	C1	*Drilonereis magna*
SPISP1	OD	*Spirosperma* sp. 1	EHLSPE	C1	*Ehlersia* sp.

(continued)

Species Code	Trophic Code	Description	Species Code	Trophic Code	Description
SPISPE	OD	*Spio* sp.	EPIANG	C1	*Epitonium angulatum*
STRBEN	OD	*Streblospio benedicti*	EPIRUP	C1	*Epitonium rupicola*
STRHAR	OD	*Streblosoma harmanae*	EPISPE	C1	*Epitonium* sp.
TAGAFF	OD	*Tagelus affinis*	ERAMAU	C1	*Erato maugeriae*
TAGDIV	OD	*Tagelus divisus*	ETEHET	C1	*Eteone heteropoda*
TAGPLE	OD	*Tagelus plebeius*	ETELAC	C1	*Eteone lactea*
TELAGI	OD	*Tellina agilis*	EUMSAN	C1	*Eumida sanguinea*
TELALT	OD	*Tellina alternata*	EUNVIT	C1	*Eunice vittata*
TELCRI	OD	*Tellidora cristata*	EUPSUL	C1	*Eupleura sulcidentata*
TELFAU	OD	*Tellina fausta*	EXODIS	C1	*Exogone dispar*
TELINA	OD	*Tellina* sp. (juvenile)	EXOSPE	C1	*Exogone* sp.
TELSPA	OD	*Tellina* sp. a	FASHUN	C1	*Fasciolaria hunteria*
TELSPB	OD	*Tellina* sp. b	FASTUL	C1	*Fasciolaria tulipa*
TELSPE	OD	*Tellina* sp.	FROSPE	C1	*Frontipoda* sp.
TELTAM	OD	*Tellina tampaensis*	GASPEN	C1	*Gastropteron pentodon*
TELTEX	OD	*Tellina texana*	GLYAME	C1	*Glycera americana*
TELVER	OD	*Tellina versicolor*	GLYSOL	C1	*Glycinde solitaria*
TEREBE	OD	Terebellidae	GONCAR	C1	*Goniadides carolinae*
TERSPA	OD	*Terebellid* sp. a	GRUMEX	C1	*Grubeulepis mexicana*
TERSTR	OD	*Terebellides stroemi*	GRUSPE	C1	*Grubeulepis* sp.
TEXSPH	OD	*Texadina sphinctostoma*	GYPBRE	C1	*Gyptis brevipalpa*
THAANN	OD	*Tharyx annulosus*	GYPSPE	C1	*Gyptis* sp.
THASPE	OD	*Tharyx* sp.	GYPVIT	C1	*Gyptis vittata*
THESET	OD	*Thelepus setosus*	HARSPE	C1	*Harmothoe* sp.
TRAHOB	OD	*Travisia hobsonae*	HELSTA	C1	*Helobdella stagnalis*
TUBBEN	OD	*Tubificoides benedeni*	HEMMIN	C1	*Hemiaegina minuta*
TUBCSY	OD	Tubificid (immature) w/ cap setae	HESION	C1	Hesionidae
TUBGAB	OD	*Tubificoides gabriellae*	HIPZOS	C1	*Hippolyte zostericola*
TUBHET	OD	*Tubificoides heterochaetus*	HYAVEL	C1	*Hyalina veliei*
TUBIFI	OD	Tubificidae	HYDDES	C1	*Hydrodromus despiciens*
TUBLIT	OD	*Tubifex litoralis*	HYDRAC	C1	Hydracarina
TUBLON	OD	*Tubificoides longipenis*	HYDRSP	C1	*Hydrodroma* sp.
TUBPSE	OD	*Tubificoides pseudogaster*	HYDSP2	C1	*Hydracarina* sp. 2
TUBSPE	OD	*Tubificoides* sp.	HYDSPE	C1	Hydridae
TUBSWI	OD	*Tubificoides swirencowi*	HYGSPE	C1	*Hygrobates* sp.
UCAMIN	OD	*Uca minax*	HYMRET	C1	*Hymanella retenuova*
UCASPE	OD	*Uca* sp.	KINBSP	C1	*Kinbergonuphis* sp.

UNIONI	OD	Unionidae	LEBERT	C1	*Lebertia* sp.
VILCHO	OD	*Villosa choctawensis*	LEPSUB	C1	*Lepidonotus sublevis*
AAAHDR	OC	**Omnivore (carnivorous)**	LEPVAR	C1	*Lepidonotus variabilis*
AMBSY2	OC	*Ambidexter symmetricus* (>10 mm)	LIMNES	C1	*Limnesia* sp.
AMBSYM	OC	*Ambidexter symmetricus*	LITANT	C1	*Litocorsa antennata*
EURCOM	OC	*Eurythoe complanta*	LUCINC	C1	*Luconacia incerta*
PALFL1	OC	*Palaemon floridanus* (20–30 mm)	LUMBRI	C1	Lumbrineridae
PODRI1	OC	*Podochela riisei* (<6 mm)	LUMBRS	C1	*Lumbrineris* sp.
PORGIB	OC	*Portunus gibbesii*	LUMLAT	C1	*Lumbrineris latreilli*
SICLA1	OC	*Sicyonia laevigata* (<20 mm)	LUMTEN	C1	*Lumbrineris tenuis*
AAAHDR	OR	**Omnivore (general)**	LUMVER	C1	*Lumbrineris verrilli*
ACAINT	OR	*Acanthohaustorius intermedius*	LYSNIN	C1	*Lysidice ninetta*
ACANTH	OR	*Acanthohaustorius* sp.	MARSAN	C1	*Marphysa sanguinea*
ACEPYG	OR	*Acerpenna pygmaeus*	MELAPP	C1	*Melita appendiculata*
ACUNAG	OR	*Acuminodeutopus naglei*	MELCOR	C1	*Melongena corona*
ALLPAR	OR	*Allonais paraguayensis*	MELINT	C1	*Melita intermedius*
ALLPEC	OR	*Allonais pectinata*	MELSPE	C1	*Melita* sp.
AMPABD	OR	*Ampelisca abdita*	METCAL	C1	*Metaporhaphis calcarata*
AMPHIL	OR	*Amphilochus* sp.	MICLEI	C1	*Micrura leidyi*
AMPHIT	OR	*Ampithoe* sp.	MICPT2	C1	*Micropthamalus* sp. 2
AMPHOL	OR	*Ampelisca holmesi*	MICPTH	C1	*Micropthamalus* sp.
AMPNEA	OR	*Amphilochus neapolitanus*	MIDSPE	C1	*Mideopsis* sp.
AMPONE	OR	*Amphipod* sp. 1	MURDIL	C1	*Murex dilectus*
AMPTWO	OR	*Amphipod* sp. 2	MURPOM	C1	*Murex pomum*
AMPVER	OR	*Ampelisca verrilli*	MURRUB	C1	*Murex rubidus*
ANTHUR	OR	Anthuridae	NATPUS	C1	*Natica pusilla*
AORIDA	OR	*Aoridae* sp. 1	NEMERT	C1	*Nemertean* sp.
APAMAG	OR	*Apanthura magnifica*	NEMHEB	C1	*Nematonereis hebes*
APEELA	OR	*Apedilum elachistus*	NEPBUC	C1	*Nephtys bucera*
APESPE	OR	*Apedilum* sp.	NEPINC	C1	*Nephtys incisa*
APSSPE	OR	*Apseudes* sp.	NEPPIC	C1	*Nephtys picta*
ARCLOM	OR	*Arcteonais lomondi*	NEPSPA	C1	*Nephtyid* sp. a
ARECRI	OR	*Arenicola cristata*	NERFRA	C1	*Nereiphylla fragilis*
ASEFOX	OR	*Asellus foxii*	NUDSPE	C1	*Nudibranch* sp.
ASELAT	OR	*Asellus laticaudatus*	OBELIA	C1	*Obelia* sp.
ASEOBT	OR	*Asellus obtusus*	OGYLIM	C1	*Ogyrides limicola*
ASERAC	OR	*Asellus racovitzai*	OLISPE	C1	*Olivella* sp.
ASESPE	OR	*Asellus* sp.	OLIVSP	C1	*Oliva* sp.

(continued)

Species Code	Trophic Code	Description	Species Code	Trophic Code	Description
ASHBEC	OR	*Asheum beckae*	ONUERE	C1	*Onuphis eremita oculata*
ASHEUM	OR	*Asheum* sp.	ONUNEB	C1	*Onuphis nebulosis*
ASSSUC	OR	*Assiminea succinea*	ONUPHD	C1	Onuphidae
ATRISP	OR	*Atrichopogon* sp.	ONUSP2	C1	*Onuphidae* sp. 2
AXASPE	OR	*Axarus* sp.	ONUSPA	C1	*Onuphis* sp. a
BAEARM	OR	*Baetis armillatus*	ONUSPE	C1	*Onuphis* sp.
BAEBEC	OR	*Baetisca becki*	OPHABS	C1	*Ophiodromus abscura*
BAEBIM	OR	*Baetis bimaculatus*	PAGBON	C1	*Pagurus bonairensis*
BAEEPH	OR	*Baetis ephippiatus*	PAGLON	C1	*Pagurus longicarpus*
BAEFRO	OR	*Baetis frondalis*	PAGPOL	C1	*Pagurus pollicaris*
BAEINT	OR	*Baetis intercalaris*	PALESP	C1	*Paleanotus* sp.
BAEOBE	OR	*Baetisca obesa*	PALFL2	C1	*Palaemon floridanus* (30–40 mm)
BAEPRO	OR	*Baetis propinquus*	PALFLO	C1	*Palaemon floridanus*
BAETID	OR	Baetidae	PANHER	C1	*Panopeus herbstii*
BAETIS	OR	*Baetis* sp.	PARAME	C1	*Parandalia americana*
BAETSC	OR	*Baetisca* sp.	PARANA	C1	*Paranaitis speciosa*
BEROSU	OR	*Berosus* sp.	PARCYP	C1	*Parametopella cypris*
BOUSPE	OR	*Bourletiella* sp.	PARLUT	C1	*Parahesione luteola*
BOWBRA	OR	*Bowmaniella brasiliensis*	PENAZ1	C1	*Penaeus aztecus* (<25 mm)
BOWDIS	OR	*Bowmaniella dissimilis*	PENDU1	C1	*Penaeus duorarum* (<25 mm)
BOWSPE	OR	*Bowmaniella* sp.	PENSE1	C1	*Penaeus setiferus* (<25 mm)
BRACHY	OR	*Brachycercus* sp.	PENSPE	C1	*Penaeus* sp.
BRAUNI	OR	*Bratislavia unidentata*	PERAME	C1	*Periclimenes americanus*
BRYOZO	OR	Bryozoa	PERLON	C1	*Periclimenes longicaudatus*
CAEHIL	OR	*Caenis hilaris*	PETSPE	C1	*Pettiboneia* sp.
CAENIS	OR	*Caenis* sp.	PHYARE	C1	*Phyllodoce arenae*
CAENIT	OR	*Caecum nitidum*	PHYFRA	C1	*Phyllodoce fragilis*
CALSPE	OR	*Callibaetis* sp.	PHYLLO	C1	Phyllodocidae
CAMBAR	OR	Cambaridae	PHYLO2	C1	*Phyllodocidae* sp. 2
CAMSPE	OR	*Cambarus* sp.	PHYMUC	C1	*Phyllodoce mucosa*
CARDSP	OR	*Cardiocladius* sp.	PILARG	C1	Pilargiidae
CARPFL	OR	*Carpias floridensis*	PIOSPE	C1	*Pionosyllis* sp.
CARSPE	OR	*Carinobatea* sp.	PLEGIG	C1	*Pleuroploca gigantea*
CARTRI	OR	*Carinobatea carinata*	PODOBS	C1	*Podarke obscura*
CERIRR	OR	*Ceratonereis irritabilis*	PODSPE	C1	*Podarke* sp.
CHAETO	OR	Chaetopteridae	POLDUP	C1	*Pollinices duplicatus*
CHEUMA	OR	*Cheumatopsyche* sp.	POLLUP	C1	*Polyodontes lupina*

CHIALM	OR	*Chiridotea almyra*	POLPUL	C1	*Pollinices pullicarpus*
CHIFLO	OR	*Chimarra florida*	POLYCL	C1	*Polycladida* sp.
CHINAE	OR	Chironominae	POLYNL	C1	*Polynoidae* larvae
CHIRID	OR	*Chiridotea* sp.	POLYNO	C1	Polynoidae
CHIRIP	OR	*Chironomus riparius*	PROGRA	C1	*Prostoma graecense*
CHIRMN	OR	Chironomini	PROSTO	C1	*Prostoma* sp.
CHIRMS	OR	*Chironomus* sp.	PROTSP	C1	*Protzia* sp.
CHIROA	OR	*Chironomini* genus A	PRUAPI	C1	*Prunum apicinum*
CHISTE	OR	*Chiridotea stenops*	PSEAMB	C1	*Pseudeurythoe ambigua*
CLADOT	OR	*Cladotanytarsus* sp.	PSECRO	C1	*Pseudoceros crozieri*
CLASPE	OR	*Cladopelma* sp.	PYCNOG	C1	Pycnogonida
CLEPLA	OR	*Cleantis planicauda*	RHIHAR	C1	*Rhithropanopeus harrisii*
COLLEM	OR	Collembola	SAGHIS	C1	*Sagitta hispida*
COLSPA	OR	*Colomastix* sp. a	SAGITT	C1	*Sagitta* sp.
CORACH	OR	*Corophium acherusicum*	SAGTEN	C1	*Sagitta tenuis*
CORLAC	OR	*Corophium lacustre*	SCHRUD	C1	*Schistomeringos rudolphi*
COROPH	OR	Corophiidae	SICLA2	C1	*Sicyonia laevigata* (>20 mm)
CORSIM	OR	*Corophium simile*	SICLAE	C1	*Sicyonia laevigata*
CORSPA	OR	*Corophium* sp. a	SIGALI	C1	Sigalionidae
CORSPE	OR	*Corophium* sp.	SIGBAS	C1	*Sigambra bassi*
CORTUB	OR	*Corophium tuberculatum*	SIGTEN	C1	*Sigambra tentaculata*
CORYNO	OR	*Corynoneura* sp.	SPERSP	C1	*Sperchonopsis* sp.
CORYS1	OR	*Corynoneura* sp. 1	SPESPE	C1	*Sperchon* sp.
CORYS2	OR	*Corynoneura* sp. 2	SPEVER	C1	*Sperchonopsis* sp. *(pr verrucosa)*
CRAFLO	OR	*Crangonyx floridanus*	SPHACI	C1	*Sphaerosyllis aciculata*
CRAPSE	OR	*Crangonyx pseudogracilis* group	STEMIN	C1	*Stenothoe minuta*
CRARIC	OR	*Crangonyx richmondensis*	STESPA	C1	*Stenothoe* sp. a
CRASPE	OR	*Crangonyx* sp.	STESPB	C1	*Stenothoe* sp. b
CRIBIC	OR	*Cricotopus bicinctus*	STHBOA	C1	*Sthenelais boa*
CRICOT	OR	*Cricotopus* sp.	STOMEL	C1	*Stomolophus meleagris*
CRIORT	OR	*Cricotopus/Orthocladius* group	STRPET	C1	*Streptosyllis pettiboneae*
CRISP1	OR	*Cricotopus* sp. 1	SYLCOR	C1	*Syllis cornuta*
CRISP2	OR	*Cricotopus* sp. 2	SYLHYA	C1	*Syllis hyalina*
CRISYL	OR	*Cricotopus sylvestris* group	SYLLSP	C1	*Syllid* sp.
CRUTRI	OR	*Crustipellis tribranchiata*	SYLSPE	C1	*Syllis* sp. e
CRYPTE	OR	*Cryptotendipes* sp.	THAHAE	C1	*Thais haemastoma*
CULICI	OR	Culicidae	THAISP	C1	*Thais* sp.
CULSPE	OR	*Culex* sp.	THODOB	C1	*Thor dobkini*

(continued)

Species Code	Trophic Code	Description	Species Code	Trophic Code	Description
DEMSPE	OR	*Demicryptochironomus* sp.	TORSPE	C1	*Torrenticola* sp.
DERDIG	OR	*Dero digitata*	TRACON	C1	*Trachypenaeus constrictus*
DERFUR	OR	*Dero furcata*	TRICLA	C1	Tricladida
DERLOD	OR	*Dero lodeni*	TRYPAR	C1	*Trypanosyllis parvidentata*
DERPEC	OR	*Dero pectinata*	TRYVIT	C1	*Trypanosyllis vittigera*
DERSPE	OR	*Dero* sp.	TUBPEL	C1	*Tubulanus pellucidus*
DERTRI	OR	*Dero trifida*	TURBEL	C1	*Turbellarian* sp.
DERVAG	OR	*Dero vaga*	UNISPE	C1	*Unionicola* sp.
DIAANT	OR	*Diadema antillarum*	UROPER	C1	*Urosalpinx perrugata*
DIAVAR	OR	*Diastoma varium*	UROTAM	C1	*Urosalpinx tampaensis*
DICLUC	OR	*Dicrotendipes lucifer*	AAAHDR	C2	**Carnivore (secondary)**
DICMOD	OR	*Dicrotendipes modestus*	CALSA2	C2	*Callinectes sapidus* (31–60 mm)
DICNEO	OR	*Dicrotendipes neomodestus*	CALSA3	C2	*Callinectes sapidus* (>60 mm)
DICNER	OR	*Dicrotendipes nervosus*	CALSAP	C2	*Callinectes sapidus*
DICROT	OR	*Dicrotendipes* sp.	ECHSPE	C2	*Echinaster* sp.
DICSIM	OR	*Dicrotendipes simpsoni*	ECHSPI	C2	*Echinaster* sp.
DICTHA	OR	*Dicrotendipes thanatogratus*	LOLBRE	C2	*Lolliguncula brevis*
DUBIAD	OR	*Dubiraphia* sp. adult	LUIALT	C2	*Luidia alternata*
DUBIRA	OR	*Dubiraphia* sp.	LUICLA	C2	*Luidia clathrata*
DUBVAD	OR	*Dubiraphia vittata* adult	LUISAG	C2	*Luidia sagamina*
DUBVIT	OR	*Dubiraphia vittata*	OCTVUL	C2	*Octopus vulgaris*
ECTSPE	OR	*Ectopria* sp.	PENAZ2	C2	*Penaeus aztecus* (25–50 mm)
EDOMON	OR	*Edotea montosa*	PENAZ3	C2	*Penaeus aztecus* (>50 mm)
EDOTRI	OR	*Edotea triloba*	PENAZT	C2	*Penaeus aztecus*
EINFEL	OR	*Einfeldia* sp.	PENDU2	C2	*Penaeus duorarum* (25–50 mm)
EINNAT	OR	*Einfeldia natchitocheae*	PENDU3	C2	*Penaeus duorarum* (>50 mm)
ELALEV	OR	*Elasmopus levis*	PENDUO	C2	*Penaeus duorarum*
ELMIDS	OR	Elmidae	PENSE2	C2	*Penaeus setiferus* (25–50 mm)
ENDNIG	OR	*Endochironomus nigricans*	PENSE3	C2	*Penaeus setiferus* (>50 mm)
ENDOCH	OR	*Endochironomus* sp.	PENSET	C2	*Penaeus setiferus*
ENOCHR	OR	*Enochrus* sp.	PROSPE	C2	*Processa* sp.
ENTSPE	OR	*Entomobrya* sp.	SQUEMP	C2	*Squilla empusa*
EPHCHO	OR	*Ephemerella choctawhatchee*	SQURUG	C2	*Squilla rugosa*
EPHEME	OR	*Ephemerella* sp.	STRALA	C2	*Strombus alatus*
EPHOPT	OR	*Ephemeroptera*	AAAHDR	C3	**Carnivore (tertiary)**
EPHPUP	OR	*Ephydridae* pupa	AAAHDR	CP	**Carnivore (parasitic)**
EPHYDR	OR	Ephydridae	ACTINE	CP	*Actinobdella inequinnulata*

ERCFRU	OR	*Erichthonius (c.f. rubricornis)*	ALBHET	CP	*Alboglossiphonia heteroclita*
ERIBRA	OR	*Erichthonius brasiliensis*	BATPHO	CP	*Batracobdella pholera*
ERIFIL	OR	*Erichsonella (c.f. filiformis)*	BRABDE	CP	*Branchiobdella*
ERISP1	OR	*Erichthonius* sp. 1	CLIARE	CP	*Climacia areolaris*
ERISP2	OR	*Erichthonius* sp. 2	CULPIP	CP	*Culex pipiens pipiens*
ERISPA	OR	*Erichthonius* sp. a	CYMEXC	CP	*Cymothoa excisa*
ERISPE	OR	*Erichsonella* sp.	ERPOBD	CP	Erpobdellidae
EUCSPE	OR	*Eucorethra* sp.	HELFUS	CP	*Helobdella fusca*
EUDHON	OR	*Eudevenopus honduranus*	HELOSP	CP	*Helobdella* sp.
EURLIT	OR	*Eurydice littoralis*	HELTRI	CP	*Helobdella triserialis*
EURTEM	OR	*Eurylophella temporalis*	HIRUDI	CP	Hirudinia
EURYSP	OR	*Eurylophella* sp.	PLAMON	CP	*Placobdella montifera*
FORCIP	OR	*Forcipomyia* sp.	PLAMUL	CP	*Placobdella multilineata*
GAMDAI	OR	*Gammarus daiberi*	PLAORN	CP	*Placobdella ornata*
GAMFAS	OR	*Gammarus fasciatus*	PLAPAP	CP	*Placobdella papillifera*
GAMSPE	OR	*Gammarus* sp.	PLAPAR	CP	*Placobdella parasitica*
GAMTIG	OR	*Gammarus tigrinus*	PLASPE	CP	*Placobdella* sp.
GITSPE	OR	*Gitanopsis* sp.	SIMADU	CP	*Simulium* sp. adult
GLOPYR	OR	*Glottidia pyramidata*	SIMVAD	CP	*Simulium vittatum* adult
GLOSOM	OR	Glossosomatidae	SISVIC	CP	*Sisyra c.f. vicaria*
GLYPTO	OR	*Glyptotendipes* sp.	AAAHDR	N	**Nonfeeding dev. stage**
GOECAR	OR	*Goeldichironomus carus*	ABLABA	N	*Ablabesmyia* sp. adult
GOEHOL	OR	*Goeldichironomus holoprasinus*	ABLPUP	N	*Ablabesmyia* sp. pupa
GONDIE	OR	*Gonielmis dietrichi* adult	APEEPU	N	*Apedilum elachistus* pupa
GONDLA	OR	*Gonielmis dietrichi*	ASHBEA	N	*Asheum beckae* adult
GONOMY	OR	*Gonomyia* sp.	ASHBPU	N	*Asheum beckae* pupa
GRASPE	OR	*Grandidierella* sp.	ASHPUP	N	*Asheum* sp. pupa
HABVIB	OR	*Habrophlebia vibrans*	CECIDO	N	Cecidomyiidae
HAEWAL	OR	*Haemonais waldvogeli*	CERATP	N	*Ceratopogonidae* sp. pupa
HAUARE	OR	*Haustorius arenarius*	CHEMAA	N	*Cheumatopsyche* sp. adult
HAUSTO	OR	Haustoridae	CHEUPU	N	*Cheumatopsyche* sp. pupa
HEDSPE	OR	*Hedriodiscus* sp.	CHIDCP	N	*Chironomus decorus* group pupa
HEPTAG	OR	Heptageniidae	CHIDEA	N	*Chironomus decorus* group adult
HETERO	OR	*Heterocloeon* sp.	CHIPUP	N	*Chironomid* pupa
HETSEC	OR	*Heterophlias seclusus*	CHIRMA	N	*Chironomini* sp. adult
HETSET	OR	*Heterophlias seticoxa*	CHIRMP	N	*Chironomus* sp. pupa
HETSPC	OR	*Heterotrissocladius* sp. C	CHISTP	N	*Chironomus stigmaterus* pupa
HETTRE	OR	*Heterotrissocladius* sp.	CLAPUP	N	*Cladotanytarsus* sp. pupa

(continued)

Species Code	Trophic Code	Description	Species Code	Trophic Code	Description
HEXAGE	OR	*Hexagenia* sp.	CLASPU	N	*Cladopelma* sp. pupa
HEXLIM	OR	*Hexagenia limbata*	COELPU	N	*Coelotanypus* sp. pupa
HOLOTH	OR	Holothuroidia	CONCPU	N	*Conchapelopia* sp. pupa
HOLSPA	OR	*Holothuroid* sp. a	CORPUP	N	*Corynoneura* sp. pupa
HOLSPB	OR	*Holothuroid* sp. b	CORYSA	N	*Corynoneura* sp. adult
HYAAZT	OR	*Hyalella azteca*	CRIBAD	N	*Cricotopus bicinctus* adult
HYDBET	OR	*Hydropsyche betteni* group	CRIBPU	N	*Cricotopus bicinctus* pupa
HYDDEC	OR	*Hydropsyche decalda*	CRIOPU	N	*Cricotopus/Orthocladius* group pupa
HYDROP	OR	Hydropsychidae	CRIPUP	N	*Cricotopus* pupa
HYDRPA	OR	*Hydropsyche* sp. a	CRISPU	N	*Cricotopus sylvestris* group pupa
HYDRPB	OR	*Hydropsyche* sp. b	CRYFUA	N	*Cryptochironomus fulvus* group adult
HYDRPS	OR	*Hydropsyche* sp.	CRYPTP	N	*Cryptotendipes* sp. pupa
HYDSIM	OR	*Hydropsyche simulans*	CRYPUP	N	*Cryptochironomus* sp. pupa
HYPOGA	OR	*Hypogastrura* sp.	CULIPU	N	Culicidae sp. pupa
IANIRO	OR	*Ianiropsis* sp.	DICLCA	N	*Dicrotendipes lucifer* complex adult
ISODIC	OR	*Isoperla dicala*	DICPUP	N	*Dicrotendipes* sp. pupa
ISOORA	OR	*Isoperla orata*	GLYPTP	N	*Glyptotendipes* sp. pupa
ISOPER	OR	*Isoperla* sp.	GOCARP	N	*Goeldichironomus carus* pupa
ISOPOD	OR	Isopoda	GOECAA	N	*Goeldichironomus carus* adult
ISOTOM	OR	*Isotomurus* sp.	GOEHOA	N	*Goeldichironomus holoprasinus* adult
ISOTSP	OR	*Isotoma* sp.	GOEHOP	N	*Goeldichironomus holoprasinus* pupa
KALSPE	OR	*Kalliapseudes* sp.	GOEPUP	N	*Goeldichironomus* sp. pupa
KIEDUX	OR	*Kiefferulus dux*	HACGCP	N	*Harnischia* complex genus C pupa
KIESPE	OR	*Kiefferulus* sp.	HARNCP	N	*Harnischia* complex sp. pupa
LAUSPE	OR	*Lauterborniella* sp.	HEMPUP	N	*Hemerodromia* sp. pupa
LEMREC	OR	*Lembos rectanulatus*	HETTPU	N	*Heterotrissocladius* sp. pupa
LEMSET	OR	*Lembos setosus*	HYDPUP	N	Hydroptilidae pupa
LEMSMI	OR	*Lembos smithi*	HYDRPU	N	*Hydroptila* sp. pupa
LEMSPE	OR	*Lembos* sp.	HYDSIA	N	*Hydropsyche simulans* adult
LEMSPI	OR	*Lembos spinicarpus*	HYDSPU	N	*Hydropsyche* sp. pupa
LEMUN2	OR	*Lembos unifasciatus*	KIEDUP	N	*Kiefferulus dux* pupa
LEMUNI	OR	*Lembos unicornis*	KIEPUP	N	*Kiefferulus* sp. pupa
LEMWEB	OR	*Lembos websteri*	LABPUP	N	*Labrundinia* sp. pupa
LEPBRA	OR	*Leptophelebia bradleyi*	LARSPU	N	*Larsia* sp. pupa
LEPSPE	OR	*Lepidactylus* sp.	LEPTPU	N	*Leptoceridae* pupa
LEPTIN	OR	*Leptophlebia intermedia*	LOPPUP	N	*Lopescladius* sp. pupa
LEPTOC	OR	Leptoceridae	MERFLP	N	*Meropelopia flavifrons* pupa

LEPTOP	OR	*Leptophlebia* sp.	MERSPU	N	*Meropelopia* sp. pupa
LEPTPH	OR	Leptophlebiidae	NANNDA	N	*Nanocladius nr. distinctus* adult
LEUSPA	OR	*Leucothoe* sp. a	NANNDP	N	*Nanocladius nr. distinctus* pupa
LEUSPI	OR	*Leucothoe spinicarpa*	NANOCA	N	*Nanocladius* sp. adult
LIMNEP	OR	Limnephilidae	NANPUP	N	*Nanocladius* sp. pupa
LIMNOP	OR	*Limnophyes* sp.	NEOTAD	N	*Neotrichia* sp. adult
LIRLIN	OR	*Lirceus lineatus*	NEOTPU	N	*Neotrichia* sp. pupa
LIRSPE	OR	*Lirceus* sp.	NILBAP	N	*Nilothauma babiyi* pupa
LISBAR	OR	*Listriella barnardi*	NILMIP	N	*Nilothauma mirabile* pupa
LISSPE	OR	*Listriella* sp. *(c.f. quintana)*	NILSPU	N	*Nilotanypus* sp. pupa
LISTSP	OR	*Listriella* sp.	NILTHP	N	*Nilothauma* sp. pupa
LOPSPE	OR	*Lopescladius* sp.	OECPUP	N	*Oecetis* sp. pupa
LYMSPE	OR	*Lymnaea* sp.	ORTAPU	N	*Orthocladius annectens* pupa
LYMSTA	OR	*Lymnea stagnalis*	ORTINA	N	*Orthocladiinae* sp. adult
LYSALB	OR	*Lysianopsis alba*	ORTPUP	N	*Orthocladius* sp. pupa
LYSHIR	OR	*Lysianopsis hirsuta*	OXYEPU	N	*Oxyethira* sp. pupa
LYTVAR	OR	*Lytechinus variegatus*	PARABP	N	*Parachaetocladius abnobaeus* pupa
MACGAD	OR	*Macronychus glabratus* adult	PARACP	N	*Parachironomus* sp. pupa
MACROS	OR	*Macrostemum* sp.	PARACU	N	*Paracladopelma* sp. pupa
MAYAYA	OR	*Mayatrichia ayama*	PARAKA	N	*Parakiefferiella* sp. adult
MEGMYE	OR	*Megaluropus myersi*	PARATP	N	*Paratanytarsus* sp. pupa
MELDEN	OR	*Melita dentata*	PARCPU	N	*Parachironomus carinatus* pupa
MELELO	OR	*Melita elongata*	PARLPU	N	*Paracladopelma loganae* pupa
MELFRE	OR	*Melita fresnelii*	PARMPU	N	*Parametriocnemus* sp. pupa
MELLON	OR	*Melita longisetosa*	PARPUP	N	*Parakiefferiella* sp. pupa
MELNIT	OR	*Melita nitida*	PENTPU	N	*Pentaneura* sp. pupa
MESFLO	OR	*Mesanthura floridensis*	POLBPU	N	*Polypedilum* sp. B pupa
METFLO	OR	*Metharpinia floridana*	POLCPU	N	*Polypedilum convictum* pupa
MICCAE	OR	*Microtendipes caelum*	POLFAP	N	*Polypedilum fallax* group pupa
MICLAR	OR	*Micrasema* sp.	POLILA	N	*Polypedilum illinoense* adult
MICRAN	OR	*Microprotopus raneyi*	POLIPU	N	*Polypedilum illinoense* pupa
MICROP	OR	*Micropsectra* sp.	POLPUP	N	*Polycentropus* sp. pupa
MICROR	OR	*Microtendipes rydalensis* group	POLSAD	N	*Polypedilum scalaenum* adult
MICROT	OR	*Microtendipes* sp.	POLSCP	N	*Polypedilum scalaenum* pupa
MICSP1	OR	*Microprotopus* sp. 1	POLTTP	N	*Polypedilum tritum* pupa
MOLANN	OR	*Molanna* sp.	POLYPA	N	*Polypedilum* sp. adult
MOLBLE	OR	*Molanna blenda*	POLYPU	N	*Polypedilum* sp. pupa
MOLTRY	OR	*Molanna tryphena*	POTAPU	N	*Potamyia* sp. pupa

(continued)

Species Code	Trophic Code	Description	Species Code	Trophic Code	Description
MONEDW	OR	*Monoculodes edwardsi*	PROCAD	N	*Procladius* sp. adult
MONNYE	OR	*Monoculodes* sp. *(c.f. nyei)*	PROPUP	N	*Probezzia* sp. pupa
MONSP2	OR	*Monoculodes* sp. 2	PROSUP	N	*Procladius sublettei* pupa
MONSP3	OR	*Monoculodes* sp. 3	PSECTA	N	*Psectrocladius* sp. adult
MYSALM	OR	*Mysidopsis almyra*	PSEEAD	N	*Psectrocladius elatus* adult
MYSBAH	OR	*Mysidopsis bahia*	PSEEPU	N	*Psectrocladius elatus* pupa
MYSBIG	OR	*Mysidopsis bigelowi*	PSEPUP	N	*Psectrocladius* sp. pupa
MYSSP4	OR	*Mysidopsis* sp. 4	PSESPU	N	*Pseudochironomus* sp. pupa
MYSSP5	OR	*Mysidopsis* sp. 5	RHEDPU	N	*Rheotanytarsus distinctissimus* group
MYSSPE	OR	*Mysid* sp.	RHEOPU	N	*Rheotanytarsus* sp. pupa
NAIBRE	OR	*Nais bretscheri*	RHEPU1	N	*Rheotanytarsus distinctissimus* group
NAICOM	OR	*Nais communis*	RHEPU2	N	*Rheotanytarsus distinctissimus* gr p
NAIDID	OR	Naididae	RHEPUP	N	*Rheocricotopus* sp. pupa
NAIELI	OR	*Nais elinguis*	RHEROA	N	*Rheocricotopus robacki* adult
NAISIM	OR	*Nais simplex*	RHERPU	N	*Rheocricotopus robacki* pupa
NAISPE	OR	*Nais* sp.	RHESPU	N	*Rheosmittia* sp. pupa
NAISSP	OR	*Nais* sp.	RHETAA	N	*Rheotanytarsus* sp. adult
NAIVAR	OR	*Nais varabilis*	RHETPU	N	*Rheocricotopus tuberculatus* pupa
NAMABI	OR	*Namalycastis abiuma*	ROBCLP	N	*Robackia claviger* pupa
NANBAL	OR	*Nanocladius balticus* group	SIMPUP	N	*Simulium* sp. pupa
NANBRA	OR	*Nanocladius (Plecopt.) nr. branchic*	SIMSPU	N	*Simulium snowi* pupa
NANNRD	OR	*Nanocladius nr. distinctus*	SIMUPU	N	*Simuliidae* pupa
NANOCL	OR	*Nanocladius* sp.	SIMVPU	N	*Simulium vittatum* pupa
NASSPE	OR	*Nassarius* sp.	STEMPU	N	*Stempellina* sp. pupa
NASVIB	OR	*Nassarius vibex*	STENPU	N	*Stenelmis* sp. pupa
NEOAME	OR	*Neomysis americana*	STEPUP	N	*Stempellinella* sp. pupa
NEOSAY	OR	*Neopanope texana sayi*	STIPUP	N	*Stictochironomus* sp. pupa
NEOSPE	OR	*Neopanope* sp.	TANCUR	N	*Tanytarsus curticornis* group pupa
NEOTE1	OR	*Neopanope texana* (5–10 mm)	TANEMI	N	*Tanytarsus eminulus* group pupa
NEOTSP	OR	*Neotrichia* sp.	TANMAD	N	*Tanytarsus mendax* adult
NERACU	OR	*Nereis acuminata*	TANMEN	N	*Tanytarsus mendax*
NEREID	OR	*Nereidae* sp.	TANNEA	N	*Tanypus neopunctipennis* adult
NERFAL	OR	*Nereis falsa*	TANNEP	N	*Tanypus neopunctipennis* pupa
NERGRA	OR	*Nereis grayi*	TANPPU	N	*Tanypodinae* sp. pupa
NERMIC	OR	*Nereis micromma*	TANYPP	N	*Tanypus* sp. pupa
NERSPA	OR	*Nereid* sp. a	TANYTP	N	*Tanytarsus* sp. pupa
NEUREC	OR	*Neureclipsis* sp.	THIEAD	N	*Thienemanniella* sp. adult

NILBAB	OR	*Nilothauma babiyi*	THIEPU	N	*Thienemanniella* sp. pupa
NILMIR	OR	*Nilothauma mirabile*	THIPUP	N	*Thienemannimyia* group pupa
NILOTH	OR	*Nilothauma* sp.	TIPPUP	N	Tipulidae pupa
NIMPIN	OR	*Nimbocera pinderi*	TRIFPU	N	*Tribelos fuscicorne* pupa
NIMSPE	OR	*Nimbocera* sp.	TRIJPU	N	*Tribelos jucundum* pupa
NYCTSP	OR	*Nyctiophylax* sp.	TRIPUP	N	*Trichoptera* pupa
OCHTHE	OR	*Ochthera* sp.	TVEDCP	N	*Tvetenia discoloripes* group pupa
ODOBIS	OR	*Odostomia bisuturalis*	TVESPU	N	*Tvetenia* sp. pupa
ODOENL	OR	*Odontosyllis enlopa*	UNNPUP	N	*Unniella multivirga* pupa
ODOLAE	OR	*Odostomia laevigata*	ZALPUP	N	*Zalutschia* sp. pupa
ODOSPE	OR	*Odontomyia* sp.			

APPENDIX III

Statistical Analyses Used in the Long-Term Studies of the Northeastern Gulf of Mexico (1971 through 2002)

1. Difference Testing

A series of statistical methods was used to analyze the long-term data. Scattergrams of the long-term field data were examined and either logarithmic or square root transformations were made, where necessary, to approximate the best fit for a normalized distribution. These transformations were used in all statistical tests of significance.

To examine spatial differences in mean values of various field characteristics, the nonparametric Wilcoxon Signed Rank test for paired differences (i.e., applied to the differences between paired observations through time at the sites being compared) was used. The Wilcoxon test (Wilcoxon, 1949) is well suited for these types of field data comparisons because the differencing uses tends to remove much of the serial correlation in the data sets, and the test does not require the normality assumption, which is often difficult to meet with field data. In all such comparisons, the null hypothesis ($\mu_1 = \mu_2$) was tested against the alternative ($\mu_1 \neq \mu_2$) using a two-tailed test with α = 0.05. Variances of the field characteristics were tested for significant differences using a modified Levene's statistic (Levene, 1960) when the underlying populations were non-normal. Both W_{10} and W_{50} statistics were examined using the recommendations of Brown and Forsythe (1974). Nonparametric testing, while avoiding problems with normality, is still subject to the assumptions of independence. Dependence was examined in both the Wilcoxon and Levene tests with an overall Q statistic (Ljung and Box, 1978).

We used a number of basic statistical tests regarding comparisons of two or more data sets. These included the following:

a. Independence Tests

Random, independent observations are integral to the comparison of two samples. The analysis can check if there are certain forms of numeric dependence in the data; if there is dependence due to problems of design or sampling methodology, for example, it cannot be detected. The independence tests used for preliminary analyses examine each sample for autocorrelation at a number of lags. For each sample, autocorrelation is computed for the lesser of 24 or $n/4$ lags, where n is the number of observations. The program then computes the Q statistic of Ljung and Box (1978) for an overall test of the autocorrelations as a group where the null hypothesis is that correlation at each lag is equal to zero. This statistic is approximately chi-square distributed under the null hypothesis with $K - m$ degrees of freedom, where K is the number of lags examined and m is the number of estimated parameters (1 if an AR1 filter has been applied to the data, 0 otherwise). The p-value of the chi-square test is checked and if it remains less than 0.05, the null hypothesis is rejected.

The output sheet also contains plots of the autocorrelation functions for the two samples. These graphs are bar charts showing the values of autocorrelation coefficients at all computed lags, and are overlain with dashed lines at the critical value (value for which any individual correlation may be considered significant) for each series. Critical values are equal to $2/\sqrt{n}$, where n is the number of observations.

b. Normality Tests

The parametric F-test for comparison of variances and t test for comparison of means assume that the data come from normally distributed populations. In support of those tests the two-sample spreadsheet provides a (chi-square) goodness-of-fit test, which compares distributions of the samples to normal distributions having the same means and variances as the samples. Because the number of observations is generally fairly small, the program is designed to divide the data into seven equal-sized intervals (frequency classes) for comparison with expected (normally distributed) frequencies. The program also examines the expected frequencies for the leftmost and rightmost classes and expands (doubles) the width of these classes if the initial expected frequency is less than one. Both these procedures help to minimize the problem of having extremely low expected frequencies in the denominator of the chi-square computation. However, very low sample sizes and sample distributions with extreme departures from normality can still interfere with the test. For this reason the program prints a warning when sample size is below 25. The chi-square test should not be considered reliable for such low sample sizes.

The spreadsheet contains a graph of the frequency distribution (with a normal curve overlain for visual comparison purposes) for each of the two samples. Additionally, the sheet contains calculated values for skewness, kurtosis, and Geary's g statistic (Geary, 1936), which is another measure of kurtosis and which may be compared with the value 0.7979 (g value for a normal distribution). All these values may be used by investigators to assist in their consideration of the normality issue.

c. Variance Testing

For independent, random samples from normally distributed populations, the F-test can be used to compare the variances of two samples. F is computed as the ratio of the variances of the samples:

$$F = \frac{s_1^2}{s_2^2}$$

where the null hypothesis is

$$H_0: \frac{\sigma_1^2}{\sigma_2^2} = 1$$

and where s_1^2 is the larger of the two sample variances. The program prints the value of F, the associated p-value from an F-table with $n_1 - 1$ numerator degrees of freedom and $n_2 - 1$ denominator degrees of freedom. It also prints a decision to reject or not to reject the null hypothesis (using an alpha level of 0.05), and a warning if it seems that one or more test assumptions have been violated. If one or both of the data series violate the independence assumption due to serial dependence at lag 1, one can use the AR(1) filter option in the two-sample menu. Running the filter will prompt the program to use a modified formula for the calculation of *F*, which computes *F* in the presence of lag 1 dependence.

For those cases where the samples do not satisfy the assumption of normality, two modifications to Levene's statistic (Levene, 1960) for variance comparisons are included in the spreadsheet. The statistics are W_{10} and W_{50}, which are both robust to departures from normality in the underlying populations (Brown and Forsythe, 1974). They use robust measures of central location (W_{10}) uses the mean of the middle 80% of the observations; W_{50} uses the median observation) and thereby decreases or eliminates the influence of extreme values in testing variance. Brown and Forsythe (1974) recommend using W_{50} for testing the variances of asymmetric distributions and using W_{10} for testing long-tailed distributions. Both formulas involve the generation of a new data series z_{ij}, which also must be free of serial dependence. If a warning message is printed with the robust tests, it applies to the new series not the original series, and it is possible to have dependence in the z_{ij} series even if the original data were independent. Dependence is examined with an overall *Q* statistic (Ljung and Box, 1978) and chi-square test developed in a fashion similar to that described above in the independence testing section. The output of the variance tests should not be used if a transformation (log, square root, arcsin, normalization) has been applied to the data. Also, performing a sequential average tends to "smooth" the data (i.e., variation is lowered in both data sets) and may have an effect on the variance test.

d. Means Testing

For independent, random samples from normally distributed populations, the parametric *t*-test can be used to compare the sample means. For unpaired samples the spreadsheet provides either a *t*-test with equal variance or a *t*-test with unequal variance (the results of the variance test above are used as the criterion for deciding which *t*-test to output). For paired samples, the sheet provides the results of a paired *t* test. For all tests the output contains the computed value of the *t*-statistic, a probability value (two-sided test, alpha = 0.05) from the *t*-distribution, a reject or do not reject decision regarding the null hypothesis, and a warning if it seems that one or more test assumptions have been violated. If one or both of the data series violate the independence assumption due to serial dependence at lag 1, one can use the AR(1) filter option from the two-sample menu. Running the filter will prompt the program to use a modified formula for the calculation of *t*, which computes *t* in the presence of lag 1 dependence.

For cases where one or both of the data sets violate the assumption of normality, a data transformation can be made to bring the data into normality (although variance testing is affected), or use one of the nonparametric tests (see appendix), which have been included in the spreadsheet. The tests are the rank sum test for unpaired comparisons (Wilcoxon, 1945) and the signed rank test for paired differences. Both tests are subject to the assumption of random independent samples. For the rank sum test both of the data series should be independent, whereas for the signed rank test only the series of differences needs to be independent. This can make the signed rank test a valuable tool for two-sample testing because, in practice, serial correlation often disappears when dependent data sets are differenced. Wilcoxon test results should not be used if an AR1 filter has been applied to the data.

Output for the Wilcoxon tests includes a description of the rejection region, the computed value for the test statistic *t* for small samples and *z* for larger samples ($n > 25$ for signed rank test, $n > 10$ for rank sum), a reject or do not reject decision regarding the null hypothesis (with a probability value based on a two-tailed test and alpha level of 0.05), and a warning if the independence assumption has been violated. Output for the signed rank Wilcoxon test also includes for the differenced data series an overall *Q* statistic and chi-square probability value, along with a graph of the autocorrelation function.

e. Distributional Shape Testing

The two-sample spreadsheet also includes a chi-square test, which compares the shapes of the distributions for the two samples. Although the samples may individually exhibit dependence, the assumption is made that the two samples are independent of each other. There are no assumptions regarding the form of the underlying distributions. To examine shape only, the program adjusts the two data sets by subtracting the respective mean value from each observation such that both series then have a common mean value of zero. Then the overall range of data in the two series is divided into the lesser of 9 or $n/4$ classes and a frequency distribution is produced for each sample. The frequency distributions are compared with a chi-square test.

Output for the shape test includes the computed values for the chi-square statistic and its associated probability value, a reject or do not reject decision regarding the null hypothesis (based on a one-tailed test and alpha level of 0.01), and a warning if the sample size is too small for a reliable test ($n < 25$) or if the independence assumption has been violated.

f. Principal Components Analysis and Regressions

Principal Component Analysis (PCA) and associated correlation matrices were determined using monthly data and the SAS™ statistical software package. A PCA was carried out as a preliminary review of the data using river flow information, sediment data, and water quality variables. The PCA was used to reduce the physicochemical variables into a smaller set of linear combinations that could account for most of the total variation of the original set. For the physicochemical variables in this study, the series were stationary after appropriate transformations; thus the sample correlation matrix of the variables was a good estimate of the population correlation matrix. Therefore, the standard PCA could be carried out based on the sample correlation matrix.

A matrix of the water and sediment quality data associated with the sampling stations in the Perdido system was prepared. The data were then grouped by station by year. Values for the dependent variables (total biomass m^{-2}, herbivore biomass m^{-2}, omnivore biomass m^{-2}, primary carnivore biomass m^{-2}, secondary carnivore biomass m^{-2}, tertiary carnivore m^{-2}) were then paired with the water/sediment quality data (independent variables) taken for stations within each sector. Unless otherwise defined, these statistics were run using SAS, Systat™, and SuperAnova™. Data analyses were run on the three independent data sets. Significant principal components were then used to run a series of regression models with the biological factors as dependent variables. Residuals were tested for independence using serial correlation (time series) analyses and the Wald–Wolfowitz (Wald and Wolfowitz, 1940) runs test. A chi-square test was run to evaluate normality.

2. Spatial Autocorrelation

The nonparametric Wilcoxon signed rank test for paired differences (Wilcoxon, 1949) was used to examine spatial differences in various field characteristics. The test was applied to the differences between paired observations through time at the sites being compared. The Wilcoxon test is well suited for field data comparisons because differencing tends to remove much of the serial correlation in the data sets, and the test does not require normally distributed data. In all such comparisons, the null hypothesis ($\mu_1 = \mu_2$) was tested against the alternative ($\mu_1 \neq \mu_2$) using a two-tailed test with $p = 0.05$. Variances of the field characteristics were tested for significant differences using a modified Levene's statistic when the underlying populations were non-normal. Both W_{10} and W_{50} statistics were examined using the recommendations of Brown and Forsythe (1974). Nonparametric testing, while avoiding problems with normality, was still subject to the assumptions of independence. Dependence was examined in both the Wilcoxon and Levene tests with an overall Q statistic (Ljung and Box, 1978). Despite attempts to satisfy the independence assumption, it was often violated, particularly in the weekly field analyses.

Spatial autocorrelation analyses were run independently on each of the nine 100-core data sets collected at stations 3, 5a, and ML to examine within-site variability. Spatial autocorrelation coefficients (ρ) were calculated based on a generalized form of Moran's statistic for the four community characteristics (i.e., total numbers of individuals, numbers of taxa and numbers of the top two dominant species). Correlation coefficients were calculated between each and all pairs of cores of increasing distance (h) apart, where h is referred to as the order of the coefficient. Correlation coefficients were calculated between each core and all surrounding cores up to order 8. Sample autocorrelation coefficients for each order were then averaged over the three 100-core replicates at each site. If $\rho_{kl}(h)$ denotes the average order-h spatial autocorrelation at the kth station for a given community characteristic l, then the average sample spatial autocorrelation $\rho_{kl}(h)$ has a mean $-1/(n-1)$ and variance $\sigma^2(h)/3$. The sample Z-score for $\rho_{kl}(h)$ is

$$Z_{kl}(h) = \left[\hat{\rho}_{kl}(h) + \frac{1}{n-1}\right] \Big/ \sqrt{\sigma^2(h)/3}$$

which has an approximate standard normal distribution. For testing the null hypothesis $\rho_{kl}(h) = 0$ vs. the alternative hypothesis $\rho_{kl}(h) \neq 0$, the rejection region occurs where the sample Z-score falls outside of the interval ± 1.96, when $p = 0.05$ (i.e., when $Z_{kl}(h) > 1.96$ or $Z_{kl}(h) < -1.96$, we conclude that $\rho_{kl}(h)$ is significantly different from zero).

To examine spatial autocorrelation in community characteristics among cores in each of the 100-core arrays, correlation coefficients were calculated and tested for significance using the following methodology. Suppose that a spatial process $\{Y(s)\}$ is observed at locations $s_1,\ldots,s_n$.. Let $\gamma(s_i,s_j)$ denote the covariance between $Y(s_i)$ and $Y(s_j)$. In this study, we assume that $\{Y(s)\}$ is an isotropic process (i.e., the mean value and variance of $Y(s)$ are independent of location s, and in addition, $\gamma(s_i,s_j)$ depends only on the distance $\|s_i - s_j\|$ between the two spatial locations s_i and s_j). When the observations $\{Y(s_i),\ldots,Y(s_n)\}$ are available, the spatial autocovariance function $\{\gamma(h), h \geq 0)$ can be estimated by

$$\hat{\gamma}(h) = \frac{1}{|N(h)|} \sum_{(i,j)\in N(h)} (Y(s_i) - \bar{Y})(Y(s_j) - \bar{Y})$$

where

$$\bar{Y} = (1/n)\sum_{i=1}^{n} Y(s_i), \quad N(h) = \{(i,j) : \| s_i - s_j \| = h\}$$

is the neighborhood set with distance h and $|N(h)|$ denotes the number of elements of $N(h)$. A typical spatial neighbor scheme for regular lattices indicates where different order neighbors are specified relative to a reference point.

For $h > 0$, let $w_{ij}(h) = 1$ for $(I,j) \in N(h)$ and $w_{ij}(h) = 0$ for $(I,j) \notin\in N(h)$. Since $w_{ij}(h) = w_{ij}(h)$, the weighting matrix $W(h) = [w_{ij}(h)]$ is symmetrical. The sample order-h spatial autocorrelation can be expressed as:

$$\hat{\rho}(h) = \frac{n}{\sum_{i,j=1}^{n} w_{ij}(h)} \left[\frac{\sum_{i,j=1}^{n} w_{ij}(h)(Y(s_i) - \bar{Y})(Y(s_j) - \bar{Y})}{\sum_{i=1}^{n} (Y(s_i) - Y)^2} \right]$$

which is a generalization of Moran's statistic (Moran, 1950).

When $\{Y(s_i),\ldots,Y(s_n)\}$ are independent, identically distributed normal variables, the spatial autocorrelation function is $\rho(h) = 0$ for $h > 0$. In this case the first two moments of the sample autocorrelation $\hat{\rho}(h)$ are

$$E(\hat{\rho}(h)) = -\frac{1}{n-1} \text{ and } \underline{E}(\hat{\rho}^2(\underline{h})) = (n^2 S_1(h) - nS_2(h)) + 3S_0(h)) / [(n^2-1)S_0^2(h)]$$

where

$$S_0(h) = \sum_{i,j=1}^{n} w_{ij}(h),\ S_1(h) = 2\sum_{ij=1}^{n} w_{ij}^2(h) = 2S_0(h),\ S_2(h) = 4\sum_{i=1}^{n} w_{i\cdot}^2(h)$$

and

$$w_i(h) = \sum_{j=1}^{n} w_{ij}(h)$$

The variance of $\hat{\rho}(h)$, denoted by $\sigma^2(h)$, can be calculated by the formula

$$\sigma^2(h) = \text{Var}(\hat{\rho}(h)) = E(\hat{\rho}^2(h)) - [E(\hat{\rho}(h))]^2$$

For large sample size n, $\hat{\rho}(h)$ is approximately normally distributed and can be defined as:

$$Z(h) = {[\hat{\rho}(h) - E(\hat{\rho}(h))]}\Big/{\sqrt{\sigma^2(h)}}$$

$Z(h)$ is called the sample Z-score for the spatial autocorrelation coefficient $\rho(h)$ and can be used to test the null hypothesis $\rho(h) = 0$ vs. the alternative hypothesis $\rho(h) \neq 0$.

a. *Listing of Common Taxa*

Below is a list of common taxa (those comprising the top 90% of the numbers of individuals) at each site along with long-term mean density, relative and cumulative abundance, and life history characteristics. This summary was derived from the longest available data set for each station (station 3: 77 collections over the period March 1975 through July 1984; station 5a: 109 collections over the period March 1975 through July 1984; station ML: 59 collections over the period November 1981 through September 1986).

Taxa	Trophic Level	Feeding Type	Mean Density (number /m^2)	Abundance (%) Relative	Abundance (%) Cumulative
		Station 3			
Mediomastus ambiseta	Detritivore	Deposit feeder (below surface)	2457	31.4	31.4
Oligochaeta spp.	Detritivore	Browser/grazer	1354	17.3	48.7
Streblospio benedicti	Detritivore	Deposit feeder (above surface)	841	10.7	59.4
Grandidierella bonnieroides	Omnivore	Browser/grazer	598	7.6	67.0

Taxa	Trophic Level	Feeding Type	Mean Density (number /m²)	Abundance (%) Relative	Cumulative
Hobsonia florida	Detritivore	Deposit feeder (above surface)	536	6.8	73.8
Cerapus sp. *(c.f. tubularis)*	Omnivore	Browser/grazer	487	6.2	80.0
Glycinde solitaria	Carnivore	Piercer	341	4.4	84.4
Laeonereis culveri	Omnivore	Piercer	216	2.8	87.2
Capitella capitata	Detritivore	Deposit feeder (below surface)	216	2.8	90.0
Parandalia americana	Carnivore	Piercer	213	2.7	92.7
All others (37 taxa)			571	7.3	100.0
Total			7830	100.0	
		Station 5A			
Mediomastus ambiseta	Detritivore	Deposit feeder (below surface)	1245	38.8	38.8
Streblospio benedicti	Detritivore	Deposit feeder (above surface)	446	13.9	52.7
Grandidierella bonnieroides	Omnivore	Browser/grazer	260	8.1	60.8
Mactra fragilis	Herbivore	Filter feeder	203	6.3	67.1
Macoma mitchelli	Detritivore	Filter feeder	158	4.9	72.0
Littoridina sphinctostoma	Omnivore	Scraper	146	4.6	76.6
Insect larva			129	4.0	80.6
Hobsonia florida	Detritivore	Deposit feeder (above surface)	123	3.8	84.4
Capitella capitata	Detritivore	Deposit feeder (below surface)	107	3.3	87.7
Parandalia americana	Carnivore	Piercer	106	3.3	91.0
All others (30 taxa)			285	9.0	100.0
Total			3206	100.0	
		Station ML			
Mediomastus ambiseta	Detritivore	Deposit feeder (below surface)	790	21.5	21.5
Oligochaeta spp.	Detritivore	Browser/grazer	754	20.6	42.1
Apoprionospio pygmaea	Detritivore	Deposit feeder (above surface)	308	8.4	50.5
Brania wellfleetensis	Carnivore	Piercer	254	6.9	57.4
Aricidea fragilis	Detritivore	Deposit feeder (below surface)	210	5.7	63.1
Tharyx sp.	Detritivore	Deposit feeder (above surface)	144	3.9	67.0
Axiothella mucosa	Detritivore	Deposit feeder (below surface)	88	2.4	69.4
Ampelisca verrilli	Omnivore	Browser/grazer	63	1.7	71.1
Carazziella hobsonae	Detritivore	Deposit feeder (above surface)	59	1.6	72.7
Notomastus hemipodus	Detritivore	Deposit feeder (below surface)	59	1.6	74.3
Tellina texana	Detritivore	Filter feeder	53	1.5	75.8
Paraonis fulgens	Detritivore	Deposit feeder (below surface)	53	1.4	77.2
Paraprionospio pinnata	Detritivore	Deposit feeder (above surface)	50	1.4	78.6
Semele prolifica	Detritivore	Filter feeder	45	1.2	79.8
Listriella barnardi	Omnivore	Browser/grazer	42	1.2	81.0
Spiophanes bombyx	Detritivore	Deposit feeder (above surface)	41	1.1	82.1
Armandia agilis	Detritivore	Deposit feeder (below surface)	39	1.1	83.2
Bivalve sp. 2	Herbivore	Filter feeder	38	1.0	84.2
Glycinde solitaria	Carnivore	Piercer	36	1.0	85.2
Chaetozone sp.	Detritivore	Deposit feeder (above surface)	34	0.9	86.1
Onuphis eremita oculata	Carnivore	Piercer	27	0.7	86.8
Glycera americana	Carnivore	Piercer	23	0.6	87.4
Scolelepis texana	Detritivore	Deposit feeder (above surface)	23	0.6	88.0
Gyptis vittata	Carnivore	Piercer	17	0.5	88.5
Lembos smithi	Omnivore	Browser/grazer	17	0.5	89.0
Corophium tuberculatum	Omnivore	Filter feeder	17	0.5	89.5
Cumacea	Detritivore	Filter feeder	17	0.5	90.0
Tanaidacea	Omnivore	filter feeder	17	0.5	90.5
All others (138 taxa)			352	9.5	100.0
Total			3669	100.0	

b. Statistical Analyses Associated with Reviews of Long-Term Databases in the Apalachicola, Perdido, Econfina, and Fenholloway Systems

Data Transformations

Scattergrams of the long-term field data were examined and either logarithmic or square root transformations were made, where necessary, to approximate the best fit for a normalized distribution. These transformations were used in all statistical tests of significance.

Analysis of Variance

Analysis of variance (ANOVA) models were run using Systat™ and SuperAnova™. The ANOVA determinations were run using year-by-season data sets for the infauna, macroinvertebrates, and fishes separately. Three-month intervals, starting with March of each year, were used to define the seasonal patterns. These seasonal definitions were based on established seasonal temperature patterns in the study area. The basic model followed that outlined by Winer (1971) as:

$$Y_{ij} = m + a_i + b_j + ab_{ij} + e_{ij}$$

where Y_{ij} is the response associated with the *i*th year and the *j*th season, *m* represents an overall effect, a_i is the effect of the *i*th year, b_j is the effect of the *j*th season, ab_{ij} is the interaction between the *i*th year and *j*th season, and e_{ij} represents the random error. Specific hypotheses were tested (e.g., years and seasons have no effect on dependent variables). A two-factor analysis was carried out with Years and Seasons as the two factors.

Major assumptions of the model were tested by studying residual distributions. The three monthly values within each season were used as replicates even though these factors were not replicates in the true sense. Therefore, an autocorrelation function (ACF) was run to check for serial correlation among the monthly residual values (positive correlations would invalidate the ANOVA test). No positive serial correlations of residuals for sequential months were noted for the various dependent variables. In fact, negative autocorrelations were usually found at lags 1 and 2, which indicated generally conservative ANOVA results (i.e., error variance was overestimated). Residuals were plotted against fitted values in addition to carrying out normal probability plots. Unless otherwise stated, ANOVA results showed random distributions of the residuals in normal probability plots. The Wald–Wolfowitz run test (with cutoff = 0) was run on residuals to determine the possibility of clumping of the positive and negative residuals (Wald and Wolfowitz, 1940). A lack of significance in the Wald–Wolfowitz run tests on residuals was usually found. *F*-values and their associated *p*-values were calculated for the final determinations with pairwise comparison of the effects/interactions via *post hoc* tests for further comparisons. *Post hoc* tests included Tukey's for the comparisons of means and Scheffe's for all comparisons of between factors only.

Principal Components Analysis

Principal Component Analysis (PCA) and associated correlation matrices were determined using monthly data and the SAS statistical software package. A PCA was carried out as a preliminary review of the data using the river and rainfall information and the water quality variables. The PCA was used to reduce the physicochemical variables into a smaller set of linear combinations that could account for most of the total variation of the original set. For the physicochemical variables in this study, the series were stationary after appropriate transformations; thus the sample correlation matrix of the variables was a good estimate of the population correlation matrix. Therefore, the standard PCA could be carried out based on the sample correlation matrix.

Time Series Analysis and Dynamic Regression

Univariate time series and dynamic regression models were developed for the five biological series, herbivores {Hb(t)}, omnivores {Om(t)}, primary carnivores {C1(t)}, secondary carnivores {C2(t)} and tertiary carnivores {C3(t)}, following the well-known Box–Jenkins modeling procedures (Box and Jenkins, 1976; Pankratz, 1983, 1991). All parameters in the models were estimated using the maximum likelihood method.

If a biological series is serially correlated (i.e., the current observation of the series depends on past observations), a univariate time series model may be fitted to the series. The response series {$Y(t)$} is modeled as an autoregressive integrated moving average model {ARIMA (p, d, q)} with the form:

$$\nabla^d Y(t) - \phi_1 \nabla^d Y(t-1) - \ldots - \phi_p \nabla^d Y(t-p) = \mu + \varepsilon(t) - \theta_1 \varepsilon(t-1) - \ldots - \theta_q \varepsilon(t-q) \quad \text{(III.1)}$$

where ∇ is the differencing operator such that $\nabla Y(t) = Y(t) - Y(t-1)$, ..., $\nabla\ dY(t) = Y(t) - Y(t-d)$ and the $\varepsilon(t)$ are assumed to be independent and normally distributed with mean zero and variance σ^2. In the model, p is the order of the autoregression (AR) term, d is the order of differencing and q is the order of the moving-average (MA) term. The parameters ϕ_p and θ_q represent the AR and MA coefficients, respectively. Each biological series was modeled using only present and past (i.e., lagged) monthly values of the series.

Dynamic regression models, also called transfer function models (Box and Jenkins, 1976), were used to characterize how the biological variables responded to changes in the environmental characteristics. A linear dynamic regression model describing the relationship between a response series {$Y(t)$} and an input series {$X(t)$} has the form:

$$Y(t) = \mu + \nu_0 X(t) + \ddot{\nu}_1 X(t-1) + \ldots + \nu_k X(t-k) + \xi(t) \quad \text{(III.2)}$$

where μ is a constant term and $\xi(t)$ is assumed to be an ARIMA process and independent of the input series {$X(t)$}. If $\xi(t)$ follows a stationary AR(p) model, such that $\xi(t) - \phi_1\xi(t-1) - \phi_2\xi(t-2) - \ldots - \phi_p\xi(t-p) = \varepsilon(t)$ and letting B be the backward shift operator such that $B\xi(t) = \xi(t-1)$, ..., $B^k\xi(t) = \xi(t-k)$, then $\varepsilon(t)$ can be expressed as $\xi(t) = (1 - \phi_1 B - \phi_2 B^2 - \ldots - \phi_p B^p)^{-1}\varepsilon(t)$ and Equation III.2 can be written as:

$$Y(t) = \mu + \nu_0 X(t) + \nu_1 X(t-1) + \ldots + \nu_k X(t-k) + (1 - \phi_1 B - \phi_2 B^2 - \ldots - \phi_p B_p)^{-1}\varepsilon(t) \quad \text{(III.3)}$$

The dynamic regression model in Equation III.3 states that the effects of {$X(t)$} on {$Y(t)$} are distributed across several time periods. The coefficients {ν_0, ν_1, ..., ν_k} in the model, called ν-weights, indicate how much the response series changes when the past and current values of the input series change one unit. Multiple inputs may be introduced into the model by adding other variables (e.g., $X_1(t)$, $X_2(t)$, $X_3(t)$) along with their appropriate weights for different time lags {$(t-k)$}.

In the current analysis, each of the five biological series was modeled using the six principal component axes (representing the environmental variables) as well as the other biological series (i.e., Hb, Om, C1, and C2). Inclusion of the other biological variables as inputs into each model allowed us to examine possible feedback mechanisms among the biological series. Although the tertiary carnivores were modeled for completeness, the C3 series was not included as an input variable in the formation of the dynamic regression models for the lower trophic groups (Hb to C2). Tertiary carnivores are highly motile organisms and were the least well represented group in our trawl collections.

An F-test was used to compare the two nested models (i.e., the univariate model and the dynamic regression model). RSS_1 and RSS_2 are the residual sum of squares from the two models, respectively. The F-statistic for comparing the two models is defined by:

$$F = \frac{(RSS_1 - RSS_2)/f_1}{RSS_2/f_2}$$

where f_1 is the difference in numbers of parameters in the two models and f_2 is the degrees of freedom for RSS_2. The F-statistic has an approximate F-distribution with degrees of freedom f_1 and f_2.

The usefulness of both the univariate time series and the dynamic regression model can be measured by R^2 which is calculated by the formula:

$$R^2 = 1 - \frac{\sum_{t=1}^{n} (Y(t) - \hat{Y}(t))^2}{\sum_{t=1}^{n} (Y(t) - \bar{Y})^2}$$

where $\hat{Y}(t)$ is the fitted value of $Y(t)$ and $\bar{Y}$ is the mean of $Y(t)$.

Estimated coefficients in the fitted univariate time series and dynamic regression models for herbivores (Hb), omnivores (Om), primary carnivores (C1), secondary carnivores (C2), and tertiary carnivores (C3) taken in East Bay and averaged over all stations (biomass/m^2) on a monthly basis from February 1975 through July 1984. P_1, P_2, P_3, and P_6 are Principal Components Axes 1, 2, 3, and 6, respectively; $(t - k)$ are the lag terms with k = lag in months. Only significant coefficients ($t = 1.98$, $n = 120$, $p < 0.05$) are included.

Variable	Lag	Coefficient Estimates	Standard Error	t Ratio
		Herbivores		
Univariate Time Series Model ($R^2 = 0.58$)				
$Hb(t) = 0.67Hb(t-1) + 0.22Hb(t-3) + \varepsilon(t)$				
Hb	1	0.665	0.080	8.31
Hb	3	0.218	0.110	1.98
Dynamic Regression Model ($R^2 = 0.82$)				
$Hb(t) = -0.10P_1(t-12) + 0.22P_2(t-2) + 0.23P_3(t-2) - 0.14P_3(t-6) + 0.21P_6(t-6) + 0.90Om(t) - 0.93C1(t-20) + (1 - 0.53B - 0.32B^3)^{-1}\varepsilon(t)$				
P_1	12	–0.097	0.041	–2.39
P_2	2	0.220	0.049	4.51
P_3	2	0.231	0.056	4.11
P_3	6	–0.144	0.060	–2.41
P_6	6	0.207	0.086	2.40
Om	0	0.900	0.455	1.98
C_1	20	–0.932	0.236	–3.94
Residuals	1	0.526	0.094	5.61
	3	0.320	0.097	3.29

Variable	Lag	Coefficient Estimates	Standard Error	*t* Ratio
		Omnivores		
Univariate Time Series Model ($R^2 = 0.33$)				
$Om(t) = 0.41 + 0.57Om(t-1) + \varepsilon(t)$				
Constant		0.410	0.038	10.79
Om	1	0.572	0.078	7.33
Dynamic Regression Model ($R^2 = 0.65$)				
$Om(t) = 0.36 - 0.02P_1(t-6) - 0.03P_1(t-12) + 0.02P_1(t-24) + 0.04Hb(t) + 0.04Hb(t-11) + 0.10C1(t-1) + (1-0.55B)^{-1}\varepsilon(t)$				
Constant		0.356	0.046	7.73
P_1	6	–0.023	0.009	–2.66
P_1	12	–0.025	0.009	–2.76
P_1	24	0.022	0.010	2.32
Hb	0	0.043	0.016	2.63
Hb	11	0.037	0.016	2.36
C_1	1	0.100	0.046	2.16
Residual	1	0.552	0.095	5.81
		Primary Carnivores		
Univariate Time Series Model ($R^2 = 0.38$)				
$C1(t) = 0.61 + 0.41C1(t-1) + 0.31C1(t-4) + \varepsilon(t)$				
Constant		0.608	0.098	6.20
C_1	1	0.414	0.084	4.93
C_1	4	0.306	0.085	3.60
Dynamic Regression Model ($R^2 = 0.51$)				
$C_1(t) = 0.66 + 0.08Hb(t-2) + 0.11Hb(t-8) + (1-0.26B-0.23B^4)^{-1}\varepsilon(t)$				
Constant		0.660	0.056	11.84
Hb	2	0.079	0.024	3.33
Hb	8	0.105	0.023	4.46
Residuals	1	0.262	0.094	2.77
	4	0.230	0.097	2.38
		Secondary Carnivores		
Univariate Time Series Model ($R^2 = 0.15$)				
$C_2(t) = 0.17 + 0.38C_2(t-1) + \varepsilon(t)$				
Constant		0.167	0.013	12.85
C_2	1	0.381	0.087	4.38
Dynamic Regression Model ($R^2 = 0.32$)				
$C_2(t) = 0.19 + 0.02Hb(t-6) - 0.12Om(t-11) + 0.06C_1(t-1) + (1-0.22B)^{-1}\varepsilon(t)$				
Constant		0.190	0.028	6.79
Hb	6	0.020	0.007	2.86
Om	11	–0.123	0.048	–2.56
C1	1	0.061	0.026	2.35
Residual	1	0.223	0.100	2.23

(continued)

Variable	Lag	Coefficient Estimates	Standard Error	*t* Ratio
		Tertiary Carnivores		
Univariate Time Series Model ($R^2 = 0.12$)				
$C_3(t) = 0.05 + 0.33C_3(t-1) + \varepsilon(t)$				
Constant		0.048	0.006	8.00
C_3	1	0.334	0.089	3.75
Dynamic Regression Model ($R^2 = 0.25$)				
$C_3(t) = 0.05 - 0.04C_1(t-1) + 0.15C_2(t) + (1 - 0.28B)^{-1}\,\varepsilon(t)$				
C_1	1	–0.043	0.012	–3.58
C_2	0	0.145	0.045	3.22
Residual	1	0.283	0.094	3.02

References

Abood, K.A. and S.G. Metzger. 1996. Comparing impacts to shallow-water habitats through time and space. Estuaries 19: 220–228.

Adams, S.M. 1976. Feeding ecology of eelgrass fish communities. Trans. Am. Fish. Soc. 105: 514–519.

Anderson, D.M. 1996. Control and mitigation of red tides, December 4, Manuscript prepared for Project Start (Solutions to Avoid Red Tide), Sarasota, Florida, 59 pp.

Anderson, D.A. and D.J. Garrison. 1997. The ecology and oceanography of harmful algal blooms. Limnol. Oceanogr. 42: 1–1305.

APHA. 1989. Standard Methods for the Examination of Water and Wastewater, 17th ed. American Public Health Association, Washington, D.C.

Arditi, R. and L.R. Ginzberg. 1989. Coupling in predator-prey dynamics: ratio dependence. J. Theor. Biol. 139: 311–326.

Armstrong, N.E. 1982. Responses of Texas estuaries to freshwater inflows. In: V.S. Kennedy, Ed., Estuarine Comparisons. Academic Press, New York, pp. 103–120.

Baird, D. and R.E. Ulanowicz. 1989. The seasonal dynamics of the Chesapeake Bay ecosystem. Ecol. Monogr. 59: 329–364.

Baird, R.C. 1996. Toward new paradigms in coastal resource management: linkages and institutional effectiveness. Estuaries 19: 337–341.

Balls, P.W., A. Macdonald, K. Pugh, and A.C. Edwards. 1995. Long-term nutrient enrichment of an estuarine system: Ythan, Scotland (1958–1993). Environ. Pollut. 90: 311–321.

Bell, T. and J. Kalff. 2001. The contribution of picophytoplankton in marine and freshwater systems of different trophic status and depth. Limnol. Oceangr. 46: 1243–1248.

Bittaker, H.F. 1975. A Comparative Study of the Phytoplankton and Benthic Macrophyte Primary Productivity in a Polluted vs. Unpolluted Coastal Area. M.S. thesis, Florida State University, Tallahassee.

Blanchet, R.H. 1979. The Distribution and Abundance of Ichthyoplankton in the Apalachicola Bay, Florida Area. M.S. thesis, Florida State University, Tallahassee.

Bobbie, R.J., S.J. Morrison, and D.C. White. 1978. Effects of substrate biodegradability on the mass and activity of the associated estuarine microbiota. Appl. Environ. Microbiol. 35: 179–184.

Blumberg, A.L. and G.L. Mellor. 1980. A coastal numerical model. In: J. Sundermann and K.P. Holz, Eds., Mathematical Modeling of Estuarine Physics. Proceedings of International Symposium, Hamburg, 24–26 August 1978. Springer-Verlag, Berlin, pp. 203–214.

Bohnsack, J.A. 1983. Resilience of reef fish communities in the Florida Keys following a January 1977 hypothermal fish kill. Environ. Biol. Fish. 9: 41–53.

Bonin, D.J., M.R. Droop, S.Y. Maestrini, and M.C. Bonin. 1986. Physiological features of six micro-algae to be used as indicators of seawater quality. Cryptogamie Algol. 7: 23–83.

Bortone, S. 1991. Sea grass mapping of Perdido Bay. Unpublished report.

Botton, M.L. 1979. Effects of sewage sludge on the benthic invertebrate community of the inshore New York Bight. Estuarine Coastal Mar. Sci. 8: 169–180.

Box, G.E.P. and G.M. Jenkins. 1976. Time Series Analysis: Forecasting and Control. Holden-Day, Inc., San Francisco, CA.

Boyer, J.N., J.W. Fourqurean, and R.D. Jones. 1999. Seasonal and long-term trends in the water quality of Florida Bay. In: J.W. Fourqurean, M.B. Robblee, and L.A. Deegan, Eds., Florida Bay: A Dynamic Subtropical Estuary. Estuaries 22: 417–430.

Boynton, W.R., W.M. Kemp, and C.W. Keefe. 1982. A comparative analysis of nutrients and other factors influencing estuarine phytoplankton production. In: V.S. Kennedy, Ed., Estuarine Comparisons. Academic Press, New York, pp. 69–90.

Brady, K. 1982. Seasonal and Spatial Distribution of Ichthyoplankton in Seagrass Beds of Apalachee Bay. M.S. thesis, Florida State University, Tallahassee.

Breitburg, D.L. 1990. Near-shore hypoxia in the Chesapeake Bay: patterns and relationships among physical factors. Estuarine Coastal Shelf Sci. 30: 593–609.

Bricelj, V.M. and S. Kuenstner. 1989. Effects of the "brown tide" on the feeding physiology and growth of juvenile and adult bay scallops and mussels. In: E.M. Cosper, E.J. Carpenter, and V.M. Bricelj, Eds., Novel Phytoplankton Blooms: Causes and Impacts of Recurrent Brown Tides and Other Unusual Blooms. Lecture Notes on Coastal and Estuarine Studies. Springer-Verlag, Berlin, pp. 491–509.

Bricker, S.B., C.G. Clement, D.E. Pirhalla, S.P. Orlando, and D.R.G. Farrow. 1999. National Estuarine Eutrophication Assessment; Effects of Nutrient Enrichment in the Nation's Estuaries. National Oceanic and Atmospheric Administration, National Ocean Service, Special Projects Office and the National Centers for Coastal Ocean Science, Silver Spring, MD, 71 pp.

Brim, M.S. 1993. Toxics characterization report for Perdido Bay, Alabama, and Florida. U.S. Fish and Wildlife Service Publ. PCFO-EC-93–94, 137 pp.

Brockman, U.H., P.M. Lane, and H. Postma. 1990. Cycling of nutrient elements in the North Sea. Neth. J. Sea Res. 26: 239–264.

Brook, I.M. 1976. Trophic relationships in a *Thalassia* community: fish diets in relation to abundance and biomass. Presented at 39th Annual Meeting, American Society of Limnology and Oceanography, Savannah, GA.

Brook, I.M. 1977. Trophic relationships in a sea grass community (*Thalassia testudinum*), in Card Sound, Florida — fish diets in relation to macrobenthic and cryptic faunal abundance. Trans. Am. Fish. Soc. 106: 219–229.

Brouard, D., C. Demers, R. Lalumiere, R. Schetagne, and R. Verdon. 1990. Evolution of mercury levels in fish of the La Grande Hydroelectric Complex, Quebec. Summary report, Hydro Quebec (Montreal) and Schooner, Inc. (Quebec), P.Q., 97 pp.

Brown, M.B. and A.B. Forsythe. 1974. Robust tests for the equality of variances. J. Am. Stat. Assoc. 69: 364–367.

Brush, G.S. 1984. Stratigraphic evidence of eutrophication in an estuary. Water Resourc. Res. 20: 531–541.

Brush, G.S. 1991. Long-term trends in Perdido Bay: a stratigraphic study. Unpublished report. Florida Department of Environmental Regulation.

Burgess, N.M., D.C. Evers, J.D. Kaplan, M. Duggan, and J.J. Kerekes. 1998. Ecological impacts of mercury. In: Mercury in the Atlantic: A Progress Report. Regional Science Coordinating Committee, Environment Canada.

Burkholder, J.A. and H.B. Glasgow, Jr. 1997. *Pfiesteria piscicidia* and other *Pfiesteria*-dinoflagellates: behaviors, impacts, and environmental controls. Limnol. Oceanogr. 42: 1052–1075.

Burkholder, J.M., E.J. Noga, C.H. Hobbs, and H.B. Glasgow, Jr. 1992. New "phantom" dinoflagellate is the causative agent of major estuarine fish kills. Nature 358: 407–410.

Buskey, E.J. and D.A. Stockwell. 1993. Effects of a persistent "brown tide" on zooplankton populations in the Laguna Madre of South Texas. In: Toxic Phytoplankton Blooms in the Sea. Proc. 5th Int. Conf. on Toxic Marine Phytoplankton. Elsevier, New York, pp. 659–665.

Buskey, E.J., P.A. Montagna, A.F. Amos, and T.E. Whitledge. 1997. Disruption of grazer populations as a contributing factor to the initiation of the Texas brown tide algal bloom. Limnol. Oceanogr. 42: 1215–1222.

Cairns, D.J. 1981. Detrital Production and Nutrient Release in a Southeastern Flood-Plain Forest. M.S. thesis, Florida State University, Tallahassee.

Campbell, K.R., C.J. Ford, and D.A. Levine. 1998. Mercury distribution in Poplar Creek, Oak Ridge, Tennessee, USA. Environ. Toxicol. Chem. 17: 1191–1198.

Camp Dresser and McKee, Inc. 1998. Site investigation report; Holtrachem Manufacturing site, Orrington, Maine. Unpublished report. Cambridge, MA.

Caraco, N., A. Tamse, O. Boutros, and I. Valiela. 1987. Nutrient limitation of phytoplankton growth in brackish coastal ponds. Can. J. Fish. Aquat. Sci. 4: 473–476.

Cardwell, R.D., S. Olsen, M.I. Carr, and E.W. Sanborn. 1979. Causes of oyster mortality in South Puget Sound. NOAA tech. mem. ERL MESA-39. Washington Department of Fisheries, Salmon Research and Development, Brinnan, WA.

Carr, W.E.S. and C.A. Adams. 1972. Food habits of juvenile marine fishes: evidence of the cleaning habit in the leatherjacket, *Oligoplites saurus*, and the spottail pinfish, *Diplodus holbrooki*. Fish. Bull. 70: 1111–1120.

Carr, W.E.S. and C.A. Adams. 1973. Food habits of juvenile marine fishes occupying seagrass beds in the estuarine zone near Crystal River, Florida. Trans. Am. Fish. Soc. 102: 511–540.

CH2M Hill. 1986. Local effects monitoring. Unpublished report. Valleho Sanitation and Flood Control District.

Chang, E.H. 1988. Distribution, abundance and size composition of phytoplankton off Westland, New Zealand, February 1982. N.Z. J. Mar. Freshwater Res. 22: 345–367.

Chang, F.J., C. Anderson, and N.C. Boustead. 1990. First record of *Heterosigma* (Raphidophyceae) bloom with associated mortality of cage-reared salmon in Big Glory Bay, New Zealand. N.Z. J. Mar. Freshwater Res. 24: 461–469.

Chanton, J. and F.G. Lewis. 1999. Plankton and dissolved inorganic carbon isotopic composition in a river-dominated estuary: Apalachicola Bay, Florida. Estuaries 22: 575–583.

Chanton, J. and F.G. Lewis. 2002. Examination of coupling between primary and secondary production in a river-dominated estuary: Apalachicola Bay, Florida, U.S.A. Limnol. Oceanogr. 47: 683–697.

Chao, L.N. and J.A. Musick. 1977. Life history, feeding habits, and functional morphology of juvenile sciaenid fishes in the York River Estuary, Virginia. Fish. Bull. 75: 657–702.

Chen, R. and R.S. Tsay. 1993. Nonlinear additive ARX models. J. Am. Stat. Assoc. 88: 955–967.

Christensen, V. and D. Pauly. 1992. A Guide to the Econpathii Software System (version 2.1). International Center for Living Aquatic Resources Management (ICLARM), Manila.

Clarke, T.A. 1978. Diel feeding patterns of 16 species of mesopelagic fishes from Hawaiian waters. Fish. Bull. 76: 495–514.

Clements, W.H. and R.J. Livingston. 1983. Overlap and pollution-induced variability in the feeding habits of filefish (Pisces: Monacanthidae) from Apalachee Bay, Florida. Copeia (1983): 331–338.

Clements, W.H. and R.J. Livingston. 1984. Prey selectivity of the fringed filefish *Monacanthus ciliatus* (Pisces: Monacanthidae): role of prey accessibility. Mar. Ecol. Prog. Ser 16: 291–295.

Clewell, A.F. 1977. Geobotany of the Apalachicola River region. In: R.J. Livingston, Ed., Proceedings of the Conference on the Apalachicola Drainage System. Marine Resources Publication 26. Florida Department of Natural Resources, St. Petersburg, pp. 6–15.

Cloern, J.E. 1979. Phytoplankton ecology of the San Francisco Bay system: the status of our current understanding. In: J.T. Conomos, Ed., San Francisco Bay: The Urbanized Estuary. Pacific Division AAAS, San Francisco, Allen Press, Lawrence, KS, pp. 247–264.

Cloern, J.E. 1996. Phytoplankton bloom dynamics in coastal ecosystems: a review with some general lessons from sustained investigations of San Francisco Bay, California. Rev. Geophys. 34: 127–168.

Cloern, J.E., A.E. Alpine, B.E. Cole, R.L.J. Wong, J.F. Arthur, and M.D. Ball. 1983. River discharge controls phytoplankton dynamics in the northern San Francisco Bay estuary. Estuarine Coastal Shelf Sci. 16: 415–429.

Coffin, R.B. and L.A. Cifuentes. 1992. A stable isotope study of carbon sources and microbial transformations in the Perdido Estuary, Fl. In: R.J. Livingston, Ed., Ecological Study of the Perdido Drainage System. Unpublished report. Florida Department of Environmental Protection.

Coffin, R.B. and L.A. Cifuentes. 1999. A stable isotope analysis of carbon cycling in the Perdido Estuary, Florida. Estuaries 22: 917–926.

Collard, S.B. 1991a. The Pensacola Bay System: Biological Trends and Current Status. Water Resources Special Report 91–3. Northwest Florida Water Management District, Havana, FL.

Collard, S.B. 1991b. Management Options for the Pensacola Bay System: The Potential Value of Seagrass Transplanting and Oyster Bed Refurbishment Programs. Water Resources Special Report 91–4. Northwest Florida Water Management District, Havana, FL.

Collette, B.B. and F.H. Talbot. 1972. Activity patterns of coral reef fishes with emphasis on diurnal-nocturnal changeover. Bull. Nat. Hist. Mus. L.A. 14: 98–125.

Cooper, S.R. 1995a. Diatoms in sediment cores from the mesohaline Chesapeake Bay, U.S.A. Diatom Res. 10: 39–89.

Cooper, S.R. 1995b. An abundant, small brackish water *Cyclotella* species in Chesapeake Bay, U.S.A. In: J.P. Kociolek and M.J. Sullivan, Eds., A Century of Diatom Research in North America: A Tribute to the Distinguished Careers of Charles W. Reimer and Ruth Patrick. Koeltz Scientific Books, Champaign, IL, pp. 133–140.

Corredor, J.E., R.W. Howarth, R.R. Twilley, and J.M. Morrell. 1999. Nitrogen cycling and anthropogenic impact in the tropical interamerican seas. Biogeochemistry 46: 163–178.

Crosby, D.O. 1981. Environmental chemistry of pentachlorophenol. J. Appl. Chem. 53: 1051–1080.

Cross, R.D. and D.L. Williams. 1981. Proceedings of the National Symposium on Freshwater Inflow to Estuaries. U.S. Fish and Wildlife Service, Office of Biological Services. FWS/OBS-81/04.

Crossland, N.O. and C.J.M. Wolff. 1985. Fate and biological effects of pentacholorophenol in outdoor ponds. Exp. Toxicol. Chem. 4: 71–86.

Danglade, E. 1917. Conditions and extent of the water level oyster beds and barren bottoms in the vicinity of Apalachicola, Florida. Appendix V, Reports of the U.S. Commissioner of Fishes for 1916, Bureau of Fishes Document 841, 75 pp.

Darnell, R.M. 1958. Food habits of fishes and larger invertebrates of Lake Pontchartrain, Louisiana, an estuarine community. Publ. Mar. Sci. (Univ. Tex.) 5: 353–416.

Darnell, R.M. 1961. Trophic spectrum of an estuarine community, based on studies of Lake Pontchartrain, Louisiana. Ecology 42: 553–568.

Darnell, R.M. 1967. Organic detritus in relation to the estuarine ecosystem. In: G.H. Lauff, Ed., Estuaries. American Society for the Advancement of Science, Washington, D.C., pp. 376–388.

Davis, W.P., M.R. Davis, and D.A. Flemer. 1999. Observations on the regrowth of subaquatic vegetation following transplantation: a potential method to assess environmental health of coastal habitats. In: S. Bortone, Ed., Seagrasses: Monitoring, Ecology, Physiology, and Management. CRC Press, Boca Raton, FL, pp. 231–238.

Dayton, P.K. 1979. Ecology: a science and a religion. In: R.J. Livingston, Ed., Ecological Processes in Coastal and Marine Systems. Plenum Press, New York, pp. 3–18.

Dayton, P.K., M.J. Tegner, P.E. Parnell, and P.B. Edwards. 1992. Temporal and spatial patterns of disturbance and recovery in a kelp forest community. Ecol. Monogr. 62: 421–445.

Deegan, L.A., J.W. Day, Jr., J.G. Gosselink, A. Yanez-Arancibia, G. Soberon Chavez, and P. Sanchez-Gil. 1986. Relationships among physical characteristics, vegetation distribution, and fisheries yield in Gulf of Mexico estuaries. In: D.A. Wolfe, Ed., Estuarine Variability. Academic Press, New York, pp. 83–100.

de Jong, V.N. 1995. The Ems Estuary, the Netherlands. In: A.J. McComb, Ed., Eutrophic Shallow Estuaries and Lagoons. CRC Press, Boca Raton, FL, pp. 81–108.

de Jong, V.N. and W. van Raaphorst. 1995. Eutrophication of the Dutch Wadden Sea (Western Europe), an estuarine area controlled by the River Rhine. In: A.J. McComb, Ed., Eutrophic Shallow Estuaries and Lagoons. CRC Press, Boca Raton, FL, pp. 129–150.

Delfino, J.J., D. Frazier, and J. Nepshinsky. 1984. Contaminants in Florida's coastal zone. Florida Sea Grant Report 62, Gainesville.

D'Elia, C.F., J.G. Sanders, and W.R. Boynton. 1986. Nutrient enrichment studies in a coastal plain estuary — phytoplankton growth in large-scale, continuous cultures. Can. J. Fish. Aquat. Sci. 43: 97–406.

Diaz, R.J., S. Luckenbach, M.H. Thornton, M. Roberts, Jr., R.J. Livingston, C.C. Koenig, G.L. Ray, and L.E. Wolfe. 1985. Field validation of multi-species laboratory test systems for estuarine benthic communities. Report CR B12053, Office of Research and Development, U.S. Environmental Protection Agency, Gulf Breeze, FL.

Dolan, D.M., A.K. Yui, and R.D. Geist. 1981. Evaluation of river load estimation methods for total phosphorus. Great Lakes Research 7: 207–214.

Dortch, Q. 1990. The interaction between ammonium and nitrate uptake in phytoplankton. Mar. Ecol. Prog. Ser. 61: 183–201.

Doudoroff, P. and D.L. Shumway. 1967. Dissolved oxygen criteria for the protection of fish. American Fisheries Society, Special Publication 4, pp. 13–19.

Drake, S. 1975. Effects of Mercuric Chloride on the Embryological Development of the Zebrafish (*Brachydanio rerio*). M.S. thesis, Florida State University, Tallahassee.

Duarte, C.M. 1995. Submerged aquatic vegetation in relation to different nutrient regimes. Ophelia 41: 87–112.

Duarte, C.M. and O. Piro. 2001. Interdisciplinary challenges and bottlenecks in the aquatic sciences. Limnol. Oceanog. Bull. 10: 57–61.

Dugan, P.J. and R.J. Livingston. 1982. Long-term variation in macroinvertebrate communities in Apalachee Bay, Florida. Estuarine Coastal Shelf Sci. 14: 391–403.

Duncan, J.L. 1977. Short-Term Effects of Storm Water Runoff on the Epibenthic Community of a North Florida Estuary (Apalachicola, Florida). M.S. thesis, Florida State University, Tallahassee.

Duxbury, A.C. 1975. Orthophosphate and dissolved oxygen in Puget Sound. Limnol. Oceanogr. 20: 270–274.

Eadie, B.J., B.A. McKee, M.B. Lansing, and S. Metz. 1994. Records of nutrient-enhanced coastal ocean productivity in sediments from the Louisiana continental shelf. Estuaries 17: 754–765.

Ebeling, A. and R.N. Bray. 1976. Day versus night activity of reef fishes in a kelp forest off Santa Barbara, California. Fish. Bull. 74: 703–717.

Edmiston, H.L. 1979. The Zooplankton of the Apalachicola Bay System. M.S. thesis, Florida State University, Tallahassee.

Elder, J.F. and D.J. Cairns. 1982. Production and decomposition of forest litter fall on the Apalachicola River floodplain, Florida. U.S. Geological Survey Water-Supply Paper 2196.

Elsner, J.B. 1992. Predicting time series using a neural network as a method to distinguish chaos from noise. J. Phys. 25: 843–850.

Elton, C. 1927. Animal Ecology. Macmillan, New York.

Elton, C. 1930. Animal Ecology and Evolution. Clarendon Press, Oxford, U.K.

Ericksson, L.O. 1978. Nocturnalism vs. diurnalism: dualism within fish individuals. In: J.E. Thorpe, Ed., Rhythmic Activity of Fishes. Academic Press, New York, pp. 69–89.

Estabrook, R.H. 1973. Phytoplankton Ecology and Hydrography of Apalachicola Bay. M.S. thesis, Florida State University, Tallahassee.

Estevez, E.D., L.K. Dixon, and M.F. Flannery. 1991. West-coastal rivers of peninsular Florida. In: R.J. Livingston, Ed., The Rivers of Florida. Springer-Verlag, New York, pp. 187–222.

Eyre, B. and P. Balls. 1999. A comparative study of nutrient behavior along the salinity gradient of tropical and temperate estuaries. Estuaries 22: 313–326.

Farmer, J.D. and J.J. Sidorowich. 1987. Predicting chaotic time series. Phys. Rev. Lett. 59: 845–848.

Federle, T.W., M.A. Hullar, R.J. Livingston, D.A. Meeter, and D.C. White. 1983a. Spatial distribution of biochemical parameters indicating biomass and community composition of microbial assemblies in estuarine mud flat sediments. Appl. Environ. Micro. 45: 58–63.

Federle, T.W., R.J. Livingston, D.A. Meeter, and D.C. White. 1983b. Modifications of estuarine sedimentary microbiota by exclusion of epibenthic predators. J. Exp. Mar. Biol. Ecol. 73: 81–94.

Federle, T.W., R.J. Livingston, L.E. Wolfe, and D.C. White. 1986. A quantitative comparison of microbial community structure of estuarine sediments for microcosms in the field. Can. J. Microbiol. 32: 319–325.

Fenchel, T.M. and B.B. Jorgensen. 1978. Detritus food chains of aquatic ecosystems: the role of bacteria. Adv. Microb. Ecol. 1: 1–57.

Fernald, E.A. and D.J. Patton. 1984. Water Resources Atlas of Florida. Institute of Science and Public Affairs, Florida State University, Tallahassee, 291 pp.

Fernald, E.A. and E.D. Purdum. 1992. Atlas of Florida. University of Florida Press, Gainesville, 280 pp.

Finlay, J.C. 2001. Stable-carbon-isotope ratios of river biota: implications for energy flow in lotic food webs. Ecology 82: 1052–1064.

Finney, B.P. and C. Huh. 1989. History of metal pollution in the Southern California Bight: an update. Environ. Sci. Technol. 23: 294–303.

Finney, B.P., I. Gregory-Eaves, J. Sweetman, M.S.V. Douglas, and J.P. Smol. 2000. Impacts of climatic change and fishing on Pacific salmon abundance over the past 300 years. Science 290: 795–799.

Fisher, T.R., L.W. Harding, Jr., D.W. Stanley, and L.G. Ward. 1988. Phytoplankton, nutrients, and turbidity in the Chesapeake, Delaware, and Hudson estuaries. Estuarine Coastal Shelf Sci. 27: 61–88.

Fisher, T.R., E.R. Peele, J.W. Ammerman, and L.W. Harding, Jr. 1992. Nutrient limitation of phytoplankton in Chesapeake Bay. Mar. Ecol. Prog. Ser. 82: 51–63.

Flemer, D.A., R.J. Livingston, and S.E. McGlynn. 1997. Phytoplankton nutrient limitation: seasonal growth responses of sub-temperate estuarine phytoplankton to nitrogen and phosphorus — an outdoor microcosm experiment. Estuaries 21: 145–159.

Flint, R.W. 1985. Long-term estuarine variability and associated biological response. Estuaries 8: 158–169.

Fourqurean, J.W., I.R.D. Jones, and J.C. Zieman. 1993. Processes influencing water column nutrient characteristics and phosphorus limitation of phytoplankton biomass in Florida Bay, FL, USA: inferences from spatial distributions. Estuarine Coastal Shelf Sci. 36: 295–314.

Fry, B. 1983. Fish and shrimp migrations in the northern Gulf of Mexico analyzed using stable C, N, and S isotope ratios. Fish. Bull. 81: 789–801.

Fryxell, G.A. and M.C. Villac. 1999. Toxic and harmful marine diatoms. In: E.F. Stoermer and J.P. Smol, Eds., The Diatoms: Applications for the Environmental and Earth Sciences. Cambridge University Press, New York, pp. 419–428.

Fuhs, G.W., S.D. Demmerle, E. Canelli, and M. Chen. 1972. Characterization of phosphorus-limited plankton algae. Limnol. Oceanogr., Spec. Symp. 1: 113–133.

Fukuyo, Y., H. Takano, M. Chihara, and K. Matsuoka. 1990. Red Tide Organisms in Japan — An Illustrated Taxonomic Guide. Uchida Rokakuho, Tokyo, 407 pp.

Fulmer, J.M. 1997. Nutrient Enrichment and Nutrient Input to Apalachicola Bay, Florida. M.S. thesis, Florida State University, Tallahassee.

Funicelli, N.A. 1984. Assessing and managing effects of reduced freshwater inflow to two Texas estuaries. In: V.S. Kennedy, Ed., The Estuary as a Filter. Academic Press. New York, pp. 435–446.

Galehouse, J.S. 1971. Sedimentary analysis. In: R.E. Carver, Ed., Procedures in Sedimentary Petrology. Wiley-Interscience, New York, pp. 69–94.

Gallagher, J.C. 1980. Population genetics of *Skeletonema costatum* (Bacillariophyceae) in Narragansett Bay. J. Phycol. 16: 464–474.

Gallegos, C.L., T.E. Jordan, and D.L. Correll. 1992. Event-scale response of phytoplankton to watershed inputs in a subestuary: timing, magnitude, and location of blooms. Limnol. Oceanogr. 37: 813–828.

Galparin, B. and G.L. Mellor. 1990a. A time-dependent, three-dimensional model of the Delaware Bay and River. Part 1: Description of the model and tidal analysis. Est. Coastal Shelf Sci. 31: 231–253.

Galparin, B. and G.L. Mellor. 1990b. A time-dependent, three-dimensional model of the Delaware Bay and River. Part 2: Three-dimensional flow fields and residual circulation. Estuarine Coastal Shelf Sci. 31: 255–281.

Galparin, B., A.L. Blumberg, and R.H. Weisberg. 1992. The importance of density driven circulation in well-mixed estuaries: the Tampa Bay experience. Estuarine Coastal Modeling, ASCE: 331–341.

Gerhart, D.Z. and G.E. Likens. 1975. Enrichment experiments for determining nutrient limitation: four methods compared. Limnol. Oceanogr. 20: 649–653.

Gihara, G. and R.M. May. 1990. Nonlinear forecasting as a way of distinguishing chaos from measurement error in time series. Nature 344: 734–741.

Gilbert, P.M., D.J. Conley, T.R. Fisher, L.W. Harding, and T.C. Malone. 1995. Dynamics of the 1990 winter/spring bloom in Chesapeake Bay. Mar. Ecol. Prog. Ser. 122: 27–43.

Gilbert, P.M., R. Magnien, M.W. Lomas, J. Alexander, C. Fan, E. Haramoto, M. Trice, and T.M. Kana. 2001. Harmful algal blooms in the Chesapeake and coastal bays of Maryland, USA: comparison of 1997, 1998, and 1999 events. Estuaries 24: 875–883.

Glass, N.R. 1971. Computer analysis of predation energetics in the largemouth bass. In: B.C. Patten, Ed., Systems Analysis and Simulation in Ecology. Academic Press, New York, Vol. I.

Graco, M., L. Farias, V. Molina, and D. Gutierrez. 2001. Massive developments of microbial mats following phytoplankton blooms in a naturally eutrophic bay: implications of nitrogen cycling. Limnol. Oceanogr. 46: 821–832.

Graham, M.H. and P.K. Dayton. 2002. On the evolution of ecological ideas: paradigns and scientific progress. Ecology 83: 1481–1489.

Graneli, E. 1987. Nutrient limitation of phytoplankton in a brackish water bay highly influenced by river discharge. Estuarine Coastal Shelf Sci. 25: 555–565.

Graneli, E. and P. Carlsson. 1997. The ecological significance of phagotrophy in photosynthetic flagellates. In: D.M. Anderson, A.D. Cembella, and G.M. Hallagraeff, Eds., Physiological Ecology of Harmful Algal Blooms. Springer-Verlag, Berlin, pp. 540–557.

Graneli, E., K. Wallstrom, U. Larsson, W. Graneli, and R. Elmgren. 1990. Nutrient limitation of primary production in the Baltic Sea area. Ambio 19: 142–151.

Greening, H.S. 1980. Seasonal and Diel Variations in the Structure of Macroinvertebrate Communities: Apalachee Bay, Florida. M.S. thesis, Florida State University, Tallahassee.

Greening, H.S. and R.J. Livingston. 1982. Diel variations in the structure of epibenthic macroinvertebrate communities of seagrass beds (Apalachee Bay, Florida). Mar. Ecol. Prog. Ser. 7: 147–156.

Greve, W. and I.R. Parsons. 1977. Photosynthesis and fish production. Hypothetical effects of climate change and pollution. Helgol. Wiss. Meeresunters. 30: 666–672.

Grossman, G.D., R. Coffin, and P.B. Moyle. 1980. Feeding ecology of the bay goby (Pisces: Gobiidae). Effects of behavioral, ontogenetic, and temporal variation in diet. J. Exp. Mar. Biol. Ecol. 44: 47–59.

Guildford, S.J. and R.E. Hecky. 2000. Total nitrogen, total phosphorus, and nutrient limitation in lakes and oceans: is there a common relationship? Limnol. Oceanogr. 45: 1213–1223.

Hackney, C.T. and E.B. Haines. 1980. Stable carbon isotope composition of fauna and organic matter collected in a Mississippi estuary. Estuarine Coastal Mar. Sci. 10: 703–708.

Hagy, J. D. 1996. Residence Times and Net Ecosystem Processes in Patuxent River Estuary. M.S. thesis, University of Maryland, College Park.

Haines, E.B. 1979. Interactions between Georgia salt marshes and coastal waters: a changing paradigm. In: R.J. Livingston, Ed., Ecological Processes in Coastal and Marine Systems. Plenum Press, New York, pp. 35–46.

Haines, E.B. and C.L. Montague. 1979. Food sources of estuarine invertebrates analyzed using $^{13}C/^{12}C$ ratios. Ecology 60: 48–56.

Hall, D.J., E.E. Werner, J.F. Gilliam, G.G. Mittelbach, D. Howard, C.G. Doner, J.A. Dickerman, and A.J. Stewart. 1979. Diel foraging behavior and prey selection in the golden shiner (*Notemigonus crysoleucas*). J. Fish. Res. Board Can. 39: 1029–1039.

Hallegreaff, G.M. 1995. Harmful algal blooms: a global overview. In: G.M. Hallegreaff, D.M. Anderson, and A.D. Cembella, Eds., Manual on Harmful Marine Microalgae, IOC Manuals and Guides 33. UNESCO, Paris, pp. 1–22.

Hallegreaff, G.M., D.M. Anderson, and A.D. Cembella, Eds. 1995. Manual on Harmful Marine Microalgae. IOC Manuals and Guides 33. UNESCO, Paris.

Hannah, R.P., A.T. Simmons, and G.A. Moshiri. 1973. Some aspects of nutrient-primary productivity relationships in a bayou estuary. J. Water Pollut. Control Fed. 45: 2508–2520.

Hansen, P.J. 1997. Phagotrophic mechanisms and prey selection in mixotrophic phytoflagellates. In: D.M. Anderson, A.D. Cembella, and G.M. Hallagraeff, Eds., Physiological Ecology of Harmful Algal Blooms. Springer-Verlag, Berlin, pp. 523–537.

Hara, Y. and M. Chihara. 1987. Morphology, ultrastructure, and taxonomy of the raphidophycean alga *Heterosigma akashiwo*. Bot. Mag. Tokyo 100: 151–163.

Hargraves, P.H. 1990. Studies on marine planktonic diatoms: V. Morphology and distribution of *Leptocylindrus minimus*. Gran. Beih. Nova Hedw. 100: 47–60.

Harmelin-Vivien, M.L. and C. Bouchon. 1976. Feeding behavior of some carnivorous fishes (Serranidae and Scorpaenidae) from Tulear (Madagascar). Mar. Biol. 37: 329–340.

Hartmann, K.J. 1993. Striped Bass, Bluefish, and Weakfish in the Chesapeake Bay: Energetics, Trophic Linkages, and Bioenergetics Model Applications. Ph.D. dissertation, University of Maryland, College Park.

Hatanaka, M. and K. Iizuka. 1962. Studies and community structure of the *Zostera* area. Trophic order in a fish group living outside of the *Zostera* area. Bull. Jpn. Soc. Sci. Fish. 28: 155–161.

Hayes, P.F. and R.W. Menzel. 1981. The reproductive cycle of early setting *Crassostrea virginica* (Gmelin) in the northern Gulf of Mexico and its implications for population recruitment. Biol. Bull. 160: 80–88.

Heard, W.H. 1977. Freshwater mollusca of the Apalachicola drainage. In: R.J. Livingston, Ed., Proceedings of the Conference on the Apalachicola Drainage System. Marine Resources Publication 26, Florida Department of Natural Resources, St. Petersburg, pp. 20–21.

Hecky, P.E. 1998. Low nitrogen:phosphorus ratios and the nitrogen fix: why watershed nitrogen removal will not improve the Baltic. In: T. Hellstrom, Ed., Effects of Nitrogen in the Aquatic Environment. Swedish National Committee for IAWQ, Royal Swedish Academy of Sciences, Stockholm.

Hecky, R.E. and P. Kilham. 1988. Nutrient limitation of phytoplankton in freshwater and marine environments. Limnol. Oceanogr. 33: 796–822.

Hein, M., M.F. Pedersen, and K. Sand-Jensen. 1995. Size-dependent nitrogen uptake in micro- and macroalgae. Mar. Ecol. Prog. Ser. 118: 247–253.

Helfman, G.S. 1978. Patterns of community structure in fishes: summary and overview. Environ. Biol. Fish. 3:129–148.

Hellstrom, T. 1996. An empirical study of nitrogen dynamics in lakes. Water Environ. Res. 68: 55–65.

Henderson-Sellers, B. and H.R. Markland. 1988. Decaying Lakes: The Origins and Control of Cultural Eutrophication. John Wiley & Sons, New York.

Hiatt, R.W. and D.W. Strasburg. 1960. Ecological relationships of fish fauna on coral reefs of the Marshall Islands. Ecol. Monogr. 30: 65–127.

Hitchcock, G.L. 1982. A comparative study of the size-dependent organic composition of marine diatoms and dinoflagellates. J. Plankton Res. 4: 363–377.

Hobson, E.S. 1965. Diurnal-nocturnal activity of some inshore fishes in the Gulf of California. Copeia (1965): 291–302.

Hobson, E.S. 1968. Predatory behavior of some shore fishes in the Gulf of California. U.S. Fish Wildlife Service Res. Rep., 73 pp.

Hobson, E.S. 1973. Diel feeding migrations in tropical reef fishes. Helgol. Wiss. Meeresunters. 24:361–370.

Hobson, E.S. 1974. Feeding relationships of teleostean fishes on coral reefs in Kona, Hawaii. Fish. Bull. 72: 915–1031.

Hobson, E.S. 1975. Feeding patterns among tropical reef fishes. Am. Sci. 63: 382–392.

Hobson, E.S. and J.R. Chess. 1976. Trophic interactions among fishes and zooplankters near shore at Santa Catalina Island, California. Fish. Bull. 74: 567–598.

Hodgkiss, I.J. and W.S. Yim. 1995. A case study of Tolo Harbour, Hong Kong. In: A.J. McComb, Ed., Eutrophic Shallow Estuaries and Lagoons. CRC Press, Boca Raton, FL, pp. 41–58.

Hoese, H.D., B.J. Copeland, F.N. Moseley, and E.D. Lane. 1968. Fauna of the Aransas Pass Inlet, Texas: diel and seasonal variations in trawlable organisms of the adjacent area. Tex. J. Sci. 20: 33–60.

Holland, A.F., N.K. Mountford, and J.A. Mihursky. 1977. Temporal variation in upper bay mesohaline benthic communities: 1. The 9-m mud habitat. Chesapeake Sci. 18: 370–378.

Holland, A.F., N.K. Mountford, M.H. Hiefel, K.R. Kaunmeyer, and J.A. Mihursky. 1980. Influence of predation on infaunal abundance in upper Chesapeake Bay, USA. Mar. Biol. 57: 221–235.

Holling, C.S. 1973. Resilience and stability of ecological systems. Annu. Rev. Ecol. Syst. 4: 1–24.

Honjo, T. 1993. Overview on bloom dynamics and physiological ecology of *Heterosigma akashiwo*. In: T.J. Smayda, and Y. Shimizu, Eds., Toxic Phytoplankton Blooms in the Sea. Proceedings of the Fifth International Conference on Toxic Marine Phytoplankton. Elsevier, New York, pp. 33–41.

Hooks, T.A., K.L. Heck, and R.J. Livingston. 1975. An inshore marine invertebrate community: structure and habitat associations in the N.E. Gulf of Mexico. Bull. Mar. Sci. 26: 99–109.

Houlihan, J. and R. Wiles. 2001. Brain food: what women should know about mercury contamination of fish. Environmental Working Group, Washington, D.C.

Howarth, R.W. 1988. Nutrient limitation of net primary production in marine ecosystems. Annu. Rev. Ecol. Syst. 19: 89–110.

Howarth, R.W. and R. Marino. 1998. A mechanistic approach to understanding why so many estuaries and brackish waters are nitrogen limited. In: T. Hellstrom, Ed., Effects of Nitrogen in the Aquatic Environment. KVA Report 1, Royal Swedish Academy of Sciences, Stockholm.

Howarth, R.W., H.S. Jensen, R. Marino, and H. Postma. 1995. Transport to and processing of phosphorus in near-shore and oceanic waters. In: H. Tiessen, Ed., Phosphorus in the Global Environment. Wiley, New York.

Howarth, R.W., D.M. Anderson, T.M. Church, H. Greening, C.S. Hopkinson, W.C. Huber, N. Marcus, R.J. Nainman, K. Segerson, A.N. Sharpley, and W.J. Wiseman. 2000. Clean Coastal Waters: Understanding and Reducing the Effects of Nutrient Pollution. Ocean Studies Board and Water Science and Technology Board, National Academy Press, Washington, D.C., 391 pp.

Hughes, T.P., A.M. Szmant, R. Steneck, R. Carpenter, and S. Miller. 1999. Algal blooms on coral reefs: what are the causes? Limnol. Oceanogr. 44: 1583–1586.

Hunter, M.D. and P.W. Price. 1992. Playing chutes and ladders: heterogeneity and the relative roles of bottom-up and top-down forces in natural communities. Ecology 73: 724–732.

Hutchinson, G.E. 1957. A Treatise on Limnology. Vol. 1, Geography, Physics and Chemistry. John Wiley & Sons, New York, 1015 pp.

Hutchinson, G.E. 1973. Eutrophication. Am. Sci. 61: 269–279.

Hydroqual, Inc. 1992. Elevenmile Creek and Perdido Bay modeling analysis. Unpublished report. Hydroqual, Inc., Mahwah, NJ.

Imai, I., M. Yamaguchi, and M. Watanabe. 1997. Ecophysiology, life cycle, and bloom dynamics of *Chattonella* in the Seto inland sea, Japan. In: D.M. Anderson, A.D. Cembella, and G.M. Hallagraeff, Eds., Physiological Ecology of Harmful Algal Blooms. Springer-Verlag, Berlin, pp. 93–112.

Isphording, W.C. and R.J. Livingston. 1989. Report on synoptic analyses of water and sediment quality in the Perdido Drainage System. Unpublished report. Florida Department of Environmental Regulation, Tallahassee.

Iverson, R.L. and H.F. Bittaker. 1986. Seagrass distribution and abundance in eastern Gulf of Mexico coastal waters. Estuarine Coastal Shelf Sci. 22: 577–602.

Iverson, R.L., W. Landing, B. Mortazawi, and J. Fulmer. 1997. Nutrient transport and primary productivity in the Apalachicola River and Bay. In: F.G. Lewis, Ed., Apalachicola River and Bay Freshwater Needs Assessment. Report to the ACF/ACT Comprehensive Study. Northwest Florida Water Management District, Havana, FL.

Ivlev, V.S. 1961. Experimental Ecology of the Feeding Fishes. Yale University Press, New Haven, CT, 302 pp.

Jackson, J.B.C, M.X. Kirby, W.H. Wolfgang, H. Berger, K.A. Bjorndal, L.W. Botsford, B.J. Bourque, R.H. Bradbury, R. Cooke, J. Erlandson, J. Estes, T.P. Hughes, S. Kidwell, C.B. Lange, H.S. Lenihan, J. M. Pandolfi, C.H. Peterson, R.S. Steneck, M.J. Tegner, and R.R. Warner. 2001. Historical overfishing and the recent collapse of coastal ecosystems. Science 293: 629–638.

Johnson, P.S. and J.NcN. Sieburth. 1979. Chrorococcoid cyanobacteria in the sea: a ubiquitous and diverse photographic biomass. Limnol. Oceanogr. 24: 928–935.

Jordan, T.E., D.L. Correll, J. Miklas, and D.E. Weller. 1991. Nutrients and chlorophyll *a* at the interface of a watershed and an estuary. Limnol. Oceanogr. 36: 251–267.

Jorgensen, B.B. and K. Richardson. 1996. Eutrophication in Coastal Marine Systems. American Geophysical Union, Washington, D.C.

Kaiser, J. 2000. Mercury report backs strict rules. Science 289: 371–372.

Kaul, L.W. and P.N. Froelich, Jr. 1984. Modeling estuarine nutrient geochemistry in a simple system. Geochim. Cosmochem. Acta 48: 1417–1433.

Keast, A. 1979. Patterns of predation in generalist feeders. In: H. Clepper, Ed., Predator-Prey Systems in Fisheries Management. Sports Fishing Institute, Washington, D.C., pp. 243–255.

Keister, J.E., E.D. Houde, and D.L. Breitburg. 2000. Effects of bottom-layer hypoxia on abundance and depth distributions of organisms in Patuxent River, Chesapeake Bay. Mar. Ecol. Prog. Ser. 205: 43–59.

Kemp, W.M. and W.R. Boynton. 1984. Spatial and temporal coupling of nutrient inputs to estuarine primary production: the role of particulate transport and decomposition. Bull. Mar. Sci. 35: 522–535.

Kemp, W.M. and W.R. Boynton. 1992. Benthic-pelagic interactions: nutrient and oxygen dynamics. In: D.E. Smith, M. Leffler, and G. Mackiernan, Eds., Oxygen Dynamics in the Chesapeake Bay. A Synthesis of Recent Research. Maryland Sea Grant College, College Park.

Kennish, M.J. 1997. Estuarine and Marine Pollution. CRC Press, Boca Raton, FL, 524 pp.

Kennish, M.J. 2001. Barnegat Bay–Little Egg Harbor, New Jersey: estuary and watershed assessment. J. Coastal Res. Spec. Iss. 32, 280 pp.

Kercher, J.R. and H.H. Shugart, Jr. 1975. Trophic structure, effective trophic position and connectivity in food webs. Am. Nat. 109: 191–206.

Kikuchi, T. 1974. Japanese contributions on consumer ecology in eelgrass (*Zostera marina*) beds with special reference to trophic relationships and resources in inshore fisheries. Aquaculture 4:15–160.

King, R.J. and B.R. Hodgson. 1995. Tuggerah Lakes system, New South Wales, Australia. 1995. In: A.J. McComb, Ed., Eutrophic Shallow Estuaries and Lagoons. CRC Press, Boca Raton, FL, pp. 19–30.

Kivi, K., S. Kaitala, H. Kuosa, J. Kuparinen, E. Leskinen, R. Lignell, B. Marcussen, and T. Tamminen. 1993. Nutrient limitation and grazing control of the Baltic plankton community during annual succession. Limnol. Oceanogr. 38: 893–905.

Klein, C.J., III and J.A. Galt. 1986. A screening model framework for estuarine assessment. In: D.A. Wolfe, Ed., Estuarine Variability. Academic Press, New York, pp. 483–501.

Koditschek, L.K. and P. Guyre. 1974. Antimicrobial-resistant coliforms in New York Bight. Mar. Pollut. Bull. 5: 71–74.

Koenig, C.C., R.J. Livingston, and C.R. Cripe. 1976. Blue crab mortality: interaction of temperature and DDT residues. Arch. Environ. Contam. Toxicol. 4: 119–128.

Komarek, J. and K. Anagnostidis. 1998. Cyanoprokaryota. 1. Chlorococcales. In: G. Fischer, Ed., Susswasser-flora von Mitteleuropa, Vol. 19. G. Fischer, Jena, 548 pp.

Konrad, P.M. 1989. Southern California Coastal Water Research Project, Annual Report.

Lam, C.W.Y. and K.C. Ho. 1989. Red tides in Tolo Harbor, Hong Kong. In: T. Okaichi, D.M. Anderson, and T. Nemoto, Eds., Red Tides: Biology. Environ. Sci. Toxicol. Elsevier, New York, pp. 49–52.

Lapointe, B.E. 1997. Nutrient thresholds for bottom-up control of macroalgal blooms on coral reefs in Jamaica and southeast Florida. Limnol. Oceanogr. 42: 1583–1586.

Lapointe, B.E. 1999. Simultaneous top-down and bottom-up forces control macroalgal blooms on coral reefs (Reply to the comment by Hughes et al.). Limnol. Oceanogr. 44: 1586–1592.

Lapointe, B.E. and P.J. Barile. 1999. Seagrass die-off in Florida Bay: an alternative interpretation. Estuaries 22: 460–470.

Lapointe, B.E., W.R. Matzie, and P.J. Barile. 2001. Biotic phase-shifts in Florida Bay and bank reef communities of the Florida Keys: linkages with historical freshwater flows and nitrogen loading from the Everglades runoff in the Florida Everglades, Florida Bay, and coral reefs of the Florida Keys: In: J.W. Porter and K.G. Porter, Eds., An Ecosystem Sourcebook. CRC Press, Boca Raton, FL.

Lassus, P. and J.P. Berthome. 1988. Status of 1987 algal blooms in IFREMER. ICES/annex III C: 5–13.

Latasa, M. and R.R. Bidagare. 1998. Comparison of phytoplankton populations of the Arabian Sea during the spring inter-monsoon and southwest monsoon of 1995 as described by HPLC-analyzed pigments. Deep-Sea Res. 11: 2133–2170.

Laughlin, R.A. 1979. Trophic Ecology and Population Distribution of the Blue Crab, *Callinectes sapidus* Rathbun in the Apalachicola Estuary (North Florida, U.S.A.). Doctoral dissertation, Florida State University, Tallahassee.

Laughlin, R.A. and R.J. Livingston. 1982. Environmental and trophic determinants of the spatial/temporal distribution of the brief squid (*Lolliguncula brevis*) in the Apalachicola Estuary (North Florida, USA). Bull. Mar. Sci. 32: 489–497.

Leber, K.M. 1983. Feeding Ecology of Decapod Crustaceans and the Influence of Vegetation on Foraging Success in a Subtropical Seagrass Meadow. Ph.D. dissertation, Florida State University, Tallahassee.

Leber, K.M. 1985. The influence of predatory decapods, refuge, and microhabitat selection on seagrass communities. Ecology 66: 1951–1964.

Legendre, L. and S. Demers. 1985. Auxiliary energy, ergoclines, and aquatic biological production. Nat. Can. 112: 5–14.

Lehman, P.W. 1992. Environmental factors associated with long-term changes in chlorophyll concentration in the Sacramento-San Joaquin delta and Suisun Bay, California. Estuaries 15: 335–348.

Leitman, H.M., J.E. Sohm, and M.A. Franklin. 1982. Wetland hydrology and tree distribution of the Apalachicola River floodplain, Florida. U.S. Geological Survey Report 82. U.S. Government Printing Office, Washington, D.C., 92 pp.

Leitman, H.M., J.E. Sohm, and M.A. Franklin. 1983. Wetland hydrology and tree distribution of the Apalachicola River flood plain, Florida. U.S. Geological Survey Water-Supply Paper 2196-A, 52 pp.

Leitman, S.F., L. Ager, and C. Mesing. 1991. The Apalachicola experience: environmental effects of physical modifications to a river for navigational purposes. In: R.J. Livingston, Ed., The Rivers of Florida. Springer-Verlag, New York, pp. 223–246.

Levene, H. 1960. Robust tests for equality of variances. In: I. Olkin, Ed., Contributions to Probability and Statistics. Stanford University Press, Palo Alto, CA, pp. 278–292.

Levin, L.A. 1984. Life history and dispersal patterns in a dense infaunal polychaete assemblage: community structure and response to disturbance. Ecology 65: 1185–1200.

Lewis, F.G., III and R.J. Livingston. 1977. Avoidance of bleached kraft pulp mill effluent by pinfish (*Lagodon rhomboides*) and gulf killifish (*Fundulus grandis*). J. Fish. Res. Board Can. 34: 568–570.

Lewis, F.G., III and A.W. Stoner. 1983. Distribution of macro-fauna within seagrass beds: an explanation for patterns of abundance, Bull. Mar. Sci. 33: 296–304.

Lindeman, R.L. 1942. The trophic-dynamic aspect of ecology. Ecology 23: 399–418.

Lirdwitayaprasit, T., S. Nishio, S. Montani, and T. Okaichi. 1990. The biochemical processes during cyst formation in *Alexandrium catenella*. In: L.E. Graneli, B. Sundstorm, L. Edler, and D.M. Anderson, Eds., Toxic Marine Phytoplankton. Elsevier, Amsterdam, pp. 294–299.

Livingston, R.J. 1975a. Impact of kraft pulp-mill effluents on estuarine and coastal fishes in Apalachee Bay, Florida, USA. Mar. Biol. 32:19–48.

Livingston, R.J. 1975b. Resource management and estuarine function with application to the Apalachicola drainage system. Estuarine Pollut. Control Assess. 1: 3–17.

Livingston, R.J. 1975c. Long-term fluctuations of epibenthic fish and invertebrate populations in Apalachicola Bay, Florida. U.S. Fish Bull. 74: 311–321.

Livingston, R.J. 1976a. Diurnal and seasonal fluctuations of organisms in a north Florida estuary. Estuarine Coastal Mar. Sci. 4: 373–400.

Livingston, R.J. 1976b. Environmental considerations and the management of barrier islands: St. George Island and the Apalachicola Bay system. In: Barrier Islands and Beaches. Technical Proceedings of the 1976 Barrier Islands Workshop. Annapolis, MD, pp. 86–102.

Livingston, R.J. 1977. The Apalachicola dilemma: wetlands development and management initiatives. In: National Wetlands Protection Symposium. Environmental Law Institute and the Fish and Wildlife Service, Washington, D.C., pp. 163–177.

Livingston, R.J. 1979. Multiple factor interactions and stress in coastal systems: a review of experimental approaches and field implications. In: F.J. Vernberg, Ed., Marine Pollution: Functional Responses. Academic Press, New York, pp. 389–413.

Livingston, R.J. 1980a. Ontogenetic trophic relationships and stress in a coastal sea grass system in Florida. In: V.S. Peterson, Ed., Estuarine Perspectives. Academic Press, New York, pp. 423–435.

Livingston, R.J. 1980b. The Apalachicola experiment: research and management. Oceanus 23: 14–28.

Livingston, R.J. 1981a. Man's impact on the sciaenid fishes. Sixth Annual Marine Recreational Fisheries Symposium, Sciaenids: Territorial Dermersal Resources. National Marine Fisheries Service, Houston, TX, pp. 189–196.

Livingston, R.J. 1981b. Man's impact on the distribution and abundance of sciaenid fishes. Sixth Annual Marine Recreational Fisheries Symposium, Sciaenids: Territorial Demersal Resources. National Marine Fisheries Service, Houston, TX, pp. 189–196.

Livingston, R.J. 1982a. Trophic organization in a coastal seagrass system. Mar. Ecol. Prog. Ser. 7: 1–12.

Livingston, R.J. 1982b. Between the idea and the reality: an essay on the problems involved in applying scientific data to research management problems. In: A. Donovan and A.L. Berge, Eds., Working Papers in Science and Technology Studies. Virginia Polytechnic Institution, Blacksburg, pp. 31–59.

Livingston, R.J. 1983a. Resource atlas of the Apalachicola estuary. Florida Sea Grant College Publication, Gainesville.

Livingston, R.J. 1983b. Identification and Analysis of Sources of Pollution in the Apalachicola River and Bay System. Final Report, Florida Department of Natural Resources, Tallahassee.

Livingston, R.J. 1984a. The ecology of the Apalachicola Bay system: an estuarine profile. U.S. Fish and Wildlife Service FWS/PBS 82/05, Gainesville, FL, 148 pp.

Livingston, R.J. 1984b. Trophic response of fishes to habitat variability in coastal seagrass systems. Ecology 65: 1258–1275.

Livingston, R.J. 1984c. River-derived input of detritus into the Apalachicola estuary. In: R.D. Cross and D.L. Williams, Eds., Proceedings of the National Symposium on Freshwater Inflow to Estuaries. Fish and Wildlife Service, Washington, D.C., pp. 320–332.

Livingston, R.J. 1984d. Long-Term Effects of Dredging and Open-Water Disposal on the Apalachicola Bay System. Final Report. National Oceanic and Atmospheric Administration, Washington, D.C.

Livingston, R.J. 1985a. The relationship of physical factors and biological response in coastal seagrass meadows. Proceedings of a Seagrass Symposium, Estuarine Research Foundation. Estuaries 7: 377–390.

Livingston, R.J. 1985b. Aquatic field monitoring and meaningful measures of stress. In: H.H. White, Ed., Concepts in Marine Pollution Measurements. Washington, D.C., pp. 681–692.

Livingston, R.J. 1985c. Application of scientific research to resource management: case history, the Apalachicola Bay system. In: N.L. Chao and W. Kirby-Smith, Eds., Proceedings of the International Symposium on Utilization of Coastal Ecosystems: Planning, Pollution, and Productivity. Fundacao Universidad do Rio Grande, Rio Grande, Brazil, pp. 103–125.

Livingston, R.J. 1986a. The Choctawhatchee River-Bay System. Final Report. Northwest Florida Water Management District, Havana, FL.

Livingston, R.J. 1986b. Analysis of Field Data Concerning Old Pass Lagoon (Choctawhatchee Bay, Florida: September, 1985–February, 1986). Final Report. Northwest Florida Water Management District, Havana, FL.

Livingston, R.J. 1987a. Historic trends of human impacts on seagrass meadows in Florida (invited paper). Symposium Proceedings: Subtropical-Tropical Seagrasses of the Southeastern U.S. Fla. Mar. Res. Publ. 42: 139–151.

Livingston, R.J. 1987b. Field sampling in estuaries: the relationship of scale to variability. Estuaries 10: 194–207.

Livingston, R.J. 1987c. Distribution of Toxic Agents and Biological Response of Infaunal Macroinvertebrates in the Choctawhatchee Bay System. Final Report. Office of Coastal Management, Florida Department of Environmental Regulation, Tallahassee.

Livingston, R.J. 1988a. Inadequacy of species-level designations for ecological studies of coastal migratory fishes. Environ. Biol. Fish. 22: 225–234.

Livingston, R.J. 1988b. Projected changes in estuarine conditions based on models of long-term atmospheric alteration. Report CR-814608-01-0, U.S. Environmental Protection Agency, Washington, D.C.

Livingston, R.J. 1989. The Ecology of the Choctawhatchee River System. Final Report. Northwest Florida Water Management District, Havana, FL.

Livingston, R.J. 1990. Inshore marine habitats. In: R.L. Myers and J.J. Ewel, Eds., Ecosystems of Florida. University of Central Florida Press, Orlando, pp. 549–573.

Livingston, R.J. 1991a. The Rivers of Florida. Springer-Verlag, New York.

Livingston, R.J. 1991b. Medium sized rivers: Gulf coastal plain. In: C.T. Hackney, Ed., Biodiversity of the Southeastern U.S. Ecological Society of America. John Wiley & Sons, New York, pp. 351–385.

Livingston, R.J. 1991c. Historical relationships between research and resource management in the Apalachicola River-estuary. Ecol. Appl. 1: 361–382.

Livingston, R.J. 1992. Ecological study of the Perdido Drainage System. Florida Department of Environmental Regulation, Tallahassee, Florida.

Livingston, R.J. 1993a. River-Gulf Study. Unpublished report. Florida Department of Environmental Regulation, Tallahassee.

Livingston, R.J. 1993b. Estuarine wetlands. In: J. Berry and M. Dennison, Eds., Wetlands. Noyes, Park Ridge, NJ, pp. 128–153.

Livingston, R.J. 1994. Ecological Study of Jack's Branch and Soldier Creek. Unpublished report. Florida Department of Environmental Regulation, Tallahassee.

Livingston, R.J. 1995. Nutrient analysis of upper Perdido Bay. Unpublished report. Florida Department of Environmental Regulation, Tallahassee.

Livingston, R.J. 1997a. Eutrophication in estuaries and coastal systems: relationships of physical alterations, salinity stratification, and hypoxia. In: F.J. Vernberg, W.B. Vernberg, and T. Siewicki, Eds., Sustainable Development in the Southeastern Coastal Zone. University of South Carolina Press, Columbus, pp. 285–318.

Livingston, R.J. 1997b. Trophic response of estuarine fishes to long-term changes of river runoff. Bull. Mar. Sci. 60: 984–1004.

Livingston, R.J. 1997c. Update on the ecological status of Perdido Bay. Unpublished report. Florida Department of Environmental Regulation, Tallahassee.

Livingston, R.J. 1997d. Final analyses of Perdido data base: the Soldier Creek System. Florida. Unpublished report. Florida Department of Environmental Regulation, Tallahassee.

Livingston, R.J. 1998. Perdido Bay Analysis: 10/88–9/98. Unpublished report. Florida Department of Environmental Regulation, Tallahassee.

Livingston, R.J. 1999. Pensacola Bay System Environmental Study: Ecology and Trophic Organization. Unpublished report. Florida Department of Environmental Regulation, Tallahassee.

Livingston, R.J. 2000. Eutrophication Processes in Coastal Systems: Origin and Succession of Plankton Blooms and Effects on Secondary Production. CRC Press, Boca Raton, FL.

Livingston, R.J. 2001. Mercury distribution in sediments and mussels in the Penobscot River-estuary. Unpublished report. Natural Resources Defense Council, New York.

Livingston, R.J. and J.L. Duncan. 1979. Climatological control of a north Florida coastal system and impact due to upland forestry management. In: R.J. Livingston, Ed., Ecological Processes in Coastal and Marine Systems. Plenum Press, New York, pp. 339–382.

Livingston, R.J. and E.A. Joyce, Jr. 1977. Proceedings of the Conference on the Apalachicola Drainage System, 23–24 April 1976, Gainesville, Florida. Florida Marine Research Publications 26, Florida Department of Natural Resources, St. Petersburg.

Livingston, R.J. and O.L Loucks. 1978. Productivity, trophic interactions, and food-web relationships in wetlands and associated systems. In: P.E. Greeson, J.R. Clark, and J.E. Clark, Eds., Wetland Functions and Values: The State of Our Understanding. American Water Resources Association, Lake Buena Vista, FL, pp. 101–119.

Livingston, R.J. and G.L. Ray. 1989. A simplified and rapid method for assessing the biological disturbances resulting from stormwater and marina discharges in estuaries. Final report. Florida Institute of Government and Florida Department of Environmental Regulation, Tallahassee.

Livingston, R.J., R.L. Iverson, R.H. Estabrook, V.E. Keys, and J. Taylor, Jr. 1974. Major features of the Apalachicola Bay system: physiography, biota, and resource management. Fla. Sci. 4: 245–271.

Livingston, R.J., G.J. Kobylinski, F.G. Lewis III, and P.F. Sheridan. 1976a. Long-term fluctuations of epibenthic fish and invertebrate populations in Apalachicola Bay, Florida. Fish. Bull. 74: 311–321.

Livingston, R.J., R.S. Lloyd, and M.S. Zimmerman. 1976b. Determination of adequate sample size for collections of benthic macrophytes in polluted and unpolluted coastal areas. Bull. Mar. Sci. 26: 569–575.

Livingston, R.J., P.S. Sheridan, B.G. McLane, F.G. Lewis III, and G.G. Kobylinski. 1977. The biota of the Apalachicola bay system: functional relationships. In: R.J. Livingston, Ed., Proceedings of the Conference on the Apalachicola Drainage System. Marine Resources Publication 26, Florida Department of Natural Resources, St. Petersburg, pp. 75–100.

Livingston, R.J., N.P. Thompson, and D.A. Meeter. 1978. Long-term variation of organochlorine residues and assemblages of epibenthic organisms in a shallow north Florida (USA) estuary. Mar. Biol. 46: 355–372.

Livingston, R.J., R.J. Diaz, and D.C. White. 1985a. Field validation of laboratory-derived multispecies aquatic test systems. Report EPA/600/SA-85/039, Gulf Breeze, FL.

Livingston, R.J., L.E. Wolfe, C.C. Koenig, and G.L. Ray. 1985b. Results of laboratory-field response to PCP. Unpublished report.

Livingston, R.J., X. Niu, F.G. Lewis, and G.C. Woodsum. 1997. Freshwater input to a Gulf estuary: long-term control of trophic organization. Ecol. Appl. 7: 277–299.

Livingston, R.J., A.W. Niedoroda, T.W. Gallagher, and A. Thurman. 1998a. Environmental Studies of Perdido Bay. Unpublished report. Florida Department of Environmental Regulation, Tallahassee.

Livingston, R.J., S.E. McGlynn, and X. Niu. 1998b. Factors controlling seagrass growth in a Gulf coastal system: water and sediment quality and light. Aquat. Bot. 60: 135–159.

Livingston, R.J., R.L. Howell, X. Niu, F.G. Lewis, and G.C. Woodsum. 1999. Recovery of oyster reefs (*Crassostrea virginica*) in a Gulf estuary following disturbance by two hurricanes. Bull. Mar. Sci. Gulf Caribbean 64: 75–94.

Livingston, R.J., F.G. Lewis III, G.C. Woodsum, X. Niu, R.L. Howell IV, G.L. Ray, J.D. Christensen, M.E. Monaco, T.A. Battista, C.J. Klein, B. Galperin, and W. Huang. 2000. Coupling of physical and biological models: response of oyster population dynamics to fresh water input in a shallow Gulf estuary. Estuarine Coastal Shelf Sci. 50: 655–672.

Ljung, R. and J. Box. 1978. On a measure of lack of fit in time series models. Biometrika 65: 297–303.

Locarnini, S.J.P. and B.J. Presley. 1996. Mercury concentrations in benthic organisms from a contaminated estuary. Mar. Environ. Res. 41: 225–239.

Long, E.R., G.M. Sloane, R.S. Carr, K.J. Scott, G.B. Thursby, E. Crecelius, C. Peven, and H.L. Windom. 1997. Magnitude and extent of sediment toxicity in four bays of the Florida Panhandle: Pensacola, Choctawhatchee, St. Andrew, and Apalachicola. NOAA Technical Memorandum NOS ORCA 117, Silver Spring, MD.

Lovegrove, T. 1960. An improved form of sedimentation apparatus for use with an inverted microscope. J. Conserv. Chem. 25: 279–284.

Lucas, L.V., J.R. Koseff, J.E. Cloern, S.G. Monismith, and J.K. Thompson. 1999a. Processes governing phytoplankton blooms in estuaries. I: The local production-loss balance. Mar. Ecol. Prog. Ser. 187: 1–15.

Lucas, L.V., J.R. Koseff, S.G. Monismith, J.E. Cloern, and J.K. Thompson. 1999b. Processes governing phytoplantion blooms in estuaries. II: The role of horizontal transport. Mar. Ecol. Prog. Ser. 187: 17–30.

Macauley, J.M., V.D. Engle, J.K. Summers, J.R. Clark, and D.A. Flemer. 1995. An assessment of water quality and primary productivity in Perdido Bay, a northern Gulf of Mexico estuary. Environ. Monitoxicol. Assess. 78:1–15.

MacPherson, E. 1981. Resource partitioning in a Mediterranean demersal fish community. Mar. Ecol. Prog. Ser. 4: 183–193.

Maestrini, S.Y., D.J. Bonin, and M.R. Droop. 1984a. Phytoplankton as indicators of seawater quality: bioassay approach and protocols. In: L.E. Shubert, Ed., Algae as Ecological Indicators. Academic Press, New York, pp. 71–131.

Maestrini, S.Y., M.R. Droop, and D.J. Bonin. 1984b. Test algae as indicators of seawater quality: prospects. In: L.E. Shubert, Ed., Algae as Ecological Indicators. Academic Press, New York, pp. 132–188.

Mahoney, B.M.S. 1982. Seasonal Fluctuations of Benthic Macrofauna in the Apalachicola Estuary, Florida. The Role of Predation and Larval Availability. Ph.D. dissertation, Florida State University, Tallahassee.

Mahoney, B.M.S. and R.J. Livingston. 1982. Seasonal fluctuations of benthic macrofauna in the Apalachicola estuary, Florida, USA. Mar. Biol. 69: 207–213.

Mahoney, J.B., F.H. Midlige, and D.G. Deuel. 1973. A fin rot disease of marine and euryhaline fishes in the New York Bight. Trans. Am. Fish. Soc. 102: 596–605.

Main, K.L. 1983. Behavioral Response of Acaridean Shrimp to a Predatory Fish. Ph.D. dissertation, Department of Biological Science, Florida State University, Tallahassee.

Maine Department of Environmental Protection. 1998. Mercury in Maine. Land and Water Resources Council, 1997 Annual Report, Appendix A.

Mallin, M.A., H.W. Paerl, J. Rudek, and P.W. Bates. 1993. Regulation of estuarine primary production by watershed rainfall and river flow. Mar. Ecol. Prog. Ser. 93: 199–203.

Malone, T.C. 1977. Environmental regulation of phytoplankton productivity in the lower Hudson estuary. Estuarine Coastal Mar. Sci. 13: 157–172.

Malone, T.C., D.J. Conley, T.R. Fisher, P.M. Gilbert, L.W. Harding, and K.G. Sellner. 1996. Scales of nutrient-limited phytoplankton productivity in Chesapeake Bay. Estuaries 19: 371–385.

Marcomini, A., A. Sfriso, B. Pavoni, and A.A. Orio. 1995. Eutrophication of the Lagoon of Venice: nutrient loads and exchanges. In: A.J. McComb, Ed., Eutrophic Shallow Estuaries and Lagoons. CRC Press, Boca Raton, FL, pp. 59–80.

Marino, D., M. Montresor, and A. Zingone. 1987. *Miraltia throndsenii* gen. et sp. nov. a planktonic diatom from the Gulf of Naples. Diatom Res. 2: 205–211.

Marino, D., G. Giuffre, M. Montresor, and A. Zingone. 1992. An electron microscope investigation on *Chaetoceros minimus* (Levander) Comb. nov. and new observations on *Chaetoceros throndsenii* (Marineo, Montresor & Zingone) comb. nov. Diatom Res. 6: 317–326.

Marsh, G.P. 1964. Man and nature: physical geography as modified by human actions. Belknap Press, Cambridge, MA.

Marshall, H.G. 1982a. Meso-scale distribution patterns for diatoms over the northeastern continental shelf of the United States. 7th Diatom-Symposium, pp. 393–400.

Marshall, H.G. 1982b. Phytoplankton distribution along the eastern coast of the USA. Part IV. Shelf waters between Cape Lookout, North Carolina, and Cape Canaveral, Florida. Proc. Biol. Soc. Washington 95: 99–113.

Marshall, H.G. 1984. Phytoplankton distribution along the eastern coast of the USA. Part V. Seasonal density and cell volume patterns for the northeastern continental shelf. J. Plankton Res. 6: 169–193.

Marshall, H.G. 1988. Distribution and concentration patterns of ubiquitous diatoms for the northeastern continental shelf of the United States. Proc. 9th Int. Diatom Symp. 1986. Otto Koeltz, Stuttgart, pp. 75–85.

Marshall, H.G. and J.A. Ranasinghe. 1989. Phytoplankton distribution along the eastern coast of the U.S.A. Part VII. Mean cell concentrations and standing crop. Cont. Shelf Res. 9: 153–164.

Marshall, N.B. 1954. Aspects of Deep Sea Biology. Hutchinson, London, 380 pp.

Mattraw, H.C. and J.F. Elder. 1982. Nutrient and detritus transport in the Apalachicola River, Florida. U.S. Geological Survey Water-Supply Paper 2196-C.

McComb, A.J. 1995. Eutrophic Shallow Estuaries and Lagoons. CRC Press, Boca Raton, FL, 240 pp.

McComb, A.J. and R.J. Lukatelich. 1995. The Peel-Harvey estuarine system, Western Australia. In: A.J. McComb, Ed., Eutrophic Shallow Estuaries and Lagoons. CRC Press, Boca Raton, FL, pp. 5–18.

McEachran, J.D., D.F. Boesch, and J.A. Musick. 1976. Food division within two sympatric species-pairs of skates (Pisces: Rajidae). Mar. Biol. 35: 301–317.

McLane, B.G. 1980. An Investigation of the Infauna of East Bay-Apalachicola Bay. M.S. thesis, Florida State University, Tallahassee.

McPherson, B.F. and K.M. Hammett. 1991. Tidal rivers of Florida. In: R.J. Livingston, Ed., The Rivers of Florida. Springer-Verlag, New York, pp. 31–46.

McQueen, D.J., M.R.S. Johannes, J.R. Post, T.J. Stewart, and D.R.S. Lean. 1989. Bottom-up and top-down impacts on freshwater pelagic community structure. Ecol. Monogr. 59: 289–309.

Means, D.B. 1977. Aspects of the significance to terrestrial vertebrates of the Apalachicola River drainage basin, Florida. In: R.J. Livingston, Ed.. Proceedings of the Conference on the Apalachicola Drainage System. Marine Resources Publication 26, Florida Department of Natural Resources, St. Petersburg, pp. 37–67.

Meeter, D.A. and R.J. Livingston. 1978. Statistical methods applied to a four-year multivariate study of a Florida estuarine system. In: J. Cairns, Jr., K. Dickson, and R.J. Livingston, Eds., Biological Data in Water Pollution Assessment: Quantitative and Statistical Analyses. American Society for Testing and Materials, Special Technical Publication 652, pp. 53–67.

Meeter, D.A., R.J. Livingston, and G.C. Woodsum. 1979. Short and long-term hydrological cycles of the Apalachicola drainage system with application to Gulf coastal populations. In: R.J. Livingston, Ed., Ecological Processes in Coastal and Marine Systems. Plenum Press, New York, pp. 315–338.

Meinesz, A. 1999. Killer Algae. University of Chicago Press, Chicago, 360 pp.

Mellor, G.L. and R. Yamada. 1982. Development of a turbulence closure model for geophysical fluid problems. Rev. Geophys. Space Phys. 20: 851–875.

Menge, B.A. 1992. Community regulation: under what conditions are bottom-up factors important on rocky shores. Ecology 73: 755–765.

Menzel, R.W. 1955a. Effects of two parasites on the growth of oysters. Proc. Natl. Shellfish Assoc. 45: 184–186.

Menzel, R.W. 1955b. The growth of oysters parasitized by the fungus *Dermocystidium marinum* and by the trematode *Bucephalus cuculus*. J. Parasitol. 41: 333–342.

Menzel, R.W. and F.E. Nichy. 1958. Studies of the distribution and feeding habits of some oyster predators in Alligator Harbor, Florida. Bull. Mar. Sci. Gulf Caribb. 8: 125–145.

Menzel, R.W., N.C. Hulings, and R.R. Hathaway. 1957. Causes of depletion of oysters in St. Vincent Bar, Apalachicola Bay, Florida. Proc. Natl. Shellfish Assoc. 48: 66–71.

Menzel, R.W., N.C. Hulings, and R.R. Hathaway. 1966. Oyster abundance in Apalachicola Bay, Florida in relation to biotic associations influenced by salinity and other factors. Gulf Res. Rep. 2: 73–96.

Menzie-Cura and Associates, Inc. 2001. Evaluation of the ecological health of the Lower Penobscot River. Report for Camp Dresser and McKee, Chelmford, MA.

Merrett, N.R. and H.J.S. Roe. 1974. Patterns and selectivity in the feeding of certain mesopelagic fishes. Mar. Biol. 28:115–126.

Meyer, M.S., D.C. Evers, J.J. Hartigan, and P.S. Rasmussen. 1998. Patterns of common loon (*Gavia immer*) mercury exposure, reproduction, and survival in Wisconsin, USA. Environ. Toxicol. Chem. 17: 184–190.

Miller, R. 1979. Relationships between habitat and feeding mechanisms in fishes. In: H. Clepper, Ed., Predator-Prey Systems in Fisheries Management. Sports Fishery Institute, Washington, D.C.

Mittlebach, G.G. 1988. Competition among refuging surf-fishes and effects of fish density on littoral zone invertebrates. Ecology 69: 614–623.

Mittlebach, G.G., C.W. Osenberg, and M.A. Leibold. 1988. Trophic relations and ontogenetic niche shifts in aquatic ecosystems. In: B. Ebenman and L. Persson, Eds., Size-structured populations. Springer-Verlag, Berlin, pp. 219–235.

Monaco, M.E. 1995. Comparative Analysis of Estuarine Biophysical Characteristics and Trophic Structure: Defining Ecosystem Function to Fishes. Ph.D. dissertation, University of Maryland, College Park.

Monaco, M.E. and R.E. Ulanowicz. 1997. Comparative ecosystem trophic structure of three U.S. mid-Atlantic estuaries. Mar. Ecol. Prog. Ser. 161: 239–254.

Montagna, P.A. and R.D. Kalke. 1992. The effect of freshwater inflow on meiofaunal and macrofaunal populations in the Guadalupe and Nuesces estuaries, Texas. Estuaries 15: 307–326.

Montgomery, R.T., B.F. McPherson, and E.T. Emmons. 1991. Effects of nitrogen and phosphorus additions on phytoplankton productivity and chlorophyll *a* in a subtropical estuary, Charlotte Harbor, Florida. Water-Resources Investigations Report, 91-4077, U.S. Geological Survey, Denver, CO.

Morrison, S.J., J.D. King, R.J. Bobbie, R.E. Bechtold, and D.C. White. 1977. Evidence of microfloral succession on allochthonous plant litter in Apalachicola Bay, Florida, USA. Mar. Biol. 41: 229–240.

Moshiri, G.A. 1976. Interrelationships between Certain Microorganisms and Some Aspects of Sediment-Water Nutrient Exchange in Two Bayou Estuaries. Phases I and II. University of Florida Water Resources Research Center Res. Ctr. Publ. 37, 45 pp.

Moshiri, G.A. 1978. Certain mechanisms affecting water column-to-sediment phosphate exchange in a bayou estuary. J. Water Pollut. Control Fed. 50: 392–394.

Moshiri, G.A. 1981. Study of Selected Water Quality Parameters in Bayou Texar. Final Report on Contract DACW01—80-0252, University of West Florida, Pensacola.

Moshiri, G.A. and W.G. Crumpton. 1978. Some aspects of redox trends in the bottom muds of a mesotrophic bayou estuary. Hydrobiologia 57:155–158.

Moshiri, G.A., W.G. Crumpton, and D.A. Blaylock. 1978. Algal metabolites and fish kills in a bayou estuary: an alternative explanation to the low dissolved oxygen controversy. J. Water Pollut. Control Fed. 50: 2043–2046.

Moshiri, G.A., W.G. Crumpton, and N.G. Aumen. 1979. Dissolved glucose in a bayou estuary, possible sources and utilization by bacteria. Hydrobiologia 62: 71–74.

Moshiri, G.A., N.G. Aumen, and W.G. Swann III. 1980. Water Quality Studies in Santa Rosa Sound, Pensacola, Florida. U.S. Environmental Protection Agency, Gulf Breeze Environmental Research Lab, Pensacola.

Moshiri, G.H., N.G. Aumen, and W.G. Crumpton. 1987. Reversal of the eutrophication process: a case study. In: B.G. Neilson and L.E. Cronin, Eds., Estuaries and Nutrients. Humana Press, Clifton, NJ, pp. 370–390.

Muessig, P.H. 1974. Acute Toxicity of Mercuric Chloride and the Accumulation and Distribution of Mercury Chloride and Methyl Mercury in Channel Catfish (*Ictalurus punctatus*). M.S. thesis, Florida State University, Tallahassee.

Muller, K. 1970. Phasenwechsel der lokomotorischen Aktivität bei der Quappe *Lota lota* L. Oikos Suppl. 13:122–129.

Muller, K. 1978. The flexibility of the circadian system of fish at different latitudes. In: J.E. Thorpe, Ed., Rhythmic Activity of Fishes. Academic Press, New York, pp. 91–104.

Murdoch, W.W. 1966. Community structure, population control and competition — a critique. Am. Nat. 100: 219–226.

Myers, V.B. 1977. Nutrient Limitation of Phytoplankton Productivity in North Florida Coastal Systems: Technical Considerations; Spatial Patterns; and Wind Mixing Effects. Ph.D. dissertation, Florida State University, Tallahassee.

Myers, V.B. and R.J. Iverson. 1977. Aspects of nutrient limitation of phytoplankton productivity in the Apalachicola Bay system. In: R.J. Livingston, Ed., Proceedings of the Conference on the Apalachicola Drainage System. Marine Resources Publication 26, Florida Department of Natural Resources, St. Petersburg, pp. 68–74.

Myers, V.B. and R.J. Iverson. 1981. Phosphorus and nitrogen limited phytoplankton productivity in Northeastern Gulf of Mexico coastal estuaries. In: B.J. Nielson and L.E. Cronin, Eds., Estuaries and Nutrients. Humana Press, Clifton, NJ, pp. 569–582.

Nakazima, M. 1965. Studies on the source of shellfish poison in Lake Hamana. III. Poisonous effects on shellfishes feeding on *Prorocentrum* sp. Bull. Jpn. Soc. Sci. Fish. 31: 281–285.

National Oceanic and Atmospheric Administration. 1978. MESA New York Bight Project Annual Report for fiscal year 1977. Boulder, CO.

National Research Council. 1999. From Monsoons to Microbes: Understanding the Ocean's Role in Human Health. National Academy Press, Washington, D.C.

Neutel, A., J.A.P. Heesterbeek, and P.C. De Rulter. 2002. Stability in real food webs: weak links in long loops. Science 296: 1120–1123.

Nichols, F.H. 1985. Increased benthic grazing: an alternative explanation for low phytoplankton biomass in northern San Francisco Bay during the 1976–1977 drought. Estuarine Coastal Shelf Sci. 21: 379–388.

Niedoroda, A.W. 1992. Hydrography and oceanography of Perdido Bay. In: R.J. Livingston, Ed., Ecological Study of the Perdido Drainage System. Unpublished report. Florida Department of Environmental Protection, Tallahassee.

Niedoroda, A.W. 1999. Pensacola Bay System Environmental Study; Physical Processes and Oceanography. Unpublished report. Florida Department of Environmental Regulation, Tallahassee.

Niu, X.-F. 1995. Asymptotic properties of maximum likelihood estimates in a class of space-time models, J. Multivariate Anal. 55: 82–104.

Niu, X.-F., H.L. Edmiston, and G.O. Bailey. 1998. Time series models for salinity and other environmental factors in the Apalachicola National Estuarine Research Reserve, Estuarine Coastal Shelf Sci. 46: 549–563.

Nixon, S.W. 1980. Between coastal marshes and coastal waters — a review of twenty years of speculation and research on the role of salt marshes in estuarine productivity and water chemistry. In: P. Hamilton and K.B. MacDonald, Eds., Estuarine and Wetland Processes. Plenum Press, New York, pp. 438–525.

Nixon, S.W. 1981a. Freshwater inputs and estuarine productivity. In: R.D. Cross and D.L. Williams, Eds., Proceedings of the National Symposium on Freshwater Inflow to Estuaries. U.S. Fish and Wildlife Service, Office of Biological Services, FWS/OBS-81/04, pp. 31–57.

Nixon, S.W. 1981b. Remineralization and nutrient cycling in coastal marine ecosystems. In: B.G. Neilson and L.E. Cronin, Eds., Estuaries and Nutrients. Humana Press, Clifton, NJ, pp. 111–138.

Nixon, S.W. 1988a. Comparative ecology of freshwater and marine systems. Limnol. Oceanogr. 33: 1–1025.

Nixon, S.W. 1988b. Physical energy inputs and the comparative ecology of lake and marine ecosystems. Limnol. Oceanogr. 33: 1005–1025.

Nixon, S.W. 1995. Coastal marine eutrophication: a definition, social causes, and future concerns. Ophelia 41:199–219.

Nixon, S.W., M.E.Q. Pilson, C.A. Oviatt, P. Donaghy, B. Sullivan, S. Seitzinger, D. Rudnick, and J. Frithsen. 1984. Eutrophication of a coastal marine ecosystem — an experimental study using the MERL microcosms. In: M.J.-R. Fasham, Ed., Flows of Energy and Materials in Marine Ecosystems: Theory and Practice. Plenum Press, New York, pp. 105–135.

Norin, L.L. 1977. Bioassays with the natural phytoplankton in the Stockholm Archipelago. Ambio Spec. Rep. 5, pp. 15–21.

O'Connors, H.B. and I.W. Duedall. 1975. The seasonal variation in sources, concentrations, and impacts of ammonium in the New York Bight apex. ACS Symposium series 18. In: Marine Chemistry in the Coastal Environment, pp. 636–663.

Odum, E.P. 1953. Fundamentals of Ecology. W.B. Saunders, Philadelphia.

Odum, W.E. and E.J. Heald. 1972. Trophic analyses of an estuarine mangrove community. Bull. Mar. Sci. 22: 671–738.

Odum, W.E., J.S. Fishes, and J.C. Pickral. 1979. Factors controlling the flux of particulate organic carbon from estuarine wetlands. In: R.J. Livingston, Ed., Ecological Processes in Coastal and Marine Systems. Plenum Press, New York, pp. 69–82.

Odum, W.E., C.C. McIvor, and T.G. Smith. 1982. The Florida mangrove zone: a community profile. FWS/OBS-82/24, U.S. Fish and Wildlife Service, Office of Biological Services, Washington, D.C., 144 pp.

Ogren, L.H. and H.A. Brusher. 1977. The distribution and abundance of fishes caught with a trawl in the St. Andrews Bay System, Florida. Northeast Gulf Sci. 1:83–105.

Okaichi, T. 1997. Red tides in the Seto Inland Sea. In: T. Okaichi and T. Yanagi, Eds., Sustainable Development in the Seto Inland Sea, Japan. From the Viewpoint of Fisheries. Terra Scientific, Tokyo, pp. 251–304.

Okey, T.A. and D. Pauly. 1999. A mass-balanced model of trophic flows in Prince William Sound: de-compartmentalizing ecosystem knowledge. In: S. Keller, Ed., Ecosystem Approaches for Fisheries Management. University of Alaska Sea Grant, Fairbanks, pp. 621–635.

Olsen, P.S. and J.B. Mahoney. 2001. Phytoplankton in the Barnegat Bay–Little Egg Harbor estuarine system: Species composition and picoplankton bloom development. In: Barnegat Bay–Little Egg Harbor, New Jersey: Estuary and Watershed Assessment. In: M.J. Kennish, Ed., J. Coastal Res. Spec. Iss. 32: 113–143.

O'Neill, R.V. 2001. Is it time to bury the ecosystem concept? (with full military honors, of course!). Ecology 82: 3275–3284.

Orth, R.J. and K.L. Heck, Jr. 1980. Structural components of eelgrass (*Zostera marina*) meadows in the lower Chesapeake Bay. II. Fishes. Estuaries 3:278–288.

O'Shea, M.L. and T.M. Brosnon. 2000. Trends in indicators of eutrophication in western Long Island Sound and the Hudson-Raritan Estuary. Estuaries 23: 877–901.

Oshima, Y., C.J. Bolch, and G.M. Hallegraeff. 1992. Acute effects of the cyanobacterium cysts of *Alexandrium tamarense* (Dinophyceae). Toxicon 30: 1539–1544.

Oviatt, C.A., P. Doering, B. Nowicki, I. Reed, J. Cole, and J. Frithsen. 1995. An ecosystem level experiment on nutrient limitation in temperate coastal marine environments. Mar. Ecol. Prog. Ser. 116: 171–179.

Paasche, E.E. and S.R. Erga. 1988. Phosphorus and nitrogen limitation of phytoplankton in the Inner Oslofjord (Norway). Sarsia 73: 229–243.

Paerl, H.W. 1997. Coastal eutrophication and harmful algal blooms: the importance of atmospheric and groundwater as "new" nitrogen and other nutrient sources. Limnol. Oceanogr. 42: 1154–1165.

Palmer, S.L. 1984. Surface water. In: E.A. Fernald and D.J. Patton, Eds., Water Resources Atlas of Florida. Florida State University, Tallahassee, pp. 54–67.

Pankratz, A. 1983. Forecasting with Univariate Box-Jenkins Models: Concepts and Cases. John Wiley & Sons, New York.

Pankratz, A. 1991. Forecasting with Dynamic Regression Models. John Wiley & Sons, New York.

Parker, C.A. and J.E. O'Reilly. 1991. Oxygen depletion in Long Island Sound: a historical perspective. Estuaries 14: 248–264.

Parks, J.W., A. Lutz, and J.A. Sutton. 1989. Water column methylmercury in the Wabigoon/English River-Lake system: factors controlling concentrations, speciation, and net production. Can. J. Fish. Aquat. Sci. 46: 2184–2202.

Parsons, T.R., Y. Maita, and C.M. Lalli. 1984. A Manual of Chemical and Biological Methods for Seawater Analysis. Pergamon Press, New York, 173 pp.

Pauly, D.V., V. Christensen, A. Dalsgaard, R. Froese, and J. Torres. 1998. Fishing down marine food webs. Science 279: 860–863.

Pauly, D.V., V. Christensen, and C. Walters. 2000. Ecopath, Ecosim, and Ecospace as tools for evaluating ecosystem impact of fisheries. ICES. J. Mar. Sci. 57: 697–706.

Pearce, J.W. 1977. Florida's environmentally endangered land acquisition program and the Apalachicola River system. In: R.J. Livingston, Ed., Proceedings of the Conference on the Apalachicola Drainage System. Marine Resources Publication 26, Florida Department of Natural Resources, St. Petersburg, pp. 141–145.

Pearson, T.H. 1980. Marine pollution effects of pulp and paper industry wastes. Helgol. Wiss. Meeresunters. 33: 340–365.

Pearson, T.H. and R. Rosenberg. 1978. Macrobenthic succession in relation to organic enrichment and pollution of the marine environment. Oceanogr. Mar. Biol. A. Rev. 16: 229–311.

Pennock, J.R. 1987. Temporal and spatial variability in plankton, ammonium, and nitrate uptake in the Delaware River. Estuarine Coastal Shelf Sci. 24: 841–857.

Pennock, J.R. and J.H. Sharp. 1994. Temporal alteration between light and nutrient limitation of phytoplankton production in a coastal plain estuary. Mar. Ecol. Prog. Ser. 111: 275–288.

Peters, R.H. 1977. The unpredictable problems of trophodynamics. Environ. Biol. Bull. 2: 97–101.

Peterson, B.J. and R.W. Howarth. 1987. Sulfur, carbon, and nitrogen isotopes used to trace organic matter flow in the salt-marsh estuaries of Sapelo Island, Georgia. Limnol. Oceanogr. 32: 1195–1213.

Peterson, C.H. 1979. Predation, competitive exclusion, and diversity in the soft-sediment benthic communities of estuaries and lagoons. In: R.J. Livingston, Ed., Ecological Processes in Coastal and Marine Systems. Plenum Press, New York, pp. 233–264.

Peterson, C.H. 1982. The importance of predation and intra- and interspecific competition in the population biology of two infaunal suspension feeding bivalves, *Protothaca staminea* and *Chione undatella*. Ecol. Monogr. 52: 437–475.

Peterson, C.H. 1991. Intertidal zonation of marine invertebrates in sand and mud. Am. Sci. 79: 236–249.

Peterson, C.H. 1992. Competition for food and its community-level implications. Benth. Res. 42: 1–11.

Philippart, C.J.M., G.C. Cadee, W. van Raaphorst, and R. Riegman. 2000. Long-term phytoplankton-nutrient interactions in a shallow coastal sea: algal community structure, nutrient budgets, and denitrification potential. Limnol. Oceanogr. 45: 131–144.

Philips, E.J., M. Cichra, F.J. Aldridge, and J. Jembeck. 2000. Light availability and variations in phytoplankton standing crops in a nutrient-rich blackwater river. Limnol. Oceanogr. 45: 916–929.

Portnoy, J.W. 1991. Summer oxygen depletion in a diked New England estuary, Estuaries 14: 122–129.

Postma, H. 1985. Eutrophication of Dutch coastal waters. Neth. J. Zool. 35: 348–359.

Potts, M. 1980. Blue-green algae (Cyanophyta) in marine coastal environments of the Sinai Peninsula; distribution, zonation, stratification and taxonomic diversity. Phycologia 19: 60–73.

Power, M.P. 1992. Top-down and bottom-up forces in food webs: do plants have primacy? Ecology 73: 733–746.

Powers, C.F., D.W. Schults, K.W. Malueg, R.M. Brice, and M.D. Schuldt. 1972. Algal responses to nutrient additions in natural waters. II. Field experiments. Limnol. Oceanogr. Spec. Symp. 1: 141–156.

Prasad, A.K.S.K. and G.A. Fryxell. 1991. Habit, frustule morphology and distribution of the Antarctic marine benthic diatom *Entopyla australis* var. *gigantea* (Greville) Fricke (Entopylaceae). Br. Phycol. J. 26: 101–122.

Prasad, A.K.S.K. and R.J. Livingston. 1995. A microbiological and systematic study of *Coscinodiscus jonesianus* (Bacillariophyta) from Florida coastal waters. Nova Hedw. Beih. 112: 247–263.

Prasad, A.K.S.K. and R.J. Livingston. 1998. Fine structure and taxonomy of *Fryxelliella* — a new genus of centric diatom (Triceratiaceae: Bacillariophyta) with a new valve feature, a circumferential marginal tube, and descriptions of *F. floridana* sp. nov. from the Atlantic coast of Florida and *F. inconspicua* (Rattray) comb. nov. from the Miocene. Phycologia 36: 305–323.

Prasad, A.K.S.K., J.A. Nienow, and R.J. Livingston. 1990. The genus *Cyclotella* (Bacillariophyta) in Choctawhatchee Bay, Florida, with special reference to *C. striata* and *C. choctawhatcheeana* sp. nov. Phycologia 29: 418–436.

Premula, V.E. and M.U. Rao. 1977. Distribution and seasonal abundance of *Oscillatoria nigroviridis* Thwaites ex. Gomant in the waters of Visakhapatnam Harbour. Ind. J. Mar. Sci. 3: 79–91.

Purcell, B.H. 1977. The Ecology of the Epibenthic Fauna Associated with *Vallisneria americana* Beds in a North Florida Estuary. M.S. thesis, Florida State University, Tallahassee.

Pyke, G.H., H.R. Pulliam, and E.L. Charnou. 1977. Optimal foraging: a selective review of theory and tests. Q. Rev. Biol. 52:137–154.

Rabalais, N.N. 1992. An updated summary of status and trends in indicators of nutrient enrichment in the Gulf of Mexico. Report to Gulf of Mexico Program, Nutrient Enrichment Subcommittee. EPA/800-R-92–004, U.S. Environmental Protection Agency, Office of Water, Gulf of Mexico Program, Stennis Space Center, MS, 421 pp.

Rabalais, N.N., J. Berg, and E. Hagmeier. 1990. Long-term changes of the annual cycles of meteorological, hydrographic, nutrient, and phytoplankton time series at Helgoland and at LV. ELBE 1 in the German Bight. Cont. Shelf Res. 10: 305–328.

Rabalais, N.N., R.E. Turner, D. Justic, Q. Dortch, W.J. Wiseman, Jr., and B.K. Sen Grupta. 1996. Nutrient changes in the Mississippi River and system responses on the adjacent continental shelf. Estuaries 19: 386–407.

Rabalais, N.N., R.E. Turner, D. Justic, Q. Dortch, and W.J. Wiseman, Jr. 1999. Characterization of Hypoxia: Topic 1 Report for the Integrated Assessment of Hypoxia in the Gulf of Mexico. NOAA Coastal Ocean Program, Silver Spring, MD, 167 pp.

Rada, R.G., J.E. Findley, and J.G. Wiener. 1986. Environmental fate of mercury discharged into the Upper Wisconsin River. Water Air Soil Pollut. 29: 57–76.

Raffaelli, D. 2002. From Elton to mathematics and back again. Science 296: 1035–1037.

Rainville, R.P., B.J. Copeland, and W.T. McKean. 1975. Toxicity of kraft mill wastes to an estuarine phytoplankton. J. Water Pollut. Control Fed. 47: 487–503.

Randall, J. 1967. Food habits of reef fishes of the West Indies. Stud. Trop. Oceanogr. Miami 5: 665–847.

Randall, J.M. and J.W. Day. 1987. Effects of river discharge and vertical circulation on aquatic primary production in a turbid Louisiana (USA) estuary. Neth. J. Sea. Res. 21: 231–242.

Reddy, P.M. and V. Venkateswarlu. 1986. Ecology of algae in the paper mill effluents and their impact on the River Tungabhadra. J. Environ. Biol. 7: 215–223.

Regnell, O. and G. Ewald. 1997. Factors controlling temporal variation in methyl mercury levels in sediment and water in a seasonally stratified lake. Limnol. Oceanogr. 42: 1784–1795.

Reid, G.K., Jr. 1967. An ecological study of the Gulf of Mexico fishes in the vicinity of Cedar Key, Florida. Bull. Mar. Sci. Gulf Caribb. 4: 1–94.

Reise, K. 1978. Experiments on epibenthic predation in the Wadden Sea. Helgol. Wiss. Meeresunters. 31: 51–101.

Reyer, A.J., D.W. Field, J.E. Cassells, C.E. Alexander, and C.L. Holland. 1988. The distribution and areal extent of coastal wetlands in estuaries of the Gulf of Mexico. National Coastal Wetlands Inventory. National Oceanic and Atmospheric Administration, Rockville, MD, 18 pp.

Reyes, E. and M. Merino. 1991. Diel dissolved oxygen dynamics and eutrophication in a shallow well-mixed tropical lagoon. Estuaries 14: 372–381.

Riegman, R. 1998. Species composition of harmful algal blooms in relation to macronutrient dynamics. In: D.M. Anderson, A.D.. Cembella, and G.M. Hallagraeff, Eds., Physiological Ecology of Harmful Algal Blooms. Springer-Verlag, Berlin, pp. 475–488.

Riegman, R., A.A.M. Noordeloos, and G.C. Cadee. 1992. *Phaeocystis* blooms and eutrophication of the continental coastal zones of the North Sea. Mar. Biol. 112: 479–484.

Rigler, F.H. 1975. The concept of energy flow and nutrient flow between trophic levels. In: W.H. van Dobben and R.H. Lowe-McConnell, Eds., Unifying Concepts in Ecology. Junk/PUDOC, The Hague, the Netherlands, pp. 15–26..

Robertson, B. 1982. Guardian of Apalachicola Bay. Oceans 5:65–67.

Roegner, G.C. and A.L. Shanks. 2001. Coastally derived chlorophyll *a* to South Slough, Oregon. Estuaries 24: 244–256.

Rosenberg, D.M., R.A. Bodaly, R.E. Hecky, and R.W. Newbury. 1987. The environmental assessment of hydroelectric impoundments and diversions in Canada. In: M.C. Healy and R.R. Wallace, Eds., Canadian Aquatic Resources. Can. Bull. Fish. Aquat. Sci. 215: 71–104.

Rosenberg, R. 1977. Benthic macrofaunal dynamics, production, and dispersion in an oxygen-deficient estuary on west Sweden. J. Exp. Mar. Biol. Ecol. 26: 107–133.

Rosenberg, R., R. Elmgren, S. Fleischer, P. Jonsson, G. Persson, and H. Dahhm. 1990. Marine eutrophication case studies in Sweden. Ambio 19: 102–108.

Ross, S.T. 1977. Patterns of resource partitioning in searobins (Pisces: Triglidae). Copeia (1977): 561–571.

Ross, S.T. 1978. Trophic ontogeny of the leopard searobin, *Prionotus scitulus* (Pisces: Triglidae). U.S. Fish. Bull. 76: 225–234.

Round, F.E. 1981. The Ecology of Algae. Cambridge University Press, Cambridge, U.K.

Rowe, G.T. and G.S. Boland. 1991. Benthic oxygen demand and nutrient regeneration on the Continental Shelf of Louisiana and Texas. Abstract for LUMCON meeting.

Rudnick, D.T., Z. Chen, D.L. Childers, J.N. Boyer, and T.D. Fontaine III. 1999. Phosphorus and nitrogen inputs to Florida Bay. The importance of the Everglades watershed. In: J.W. Fourquerean, M.B. Robblee, and L.A. Deegan, Eds., Florida Bay: A Dynamic Subtropical Estuary. Estuaries 22: 398–416.

Ryan, J.R. 1981. Trophic Analysis of Nocturnal Fishes in Seagrass Beds in Apalachee Bay, Florida. M.S. thesis, Florida State University, Tallahassee.

Ryther, J.H. 1954. The ecology of phytoplankton blooms in Moriches Bay and Great South Bay, Long Island, New York. Biol. Bull. 106: 198–209.

Ryther, J.H. and W.M. Dunstan. 1971. Nitrogen, phosphorus, and eutrophication in the coastal marine environment. Science 171: 1008–1013.

Sabatier, R., J.D. Lebreton, and D. Chessel. 1989, Principal component analysis with instrumental variables as a tool for modeling composition data. In: R. Coppi and S. Bolasco, Eds., Multiway Data Analysis. Elsevier/North Holland, Amsterdam, pp. 341–352.

Sakshaug, E. and Y. Olsen. 1986. Nutrient status of phytoplankton blooms in Norwegian waters and algal strategies for nutrient competition. Can. J. Fish. Aquat. Sci. 43: 389–396.

Santos, S.H. and S.A. Bloom. 1980. Stability in an annually defaunated estuarine soft-bottom community. Oecologia 46: 290–294.

Santos, S.L. and J.L. Simon. 1980a. Marine soft-bottom community establishment following annual defaunation: larval or adult recruitment? Mar. Ecol. 2: 235–241.

Santos, S.L. and J.L. Simon. 1980b. Response of soft-bottom benthos to annual catastrophic disturbance in a south Florida estuary. Mar. Ecol. 3: 347–355.

Schnable, J.E. 1966. The Evolution and Development of Part of the Northwest Florida Coast. Ph.D. dissertation, Florida State University, Tallahassee.

Schoenly, K. and J.E. Cohen. 1991. Temporal variation in food web structure: 16 empirical cases. Ecol. Monogr. 61: 267–298.

Schroeder, W.W. 1978. Riverine influence on estuaries: a case study. In: V.S. Kennedy, Ed., Estuarine Interactions. Academic Press, New York, pp. 347–364.

Schroeder, W.W. and W.J. Wiseman, Jr. 1985. An analysis of the winds (1974–1984) and sea level elevations (1973–1983) in coastal Alabama. Publ. MASGP-84–024, Mississippi-Alabama Sea Grant Consortium, Jackson, MS.

Schroeder, W.W. and W.J. Wiseman. 1986. Low-frequency shelf-estuarine exchange processes in Mobile Bay and other estuarine systems on the northern Gulf of Mexico. In: D.A. Wolfe, Ed., Estuarine Variability. Academic Press, New York.

Schropp, S.J., F.D. Calder, G.M. Sloane, J.C. Carlton, G.L. Holcomb, H.L. Windom, F. Huan, and R.B. Taylor. 1991. A report on physical and chemical processes affecting the management of Perdido Bay: results of Perdido Bay Interstate Project, Alabama Department of Environmental Management and Florida Department of Environmental Regulation, Tallahassee.

Sciarotta, T.C. and D.R. Nelson. 1977. Diel behavior of the blue shark, *Prionace glauca*, near Santa Catalina Island, California. Fish. Bull. 75: 519–529.

Seal, T.L., F.D. Calder, G.M. Sloane, S.J. Schropp, and H.L. Windom. 1994. Florida Coastal Sediment Contaminants Atlas: A Summary of Coastal Sediment Quality Surveys. Florida Department of Environmental Protection, Tallahassee.

Seliger, H.H. and J.A. Boggs. 1988. Long term pattern of anoxia in the Chesapeake Bay. In: M.P. Lynch and E.C. Krome, Eds., Understanding the Estuary: Advances in Chesapeake Bay Research. Chesapeake Research Consortium, Solomons, MD, pp. 570–583.

Sellner, K.G., S.E. Shumway, M.W. Luckenbach, and T.L. Cucci. 1995. The effects of dinoflagellate blooms on the oyster *Crassostrea virginica* in Chesapeake Bay. In: P. Lassus, G. Arzul, E. Erard, P. Gentien, and C. Marcalliou, Eds., Harmful Marine Algal Blooms. Intercept Ltd., Lavoisier, Belgium, pp. 505–511.

Sheridan, P.F. 1978. Trophic Relationships of Dominant Fishes in the Apalachicola Bay System (Florida). Dissertation, Florida State University, Tallahassee.

Sheridan, P.F. 1979. Trophic resource utilization by three species of sciaenid fishes in a northwest Florida estuary. Northeast Gulf Sci. 3: 1–15.

Sheridan, P.F. and R.J. Livingston. 1979. Cyclic trophic relationships of fishes in an unpolluted, river-dominated estuary in North Florida. In: R.J. Livingston, Ed., Ecological Processes in Coastal and Marine Systems. Plenum Press, New York, pp. 143–161.

Sheridan, P.F. and R.J. Livingston. 1983. Abundance and seasonality of infauna and epifauna inhabiting a *Halodule wrightii* meadow in Apalachicola Bay, Florida. Estuaries 6: 407–419.

Sherman, K., L.M. Alexander, and B.D. Gold. 1991. Food Chains, Yields, Models and Management of Large Marine Ecosystems. Westview Press, Boulder, CO, 320 pp.

Shimada, H., T. Hayashi, and T. Mizushima. 1996. Spatial distribution of *Alexandrium tamarencse* "species complex" (Dinophyceae): dispersal in the North American and West Pacific regions. Phycologia 34: 472–485.

Shimada, M., T.H. Murakami, T. Imahayashi, H.S. Ozaki, T. Toyashima, and T. Okaichi. 1983. Effects of sea bloom, *Chattonella antiqua*, on grill primary lamellae of the young yellowtail, *Seriola quinqueradiata*. Acta Histochem. Cytochem. 16: 232–244.

Shumway, S.E. 1990. A review of the effects of algal blooms on shellfish and aquaculture. J. World Aquacult. Soc. 21: 65–104.

Shumway, S.E., J. Barter, and S. Sherman-Caswell. 1990. Auditing the impact of toxic algal blooms on oysters. Environ. Auditor 2: 41–56.

Sikora, W.B., R.W. Heard, and M.D. Dahlberg. 1972. The occurrence and food habits of two species of hake, *Urophycis regius* and *U. floridanus* in Georgia estuaries. Trans. Am. Fish. Soc. 101:513–525.

Simmons, E.G. and W.H. Thomas. 1962. Phytoplankton of the eastern Mississippi delta. Publ. Inst. Mar. Sci. (Univ. Texas) 8: 269–298.

Simpson, R.L., R.E. Good, M.A. Leck, and D.F. Whigham. 1983. The ecology of freshwater tidal wetlands. Bioscience 33: 255–259.

Sin, Y., R.L. Wetzel, and I.C. Anderson. 1999. Spatial and temporal characteristics of nutrient and phytoplankton dynamics in the York River Estuary, Virginia: analyses of long-term data. Estuaries 22: 260–275.

Sinclair, M., M. El-Sabh, and J. Brindle. 1976. Seaward nutrient transport in the Lower St. Lawrence estuary. J. Fish. Res. Board Can. 33: 1271–1277.

Smayda, T.A. 1978. From phytoplankters to biomass. In: A. Sournia, Ed., Phytoplankton Manual. UNESCO, Paris, pp. 273–279.

Smayda, T.A. 1990. Novel and nuisance phytoplankton blooms in the sea. Evidence for a global epidemic. In: E. Granneli, B. Sundstrom, R. Edler, and D.M. Anderson, Eds., Toxic Marine Phytoplankton: Proceedings of the Fourth International Conference. Elsevier, New York, pp. 29–40.

Smayda, T.A. 1997a. What is a bloom? A commentary. Limnol. Oceanogr. 42: 1132–1136.

Smayda, T.A. 1997b. Harmful algal blooms. Their ecophysiology and general relevance to phytoplankton blooms in the sea. Limnol. Oceanogr. 42: 1137–1153.

Smayda, T.A. 1997c. Ecophysiology and bloom dynamics of *Heterosigma akashiwo* (Raphidophyceae). In: D.M. Anderson, A.D. Cembella, and G.M. Hallagraeff, Eds., Physiological Ecology of Harmful Algal Blooms. Springer-Verlag, Berlin, pp. 111–131.

Smayda, T.J. 1965. A quantitative analysis of the phytoplankton of the Gulf of Panama. II. On the relationship between ^{14}C assimilation and the diatom standing crop. Bull. Inter-Am. Trop. Comm. 9: 46–531.

Smayda, T.J. 1980. Phytoplankton species succession. In: I. Morris, Ed., The Physiological Ecology of Phytoplankton. University of California Press, Berkeley, pp. 493–570.

Smayda, T.J. 1989. Primary production and the global epidemic of phytoplankton blooms in the sea: a linkage? In: E.M. Cosper, J. Carpenter, and V.M. Bricelj, Eds., Novel Phytoplankton Blooms: Causes and Impacts of Recurrent Brown Tides and Other Unusual Blooms. Springer-Verlag, Berlin.

Smayda, T.J. and Y. Shimizu. 1993. Toxic Phytoplankton Blooms in the Sea. Proceedings of the Fifth International Conference on Toxic Marine Phytoplankton. Elsevier, New York, 952 pp.

Smith, G.A., J.S. Nickels, W.M. Davis, R.F. Martz, R.H. Findlay, and D.C. White. 1982. Perturbations in the biomass, metabolic activity, and community structure of the estuarine detrital microbiota: resource partitioning in amphipod grazing. J. Exp. Mar. Biol. Ecol. 64: 125–143.

Smith, S.M. and G.L. Hitchcock. 1994. Nutrient enrichment and phytoplankton growth in the surface waters of the Louisiana Bight. Estuaries 17: 740–753.

Snedaker, S., D. de Sylva, and D. Cottrell. 1977. A review of the role of freshwater in estuarine ecosystems. Final Report, Southwest Florida Water Management District, Brooksville, 126 pp.

Sorokin, Y.I., F. Dallocchio, F. Gelli, and L. Pregnollato. 1996. Phosphorus metabolism in anthropogenically transformed lagoon ecosystems: the Comacchio Lagoons (Ferrar, Italy). J. Sea Res. 35: 243–250.

Sournia, A. 1978. Phytoplankton Manual. Monographs on Oceanic Methodology 6. UNESCO, Paris, 337 pp.

Sournia, A., M.J. Chretiennot-Dinet, and M. Ricard. 1991. Marine phytoplankton: how many species in the world ocean? J. Plankton Res. 13: 1093–1099.

South Florida Water Management District. 1994. An update of the surface water improvement and management plan for Biscayne Bay. Draft Final Report, West Palm Beach.

South Florida Water Management District. 2000. Everglades Consolidated Report, West Palm Beach.

Sowles, J.W. 1997a. Memo to Stacey Ladner. Subject: Interpretation of Hg in Penobscot Sediments. Maine Department of Environmental Protection, Augusta.

Sowles, J.W. 1997b. Water resources survey — HoltraChem (Part II). September 29, 1997 memorandum to Stacy Ladner, Maine Department of Environmental Protection, Augusta.

Sowles, J.W. 1999. Mercury contamination in the Penobscot River estuary at HoltraChem Manufacturing Company — an evaluation of monitoring data and interpretation of toxic potential and ecological implications. April 23, 1999 data summary report. Maine Department of Environmental Protection, Augusta.

Squires, L.E. and N.A. Sinnu. 1982. Seasonal changes in the diatom flora in the estuary of the Damour River, Lebanon. 7th Diatom Symposium, pp. 359–372.

Stanley, D.W. and S.W. Nixon. 1992. Stratification and bottom-water hypoxia in the Pamlico River Estuary. Estuaries 15: 270–281.

Starck, W.A. II and Davis, W.P. 1966. Night habits of fishes of Alligator Reef, Florida. Ichthyologia (1966): 313–357.

Stefanou, P., G. Tsirtsis, and M. Karydis. 2000. Nutrient scaling for assessing eutrophication: the development of a simulated normal distribution. Ecol. Appl. 10: 303–309.

Steidinger, K.A., J.T. Davis, and J. Williams. 1966. Observations of *Gymnodinium breve* Davis and other dinoflagellates. In: Observations of an Unusual Red Tide. A Symposium. Prof. Pap. Ser. 8. Florida Board of Conservation, Marine Laboratory, St. Petersburg, pp. 9–15.

Steidinger, K.A., J.H. Landsberg, D.W. Trudy, and B.S. Roberts. 1998a. First report of *Gymnodinium breve* Larson 1994 (Dinophyceae) in North America and associated fish kills in the Indian River. J. Phycol. 37: 58–63.

Steidinger, K.A., P. Carlson, D. Baden, C.D. Rodriguez, and J. Seagle. 1998b. Neurotoxic shellfish poisoning due to toxin retention in the clam *Chione cancellata*. In: B. Reguera, J. Blanco, M.L. Fernandez, and T. Wyatt, Eds., Harmful Algae, Xunta de Galicia and IOC.

Stockner, J.G. and D.D. Cliff. 1976. Effects of pulp mill effluent on phytoplankton production in coastal marine waters of British Columbia. J. Fish. Res. Board Can. 33: 2433–2442.

Stockner, J.G. and D.D. Costella. 1976. Marine phytoplankton growth in high concentrations of pulp mill effluent. J. Fish. Res. Board Can. 33: 2758–2765.

Stoner, A.W. 1976. Growth and Food Conversion Efficiency of Pin-fish (*Lagodon rhomboides*) Exposed to Sublethal Concentrations of Bleached Kraft Mill Effluents. M.S. thesis, Florida State University, Tallahassee.

Stoner, A.W. 1979a. The Macrobenthos of Seagrass Meadows in Apalachee Bay, Florida, and the Feeding Ecology of *Lagodon rhomboides* (Pisces: Sparidae). Ph.D. dissertation, Florida State University, Tallahassee.

Stoner, A.W. 1979b. Species-specific predation on amphipod Crustacea by the pinfish, *Lagodon rhomboides*: mediation by macrophyte standing crop. Mar. Biol. 55: 201.

Stoner, A.W. 1980a. The role of seagrass biomass in the organization of benthic macrophyte assemblages. Bull. Mar. Sci. 30:537–551.

Stoner, A.W. 1980b. Abundance, reproductive seasonality and habitat preferences of amphipod crustaceans in seagrass meadows of Apalachee Bay, Florida. Contrib. Mar. Sci. 23: 63–77.

Stoner, A.W. 1980c. Feeding ecology of the *Lagodon rhomboides* (Pisces: Sparidae): variation and functional responses. Fish. Bull. 78: 337–352.

Stoner, A.W. 1982. The influence of benthic macrophytes on the foraging behavior of pinfish *Lagodon rhomboides* (Linnaeus). J. Exp. Mar. Biol. Ecol. 58: 271–284.

Stoner, A.W. and R.J. Livingston. 1980. Distributional ecology and food habits of the banded blenny *Paraclinus fasciatus* (Clinidae): a resident in a mobile habitat. Mar. Biol. 56: 239–246.

Stoner, A.W. and R.J. Livingston. 1984. Ontogenetic patterns in diet and feeding morphology in sympatric sparid fishes from seagrass meadows. Copeia (1984): 174–187.

Stoner, A.W., H.S. Greening, J.D. Ryan, and R.J. Livingston. 1982. Comparison of macrobenthos collected with cores and suction dredge. Estuaries 6: 76–82.

Suttle, C.A. and P.J. Harrison. 1988. Ammonium and phosphate uptake rates, N:P supply ratios, and evidence for N and P limitation in some oligotrophic lakes. Limnol. Oceanogr. 33: 186–202.

Swift, F. 1896. Report of a survey of the oyster region of St. Vincent Sound, Apalachicola Bay, and St. George Sound, Florida. Report of the U.S. Commission for 1896. U.S. Commission of Fish and Fisheries. Appendix 4, pp. 187–221.

Sykes, J.E. and C.S. Manooch III. 1979. Predator-prey systems in fisheries management. In: R.H. Clepper, Ed., International Symposium on Predator-Prey Systems in Fish Communities and Their Role in Fisheries Management. Sport Fishing Institute, Washington, D.C., pp. 93–101.

Tanner, W.F. 1960. Florida coastal classification. Trans. Gulf Coast Assoc. Geol. Soc. 10: 259–266.

Tansley, A.G. 1935. The use and abuse of vegetational concepts and terms. Ecology 16: 284–307.

Thienemann, A. 1918. Lebensgemeinschaft und Lebensraum. Naturw. Wochenschrift 17: 282–290, 297–303.

Thistle, D. 1981. Natural physical disturbances and communities of marine soft bottoms. Mar. Ecol. Prog Ser. 6: 223–228.

Thomas, D.L. 1971. The early life history and ecology of six species of drum (Sciaenidae) in the lower Delaware River, a brackish tidal estuary. Ichthyol. Assoc. Bull. 3: 247.

Thornton, J.A., H. Beekman, G. Boddington, R. Dick, W.R. Harding, M. Lief, I.R. Morrison, and A.J.R. Quick. 1995b. The ecology and management of Zandvlei (Cape Province, South Africa), an enriched shallow African estuary. In: A.J. McComb, Ed., Eutrophic Shallow Estuaries and Lagoons. CRC Press, Boca Raton, FL, pp. 109–128.

Thornton, J.A., J. McComb, and S.O. Ryding. 1995a. The role of sediments. In: A.J. McComb, Ed., Eutrophic Shallow Estuaries and Lagoons. CRC Press, Boca Raton, FL, pp. 205–224.

Thorpe, P., R. Bartel, P. Ryan, K. Albertson, T. Pratt, and D. Cairns. 1997. The Pensacola Bay system surface water improvement and management plan. Unpublished report. Northwest Florida Water Management District, Havana, FL.

Tichken, E.J. 1991. Santa Monica Bay restoration project: assessment of geophysical properties of ocean sediments off Palos Verdes Peninsula Los Angeles County, California. Unpublished report.

Thompson, P.A. 1998. Spatial and temporal patterns of factors influencing phytoplankton in a salt wedge estuary, the Swan River, Western Australia. Estuaries 21: 801–817.

Throndsen, Y. 1993. The planktonic marine flagellates. In: C.R. Tomas, Ed., Marine Phytoplankton. Academic Press, New York, pp. 7–145.

Tomas, C.R., B. Bendis, and D.K. Johns. 1999. Role of nutrients in regulating plankton blooms in Florida Bay. In: H. Kumpf, K. Steidinger, and K. Sherman, Eds., The Gulf of Mexico: Large Marine Ecosystem. Blackwell Science, New York, pp. 323–337.

Toner, W. 1975. Oysters and the good ol' boys. Planning 41: 10–15.

Tong, H. 1990. Non-Linear Time Series: A Dynamical System Approach. Clarendon Press, Oxford.

Tracey, G.A. 1985. Feeding reduction, reproductive failure, and mortality in *Mytilus edulis* during the 1985 "brown tide" in Narragansett Bay, Rhode Island. Mar. Ecol. Prog. Ser. 50: 73–81.

Turner, R.E., W.W. Schroeder, and W.J. Wiseman, Jr. 1987. The role of stratification in the deoxygenation of Mobile Bay and adjacent shelf bottom waters. Estuaries 10: 13–20.

Turner, S.J., S.F. Thrush, R.D. Pridmore, J.E. Hewitt, V.J. Cummings, and M. Maskery. 1995. Are soft-sediment communities stable? An example from a windy harbour. Mar. Ecol. Prog. Ser. 120: 219–230.

Tyrrell, T. 1999. The relative influences of nitrogen and phosphorus on oceanic primary production. Nature 368: 619–621.

Ulanowicz, R.E. 1987. NETWRK: a package of computer algorithms to analyze ecological flow networks. University of Maryland Chesapeake Biological Laboratory, Solomons, MD.

Ulanowicz, R.E. and J.J. Kay. 1991. A package for the analysis of ecosystem flow networks. Environ. Software 6: 131–142.

U.S. Army Corps of Engineers. 1976. Statement of Findings: Perdido Pass Channel (maintenance dredging), Baldwin County, AL.

U.S. Army Corps of Engineers. 1980. Water Resources Study: Escambia-Yellow Rivers Basins. Mobile District, Mobile, AL.

U.S. Environmental Protection Agency. 1971. Conference in the matter of pollution of the interstate waters of the Escambia River basin (Alabama-Florida) and the intrastate portions of the Escambia basin within the state of Florida. Pensacola, FL.

U.S. Environmental Protection Agency. 1983. Methods for the chemical analysis of water and wastes. EPA-600/4-79-020.

U.S. Environmental Protection Agency. 1990. Analysis of the section 301(h) secondary treatment variance application by Los Angeles County Sanitation Districts for Joint Water Pollution Control Plant. Region 9, Water Management Division.

U.S. Environmental Protection Agency. 1993. Water-quality protection program for the Florida Keys National Marine Sanctuary. Phase II report. U.S. Environmental Protection Agency, Washington, D.C.

U.S. Environmental Protection Agency. 1997. Mercury Report to Congress. An Assessment of Exposure to Mercury in the United States. Office of Air Quality Planning and Standards, and Office of Research and Development, EPA452/R-97-006.

U.S. Environmental Protection Agency. 1998. National strategy for the development of regional nutrient criteria. U.S. Environmental Protection Agency, Washington, D.C.

U.S. Environmental Protection Agency. 2000. Fenholloway Nutrient Study, Perry FL. U.S. Environmental Protection Agency, Region 4, Athens, GA.

U.S. Fish and Wildlife Service. 1990. Aerial photography of the seagrass beds of Perdido Bay. Unpublished report. Pensacola, FL.

Valiela, L., J. McClelland, J. Hauxwell, P.J. Behr, D. Hersh, and K. Foreman. 1997. Macroalgal blooms in shallow estuaries: controls and ecophysiological and ecosystem consequences. Limnol. Oceanogr. 42: 1105–1118.

van den Hoek, C., D.G. Mann, and H.M. Jahns. 1995. Algae: An Introduction to Phycology. Cambridge University Press, Cambridge, U.K.

Van Dolah, R.F. 1978. Factors regulating the distribution and population dynamics of the amphipod *Gammarus palustris* in an intertidal salt marsh community. Ecol. Monogr. 48: 191–217.

Van Valkenburg, S.D. and D.A. Flemer. 1974. The distribution and productivity of nanoplankton in a temperate estuarine area. Estuarine Coastal Mar. Sci. 2: 311–322.

Vince, S.I., I. Valiela, N. Backus, and J. Teal. 1976. Predation by the salt marsh killifish *Fundulus heteroclitus* (L.) in relation to prey size and habitat structure; consequences of prey distribution and abundance. J. Exp. Mar. Biol. Ecol. 23:255–266.

Virnstein, R.W. 1977. The importance of predation by crabs and fishes on benthic infauna in Chesapeake Bay. Ecology 58: 1199–1217.

Wald, A. and J. Wolfowitz. 1940. On a test whether two samples are from the same population. Ann. Math. Stat. 11: 147–162.

Wales, D.J. 1991. Calculating the rate of loss of information from chaotic time series by forecasting. Nature 350: 485–488.

Ward, C.H., M.E. Bender, and D.J. Reish, Eds. 1979. The Offshore Ecology Investigation: effects of oil drilling and production in a coastal environment. Rice Univ. Stud. 65: 1–589.

Ware, D.H. 1972. Predation of rainbow trout (*Salmo gairdneri*): the influence of hunger, prey density, and prey size. J. Fish. Res. Board Can. 29: 1193–1201.

Waterbury, J., S. Watson, R. Guillard, and L. Brand. 1979. Widespread occurrence of a unicellular, marine, planktonic, cyanobacterium. Nature 277: 293–294.

Weinstein, M. and K.L. Heck. 1979. Ichthyofauna of seagrass meadows along the Caribbean coast of Panama and in the Gulf of Mexico: composition, structure and community ecology. Mar. Biol. 50: 97–107.

Welsh, B.L. and F.C. Eller. 1991. Mechanisms controlling summertime oxygen depletion in Western Long Island Sound. Estuaries 14: 265–278.

Welsh, B.L., R.B. Whitlach, and W.F. Bohlen. 1982. Relationship between physical characteristics and organic carbon sources as a basis for comparing estuaries in southern New England. In: V.S. Kennedy, Ed., Estuarine Comparisons. Academic Press, New York, pp. 53–67.

Werner, E.E. and D.J. Hall. 1974. Optimal foraging and the size selection of prey by bluegill sunfish (*Lepomis macrochirus*). Ecology 55: 1042–1052.

Westman, W.E. 1978. Measuring the inertia and resilience of ecosystems. Bioscience 28: 705–710.

Wetzel, R.G. 1984. Detrital dissolved and particulate organic carbon functions in aquatic ecosystems. Bull. Mar. Sci. 35: 503–509.

Wetzel, R.G. and G.E. Likens. 1990. Limnological Analyses, 2nd ed. Springer-Verlag, Berlin, 212 pp.

White, D.C. 1983. Analysis of microorganisms in terms of quantity and activity in natural environments. Microbes in their natural environments. Soc. Gen. Microbiol. Symp. 34: 37–66.

White, D.C., R.J. Bobbie, S.J. Morrison, D.K. Oesterhof, C.W. Taylor, and D.A. Meeter. 1977. Determination of microbial activity of lipid biosynthesis. Limnol. Oceanogr. 22: 1089–1099.

White, D.C., R.J. Livingston, R.J. Bobbie, and J.S. Nickels. 1979a. Effects of surface composition, water column chemistry, and time of exposure on the composition of the detrital microflora and associated macrofauna in Apalachicola Bay, Florida. In: R.J. Livingston, Ed., Ecological Processes in Coastal and Marine Systems. Plenum Press, New York, pp. 53–67.

White, D.C., W.M. Davis, J.S. Nickels, J.D. King, and R.J. Bobbie. 1979b. Determination of the sedimentary microbial biomass by extractable lipid phosphate. Oecologia 40: 51–62.

Whitfield, W.K., Jr. and D.S. Beaumariage. 1977. Shellfish management in Apalachicola Bay: past, present, and future. In: R.J. Livingston and E.A. Joyce, Jr., Eds., Proceedings of the Conference on the Apalachicola Drainage System. Marine Resources Publication 26, Florida Department of Natural Resources, St. Petersburg, pp. 130–140.

Whitlach, R.B. and R.N. Zajac. 1985. Biotic interactions among estuarine infaunal opportunistic species. Mar. Ecol. Prog. Ser. 21: 299–311.

Wiener, J.G. and P.J. Shields. 2000. Mercury in the Sudbury River (Massachusetts, U.S.A.): pollution history and a synthesis of recent research. Can. J. Fish. Aquat. Sci. 57: 1053–1061.

Wikfors, G.H. and R.M. Smolowitz. 1993. Detrimental effects of a *Prorocentrum* isolate upon hard clams and bay scallops in laboratory feeding studies. In: T.J. Smayda and Y. Shimizu, Eds., Toxic Phytoplankton Blooms in the Sea. Proceedings of the Fifth International Conference on Toxic Marine Phytoplankton. Elsevier, New York, pp. 447–452.

Wilber, D.H. 1992. Associations between freshwater inflows and oyster productivity in Apalachicola Bay, Florida. Estuarine Coastal Shelf Sci. 35: 179–190.

Wilcoxon, F. 1949. Some Rapid Approximate Statistical Procedures. American Cyanamid Company, Stamford, CT.

Williams, J.E., J.E. Johnson, D.A. Hendrickson, S. Contreras-Balderas, J.D. Williams, M. Navarro-Mendoza, D.E. McAllister, and J.E. Deacon. 1989. Fishes of North America, endangered, threatened, or of special concern: 1989. Fisheries 14: 2–20.

Winemiller, K.O. 1990. Spatial and temporal variation in tropical fish trophic networks. Ecol. Monogr. 60: 331–367.

Winer, B.J. 1971. Statistical Principles in Experimental Design. McGraw-Hill, New York.

Woelke, C.E. 1961. Pacific oyster *Crassostrea gigas* mortalities with notes on common oyster predators in Washington waters. Proc. Natl. Shellfish. Assoc. 50: 53–66.

Wolfe, D.A., M.A. Champ, D.A. Flemer, and A.J. Mearns. 1987. Long-term biological data sets: their role in research, monitoring, and management of estuarine and coastal marine systems. Estuaries 10: 181–193.

Wolfe, M.F., S. Schwarzbach, and R.A. Sulaiman. 1998. Effects of mercury on wildlife: a comprehensive review. Environ. Toxicol. Chem. 17: 146–160.

Wolfe, S.H., J.A. Reidenauer, and D.B. Means. 1988. An ecological characterization of the Florida Panhandle. U.S. Fish and Wildlife Service, Biol. Rep. 88, 277 pp.

Woodhead, P.M.J. 1966. The behavior of fish in relation to light in the sea. Oceanogr. Mar. Biol. Annu. Rev. 4: 337–403.

Woodwell, G.M. and D.E. Whitney. 1977. Flax Pond ecosystem study: exchanges of phosphorous between a salt marsh and the coastal waters of Long Island Sound. Mar. Biol. 41: 1–6.

Woodwell, G.M., D.E. Whitney, C.A.S. Hall, and R.A. Houghton. 1977. The Flax Pond ecosystem study: exchanges of carbon in water between a salt marsh and Long Island Sound. Limnol. Oceanogr. 22: 833–838.

Yerger, R.W. 1977. Fishes of the Apalachicola River. In: R.J. Livingston, Ed., Proceedings of the Conference on the Apalachicola Drainage System. Marine Resources Publication 26, Florida Department of Natural Resources, St. Petersburg, pp. 22–33.

Young, D.L. and R.T. Barber. 1973. Effects of waste dumping in New York Bight on the growth of natural populations of phytoplankton. Environ. Pollut. 5: 237–252.

Young, P.H. 1964. Some effects of sewer effluent on marine life. Calif. Fish Game 50: 33–37.

Zajac, R.N. and R.B. Whitlach. 1982. Responses of estuarine infauna to disturbance. II. Spatial and temporal variation of succession. Mar. Ecol. Prog. Ser. 10: 15–27.

Zaret, T.M. and A.S. Rand. 1971. Competition in tropical stream fishes: support for the competitive exclusion principle. Ecology 52: 336–342.

Zeeman, C. 2001. Memo to S. Ladner regarding Media Protection Goals for Contaminants of Concern Released by the HoltraChem Manufacturing Facility, Orrington, Maine. Maine Department of Environmental Protection, Augusta.

Zillioux, E.J., D.B. Porcella, and J.M. Benoit. 1993. Mercury cycling and effects in freshwater wetland ecosystems. Environ. Toxicol. Chem. 12: 2245–2264.

Zimmerman, M.S. and R.J. Livingston. 1976a. The effects of kraft mill effluents on benthic macrophyte assemblages in a shallow bay system (Apalachee Bay, North Florida, U.S.A.). Mar. Biol. 34: 297–312.

Zimmerman, M.S. and R.J. Livingston. 1976b. Seasonality and physico-chemical ranges of benthic macrophytes from a north Florida estuary (Apalachee Bay). Contrib. Mar. Sci. (Univ. Tex.) 20: 34–45.

Zimmerman, M.S. and R.J. Livingston. 1979. Dominance in benthic macrophyte assemblages from a north Florida estuary (Apalachee Bay). Bull. Mar. Sci. 29: 27–40.

Zimmerman, R., R. Gibson, and J. Harrington. 1979. Herbivory and detrivory among gammaridean amphipods from a Florida seagrass community. Mar. Biol. 54: 41–47.

Index

A

B

C

D

E

F

G

H

I

J

L

M

N

O

P

R

S

T

U